# Lecture Notes in Artificial Intelligence 12167

Subseries of Lecture Notes in Computer Science

More information about this series at http://www.springer.com/series/1244

Nicolas Peltier · Viorica Sofronie-Stokkermans (Eds.)

# Automated Reasoning

10th International Joint Conference, IJCAR 2020
Paris, France, July 1–4, 2020
Proceedings, Part II

*Editors*
Nicolas Peltier
CNRS, LIG, Université Grenoble Alpes
Saint Martin d'Hères, France

Viorica Sofronie-Stokkermans
University Koblenz-Landau
Koblenz, Germany

ISSN 0302-9743 ISSN 1611-3349 (electronic)
Lecture Notes in Artificial Intelligence
ISBN 978-3-030-51053-4 ISBN 978-3-030-51054-1 (eBook)
https://doi.org/10.1007/978-3-030-51054-1

LNCS Sublibrary: SL7 – Artificial Intelligence

# Preface

These volumes contain the papers presented at the 10th International Joint Conference on Automated Reasoning (IJCAR 2020) initially planned to be held in Paris, but – due to the COVID-19 pandemic – held by remote conferencing during July 1-4, 2020.

IJCAR is the premier international joint conference on all aspects of automated reasoning, including foundations, implementations, and applications, comprising several leading conferences and workshops. IJCAR 2020 united CADE, the Conference on Automated Deduction, TABLEAUX, the International Conference on Automated Reasoning with Analytic Tableaux and Related Methods, FroCoS, the International Symposium on Frontiers of Combining Systems, and ITP, the International Conference on Interactive Theorem Proving. Previous IJCAR conferences were held in Siena (Italy) in 2001, Cork (Ireland) in 2004, Seattle (USA) in 2006, Sydney (Australia) in 2008, Edinburgh (UK) in 2010, Manchester (UK) in 2012, Vienna (Austria) in 2014, Coimbra (Portugal) in 2016, and Oxford (UK) in 2018.

150 papers were submitted to IJCAR: 105 regular papers, 21 system description, and 24 short papers, describing interesting work in progress. Each submission was assigned to three Program Committee (PC) members; in a few cases, a fourth additional review was requested. A rebuttal phase was added for the authors to respond to the reviews before the final deliberation. The PC accepted 62 papers, resulting in an acceptance rate of about 41%: 46 regular papers (43%), 11 system descriptions (52%), and 5 short papers (20%).

In addition, the program included three invited talks, by Clark Barrett, Elaine Pimentel, and Ruzica Piskac, plus two additional invited talks shared with the conference on Formal Structures for Computation and Deduction (FSCD), by John Harrison and René Thiemann (the abstract of the invited talk by René Thiemann is available in the proceedings of FSCD 2020).

The Best Paper Award was shared this year by two papers: "Politeness for The Theory of Algebraic Datatypes" by Ying Sheng, Yoni Zohar, Christophe Ringeissen, Jane Lange, Pascal Fontaine, and Clark Barrett, and "The Resolution of Keller's Conjecture" by Joshua Brakensiek, Marijn Heule, John Mackey, and David Narvaez.

IJCAR acknowledges the generous sponsorship of the CNRS (French National Centre for Scientific Research), Inria (French Institute for Research in Computer Science and Automation), the Northern Paris Computer Science (LIPN: Laboratoire d'Informatique de Paris Nord) at the University of Paris North (Université Sorbonne Paris Nord), and of the Computer Science Laboratory of Ecole Polytechnique (LIX: Laboratoire d'Informatique de l'École Polytechnique) in the École Polytechnique.

The EasyChair system was extremely useful for the reviewing and selection of papers, the organization of the program, and the creation of this proceedings volume. The PC chairs also want to thank Springer for their support of this publication.

We would like to thank the organizers of IJCAR, the members of the IJCAR PC, and the additional reviewers, who provided high-quality reviews, as well as all authors, speakers, and attendees.

The COVID-19 pandemic had a strong impact on the organization of IJCAR and significantly weighted the burden on authors, reviewers, and organizers. We are very grateful to all of them for their hard work under such difficult and unusual circumstances.

April 2020                                                    Nicolas Peltier
                                            Viorica Sofronie-Stokkermans

# Organization

## Program Committee Chairs

Nicolas Peltier       Université Grenoble Alpes, CNRS, LIG, France
Viorica       University of Koblenz-Landau, Germany
  Sofronie-Stokkermans

## Program Committee

| | |
|---|---|
| Takahito Aoto | Niigata University, Japan |
| Carlos Areces | Universidad Nacional de Córdoba (FaMAF), Argentina |
| Jeremy Avigad | Carnegie Mellon University, USA |
| Franz Baader | TU Dresden, Germany |
| Peter Baumgartner | CSIRO, Australia |
| Christoph Benzmüller | Freie Universität Berlin, Germany |
| Yves Bertot | Inria, France |
| Armin Biere | Johannes Kepler University Linz, Austria |
| Nikolaj Bjørner | Microsoft, USA |
| Jasmin Blanchette | Vrije Universiteit Amsterdam, The Netherlands |
| Maria Paola Bonacina | Università degli Studi di Verona, Italy |
| James Brotherston | University College London, UK |
| Serenella Cerrito | IBISC, Université d'Évry, Université Paris-Saclay, France |
| Agata Ciabattoni | Vienna University of Technology, Austria |
| Koen Claessen | Chalmers University of Technology, Sweden |
| Leonardo de Moura | Microsoft, USA |
| Stéphane Demri | LSV, CNRS, ENS Paris-Saclay, France |
| Gilles Dowek | Inria, ENS Paris-Saclay, France |
| Marcelo Finger | University of São Paulo, Brazil |
| Pascal Fontaine | Université de Liège, Belgium |
| Didier Galmiche | Université de Lorraine, CNRS, LORIA, France |
| Silvio Ghilardi | Università degli Studi di Milano, Italy |
| Martin Giese | University of Oslo, Norway |
| Jürgen Giesl | RWTH Aachen University, Germany |
| Valentin Goranko | Stockholm University, Sweden |
| Rajeev Gore | The Australian National University, Australia |
| Stefan Hetzl | Vienna University of Technology, Austria |
| Marijn Heule | Carnegie Mellon University, USA |
| Cezary Kaliszyk | University of Innsbruck, Austria |
| Deepak Kapur | University of New Mexico, USA |
| Laura Kovacs | Vienna University of Technology, Austria |
| Andreas Lochbihler | Digital Asset GmbH, Switzerland |

| | |
|---|---|
| Christopher Lynch | Clarkson University, USA |
| Assia Mahboubi | Inria, France |
| Panagiotis Manolios | Northeastern University, USA |
| Dale Miller | Inria, LIX/Ecole Polytechnique, France |
| Cláudia Nalon | University of Brasília, Brazil |
| Tobias Nipkow | Technical University of Munich, Germany |
| Albert Oliveras | Universitat Politècnica de Catalunya, Spain |
| Jens Otten | University of Oslo, Norway |
| Lawrence Paulson | University of Cambridge, UK |
| Nicolas Peltier | Université Grenoble Alpes, CNRS, LIG, France |
| Frank Pfenning | Carnegie Mellon University, USA |
| Andrei Popescu | Middlesex University London, UK |
| Andrew Reynolds | University of Iowa, USA |
| Christophe Ringeissen | Inria, France |
| Christine Rizkallah | The University of New South Wales, Australia |
| Katsuhiko Sano | Hokkaido University, Japan |
| Renate A. Schmidt | The University of Manchester, UK |
| Stephan Schulz | DHBW Stuttgart, Germany |
| Roberto Sebastiani | University of Trento, Italy |
| Viorica Sofronie-Stokkermans | University of Koblenz-Landau, Germany |
| Matthieu Sozeau | Inria, France |
| Martin Suda | Czech Technical University in Prague, Czech Republic |
| Geoff Sutcliffe | University of Miami, USA |
| Sofiène Tahar | Concordia University, Canada |
| Cesare Tinelli | The University of Iowa, USA |
| Christian Urban | King's College London, UK |
| Josef Urban | Czech Technical University in Prague, Czech Republic |
| Uwe Waldmann | Max Planck Institute for Informatics, Germany |
| Christoph Weidenbach | Max Planck Institute for Informatics, Germany |

## Conference Chair

| | |
|---|---|
| Kaustuv Chaudhuri | Inria, Ecole Polytechnique Paris, France |

## Workshop Chairs

| | |
|---|---|
| Giulio Manzonetto | Université Paris-Nord, France |
| Andrew Reynolds | University of Iowa, USA |

## IJCAR Steering Committee

| | |
|---|---|
| Franz Baader | TU Dresden, Germany |
| Kaustuv Chaudhuri | Inria, Ecole polytechnique Paris, France |
| Didier Galmiche | Université de Lorraine, CNRS, LORIA, France |
| Ian Horrocks | University of Oxford, UK |

Jens Otten                     University of Oslo, Norway
Lawrence Paulson               University of Cambridge, UK
Pedro Quaresma                 University of Coimbra, Portugal
Viorica                        University of Koblenz-Landau, Germany
  Sofronie-Stokkermans
Christoph Weidenbach           Max Planck Institute for Informatics, Germany

# Additional Reviewers

Abdelghany, Mohamed
Abdul Aziz, Mohammad
Ahmad, Waqar
Ahmed, Asad
Alassaf, Ruba
Baanen, Anne
Baldi, Paolo
Balyo, Tomas
Barbosa, Haniel
Barnett, Lee
Bartocci, Ezio
Berardi, Stefano
Bobillo, Fernando
Bofill, Miquel
Boutry, Pierre
Boy de La Tour, Thierry
Brain, Martin
Bromberger, Martin
Brown, Chad
Casal, Filipe
Ceylan, Ismail Ilkan
Chaudhuri, Kaustuv
Chen, Joshua
Cialdea Mayer, Marta
Cohen, Cyril
D'Agostino, Marcello
Dahmen, Sander
Dalmonte, Tiziano
Dang, Thao
Dawson, Jeremy
De, Michael
Del-Pinto, Warren
Dumbrava, Stefania
Eberl, Manuel
Ebner, Gabriel

Echenim, Mnacho
Egly, Uwe
Esen, Zafer
Fleury, Mathias
Fuhs, Carsten
Gammie, Peter
Gauthier, Thibault
Genestier, Guillaume
Gianola, Alessandro
Girlando, Marianna
Graham-Lengrand, Stéphane
Hirokawa, Nao
Holliday, Wesley
Hustadt, Ullrich
Jaroschek, Maximilian
Kaminski, Benjamin Lucien
Kappé, Tobias
Ketabii, Milad
Kiesel, Rafael
Kim, Dohan
Kocsis, Zoltan A.
Kohlhase, Michael
Konev, Boris
Koopmann, Patrick
Korovin, Konstantin
Lammich, Peter
Lange, Martin
Larchey-Wendling, Dominique
Leitsch, Alexander
Lellmann, Bjoern
Lewis, Robert Y.
Li, Wenda
Lin, Hai
Litak, Tadeusz
Liu, Qinghua

Lyon, Tim
Magaud, Nicolas
Mahmoud, Mohamed Yousri
Manighetti, Matteo
Mansutti, Alessio
Marti, Johannes
Martins, Ruben
Mazzullo, Andrea
McKeown, John
Mclaughlin, Craig
Metcalfe, George
Mineshima, Koji
Méry, Daniel
Nagashima, Yutaka
Negri, Sara
Nishida, Naoki
Norrish, Michael
Nummelin, Visa
Olarte, Carlos
Omori, Hitoshi
Overbeek, Roy
Pattinson, Dirk
Peñaloza, Rafael
Preiner, Mathias
Preto, Sandro
Ramanayake, Revantha
Rashid, Adnan
Robillard, Simon
Roux, Pierre

Roveri, Marco
Rowe, Reuben
Sakr, Mostafa
Samadpour Motalebi, Kiana
Sato, Haruhiko
Sefidgar, Seyed Reza
Thiemann, René
Théry, Laurent
Toluhi, David
Traytel, Dmitriy
Trevizan, Felipe
van Oostrom, Vincent
Vierling, Jannik
Vigneron, Laurent
Virtema, Jonni
Wahl, Thomas
Wansing, Heinrich
Wassermann, Renata
Weber, Tjark
Wiedijk, Freek
Wies, Thomas
Winkler, Sarah
Wojtylak, Piotr
Wolter, Frank
Yamada, Akihisa
Zamansky, Anna
Zivota, Sebastian
Zohar, Yoni

# Contents – Part II

# Contents – Part I

## Superposition

## Proof Procedures

## Non Classical Logics

**Interactive Theorem Proving/HOL**

# Competing Inheritance Paths
# in Dependent Type Theory:
# A Case Study in Functional Analysis

Reynald Affeldt[1]([✉]), Cyril Cohen[2], Marie Kerjean[3], Assia Mahboubi[3],
Damien Rouhling[4], and Kazuhiko Sakaguchi[5]

[1] National Institute of Advanced Industrial Science and Technology (AIST),
Tsukuba, Japan
reynald.affeldt@aist.go.jp
[2] Université Côte d'Azur, Inria, Sophia Antipolis, France
[3] Inria, Rennes-Bretagne Atlantique, Rennes, France
[4] Inria & Université de Strasbourg, CNRS, ICube, Nancy-Grand Est,
Villers-lès-Nancy, France
[5] University of Tsukuba, Tsukuba, Japan

**Abstract.** This paper discusses the design of a hierarchy of structures
which combine linear algebra with concepts related to limits, like topology and norms, in dependent type theory. This hierarchy is the backbone
of a new library of formalized classical analysis, for the Coq proof assistant. It extends the Mathematical Components library, geared towards
algebra, with topics in analysis. Issues of a more general nature related
to the inheritance of poorer structures from richer ones arise due to this
combination. We present and discuss a solution, coined forgetful inheritance, based on packed classes and unification hints.

**Keywords:** Formalization of mathematics · Dependent type theory ·
Packed classes · Coq

## 1 Introduction

*Mathematical structures* are the backbone of the axiomatic method advocated
by Bourbaki [8,9] to spell out mathematically relevant abstractions and establish the corresponding vocabulary and notations. They are instrumental in making the mathematical literature more precise, concise, and intelligible. Modern libraries of formalized mathematics also rely on hierarchies of mathematical
structures to achieve modularity, akin to interfaces in generic programming. By
analogy, we call *instance* a witness of a mathematical structure on a given carrier. Mathematical structures, as interfaces, are essential to factor out the shared
*vocabulary* attached to their instances. This vocabulary comes in the form of formal definitions and generic theorems, but also parsable and printable notations,
and sometimes delimited automation. Some mathematical structures are *richer*
than others in the sense that they extend them. Like in generic programming,

N. Peltier and V. Sofronie-Stokkermans (Eds.): IJCAR 2020, LNAI 12167, pp. 3–20, 2020.
https://doi.org/10.1007/978-3-030-51054-1_1

rich structures *inherit* the vocabulary attached to poorer structures. Working out the precise meaning of symbols of this shared vocabulary is usually performed by enhanced type inference, which is implemented using type classes [5,27,28] or unification hints [3,17,20,24]. In particular, these mechanisms must automatically identify inheritance relations between structures.

This paper discusses the design of a hierarchy of mathematical structures supporting a Coq [30] formal library for functional analysis, i.e., the study of spaces of functions, and of structure-preserving transformations on them. The algebraic vocabulary of linear algebra is complemented with a suitable notion of "closeness" (e.g., topology, distance, norm), so as to formalize convergence, limits, size, etc. This hierarchy is based on the *packed classes* methodology [17, 20], which represents structures using dependent records. The library strives to provide notations and theories that are as generic as they would be on paper. It is an extension of the "Mathematical Components" library [32] (hereafter MathComp), which is geared towards algebra. This extension is inspired by the Coquelicot real analysis library [7], which has its own hierarchy.

Fusing these two hierarchies, respectively from the MathComp and Coquelicot libraries, happens to trigger several interesting issues, related to inheritance relations. Indeed, when several constructions compete to infer an instance of the poorer structure, the proof assistant displays indiscernible notations, or keywords, for constructions that are actually different. This issue is not at all specific to the formalization of functional analysis: actually, the literature reports examples of this exact problem but in different contexts, e.g., in Lean's math-lib [11,33]. It is however more likely to happen when organizing different flavors of mathematics in a same coherent corpus, as this favors the presence of problematic competing constructions. Up to our knowledge, the problem of competing inheritance paths in hierarchies of dependent records was never discussed per se, beyond some isolated reports of failure, and of ad hoc solutions. We thus present and discuss a general methodology to overcome this issue, coined *forgetful inheritance*, based on packed classes and unification hints.

The paper is organized as follows: in Sect. 2, we recall the packed classes methodology, using a running example. Section 3 provides two concrete examples of the typical issues raised by the presence of competing inheritance paths, before describing the general issue, drawing its solution, and comparing with other type-class-like mechanisms. Finally, Sect. 4 describes the design of our hierarchy of structures for functional analysis, and its features, before Sect. 5 concludes.

## 2    Structures, Inheritance, Packed Classes

We recall some background on the representation of mathematical structures in dependent type theory, and on the construction of hierarchies using packed classes. For that purpose, we use a toy running example (see the accompanying file `packed_classes.v` [1]), loosely based on the case study presented in Sect. 4.

### 2.1   Dependent Records

In essence, a mathematical structure attaches to a carrier set some data (e.g., operators of the structure, collections of subsets of the carrier) and prescribed properties about these data, called the *axioms* of the structure. The Calculus of Inductive Constructions [16], as implemented, e.g., by Coq [30], Agda [29], or Lean [21,22], provides a *dependent record* construct, which allows to represent a given mathematical structure as a type, and its instances as terms of that type. A dependent record is an inductive type with a single constructor, which generalizes dependent pairs to dependent tuples. The elements of such a tuple are the arguments of the single constructor. They form a *telescope* [10], i.e., collection of terms, whose types can depend on the previous items in the tuple.

For example, the `flatNormModule` record type formalizes a structure with two operators, `fminus` and `fnorm`, and one axiom `fnormP`. This structure is a toy generalisation for the mathematical notion of normed module. Its purpose is to simulate one basic axiom of norms via a minimal amount of constructors. Thus, the `flatNormModule` has a single constructor named `FlatNormModule`, and four projections (also called fields) `carrier`, `fminus`, `fnorm`, and `fnormP`, onto the respective components of the tuple:

```
Structure flatNormModule := FlatNormModule {
  carrier : Type ;
  fminus : carrier → carrier → carrier;
  fnorm : carrier → nat;
  fnormP : ∀ x : carrier, fnorm (fminus x x) = 0 }.
```

The `fnormP` axioms makes use of `fminus x x` to avoid the introduction of a 0 of carrier type. Fields have a dependent type, parameterized by the one of the structure:

```
fminus : ∀ f : flatNormModule, carrier f → carrier f → carrier f
fnormP : ∀ (f : flatNormModule) (x : carrier f), fnorm f (fminus f x x) = 0.
```

In this case, declaring an instance of this structure amounts to defining a term of type `flatNormModule`, which packages the corresponding instances of carrier, data, and proofs. For example, here is an instance on carrier Z (using the `Z.sub` and `Z.abs_nat` functions from the standard library resp. as the `fminus` and the `fnorm` operators):

```
Lemma Z_normP (n : Z) : Z.abs_nat (Z.sub n n) = 0. Proof. ... Qed.
Definition Z_flatNormModule := FlatNormModule Z Z.sub Z.abs_nat Z_normP.
```

## 2.2   Inference of Mathematical Structures

Hierarchies of mathematical structures are formalized by nesting dependent records but naive approaches quickly incur scalability issues. *Packed classes* [17, 20] provide a robust and systematic approach to the organization of structures into hierarchies. In this approach, a *structure* is a two-field record, which associates a carrier with a *class*. A class encodes the inheritance relations of the structure and packages various *mixins*. Mixins in turn provide the data, and their properties. In Coq, `Record` and `Structure` are synonyms, but we reserve the latter for record types that represent actual structures. Let us explain the basics of inference of mathematical structures with packed classes by replacing the structure of Sect. 2.1 with two structures represented as packed classes. The first one provides just a binary operator:

```
1 Record isModule T := IsModule { minus_op : T → T → T }.
2 Structure module := Module {
3    module_carrier  : Type;
4    module_isModule : isModule module_carrier }.
```

Since the module structure is expected to be the bottom of the hierarchy, we are in the special class where the class is the same as the mixin (here, the class would be *equal* to isModule). To endow the operator minus_op with a generic infix notation, we introduce a definition minus, parameterized by an instance of module. In the definition of the corresponding notation, the wildcard _ is a placeholder for the instance of module to be inferred from the context.

```
Definition minus (M : module) :
    module_carrier M → module_carrier M → module_carrier M :=
  minus_op _ (module_isModule M).
Notation "x - y" := (minus _ x y).
```

We can build an instance of the module structure with the type of integers as the carrier and the subtraction of integers for the operator:

```
Definition Z_isModule : isModule Z := IsModule Z Z.sub.
Definition Z_module              := Module   Z Z_isModule.
```

But defining an instance is not enough to make the _ - _ notation available:

```
Fail Check ∀ x y : Z, x - y = x - y.
```

To type-check the expression just above, Coq needs to fill the wildcard in the _ - _ notation, which amounts to solving the equation module_carrier ?M ≡ Z, where _ ≡ _ is the definitional equality (i.e., equality up to the conversion rules of Coq type theory [30, Section "Conversion Rules" in Chapter "Calculus of Constructions"]). One can indicate that the instance Z_module is a *canonical* solution by declaring it as a *canonical instance*:

```
Canonical Z_module.
Check ∀ x y : Z, x - y = x - y.
```

This way, Coq fills the wildcard in minus _ x y with Z_module and retrieves as expected the subtraction for integers.

## 2.3   Inheritance and Packed Classes

We introduce a second structure class to illustrate how inheritance is implemented. This structure extends the module structure of Sect. 2.2 with a norm operator and a property (the fact that $x - x$ is 0 for any $x$):

```
1 Record naiveNormMixin (T : module) := NaiveNormMixin {
2    naive_norm_op  : T → nat ;
3    naive_norm_opP : ∀ x : T, naive_norm_op (x - x) = 0 }.
4 Record isNaiveNormModule (T : Type) := IsNaiveNormModule {
```

```
 5    nbase : isModule T ;
 6    nmix  : naiveNormMixin (Module _ nbase) }.
 7  Structure naiveNormModule := NaiveNormModule {
 8    naive_norm_carrier              :> Type;
 9    naive_normModule_isNormModule : isNaiveNormModule naive_norm_carrier
          }.
10  Definition naive_norm (N : naiveNormModule) :=
11    naive_norm_op _ (nmix _ (naive_normModule_isNormModule N)).
12  Notation "| x |" := (naive_norm _ x).
```

The new mixin for the norm appears at line 1 (it takes a `module` structure as
parameter), the new class appears at line 4, and the structure[1] at line 7. It is the
class that defines the inheritance relation between `module` and `naiveNormModule`
(at line 6 precisely). The definitions above are however not enough to achieve
proper inheritance. For example, `naiveNormModule` does not yet enjoy the _ - _
notation coming with the `module` structure:

```
Fail Check ∀ (N : naiveNormModule) (x y : N), x - y = x - y.
```

Here, Coq tries to solve the following equation[2]:

```
module_carrier ?M ≡ naive_norm_carrier N
```

The solution consists in declaring a canonical way to build a `module` structure
out of a `naiveNormModule` structure in the form of a function that Coq can use to
solve the equation above (using `naiveNorm_isModule N` in this particular case):

```
Canonical naiveNorm_isModule (N : naiveNormModule) :=
  Module N (nbase _ (naive_normModule_isNormModule N)).
Check ∀ (N : naiveNormModule) (x y : N), x - y = x - y.
```

# 3    Inheritance with Packed Classes and type classes

When several inheritance paths compete to establish that one structure extend
the other, the proof assistant may display misleading information to the user
and prevent proofs.

## 3.1    Competing Inheritance Paths

We extend the running example of Sect. 2 with a third and last structure. Our
objective is to implement a toy generalisation of the mathematical notion of
pseudometric space. This is done by introducing a reflexive relation mimicking
the belonging of one of the argument to a unit ball around the other argument.
The `hasReflRel` structure below provides a binary relation (line 2) together with
a property of reflexivity (line 3):

---

[1] The notation `:>` in the structure declares the carrier as a *coercion* [30, Chapter
"Implicit Coercions"], which means that Coq has the possibility to use the function
`naive_norm_carrier` to fix type-mismatches, transparently for the user.

[2] The application of `naive_norm_carrier` is not necessary in our case thanks to the
coercion explained in footnote (see footnote 1).

```
1 Record isReflRel T := IsReflRel {
2   ball_op           : T → T → Prop ;
3   ball_opP          : ∀ x : T, ball_op x x }.
4 Structure hasReflRel    := HasReflRel {
5   hasReflRel_carrier    :> Type;
6   hasReflRel_isReflRel  : isReflRel hasReflRel_carrier}.
7 Definition ball {N : hasReflRel} := ball_op _ (hasReflRel_isReflRel N).
8 Notation "x ~~ y" := (ball x y).
```

For the sake of the example, we furthermore declare a canonical way of building a `hasReflRel` structure out of a `naiveNormModule` structure:

```
Variable (N : naiveNormModule).
Definition norm_ball (x : N) := fun y : N => |x - y| ≤ 1.
(* details about naiveNormModule_isReflRel omitted *)
Canonical nnorm_hasReflRel := HasReflRel N naiveNormModule_isReflRel.
```

We first illustrate the issue using a construction (here the Cartesian product) that preserves structures, and that is used to build canonical instances. First, we define the product of `module` structures, and tag it as canonical:

```
Variables (M M' : module).
Definition prod_minus (x y : M * M') := (fst x - fst y, snd x - snd y).
Definition prod_isModule         := IsModule (M * M') prod_minus.
Canonical  prod_Module           := Module   (M * M') prod_isModule.
```

Similarly, we define canonical products of `hasReflRel` and `naiveNormModule`:

```
1 Variables (B B' : hasReflRel) (N N' : naiveNormModule).
2 Definition prod_ball (x y : B * B') := fst x ~~ fst y ∧ snd x ~~ snd y.
3 (* definition of prod_isReflRel omitted from the paper *)
4 Canonical prod_hasReflRel := HasReflRel (B * B') prod_isReflRel.
5
6 Definition prod_nnorm (x : N * N') := max (|fst x|) (|snd x|).
7 (* definition of prod_isNNModule omitted from the paper *)
8 Canonical prod_naiveNormModule := NaiveNormModule (N * N')
      prod_isNNModule.
```

The problem is that our setting leads Coq's type-checker to fail in unexpected ways, as illustrated by the following example:

```
Variable P : ∀ {T}, (T → Prop) → Prop.
Example failure (Pball : ∀ V : naiveNormModule, ∀ v : V, P (ball v))
  (W : naiveNormModule) (w : W * W): P (ball w).
Proof. Fail apply Pball. Abort.
```

The hypothesis `Pball` applies to any goal `P (ball v)` where the type of `v` is of type `naiveNormModule`, so that one may be led to think that it should also apply in the case of a product of `naiveNormModules`, since there is a canonical way to build one. What happens is that the type-checker is looking for an instance of a normed module that satisfies the following equation:

```
nnorm_hasReflRel ?N ≡
    prod_hasReflRel (nnorm_hasReflRel W) (nnorm_hasReflRel W)
```

while the canonical instance Coq infers is `?N := prod_naiveNormModule`
W W, which does not satisfy the equation. In particular, (`ball_op x`
`y`) is definitionally equal to $|x - y| \leq 1$ on the left-hand side and
(`fst x ~~ fst y` $\wedge$ `snd x ~~ snd y`) on the right-hand side: the two are not
definitionally equal. One can describe the problem as the fact that the diagram
in Fig. 1 *does not* commute definitionally.

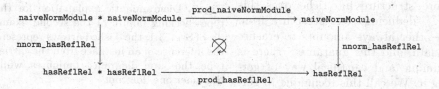

**Fig. 1.** Diagrammatic explanation for the failure of the first example of Sect. 3.1

This is of course not specific to Cartesian products and similar problems
would also occur when lifting dependent products, free algebras, closure, com-
pletions, etc., on metric spaces, topological groups, etc. as well as in simpler
settings without generic constructions as illustrated by our last example.

As a consequence of the definition of `nnorm_hasReflRel`, the following lemma
about balls is always true for any `naiveNormModule`:

```
Lemma ball_nball (N : nNormModule) (x y : N) : x ~~ y ↔ |x - y| ≤ 1.
Proof. reflexivity. Qed.
```

For the sake of the example, we define canonical instances of the `hasReflRel`
and `naiveNormModule` structures with integers:

```
Definition Z_ball (m n : Z) := (m = n ∨ m = n + 1 ∨ m = n - 1)%Z.
(* definition of Z_isReflRel omitted *)
Canonical Z_hasReflRel := HasReflRel Z Z_isReflRel.

Definition Z_naiveNormMixin := NaiveNormMixin Z_module Z.abs_nat Z_normP.
Canonical Z_naiveNormModule :=
    NaiveNormModule Z (IsNaiveNormModule Z_naiveNormMixin).
```

Since the generic lemma `ball_nball` holds, the user might expect to use it
to prove a version specialized to integers. This is however not the case as the
following script shows:

```
Example failure (x y : Z) : x ~~ y ↔ |x - y| ≤ 1.
rewrite -ball_nball. (* the goal is: x ~~ y ↔ x ~~ y *)
Fail reflexivity. (* !!! *)
```

The problem is that on the left-hand side Coq infers the instance `Z_hasReflRel`
with the `Z_ball` relation, while on the right-hand side it infers the instance
`nnorm_hasReflRel Z_naiveNormModule` whose `ball x y` is definitionally equal to
$|x - y| \leq 1$, which is not definitionally equal to the newly defined `Z_ball`.

In other words, the problem is the multiple ways to construct a "canonical instance" of hasReflRel with carrier Z, as in Fig. 2.

The solution to the problems explained in this section is to ensure definitional equality by including poorer structures into richer ones; this way, "deducing" one structure from the other always amounts to erasure of data, and this guarantees there is a unique and canonical way of getting it. We call this technique *forgetful inheritance*, as it is reminiscent of forgetful functors in category theory.

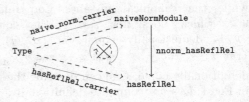

**Fig. 2.** Diagrammatic explanation for the type-checking failure of the second example of Sect. 3.1: the dashed arrows represent the inference of an instance from the carrier type; the outer diagrams commutes, while the inner one does not

## 3.2 Forgetful Inheritance with Packed Classes

When applied to the first problem exposed in Sect. 3.1, forgetful inheritance makes the diagram of Fig. 1 commute *definitionally*. Indeed, the only way to achieve commutation is to have nnorm_hasReflRel be a mere erasure. This means that one needs to include inside each instance of normModule a canonical hasReflRel (line 7 below). Furthermore the normMixin must record the compatibility between the operators ball_op and norm_op (line 4 below):

```
1 Record normMixin (T : module) (m : isReflRel T) := NormMixin {
2   norm_op         : T → nat;
3   norm_opP        : ∀ x,   norm_op (x - x) = 0;
4   norm_ball_opP : ∀ x y, ball_op _ m x y ↔ norm_op (x - y) ≤ 1 }.
5 Record isNormModule (T : Type) := IsNormModule {
6   base : isModule T;
7   bmix : isReflRel T;
8   mix  : normMixin (Module _ base) bmix }.
9 Structure normModule := NormModule {
10   norm_carrier            :> Type;
11   normModule_isNormModule : isNormModule norm_carrier }.
12 Definition norm (N : normModule) :=
13   norm_op _ _ (mix _ (normModule_isNormModule N)).
```

Since every normModule includes a canonical hasReflRel, the construction of the canonical hasReflRel given a normModule is exactly a projection:

```
Canonical norm_hasReflRel (N : normModule) :=
  HasReflRel N (bmix _ (normModule_isNormModule N)).
```

As a consequence, the equation

```
norm_hasReflRel ?N ≡
    prod_hasReflRel (norm_hasReflRel W) (norm_hasReflRel W)
```

holds with `prod_normModule W W` and the diagram in Fig. 1 (properly updated with the new `normModule` structure) commutes definitionally, and so does the diagram in Fig. 2, for the same reasons.

*Factories.* Because of the compatibility axioms required by forgetful inheritance, the formal definition of a structure can depart from the expected presentation. In fact, with forgetful inheritance, the very definition of a mathematical structure should be read in *factories*, i.e., functions that construct the mixins starting from only the expected axioms. And `Structure` records are rather interfaces, in a software engineering terminology. Note that just like there can be several equivalent presentations of a same mathematical stuctures, several mixins can be associated with a same target `Structure`.

In our running example, one can actually derive, from the previously defined `naiveNormMixin`, two mixins for both `hasReflRel`:

```
Variable (T : module) (m : naiveNormMixin T).
Definition fact_ball (x y : T) := naive_norm_op T m (x - y) ≤ 1.
Lemma fact_ballP (x : T) : fact_ball x x. Proof. (* omitted *). Qed.
Definition nNormMixin_isReflRel := IsReflRel T fact_ball fact_ballP.
```

(where the `ball` relation is the one induced by the norm, by construction) and `normModule`:

```
(* details for fact_normP and fact_norm_ballP omitted from the paper *)
Definition nNormMixin_normMixin :=
  NormMixin T nNormMixin_isReflRel (naive_norm_op T m)
          fact_normP fact_norm_ballP.
```

These two mixins make `naiveNormMixin` the source of two factories we mark as coercions, in order to help building two structures:

```
Coercion nNormMixin_isReflRel : naiveNormMixin ↣ isReflRel.
Coercion nNormMixin_normMixin  : naiveNormMixin ↣ normMixin.

Canonical alt_Z_hasReflRel := HasReflRel Z Z_naiveNormMixin.
Canonical alt_Z_normModule :=
  NormModule Z (IsNormModule Z_naiveNormMixin).
```

The second part of this paper provides concrete examples of factories for our hierarchy for functional analysis.

## 3.3  Forgetful Inheritance with type classes

Type class mechanisms [5, 27, 28] propose an alternative implementation of hierarchies. Inference relations are coded using parameters rather than projections, and proof search happens by enhancing the resolution of implicit arguments. But the issue of competing inheritance paths does not pertain to the inference mechanism at stake, nor to the prover which implements them. Its essence rather lies in the non definitionally commutative diagrams as in Fig. 1 and Fig. 2.

We illustrate this with a type classes version of our examples, in both Coq and Lean, using a semi-bundled approach (see the accompanying files `type_classes.v` and `type_classes.lean` [1]). Compared to the packed class approach, hierarchies implemented using type classes remove the *structure* layer, which packages the carrier and the *class*. Hence our example keeps only the records whose name starts with `is`, declares them as type classes, and substitutes `Canonical` declarations with appropriate `Instance` declarations.

The choice on the level of bundling in the resulting classes, i.e., what are parameters of the record, and what are its fields, is not unique. For example, one may choose to formalize rings as groups extended with additional operations and axioms, or as a class on a type which is also a group.

```
Class isGroup T := IsGroup { ... };
Class isRing_choice1 T := IsRing { ring_isGroup : isGroup T; ... }.
Class isRing_choice2 T '{isGroup T} := IsRing { ... }.
```

By contrast, a structure in the packed class approach must always package with its carrier every mixins and classes that characterize the structure. The transposition of forgetful inheritance to type class would apply to fully bundled classes (where all the operations and axioms are bundled but not the carrier).

Because it triggers no "backtracking search", the use of packed classes and unification hints described in this paper avoids some issues encountered in mathlib [33, Sect. 4.3], which are more explicitly detailed in the report on the implementation of type classes in Lean 4 [26]. We do not know either how a type class version of forgetful inheritance would interact with the performance issues described in the latter paper, or whether tabling helps. Moreover, with the current implementations of type classes in both Coq and Lean, different choices on bundling may have dramatic consequences on resolution feasibility and performance. For example, former experiments in rewriting MathComp with type classes in Coq did not scale up to modules on a ring. Incidentally, our companion file `type_classes.v` illustrates some predictability issues of the current implementation of Coq type classes.

## 4    The Mathematical Components Analysis Library

The Coquelicot library comes with its own hierarchy of mathematical structures and the intent of the MathComp-Analysis library is to improve it with the algebraic constructs of the MathComp library, for the analysis of multivariate functions for example. This section explains three applications of forgetful inheritance that solve three design issues of a different nature raised by merging MathComp and Coquelicot, as highlighted in Fig. 3.

We begin by an overview of the mathematical notions we deal with in this section. A *topological space* is a set endowed with a topology, i.e., a total collection of open sets closed under finite intersection and arbitrary unions. Equivalently, a topology can be described by the *neighborhood filter* of each point. A

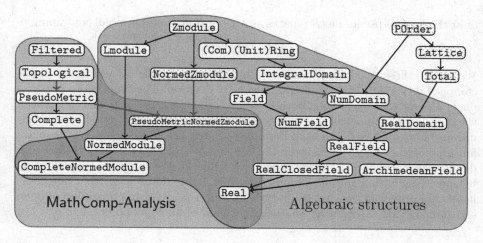

**Fig. 3.** Excerpt of MathComp and MathComp-Analysis hierarchies. Each rounded box corresponds to a mathematical structure except for `(Com)(Unit)Ring` that corresponds to several structures [17]. Dotted boxes correspond to mathematical structures introduced in Sect. 4.2 and Sect. 4.3. Thick, red arrows correspond to forgetful inheritance.

*neighborhood* of a point $x$ is a set containing an open set around $x$; the neighborhood filter of a point $x$ is the set of all neighborhoods of $x$. In MathComp-Analysis, neighborhood filters are the primary component of topological spaces. *Pseudometric spaces* are intermediate between topological and metric spaces. They were introduced as the minimal setting to handle Cauchy sequences. In Coquelicot, pseudometric spaces are called "uniform spaces" and are formalized as spaces endowed with a suitable `ball` predicate. This is the topic of Sect. 4.1. Coquelicot also provides *normed spaces*, i.e., $\mathbb{K}$-vector spaces $E$ endowed with a suitable norm. On the other hand, in MathComp, the minimal structure with a norm operator corresponds to *numerical domains* [12, Chap. 4][13, Sect. 3.1], i.e., integral domains with order and absolute value. This situation led to a generalization of MathComp described in Sect. 4.2. Finally, in Sect. 4.3, we explain how to do forgetful inheritance across the two distinct libraries MathComp and MathComp-Analysis.

## 4.1 Forgetful Inheritance from Pseudometric to Topological Spaces

When formalizing topology, we run into a problem akin to Sect. 3.1 because we face several competing notions of neighborhoods; we solve this issue with forgetful inheritance as explained in Sect. 3.

A neighborhood of a point $p$ can be defined at the level of topological spaces using the notion of open as a set $A$ that contains an open set containing $p$:

$$\exists B. \ B \text{ is open}, \ p \in B \text{ and } B \subseteq A. \tag{1}$$

or at the level of pseudometric spaces as a set $A$ that contains a ball containing $p$:

$$\exists \varepsilon > 0. \ B_\varepsilon(p) \subseteq A. \tag{2}$$

We ensure these two definitions of neighborhoods coincide by adding to mixins compatibility axioms that constrain a shared function. The function in question maps a point to a set of neighborhoods (hereafter `locally`), it is shared between the mixins for topological and pseudometric spaces, and constrained by the definitions of open set and ball as in Formulas (1) and (2). More precisely, the mixin for topological spaces introduces the set of open sets (see line 3 below) and defines neighborhoods as in Formula (1) (at line 5). We complete the definition by specifying with a specific axiom (not explained in detail here) that neighborhoods are proper filters (line 4) and with an alternative characterization of open set (namely that an open set is a neighborhood of all of its points, line 6).

```
1 (* Module Topological. *)
2 Record mixin_of (T : Type) (locally : T → set (set T)) := Mixin {
3   open : set (set T) ;
4   ax1 : ∀ p : T, ProperFilter (locally p) ;
5   ax2 : ∀ p : T, locally p = [set A | ∃ B, open B ∧ B p ∧ B ⊆ A] ;
6   ax3 : open = [set A : set T | A ⊆ (fun x => locally x A) ] }.
```

The mixin for pseudometric spaces introduces the notion of balls (line 10) and defines neighborhoods as in Formula (2) (at line 12, `locally_ ball` corresponds to the set of supersets of balls). The rest of the definition (line 11) are axioms about `ball` which are omitted for lack of space.

```
7  (* Module PseudoMetric. *)
8  Record mixin_of (R : numDomainType) (M : Type)
9     (locally : M → set (set M)) := Mixin {
10   ball : M → R → M → Prop ;
11   ax1 : ... ; ax2 : ... ; ax3 : ... ;
12   ax4 : locally = locally_ ball }.
```

Here, our definition of topological space departs from the standard definition as a space endowed with a family of subsets containing the full set and the empty set and closed under union and by finite intersection. However, the latter definition can be recovered from the former. Factories (see Sect. 3.2) are provided for users who want to give only `open` and to infer `locally` (using [31], file `topology.v`, definition `topologyOfOpenMixin`]), or the other way around.

## 4.2 Forgetful Inheritance from Numerical Domain to Normed Abelian Group

The second problem we faced when developing the MathComp-Analysis library is the competing formal definitions of norms and absolute values. The setting is more complicated than Sect. 4.1 as it involves amending the hierarchy of mathematical structures of the MathComp library.

While the codomain of a norm is always the set of (non-negative) reals, an absolute value on a `numDomainType` is always an endofunction `norm` of type

$\forall$ (R : numDomainType), R $\rightarrow$ R. Thanks to this design choice, the absolute value preserves some information about its input, e.g., the integrality of an integer. On the other hand, the Coquelicot library also had several notions of norms: the absolute value of the real numbers (from the Coq standard library), the absolute value of a structure for rings equipped with an absolute value, and the norm operator of normed modules (the latter two are Coquelicot-specific).

We hence generalize the norm provided by the MathComp library to encompass both absolute values on numerical domains and norms on vector spaces, and share notation and lemmas. This is done by introducing a new structure in MathComp called normedZmodType, for normed Abelian groups, since Abelian groups are called $\mathbb{Z}$-modules in MathComp. This structure is now the poorest structure with a norm, which every subsequent normed type will inherit from.

The definition of the normedZmodType structure requires to solve a mutual dependency problem. Indeed, to state the fundamental properties of norms, such as the triangle inequality, the codomain of the norm function should be at least an ordered and normed Abelian group, requiring normedZmodType to be parameterized by such a structure. However the codomain should also inherit from normedZmodType to share the notations for norm and absolute value.

Our solution is to dispatch the order and the norm originally contained in numDomainType between normed Abelian groups normedZmodType and partially ordered types porderType as depicted in Fig. 3. We define the two following mixins for normedZmodType and numDomainType.

```
Record normed_mixin_of (R T : zmodType) (Rorder : lePOrderMixin R) :=
  NormedMixin { norm_op : T → R ; (* properties of the norm omitted *) }.
```

```
Record num_mixin_of (R : ringType) (Rorder : lePOrderMixin R)
  (normed : @normed_mixin_of R R Rorder) := Mixin { (* omitted *) }.
```

Now we define numDomainType (which is an abbreviation for NumDomain.type) using these two mixins but *without* declaring inheritance from normedZmodType (yet to be defined). More precisely, the class of numDomainType includes the mixin for normedZmodType (at line 5 below), which will allow for forgetful inheritance:

```
1 (* Module NumDomain. *)
2 Record class_of T := Class {
3   base : GRing.IntegralDomain.class_of T ;
4   order_mixin : lePOrderMixin (ring_for T base) ;
5   normed_mixin : normed_mixin_of (ring_for T base) order_mixin ;
6   num_mixin : num_mixin_of normed_mixin }.
7 Structure type := Pack { sort :> Type; class : class_of sort }.
```

Finally, we define the class of normedZmodType, parameterized by a numDomainType:

```
(* Module NormedZmodule. *)
Record class_of (R : numDomainType) (T : Type) := Class {
  base : GRing.Zmodule.class_of T ;
  normed_mixin : @normed_mixin_of R (@GRing.Zmodule.Pack T base)
          (NumDomain.class R) }.
```

It is only then that we declare inheritance from the `normedZmodType` structure to `numDomainType`, effectively implementing forgetful inheritance. We finally end up with a norm of the general type

```
norm : ∀ (R : numDomainType) (V : normedZmodType R), V → R.
```

*Illustration: sharing of norm notation and lemmas.* As an example, we explain the construction of two norms and show how they share notation and lemmas. In MathComp, the type of matrices is `'M[K]_(m,n)` where `K` is the type of coefficients. The norm `mx_norm` takes the maximum of the absolute values of the coefficients:

```
Variables (K : numDomainType) (m n : nat).
Definition mx_norm (x : 'M[K]_(m, n)) : K := \big[maxr/0]_i '|x i.1 i.2|.
```

This definition uses the generic big operator [6] to define a "big max" operation out of the binary operation `maxr`. Similarly, we define a norm for pairs of elements by taking the maximum of the absolute value of the two projections[3]:

```
Variables (R : numDomainType) (U V : normedZmodType R).
Definition pair_norm (x : U * V) : R := maxr '|x.1| '|x.2|.
```

We then go on proving that these definitions satisfy the properties of the norm and declare canonical instances of `normedZmodType` for matrices and pairs (see [31] for details). All this setting is of course carried out in advance and the user only sees one notation and one set of lemmas (for example `ler_norm_add` for the triangle inequality), so that (s)he can mix various norms transparently in the same development, as in the following two examples:

```
Variable (K : numDomainType).
Example mx_triangle m n (M N : 'M[K]_(m, n)) : '|M + N| ≤ '|M| + '|N|.
Proof. apply ler_norm_add. Qed.
Example pair_triangle (x y : K * K) : '|x + y| ≤ '|x| + '|y|.
Proof. apply ler_norm_add. Qed.
```

One could fear that the newly introduced structures make the library harder to use since that, to declare a canonical `numDomainType` instance, a user also needs now to declare canonical `porderType` and `normedZmodType` instances of the same type. Here, the idea of factories (Sect. 3.2) comes in handy for the original `numDomainType` mixin has been re-designed as a factory producing `porderType`, `normedZmodType`, and `numDomainType` mixins in order to facilitate their declaration.

## 4.3 Forgetful Inheritance from Normed Modules to Pseudometric Spaces

The combination of the MathComp library with topological structures ultimately materializes as a mathematical structure for *normed modules*. It is made possible by introducing an intermediate structure that combines norm (from algebra)

---

[3] The actual code of `mx_norm` and `pair_norm` is slightly more complicated because it uses a type for non-negative numeric values, see [31, file `normedtype.v`].

with pseudometric (from topology) into normed Abelian groups. The precise justification for this first step is as follows.

Since normed Abelian groups have topological and pseudometric space structures induced by the norm, `NormedZmodType` should inherit from `PseudoMetricType`. To do so, we can (1) insert a new structure above `NormedZmodType`, or (2) create a common extension of `PseudoMetricType` and `NormedZmodType`. We choose (2) to avoid amending the MathComp library. This makes both `NormedZmodType` and its extension `PseudoMetricNormedZmodType` normed Abelian groups, where the former is inadequate for topological purposes.

The only axiom of this extension is the compatibility between the pseudometric and the norm, as expressed line 5 below, where `PseudoMetric.ball` has been seen in Sect. 4.1 and the right-hand side represents all the ternary relations $\lambda x, \varepsilon, y. |x - y| < \varepsilon$:

```
1 (* Module PseudoMetricNormedZmodule. *)
2 Variable R : numDomainType.
3 Record mixin_of (T : normedZmodType R) (locally : T → set (set T))
4    (m : PseudoMetric.mixin_of R locally) :=
5  Mixin { _ : PseudoMetric.ball m = ball_ (fun x => '|x|) }.
```

The extension is effectively performed by using this mixin in the following class definition at line 12 (see also Fig. 3):

```
6 Record class_of (T : Type) := Class {
7  base : Num.NormedZmodule.class_of R T ; ... ;
8  locally_mixin : Filtered.locally_of T T ;
9  topological_mixin : @Topological.mixin_of T locally_mixin ;
10 pseudometric_mixin : @PseudoMetric.mixin_of R T locally_mixin ;
11 mixin :
12   @mixin_of (Num.NormedZmodule.Pack _ base) _ pseudometric_mixin }.
```

Finally, the bridge between algebraic and topological structures is completed by a common extension of a normed Abelian group (`PseudoMetricNormedZmodType`) with a left-module (`lmodType` from the MathComp library, which provides scalar multiplication), extended with the axiom of linearity of the norm for the scalar product (line 5 below).

```
1 (* Module NormedModule. *)
2 Variable (K : numDomainType).
3 Record mixin_of
4    (V : pseudoMetricNormedZmodType K) (scale : K → V → V) :=
5  Mixin { _ : ∀ (l : K) (x : V), '|scale l x| = '|l| * '|x| }.
```

One can again observe here the overloaded notation for norms explained in Sect. 4.2. The accompanying file `scalar_notations.v` [1] provides an overview of MathComp-Analysis operations regarding norms and scalar notations.

We ensured that the structure of normed modules indeed serves its intended purpose of enabling multivariate functional analysis by generalizing existing theories of Bachmann-Landau notations and of differentiation [2, Sect. 4].

# 5   Conclusion and Related Work

This paper has two main contributions: forgetful inheritance using packed classes, and the hierarchy of the MathComp-Analysis library. The latter library is still in its infancy and covers far less real and complex analysis than the libraries available in HOL Light and Isabelle/HOL [19,23]. However, differences in foundations matter here, and the use of dependent types in type-class-like mechanisms is instrumental in the genericity of notations illustrated in this paper. Up to our knowledge, no other existing formal library in analysis has comparable sharing features.

The methodology presented in this paper to tame competing inheritance paths in hierarchies of dependent records is actually not new. The original description of packed classes [17, end of Sect. 3.1] already mentions that a choice operator (in fact, a mixin) should be included in the definition of a structure for countable types, even if choice operators can be defined for countable types in Coq without axiom. Yet, although the MathComp library uses forgetful inheritance at several places in its hierarchy, this solution was never described in earlier publications, nor was the issue precisely described. Proper descriptions, as well as the comparison with other inference techniques, are contributions of the present paper.

As explained in Sect. 3.3, type classes based on augmented inference of implicit arguments also allow for a variant of forgetful inheritance. For instance, Buzzard et al. mention that this pattern is used for defining metric spaces both in Lean and Isabelle/HOL's libraries [11, Sect. 3]. In the same paper, the authors describe another formalization issue, about completing abelian groups and rings, pertaining to the same problem [11, Sect. 5.3], and which can be solved as well using forgetful inheritance.

Extensional flavors of dependent type theory could make the problematic diagram in Fig. 1 commute judgmentally. However, to the best of our knowledge, the related tools [4,15] available at the time of writing do not implement any of the type class mechanisms discussed here.

Packed classes, and forgetful inheritance, already proved robust and efficient enough to formalize and populate large hierarchies [18], where "large" applies both to the number of structures and to the number of instances. Arguably, this approach also has drawbacks: defining deep hierarchies becomes quite verbose, and inserting new structures is tedious and error-prone. We argue that, compared to their obvious benefits in control and efficiency of the proof search, this is not a fundamental issue. As packed classes are governed by systematic patterns and invariants, this rather calls for more inspection [25] and automated generation [14] tooling, which is work in progress.

**Acknowledgments.** The authors are grateful to Georges Gonthier for the many fruitful discussions that helped rewriting parts of MathComp and MathComp-Analysis library. We also thank Arthur Charguéraud and all the anonymous reviewers for their comments on the successive versions of this paper.

# References

1. Affeldt, R., Cohen, C., Kerjean, M., Mahboubi, A., Rouhling, D., Sakaguchi, K.: Formalizing functional analysis structures in dependent type theory (accompanying material). https://math-comp.github.io/competing-inheritance-paths-in-dependent-type-theory (2020), contains the files packed_classes.v, packed_classes.v, type_classes.lean, and scalar_notations.v, and a stand-alone archive containing snapshots of the Mathematical Components [32] and Analysis [31] libraries
2. Affeldt, R., Cohen, C., Rouhling, D.: Formalization techniques for asymptotic reasoning in classical analysis. J. Formaliz. Reason. **11**, 43–76 (2018)
3. Asperti, A., Ricciotti, W., Sacerdoti Coen, C., Tassi, E.: Hints in unification. In: Berghofer, S., Nipkow, T., Urban, C., Wenzel, M. (eds.) TPHOLs 2009. LNCS, vol. 5674, pp. 84–98. Springer, Heidelberg (2009). https://doi.org/10.1007/978-3-642-03359-9_8
4. Bauer, A., Gilbert, G., Haselwarter, P.G., Pretnar, M., Stone, C.A.: Design and implementation of the andromeda proof assistant. CoRR abs/1802.06217 (2018). http://arxiv.org/abs/1802.06217
5. Bauer, A., Gross, J., Lumsdaine, P.L., Shulman, M., Sozeau, M., Spitters, B.: The HoTT library: a formalization of homotopy type theory in Coq. In: 6th ACM SIGPLAN Conference on Certified Programs and Proofs (CPP 2017), Paris, France, 16–17 January 2017, pp. 164–172. ACM (2017)
6. Bertot, Y., Gonthier, G., Ould Biha, S., Pasca, I.: Canonical big operators. In: Mohamed, O.A., Muñoz, C., Tahar, S. (eds.) TPHOLs 2008. LNCS, vol. 5170, pp. 86–101. Springer, Heidelberg (2008). https://doi.org/10.1007/978-3-540-71067-7_11
7. Boldo, S., Lelay, C., Melquiond, G.: Coquelicot: a user-friendly library of real analysis for Coq. Math. Comput. Sci. **9**(1), 41–62 (2015)
8. Bourbaki, N.: The architecture of mathematics. Am. Math. Mon. **57**(4), 221–232 (1950), http://www.jstor.org/stable/2305937
9. Bourbaki, N.: Théorie des ensembles. Éléments de mathématique. Springer (2006). Original Edition published by Hermann, Paris (1970)
10. de Bruijn, N.G.: Telescopic mappings in typed lambda calculus. Inf. Comput. **91**(2), 189–204 (1991)
11. Buzzard, K., Commelin, J., Massot, P.: Formalising perfectoid spaces. In: 9th ACM SIGPLAN International Conference on Certified Programs and Proofs (CPP 2020), New Orleans, LA, USA, 20–21 January 2020, pp. 299–312. ACM (2020)
12. Cohen, C.: Formalized algebraic numbers: construction and first-order theory. (Formalisation des nombres algébriques : construction et théorie du premier ordre). Ph.D. thesis, Ecole Polytechnique X (2012). https://pastel.archives-ouvertes.fr/pastel-00780446
13. Cohen, C., Mahboubi, A.: Formal proofs in real algebraic geometry: from ordered fields to quantifier elimination. Logic. Methods Comput. Sci. **8**(1) (2012). https://doi.org/10.2168/LMCS-8(1:2)2012
14. Cohen, C., Sakaguchi, K., Tassi, E.: Hierarchy Builder: algebraic hierarchies made easy in Coq with Elpi (2020). Accepted in the proceedings of FSCD 2020. https://hal.inria.fr/hal-02478907
15. Constable, R.L., et al.: Implementing Mathematics with the NuPRL Proof Development System. Prentice-Hall, Upper Saddle River (1986)

16. Coquand, T., Paulin, C.: Inductively defined types. In: Martin-Löf, P., Mints, G. (eds.) COLOG 1988. LNCS, vol. 417, pp. 50–66. Springer, Heidelberg (1990). https://doi.org/10.1007/3-540-52335-9_47

17. Garillot, F., Gonthier, G., Mahboubi, A., Rideau, L.: Packaging mathematical structures. In: Berghofer, S., Nipkow, T., Urban, C., Wenzel, M. (eds.) TPHOLs 2009. LNCS, vol. 5674, pp. 327–342. Springer, Heidelberg (2009). https://doi.org/10.1007/978-3-642-03359-9_23

18. Gonthier, G., et al.: A machine-checked proof of the odd order theorem. In: Blazy, S., Paulin-Mohring, C., Pichardie, D. (eds.) ITP 2013. LNCS, vol. 7998, pp. 163–179. Springer, Heidelberg (2013). https://doi.org/10.1007/978-3-642-39634-2_14

19. Harrison, J.: The HOL Light System REFERENCE (2017). https://www.cl.cam.ac.uk/~jrh13/hol-light/index.html

20. Mahboubi, A., Tassi, E.: Canonical structures for the working Coq user. In: Blazy, S., Paulin-Mohring, C., Pichardie, D. (eds.) ITP 2013. LNCS, vol. 7998, pp. 19–34. Springer, Heidelberg (2013). https://doi.org/10.1007/978-3-642-39634-2_5

21. Microsoft Reasearch: L∃∀N THEOREM PROVER (2020). https://leanprover.github.io

22. de Moura, L., Kong, S., Avigad, J., van Doorn, F., von Raumer, J.: The lean theorem prover (System Description). In: Felty, A.P., Middeldorp, A. (eds.) CADE 2015. LNCS (LNAI), vol. 9195, pp. 378–388. Springer, Cham (2015). https://doi.org/10.1007/978-3-319-21401-6_26

23. Nipkow, T., Paulson, L.C., Wenzel, M.: Isabelle HOL: A Proof Assistant for Higher-Order Logic. Springer (2019). https://isabelle.in.tum.de/doc/tutorial.pdf

24. Saïbi, A.: Outils Génériques de Modélisation et de Démonstration pour la Formalisation des Mathématiques en Théorie des Types. Application à la Théorie des Catégories. (Formalization of Mathematics in Type Theory. Generic tools of Modelisation and Demonstration. Application to Category Theory). Ph.D. thesis, Pierre and Marie Curie University, Paris, France (1999)

25. Sakaguchi, K.: Validating mathematical structures. In: Peltier, N., Sofronie-Stokkermans, V. (eds.) IJCAR 2020. LNAI, vol. 12167, pp. 138–157. Springer, Heidelberg (2020)

26. Selsam, D., Ullrich, S., de Moura, L.: Tabled typeclass resolution (2020). https://arxiv.org/abs/2001.04301

27. Sozeau, M., Oury, N.: First-class type classes. In: Mohamed, O.A., Muñoz, C., Tahar, S. (eds.) TPHOLs 2008. LNCS, vol. 5170, pp. 278–293. Springer, Heidelberg (2008). https://doi.org/10.1007/978-3-540-71067-7_23

28. Spitters, B., van der Weegen, E.: Type classes for mathematics in type theory. Math. Struct. Comput. Sci. **21**(4), 795–825 (2011)

29. The Agda Team: The Agda User Manual (2020). https://agda.readthedocs.io/en/v2.6.0.1. Version 2.6.0.1

30. The Coq Development Team: The Coq Proof Assistant Reference Manual. Inria (2019). https://coq.inria.fr. Version 8.10.2

31. The Mathematical Components Analysis Team: The Mathematical Components Analysis library (2017). https://github.com/math-comp/analysis

32. The Mathematical Components Team: The Mathematical Components library (2007). https://github.com/math-comp/math-comp

33. The mathlib Community: The lean mathematical library. In: 9th ACM SIGPLAN International Conference on Certified Programs and Proofs (CPP 2020), New Orleans, LA, USA, 20–21 January 2020, pp. 367–381. ACM (2020)

# A Lean Tactic for Normalising Ring Expressions with Exponents (Short Paper)

Anne Baanen$^{(\boxtimes)}$ [ID]

Vrije Universiteit Amsterdam, Amsterdam, The Netherlands
t.baanen@vu.nl

**Abstract.** This paper describes the design of the normalising tactic ring_exp for the Lean prover. This tactic improves on existing tactics by extending commutative rings with a binary exponent operator. An inductive family of types represents the normal form, enforcing various invariants. The design can also be extended with more operators.

## 1  Introduction

In interactive theorem proving, normalising tactics are powerful tools to prove equalities. Given an expression $a$, these tactics return an expression $a'$ in normal form together with a proof that $a = a'$. For instance, in mathlib [10], the mathematical library for the Lean theorem prover [7], the ring tactic normalises expressions in a commutative (semi)ring. Analogous tactics or conversions exist in many theorem provers [5,8,9]. The ring tactic in Lean can be directly invoked by the user and is called by the decision procedure linarith. The utility of ring is evident from the fact that it is invoked over 300 times in mathlib.

The ring tactic in Lean, and the tactic in Coq it is based on, use a Horner normal form representation of polynomials [4]. The Horner form represents a polynomial $f(x)$ with one of two cases: either it is constant ($f(x) = c$) or it is of the form $f(x) = c + x * g(x)$. This representation allows ring to uniquely and efficiently represent any polynomial, i.e. any expression consisting of the operators $+$ and $*$, numerals and variables. Problems arise when expressions include other operators than $+$ and $*$, such as the exponentiation operator $\wedge$. The Horner form fundamentally assumes the degree of a term is a constant integer, so it cannot be simply modified to represent variable exponents, or more generally to represent $\wedge$ applied to compound expressions. The analogous procedures in other theorem provers have the same restriction. Adding rewrite rules such as $x^{n+1} \mapsto x * x^n$ is not a universal solution. This rule would unfold the expression $x^{100}$ into a large term composed of repeated multiplications, reducing the performance of the procedure significantly. The result is that ring cannot prove that $2^{n+1} - 1 = 2 * 2^n - 1$ for a free variable $n : \mathbb{N}$.

The ring_exp tactic uses a new extensible normal form, currently supporting the operators $+$, $*$ and $\wedge$, numerals and variables. Its domain is a strict superset

N. Peltier and V. Sofronie-Stokkermans (Eds.): IJCAR 2020, LNAI 12167, pp. 21–27, 2020.
https://doi.org/10.1007/978-3-030-51054-1_2

of the domain of previous semiring tactics, without sacrificing too much of the efficiency of `ring`. This paper describes the design and engineering challenges encountered in implementing `ring_exp`.

The version of `ring_exp` discussed in this paper was merged into `mathlib` in commit 5c09372658.[1] Additional code and setup instructions are available online.[2]

## 2    Design Overview

The `ring_exp` tactic uses a normalisation scheme similar to the original `ring` tactic. The input from the tactic system is an abstract syntax tree representing the expression to normalise. An `eval` function maps inputs to a type `ex` of normalised expressions. The normal form should be designed in such a way that values of type `ex` are equal if and only if the input expressions can be proved equal using the axioms of commutative semirings. From the `ex` representation, the normalised output expression is constructed by a function `simple`. Both `eval` and `simple` additionally return a proof showing that the input and output expressions are equal.

The `ring_exp` tactic does not use reflection but directly constructs proof terms to be type checked by Lean's kernel, as is typical for tactics in `mathlib` [10]. Reflective tactics avoid the construction and checking of a large proof term by performing most computation during proof checking, running a verified program [2]. If the proof checker performs efficient reduction, this results in a significant speed-up of the tactic, at the same time as providing more correctness guarantees. Unfortunately, the advantages of reflection do not translate directly to Lean. Tactic execution in Lean occurs within a fast interpreter, while the kernel used in proof checking is designed for simplicity instead of efficient reduction [3]. Achieving an acceptable speed for `ring_exp` requires other approaches to the benefits that reflection brings automatically.

The language of semirings implemented by `ring`, with binary operators $+$, $*$ and optionally $-$ and $/$, is extended in `ring_exp` with a binary exponentiation operator $^\wedge$. The input expression can consist of these operators applied to other expressions, with two base cases: natural numerals such as 0 and 37, and *atoms*. An atom is any expression which is not of the above form, e.g. a variable name $x$ or a function application $\sin(y - z)$. It is treated as an opaque variable in the expression. Two such expressions are considered equal if in every commutative semiring they evaluate to equal values, for any assignment to the atoms.

Using a suitable representation of the normal form is crucial to easily guarantee correctness of the normaliser. Since there is no clear way to generalise the Horner form, `ring_exp` instead represents its normal form `ex` as a tree with operators at the nodes and atoms at the leaves. Certain classes of non-normalised expressions are prohibited by restricting which sub-node can occur for each node. The `ex` type captures these restrictions through a parameter in

---

[1] https://github.com/leanprover-community/mathlib/tree/5c09372658.
[2] https://github.com/lean-forward/ring_exp.

```
inductive ex_type : Type
| sum | prod | exp | base
inductive ex : ex_type → Type
| zero  : ex_info →                          ex sum      -- 0
| sum   : ex_info → ex prod →  ex sum → ex sum      -- +
| coeff : ex_info →                  coeff → ex prod   -- numerals
| prod  : ex_info → ex exp  → ex prod → ex prod   -- *
| exp   : ex_info → ex base → ex prod → ex exp    -- ^
| var   : ex_info →                   atom → ex base   -- atoms
| sum_b : ex_info →                 ex sum → ex base
```

**Fig. 1.** Definition of ex_type and ex

the enum ex_type, creating an inductive family of types. Each constructor allows specific members of the ex family in its arguments and returns a specific type of ex. The full definition is given in Fig. 1. The additional ex_info record passed to the constructors contains auxiliary information used to construct correctness proofs. The sum_b constructor allows sums as the base of a power, analogously to the parentheses in $(a+b)^c$.

For readability, we will write the ex representation in symbols instead of the constructors of ex. Thus, the term sum (prod (exp (var n) (coeff 1)) (coeff 1)) zero (with ex_info fields omitted) is written as $n^1 * 1 + 0$, and the normalised form of $2^n - 1$ is written $(2+0)^{n^1*1} * 1 + (-1) + 0$.

**Table 1.** Associativity and distributivity properties of the $+$, $*$ and $\wedge$ operators

|   | $+$ | $*$ | $\wedge$ |
|---|---|---|---|
| $+$ | $(a+b)+c = a+(b+c)$ | — | — |
| $*$ | $(a+b)*c = a*c+b*c;$ <br> $a*(b+c) = a*b+a*c$ | $(a*b)*c = a*(b*c)$ | — |
| $\wedge$ | $a^{b+c} = a^b * a^c$ | $(a*b)^c = a^c * b^c$ | $(a^b)^c = a^{b*c}$ |

The types of the arguments to each constructor are determined by the associativity and distributivity properties of the operators involved, summarised in Table 1. Since addition does not distribute over either other operator (as seen from the empty entries on the $+$ row), an expression with a sum as outermost operator cannot be rewritten so that another operator is outermost. Thus, the set of all expressions should be represented by ex sum. Since $*$ distributes over $+$ but not over $\wedge$, the next outermost operator after $+$ will be $*$. By associativity (the diagonal entries of the table) the left argument to $+$ should have $*$ as outermost operator; otherwise we can apply the rewrite rule $(a+b)+c \mapsto a+(b+c)$. Analogously, the left argument to the prod constructor is not an ex prod but an ex exp, and the left argument to exp is an ex base.

The `eval` function interprets each operator in the input expression as a corresponding operation on `ex`, building a normal form for the whole expression out of normalised subexpressions. The operations on `ex` build the correctness proof of normalisation out of the proofs for subexpressions using a correctness lemma: for example, the lemma `add_pf_z_sum : ps = 0 → qs = qs' → ps + qs = qs'` is used on the input expression `ps + qs` when `ps` normalises to 0.

Adding support for a new operator would take relatively little work: after extending the table of associativity and distributivity relations, one can insert the constructor in `ex` using the table to determine the relevant `ex_type`, and add an operation on `ex` that interprets the operator.

## 3   Intricacies

The `ex` type enforces that distributivity and associativity rules are always applied, but commutative semirings have more equations. In a normal form, arguments to commutative operators should be sorted according to some linear order $\prec$: if $a \prec b$, then $a + (b + 0)$ is normalised and $b + (a + 0)$ is not. Defining a linear order on `ex` requires an order on atoms; definitional equality of atoms is tested (with user control over the amount of definitional unfolding) in the `tactic` monad [3], so a well-defined order on atoms cannot be easily expressed on the type level. Additionally, the recursive structure of expressions means any expression $a$ can also be represented as $(a)^1 * 1 + 0$; if the left argument to $\wedge$ is 0 or $a * b + 0$, the expression is not in normal form. Although these invariants can also be encoded in a more complicated `ex` type, they are instead maintained by careful programming. A mistake in maintaining these invariants is not fatal: invariants only protect completeness, not soundness, of `ring_exp`.

Efficient handling of numerals in expressions, using the `coeff` constructor, is required for acceptable running time without sacrificing completeness. The tactic should not unfold expressions like $x * 1000$ as 1000 additions of the variable $x$. Representing numerals with the `coeff` constructor requires an extra step to implement addition. When terms overlap, differing only in the coefficients as for $a*b^2*1 + a*b^2*2$, their sum is given by adding their coefficients: $a*b^2*3$. Moreover, when the coefficients add up to 0, the correct representation is not $a * b^2 * 0$ : `ex prod` but 0 : `ex sum`. Coefficients must be treated similarly in exponents: $x^{a*b^2*1} * x^{a*b^2*2} = x^{a*b^2*3}$. Both cases are handled by a function `add_overlap` which returns the correct sum if there is overlap, or indicates that there is no such overlap. By choosing the order on expressions such that overlapping terms will appear adjacent in a sum, `add_overlap` can be applied in one linear scan.

A subtle complication arises when normalising in the exponent of an expression a $\wedge$ b: the type of a is an arbitrary commutative semiring, but b must be a natural number. To correctly compute a normalised expression for b, the tactic needs to keep track of the type of b. The calculations of the `eval` function are thus done in an extension of the `tactic` monad, called the `ring_exp_m` monad. Using a reader monad transformer [6], `ring_exp_m` stores the type of the current expression as a variable which can be replaced locally when operating on exponents.

Implementing subtraction and division also requires more work, since semirings in general do not have well-defined − or / operators. The tactic uses typeclass inference to determine whether the required extra structure exists on the type. When this is the case, the operators can be rewritten: $a - b$ becomes $a + (-1) * b$ in a ring and $a/b$ becomes $a * b^{-1}$ in a field. Otherwise, the subtraction or division is treated as an atom. Conditionally rewriting avoids the need for an almost-ring concept to treat semirings and rings uniformly [4]. Cancellation of multiplication and division, such as $a * b/a = b$, is not supported by the tactic, since such lemmas require an $a \neq 0$ side condition. In future work, extending the ex type with a negation or multiplicative inverse constructor could allow for handling of these operators in more general cases.

For completeness, atoms should be considered up to definitional equality: $(\lambda$ x, x$)$ a and $(\lambda$ x y, x$)$ a b reduce to the same value a, so they should be treated as the same atom. The ring_exp_m monad contains a state monad transformer to keep track of which atoms are definitionally equal. The state consists of a list of all distinct atoms encountered in the whole input expression, and any comparisons between atoms are instead made by comparing their indices in the list. As an additional benefit, the indices induce an order on atoms, which is used to sort arguments to commutative operators. Within atoms, there may be subexpressions that can be normalised as well. Instead of running the normaliser directly, ring_exp calls the built-in tactic simp with the normaliser as an argument. The simp tactic calls a given normaliser on each subexpression, rewriting it when the normaliser succeeds.

# 4  Optimisations

An important practical consideration in implementing ring_exp is its efficiency, especially running time. Among the approximately 300 calls to ring in mathlib, about half are invocations on linear expressions by the tactic linarith. Since ring_exp is intended to work as a drop-in replacement for ring, its performance characteristics, especially for linear expressions, should be comparable.

Optimising the code was a notable part of the implementation of ring_exp. Profiling revealed that up to 90% of running time could be spent on inferring implicit arguments and typeclass instances. The solution was to pass all arguments explicitly and maintain a cache of typeclass instances, also caching the expressions for the constants 0 and 1. It was possible to apply this solution without large changes to the codebase, because the required extra fields were hidden behind the ring_exp_m and ex_info types.

The result of these optimisations can be quantified by comparing the running time of ring and ring_exp on randomly generated expressions[3]. The performance measure is the tactic execution time reported by the Lean profiler, running on a 3 GHz Intel® Core™ i5-8500 CPU with 16 GB of RAM. On arbitrary expressions, the benchmark indicates that ring_exp is a factor of approximately

---

[3] The benchmark program and analysis scripts are available at https://github.com/lean-forward/ring_exp.

3.9 times slower than `ring`; on linear expressions such as are passed by `linarith`, `ring_exp` is 1.7 times slower than `ring`.

Compared to a constant factor difference in the average cases, `ring_exp` has an advantage on problems requiring efficient handling of numeric exponents. The `ring_exp` tactic is a factor 20 faster than `ring` when showing $x^{50} * x^{50} = x^{100}$ in an arbitrary ring. A similar speedup for `ring_exp` was found in practice, for the goal $(1 + x^2 + x^4 + x^6) * (1 + x) = 1 + x + x^2 + x^3 + x^4 + x^5 + x^6 + x^7$. The Horner normal form used by `ring` is optimal for representing expressions with additions and multiplications, so a constant-factor slowdown compared to `ring` on simpler goals is traded off for faster and more powerful handling of more complicated goals.

## 5    Discussion

The `ring` tactic for Coq and Lean can efficiently convert expressions in commutative semirings to normal form. A normalizing procedure for polynomials is also included with the Agda standard library [9], HOL Light [5] and Isabelle/HOL [8], and decision procedures exist that support exponential functions [1]; there is no single normalisation procedure supporting compound expressions in exponents.

Compared with the `ring` tactic, the `ring_exp` tactic can deal with a strict superset of expressions, and can do so without sacrificing too much speed. The extensible nature of the `ex` type should make it simple to add support for more operators to `ring_exp`. Independently, it should be possible to adapt the `ex` type to other algebraic structures such as lattices or vector spaces. Although more optimisations are needed to fully equal `ring` in average case efficiency, the `ring_exp` tactic already achieves its goal of being a useful, more general normalisation tactic. These results are as much a consequence of engineering effort as of theoretical work.

**Acknowledgements.** The author has received funding from the NWO under the Vidi program (project No. 016.Vidi.189.037, Lean Forward).

Floris van Doorn, Mario Carneiro and Robert Y. Lewis reviewed the code and suggested improvements. Brian Gin-Ge Chen, Gabriel Ebner, Jasmin Blanchette, Kevin Buzzard, Robert Y. Lewis, Sander Dahmen and the anonymous reviewers read this paper and gave useful suggestions. Many thanks for the help!

## References

1. Akbarpour, B., Paulson, L.C.: Extending a resolution prover for inequalities on elementary functions. In: Dershowitz, N., Voronkov, A. (eds.) LPAR 2007. LNCS (LNAI), vol. 4790, pp. 47–61. Springer, Heidelberg (2007). https://doi.org/10.1007/978-3-540-75560-9_6
2. Boutin, S.: Using reflection to build efficient and certified decision procedures. In: Abadi, M., Ito, T. (eds.) TACS 1997. LNCS, vol. 1281, pp. 515–529. Springer, Heidelberg (1997). https://doi.org/10.1007/BFb0014565

3. Ebner, G., Ullrich, S., Roesch, J., Avigad, J., de Moura, L.: A metaprogramming framework for formal verification. In: ICFP 2017, PACMPL. ACM (2017). https://doi.org/10.1145/3110278
4. Grégoire, B., Mahboubi, A.: Proving equalities in a commutative ring done right in Coq. In: Hurd, J., Melham, T. (eds.) TPHOLs 2005. LNCS, vol. 3603, pp. 98–113. Springer, Heidelberg (2005). https://doi.org/10.1007/11541868_7
5. Harrison, J.: HOL light: a tutorial introduction. In: Srivas, M., Camilleri, A. (eds.) FMCAD 1996. LNCS, vol. 1166, pp. 265–269. Springer, Heidelberg (1996). https://doi.org/10.1007/BFb0031814
6. Liang, S., Hudak, P., Jones, M.: Monad transformers and modular interpreters. In: POPL 1995, pp. 333–343. ACM (1995). https://doi.org/10.1145/199448.199528
7. de Moura, L., Kong, S., Avigad, J., van Doorn, F., von Raumer, J.: The lean theorem prover (System Description). In: Felty, A.P., Middeldorp, A. (eds.) CADE 2015. LNCS (LNAI), vol. 9195, pp. 378–388. Springer, Cham (2015). https://doi.org/10.1007/978-3-319-21401-6_26
8. Nipkow, T., Wenzel, M., Paulson, L.C. (eds.): Isabelle/HOL: A Proof Assistant for Higher-Order Logic. LNCS, vol. 2283. Springer, Heidelberg (2002). https://doi.org/10.1007/3-540-45949-9
9. The Agda Team: Agda standard library, version 1.3 for Agda 2.6.1. https://wiki.portal.chalmers.se/agda/Libraries/StandardLibrary
10. The mathlib Community: The Lean Mathematical Library. In: Blanchette, J., Hriţcu, C. (eds.) CPP 2020, pp. 367–381. ACM (2020). https://doi.org/10.1145/3372885.3373824

# Practical Proof Search for Coq by Type Inhabitation

Łukasz Czajka[(⊠)] [iD]

TU Dortmund University, Dortmund, Germany
lukasz.czajka@tu-dortmund.de

**Abstract.** We present a practical proof search procedure for Coq based on a direct search for type inhabitants in an appropriate normal form. The procedure is more general than any of the automation tactics natively available in Coq. It aims to handle as large a part of the Calculus of Inductive Constructions as practically feasible.

For efficiency, our procedure is not complete for the entire Calculus of Inductive Constructions, but we prove completeness for a first-order fragment. Even in pure intuitionistic first-order logic, our procedure performs competitively.

We implemented the procedure in a Coq plugin and evaluated it on a collection of Coq libraries, on CompCert, and on the ILTP library of first-order intuitionistic problems. The results are promising and indicate the viablility of our approach to general automated proof search for the Calculus of Inductive Constructions.

**Keywords:** Proof search · Inhabitation · Coq · Proof automation · Intuitionistic logic · Dependent type theory

## 1   Introduction

The Curry-Howard isomorphism [39] is a correspondence between systems of formal logic and computational lambda-calculi, interpreting propositions as types and proofs as programs (typed lambda-terms). Coq [10] is an interactive proof assistant based on this correspondence. Its underlying logic is the Calculus of Inductive Constructions [10,33,46] – an intuitionistic dependent type theory with inductive types.

Because of the complexity and constructivity of the logic, research on automated reasoning for Coq has been sparse so far, limited mostly to specialised tactics for restricted fragments or decidable theories. The automation currently available in Coq is weaker than in proof assistants based on simpler classical foundations, like Isabelle/HOL [29].

We present a practical general fully automated proof search procedure for Coq based on a direct search for type inhabitants. We synthesise Coq terms in an appropriate normal form, using backward-chaining goal-directed search. To make this approach practical, we introduce various heuristics including hypothesis

© Springer Nature Switzerland AG 2020
N. Peltier and V. Sofronie-Stokkermans (Eds.): IJCAR 2020, LNAI 12167, pp. 28–57, 2020.
https://doi.org/10.1007/978-3-030-51054-1_3

simplification, limited forward reasoning, ordered rewriting and loop checking. For efficiency, we sacrifice completeness for the entire logic of Coq, though we give a completeness proof for a first-order fragment.

We evaluated our procedure on a collection of Coq libraries (40.9% success rate), on CompCert [27] (17.1%) and on the ILTP library [36] (29.5%) of first-order intuitionistic problems. The percentages in brackets denote how many problems were solved fully automatically by the standalone tactic combined with heuristic induction. These results indicate the viability of our approach.

The procedure can be used as a standalone Coq tactic or as a reconstruction backend for CoqHammer [14] – a hammer tool [7] which invokes external automated theorem provers (ATPs) on translations of Coq problems and then reconstructs the found proofs in Coq using the information obtained from successful ATP runs. With our procedure used for reconstruction, CoqHammer achieves a 39.1% success rate on a collection of Coq libraries and 25.6% on CompCert. The reconstruction success rates (i.e. the percentage of problems solved by the ATPs that can be re-proved in Coq) are 87–97%.

## 1.1 Related Work

A preliminary version of a proof search procedure partly based on similar ideas was described in [14]. That procedure is less complete (not complete for the first-order fragment), slower, much more heuristic, and it performs visibly worse as a standalone tactic (see Sect. 5). It partially includes only some of the actions, restrictions and heuristic improvements described here. In particular, the construction and the unfolding actions are absent, and only special cases of the elimination and the rewriting actions are performed. See Sect. 3.

From a theoretical perspective, a complete proof search procedure for the Cube of Type Systems, which includes the pure Calculus of Constructions without inductive types, is presented in [15]. It is also based on an exhaustive search for type inhabitants. Sequent calculi suitable for proof search in Pure Type Systems are described in [22, 26].

In practice, Chlipala's `crush` tactic [9] can solve many commonly occurring Coq goals. However, it is not a general proof search procedure, but an entirely heuristic tactic. In the evaluations we performed, the `crush` tactic performed much worse than our procedure. For Agda [30] there exists an automated prover Agsy [28] which is, however, not much stronger than Coq's `auto`.

Proof search in intuitionistic first-order logic has received more attention than inhabitation in complex constructive dependent type theories. We do not attempt here to provide an overview, but only point the reader to [36] for a comparison of intuitionistic first-order ATP systems. The most promising approaches to proof search in intuitionistic first-order logic seem to be connection-based methods [6, 24, 32, 45]. Indeed, the connection-based ileanCoP [31] prover outperforms other intuitionistic ATPs by a wide margin [36].

An automated intuitionistic first-order prover is available in Coq via the `firstorder` tactic [11], which is based on a contraction-free sequent calculus extending the LJT system of Dyckhoff for intuitionistic propositional logic [17–19]. There also exists a Coq plugin for the connection-based JProver [37]. However, the plugin is not maintained and not compatible with new versions of Coq.

Coq's type theory may be viewed as an extension of intuitionistic higher-order logic. There exist several automated provers for classical higher-order logic, like Leo [40] or Satallax [8]. Satallax can produce Coq proof terms which use the excluded middle axiom.

The approach to proof search in intuitionistic logic via inhabitation in the corresponding lambda-calculus has a long tradition. It is often an easy way to establish complexity bounds [23,38,42]. This approach can be traced back to Ben-Yelles [5,23] and Wajsberg [43,44].

One of the motivations for this work is the need for a general automated reasoning procedure in a CoqHammer [14] reconstruction backend. CoqHammer links Coq with general classical first-order ATPs, but tries to find intuitionistic proofs with no additional assumptions and to handle as much of Coq's logic as possible. A consequence is that the reconstruction mechanism of CoqHammer cannot rely on a direct translation of proofs found by classical ATPs, in contrast to e.g. SMTCoq [2,21] which integrates external SAT and SMT solvers into Coq.

## 2    Calculus of Inductive Constructions

In this section, we briefly and informally describe the Calculus of Inductive Constructions (CIC) [10,33,46]. For precise definitions and more background, the reader is referred to the literature. Essentially, CIC is a typed lambda calculus with dependent products $\forall x : \tau.\sigma$ and inductive types.

An inductive type is given by its constructors, presented as, e.g.,

```
Inductive List (A : Type) : Type :=
  nil : List A | cons : A -> List A -> List A
```

This declares list $A$ to be a type of sort Type for any parameter $A$ of sort Type. Above $A$ is a *parameter* and Type $\rightarrow$ Type is the *arity* of list. The types of constructors implicitly quantify over the parameters, i.e., the type of cons above is $\forall A : \text{Type}.A \rightarrow \text{list } A \rightarrow \text{list } A$. In the presentation we sometimes leave the parameter $A$ implicit.

Propositions (logical formulas) are represented by dependent types. Inductive predicates are represented by dependent inductive types, e.g., the inductive type

```
Inductive Forall (A : Type) (P : A -> Prop) : List A -> Prop :=
| fnil : Forall P nil
| fcons : forall (x : A) (l : List A),
          P x -> Forall P l -> Forall P (cons x l)
```

defines a predicate Forall on lists, parameterised by a type $A$ and a predicate $P : A \rightarrow \text{Prop}$. Then Forall $A\,P\,l$ states that $Px$ holds for every element $x$ of $l$.

All intuitionistic connectives may be represented using inductive types:

```
Inductive ⊤ : Prop := I : ⊤.
Inductive ⊥ : Prop := .
Inductive ∧ (A : Prop) (B : Prop) : Prop := conj : A -> B -> A ∧ B.
Inductive ∨ (A : Prop) (B : Prop) : Prop :=
  inl : A -> A ∨ B | inr : B -> A ∨ B.
Inductive ∃ (A : Type) (P : A -> Prop) : Prop :=
  exi : forall x : A, P x -> ∃ A P.
```

where $\wedge$ and $\vee$ are used in infix notation. All the usual introduction and elimination rules are derivable. Equality can also be defined inductively.

Below by $t$, $u$, $w$, $\tau$, $\sigma$, etc., we denote terms, by $c$, $c'$, etc., we denote constructors, and by $x$, $y$, $z$, etc., we denote variables. We use $\vec{t}$ for a sequence of terms $t_1 \ldots t_n$ of an unspecified length $n$, and analogously for a sequence of variables $\vec{x}$. For instance, $t\vec{y}$ stands for $ty_1 \ldots y_n$, where $n$ is not important or implicit. Analogously, we use $\lambda\vec{x} : \vec{\tau}.t$ for $\lambda x_1 : \tau_1.\lambda x_2 : \tau_2. \ldots .\lambda x_n : \tau_n.t$, with $n$ implicit or unspecified. We write $t[\vec{u}/\vec{x}]$ for $t[u_1/x_1] \ldots [u_n/x_n]$.

The logic of Coq includes over a dozen term formers. The ones recognised by our procedure are: a sort $s$ (e.g. Type, Set or Prop), a variable $x$, a constant, a constructor $c$, an inductive type $I$, an application $t_1 t_2$, an abstraction $\lambda x : t_1.t_2$, a dependent product $\forall x : t_1.t_2$ (written $t_1 \rightarrow t_2$ if $x \notin \mathrm{FV}(t_2)$), and a case expression $\mathtt{case}(t; \lambda\vec{a} : \vec{\alpha}.\lambda x : I\vec{q}\vec{a}.\tau; \vec{x_1} : \vec{\sigma_1} \Rightarrow t_1 \mid \ldots \mid \vec{x_k} : \vec{\sigma_k} \Rightarrow t_k)$.

In a case expression: $t$ is the term matched on; $I$ is an inductive type with constructors $c_1, \ldots, c_k$; the type of $c_i$ is $\forall\vec{p} : \vec{\rho}.\forall\vec{x_i} : \vec{\tau_i}.I\vec{p}\vec{u_i}$ where $\vec{p}$ are the parameters of $I$; the type of $t$ has the form $I\vec{q}\vec{u}$ where $\vec{q}$ are the values of the parameters; the type $\tau[\vec{u}/\vec{a}, t/x]$ is the return type, i.e., the type of the whole case expression; $t_i$ has type $\tau[\vec{w_i}/\vec{a}, c_i\vec{q}\vec{x_i}/x]$ in $\vec{x_i} : \vec{\sigma_i}$ where $\vec{\sigma_i} = \vec{\tau_i}[\vec{q}/\vec{p}]$ and $\vec{w_i} = \vec{u_i}[\vec{q}/\vec{p}]$; $t_i[\vec{v}/\vec{x}]$ is the value of the case expression if the value of $t$ is $c_i\vec{q}\vec{v}$.

Note that some equality information is "forgotten" when typing the branches of a case expression. We require $t_i$ to have type $\tau[\vec{w_i}/\vec{a}, c_i\vec{q}\vec{x_i}/x]$ in context $\vec{x_i} : \vec{\sigma_i}$. We know that "inside" the $i$th branch $t = c_i\vec{q}\vec{x_i}$ and $\vec{u} = \vec{w_i}$, but this information cannot be used when checking the type of $t_i$. A consequence is that permutative conversions [41, Chapter 6] are not sound for CIC and this is one reason for the incompleteness of our procedure outside the restricted first-order fragment.

Coq's notation for case expressions is

```
match t as x in I _ ā return τ
with c₁ _ x⃗₁ => t₁ | ... | cₖ _ x⃗ₖ => tₖ end
```

where $c_1, \ldots, c_k$ are all constructors of $I$, and $\_$ are the wildcard patterns matching the inductive type parameters $\vec{q}$. For readability, we often use Coq match notation. When $x$ (resp. $\vec{a}$) does not occur in $\tau$ then we omit as $x$ (resp. in $I$ _ $\vec{a}$) from the match. If $\tau$ does not on either $x$ or $\vec{a}$, we also omit the return $\tau$.

A *typing judgement* has the form $E; \Gamma \vdash t : \tau$ where $E$ is an *environment* consisting of declarations of inductive types and constant definitions, $\Gamma$ is a *context* - a list of variable type declarations $x : \sigma$, and $t, \tau$ are terms. We refer to the Coq manual [10] for a precise definition of the typing rules.

Coq's definitional equality (conversion rule) includes $\beta$- and $\iota$-reduction:

$$(\lambda x : \tau.t_1)t_2 \rightarrow_\beta t_1[t_2/x]$$
$$\mathsf{case}(c_i\vec{p}\vec{v}; \lambda\vec{a} : \vec{\alpha}.\lambda x : I\vec{p}\vec{a}.\tau; \vec{x_1} : \vec{\tau_1} \Rightarrow t_1 \mid \ldots \mid \vec{x_k} : \vec{\tau_k} \Rightarrow t_k) \rightarrow_\iota t_i[\vec{v}/\vec{x_i}]$$

An inductive type $I$ is *non-recursive* if the types of constructors of $I$ do not contain $I$ except as the target. We assume the well-foundedness of the relation $\succ$ defined by: $I_1 \succ I_2$ iff $I_2 \neq I_1$ occurs in the arity of $I_1$ or the type of a constructor of $I_1$. We write $I \succ t$ if $I \succ I'$ for every inductive type $I'$ occurring in the term $t$.

# 3   The Proof Search Procedure

In this section, we describe our proof search procedure. Our approach is based on a direct search for type inhabitants in appropriate normal form [42]. For the sake of efficiency, the normal forms we consider are only a subset of possible CIC normal forms. This leads to incompleteness outside the restricted first-order fragment (see Sects. 3.6 and 4).

More precisely, the inhabitation problem is: given an environment $E$, a context $\Gamma$ and a $\Gamma$-type $\tau$ (i.e. $\Gamma \vdash \tau : s$ for a sort $s$), find a term $t$ such that $E; \Gamma \vdash t : \tau$. The environment $E$ will be kept fixed throughout the search, so we omit it from the notation.

A *goal* is a pair $(\Gamma, \tau)$ with $\tau$ a $\Gamma$-type, denoted $\Gamma \vdash ? : \tau$, where $\Gamma$ is the *context* and $\tau$ is the *conjecture*. A *solution* of the goal $\Gamma \vdash ? : \tau$ is any term $t$ such that $\Gamma \vdash t : \tau$.

## 3.1   Basic Procedure

The basic inhabitation procedure is to nondeterministically perform one of the following actions, possibly generating new subgoals to be solved recursively. If the procedure fails on one of the subgoals then the action fails. If each possible action fails then the procedure fails. The choices in the actions (e.g. of specific subgoal solutions) are nondeterministic, i.e., we consider all possible choices, each leading to a potentially different solution.

The actions implicitly determine an and-or proof search tree. We leave the exact order in which this tree is traversed unspecified, but a complete search order is to be used, e.g., breadth-first or iterative deepening depth-first.

The procedure supports five term formers in synthesised solutions: variables, constructors, applications, lambda-abstractions, case expressions. These are built with the four actions below.

1. **Introduction.** If $\Gamma \vdash ? : \forall x : \alpha.\beta$ then:
   - recursively search for a solution $t$ of the subgoal $\Gamma, x : \alpha \vdash ? : \beta$;
   - return $\lambda x : \alpha.t$ as the solution.

2. **Application.** If $\Gamma \vdash ? : \tau$ and $x : \forall \vec{y} : \vec{\sigma}.\rho$ is in $\Gamma$ then:
   - for $i = 1, \ldots, n$, recursively search for a solution $t_i$ of the subgoal $\Gamma \vdash ? :$ $\sigma_i[t_1/y_1] \ldots [t_{i-1}/y_{i-1}]$;
   - if $\rho[\vec{t}/\vec{y}] =_{\beta\iota} \tau$ then return $x\vec{t}$ as the solution.
3. **Construction.** If $\Gamma \vdash ? : I\vec{q}\vec{w}$ with $\vec{q}$ the parameters and $c : \forall \vec{p} : \vec{\rho}.\forall \vec{y} : \vec{\sigma}.I\vec{p}\vec{u}$ is a constructor of $I$ then:
   - for $i = 1, \ldots, n$, recursively search for a solution $t_i$ of the subgoal $\Gamma \vdash ? :$ $\sigma_i[\vec{q}/\vec{p}][t_1/y_1] \ldots [t_{i-1}/y_{i-1}]$;
   - if $\vec{u}[\vec{q}/\vec{p}][\vec{t}/\vec{y}] =_{\beta\iota} \vec{w}$ then return $c\vec{q}\vec{t}$ as the solution.
4. **Elimination.** If $\Gamma \vdash ? : \tau$, and $x : \forall \vec{y} : \vec{\sigma}.I\vec{q}\vec{u}$ is in $\Gamma$, and $I : \forall \vec{p} : \vec{\rho}.\forall \vec{a} : \vec{\alpha}.s$ is in $E$ with $\vec{p}$ the parameters, and $c_j : \forall \vec{p} : \vec{\rho}.\forall \vec{z_j} : \vec{\gamma_j}.I\vec{p}\vec{u_j}$ for $j = 1, \ldots, m$ are all constructors of $I$, then:
   - for $i = 1, \ldots, n$, recursively search for a solution $t_i$ of the subgoal $\Gamma \vdash ? :$ $\sigma_i[t_1/y_1] \ldots [t_{i-1}/y_{i-1}]$;
   - let $\vec{v} = \vec{u}[\vec{t}/\vec{y}]$ and $\vec{r} = \vec{q}[\vec{t}/\vec{y}]$;
   - choose $\tau'$ such that $\tau'[\vec{v}/\vec{a}, x\vec{t}/z] =_{\beta\iota} \tau$;
   - for $j = 1, \ldots, m$, recursively search for a solution $b_j$ of $\Gamma, \vec{z_j} : \vec{\delta_j} \vdash ? :$ $\tau'[\vec{w_j}/\vec{a}, c_j\vec{r}\vec{z_j}/z]$ where $\vec{\delta_j} = \vec{\gamma_j}[\vec{r}/\vec{p}]$ and $\vec{w_j} = \vec{u_j}[\vec{r}/\vec{p}]$;
   - return $\mathsf{case}(x\vec{t}; \lambda \vec{a} : \vec{\alpha}.\lambda z : I\vec{r}\vec{a}.\tau'; \vec{z_1} : \vec{\gamma_1} \Rightarrow b_1 \mid \ldots \mid \vec{z_m} : \vec{\gamma_m} \Rightarrow b_m)$ as the solution.

The intuition is that we search for *normal* inhabitants of a *type*. For instance, if $\Gamma \vdash (\lambda x : \alpha.t)u : \tau$ then also $\Gamma \vdash t[u/x] : \tau$, so it suffices to consider solutions without $\beta$-redexes. Assuming $\tau$ is not a sort, it suffices to consider only variables and constructors at the head of the solution term, because $I : \forall \vec{x} : \vec{\sigma}.s$ with $s$ a sort for any inductive type $I$. This of course causes incompleteness because it may be necessary to search for inhabitants of a sort $s$ in a subgoal.

It is straightforward to check (by inspecting the typing rules of CIC) that the described procedure is sound, i.e., any term obtained using the procedure is indeed a solution.

## 3.2   Search Restrictions

We now introduce some restrictions on the search procedure, i.e., on when each action may be applied. Note that this may compromise completeness, but not soundness. For a first-order fragment completeness is in fact preserved (Sect. 4).

- **Eager introduction.** Perform introduction eagerly, i.e., if $\Gamma \vdash ? : \forall x : \alpha.\beta$ then immediately perform introduction without backtracking.
  This is justified by observing that we may restrict the search to solutions in $\eta$-long normal form. However, in general $\eta$-long normal forms may not exist.
- **Elimination restriction.** Perform elimination only immediately after introduction or another elimination.

The intuitive justification is that in a term of the form

u (match t as x in I _ $\vec{a}$ return $\tau$
  with $c_1$ _ $\vec{x_1}$ => $t_1$ | ... | $c_k$ _ $\vec{x_k}$ => $t_k$ end)

we may usually move $u$ inside the match while preserving the type:

match t as x in I _ $\vec{a}$ return $\tau$
with $c_1$ _ $\vec{x_1}$ => u $t_1$ | ... | $c_k$ _ $\vec{x_k}$ => u $t_k$ end

However, this is not always possible in CIC (see Sect. 2).

- **Eager simple elimination.** Immediately after adding $x : I\vec{q}\vec{u}$ with parameters $\vec{q}$ into the context $\Gamma$ (by the introduction or the elimination action), if $I$ is a non-recursive inductive type and $I \succ \vec{q}$, then perform elimination of $x$ eagerly and remove the declaration of $x$ from the context.
  If $\Gamma \vdash (\lambda x : I\vec{q}.t) : \tau$ then usually $\Gamma \vdash (\lambda x : I\vec{q}.t') : \tau$ where $t'$ is

  match x with $c_1\vec{x_1}$ => $t[c_1\vec{x_1}/x]$ | ... | $c_k\vec{x_k}$ => $t[c_k\vec{x_k}/x]$ end

  However, in general replacing a subterm $u'$ of a term $u$ with $u''$ may change the type of $u$, even if $u', u''$ have the same type. See Sect. 4.
- **Loop checking.** If the same conjecture is encountered for the second time on the same proof search tree branch without performing the introduction action in the meantime, then fail.
  This is justified by observing that if $\Gamma \vdash t[u/x] : \tau$ and $\Gamma \vdash u : \tau$, then we can just use $u$ instead of $t$ as the solution. In general, this restriction also causes incompleteness, for the same reason as the previous one.

It is instructive to observe how the elimination restrictions specialise to inductive definitions of logical connectives. For example, the eager simple elimination restriction for conjunction is that a goal $\Gamma, x : \alpha \wedge \beta \vdash ? : \tau$ should be immediately replaced by $\Gamma, x_1 : \alpha, x_2 : \beta \vdash ? : \tau$.

### 3.3 Heuristic Improvements

The above presentation of the proof search procedure does not yet directly lead to a practical implementation. We thus introduce "heuristic" improvements. All of them preserve soundness, but some may further compromise completeness. In fact, we believe several of the "heuristics" (e.g. most context simplifications and forward reasoning) actually do preserve completeness (under certain restrictions), but we did not attempt to rigorously prove it[1].

- In the application action, instead of checking $\rho[\vec{t}/\vec{y}] =_{\beta\iota} \tau$ a posteriori, use unification modulo simple heuristic equational reasoning to choose an appropriate $(x : \tau) \in \Gamma$, possibly introducing existential metavariables to be instantiated later (like with Coq's eapply tactic). Analogously, we use unification in the construction action.

---

[1] It is actually clear that limited forward reasoning (4th point) preserves completeness in general, because it corresponds to performing $\beta$-expansions on the proof term.

– In the elimination action, the choice of $\tau'$ is done heuristically without back-tracking. In practice, we use either Coq's **destruct** or **inversion** tactic, depending on the form of the inductive type $I$.
– Immediately after the introduction action, simplify the context:
   - replace $h : \forall \vec{x} : \vec{\sigma}.\tau_1 \wedge \tau_2$ with $h_1 : \forall \vec{x} : \vec{\sigma}.\tau_1$ and $h_2 : \forall \vec{x} : \vec{\sigma}.\tau_2$;
   - replace $h : \forall \vec{x} : \vec{\sigma}.\tau_1 \vee \tau_2 \rightarrow \rho$ with $h_1 : \forall \vec{x} : \vec{\sigma}.\tau_1 \rightarrow \rho$ and $h_2 : \forall \vec{x} : \vec{\sigma}.\tau_2 \rightarrow \rho$;
   - replace $h : \forall \vec{x} : \vec{\sigma}.\tau_1 \wedge \tau_2 \rightarrow \rho$ with $h' : \forall \vec{x} : \vec{\sigma}.\tau_1 \rightarrow \tau_2 \rightarrow \rho$;
   - replace $h : \exists x : \sigma.\tau$ with $h' : \tau$ (assuming $x$ fresh);
   - remove some intuitionistic tautologies;
   - perform invertible forward reasoning, i.e., if $h : \sigma$ and $h' : \sigma \rightarrow \tau$ are in $\Gamma$ then we replace $h'$ with $h'' : \tau$.
   - use Coq's **subst** tactic to rewrite with equations on variables;
   - perform rewriting with some predefined lemmas from a hint database.
– Immediately after simplifying the context as above, perform some limited forward reasoning. For instance, if $h : Pa$ and $h' : \forall x.Px \rightarrow \varphi$ are in $\Gamma$, then add $h'' : Pa \rightarrow \varphi[a/x]$ to $\Gamma$. To avoid looping, we do not use newly derived facts for further forward reasoning.
– Elimination on terms matched in case expressions is done eagerly. In other words, if **match** $t$ **with** ... **end** occurs in the conjecture or the context, with $t$ closed, then we immediately perform the elimination action on $t$.
– After performing each action, simplify the conjecture by reducing it to (weak) normal form (using Coq's **cbn** tactic) and rewriting with some predefined lemmas from a hint database.
– We use a custom leaf solver at the leaves of the search tree. The leaf solver eagerly splits the disjunctions in the context (including quantified ones), uses Coq's **eauto** with depth 2, and tries the Coq tactics **congruence** (congruence closure) and **lia** (linear arithmetic).
– We extend the search procedure with two more actions (performed non-eagerly with backtracking):

1. **Unfolding.** Unfold a Coq constant definition, provided some heuristic conditions on the resulting unfolding are satisfied.
2. **Rewriting.** The order $>$ on constants is defined to be the transitive closure of $\{(c_1, c_2) \mid c_2$ occurs in the definition of $c_1\}$. By $\mathrm{lpo}_>$ we denote the lifting of $>$ to the lexicographic path order (LPO) on terms [3, Section 5.4.2]. For the LPO lifting, we consider only terms which have obvious first-order counterparts, e.g., $fxyz$ with $f$ a constant corresponds to a first-order term $f(x, y, z)$. The action is then as follows. Assume $h : \forall \vec{x} : \vec{\sigma}.t_1 = t_2$ is in $\Gamma$.

   - If $\mathrm{lpo}_>(t_1, t_2)$ then rewrite with $h$ from left to right, in the conjecture and the hypotheses, generating new subgoals and introducing existential metavariables for $\vec{x}$ as necessary.
   - If $\mathrm{lpo}_>(t_2, t_1)$ then rewrite with $h$ from right to left.
   - If $t_1, t_2$ are incomparable with $\mathrm{lpo}_>$, then rewrite heuristically from left to right, or right to left if that fails.

For heuristic rewriting in the last point, we use the leaf solver to discharge the subgoals and we track the hypotheses to avoid unordered heuristic rewriting with the same hypothesis twice.

- Immediately after forward reasoning, eagerly perform rewriting with those hypotheses which satisfy: (1) the target can be ordered with the lexicographic path order described above, and (2) the generated subgoals can be solved with the leaf solver.

### 3.4  Search Strategy

The proof search strategy is based on (bounded or iterative deepening) depth-first search. We put a bound on the cost of proof search according to one of the following two cost models.

- **Depth cost model.** The depth of the search tree is bounded, with the leaf solver tactic tried at the leaves.
- **Tree cost model.** The size of the entire search tree is bounded, but not the depth directly. The advantages of this approach are that (1) it allows to find deep proofs with small branching, and (2) it is easier to choose a single cost bound which performs well in many circumstances. However, this model performs slightly worse on pure first-order problems (see Sect. 5).

### 3.5  Soundness

Our proof search procedure, including all heuristic improvements, is sound. For the basic procedure (Sects. 3.1 and 3.2) this can be shown by straightforward induction, noting that the actions essentially directly implement CIC typing rules. For the heuristic improvements (Sect. 3.3), one could show soundness by considering the shapes of the proof terms. This is straightforward but tedious. The implementation in the Coq tactic monad guarantees soundness as only well-typed proof terms can be produced by standard Coq tactics.

### 3.6  Incompleteness

The inhabitation procedure presented above is not complete for the full logic of Coq. The reasons for incompleteness are as follows.

1. Higher-order unification in the application action is not sufficient for completeness in the presence of impredicativity. A counterexample (from [15]) is

$$Q : * \to *, u : \forall P : *.QP \to P, v : Q(A \to B), w : A \vdash ? : B.$$

The solution is $u(A \to B)vw$. The target $P$ of $u$ unifies with $B$, but this does not provide the right instantiation (which is $A \to B$) and leaves an unsolvable subgoal $? : QB$. It is worth noting, however, that this particular example can be solved thanks to limited forward reasoning (see Sect. 3.3).

2. For efficiency, most variants of our tactics do not perform backtracking on instantiations of existential metavariables.
3. The normal forms we implicitly search for do not suffice for completeness. One reason is that permutative conversions [41, Chapter 6] are not sound for dependent case expressions `case` in CIC if the return type $\tau$ depends on $\vec{a}$ or $x$. We elaborate on this in the next section.
4. The full logic of Coq contains more term formers than the five supported by our procedure: `fix`, `cofix`, `let`, ... In particular, our inhabitation procedure never performs induction over recursive inductive types, which requires `fix`. It does reasoning by cases, however, via the elimination action.

We believe the compromises on completeness are in practice not very severe and our procedure may reasonably be considered a general automated proof search method for CIC without fixpoints. In fact, many of the transformations on proof terms corresponding to the "restrictions" and "heuristics" above would preserve completeness in the presence of definitional proof irrelevance.

## 4   Normal Forms and Partial Completeness

The basic inhabitation procedure (Sects. 3.1 and 3.2) with restricted looping check is complete for a first-order fragment of CIC. We conjecture that a variant of our procedure is also complete for CIC with definitional proof irrelevance, with only non-dependent elimination, and without fixpoints. First, we describe the subset of normal forms our procedure searches for.

Permutative conversions are the two reductions below. They "move around" case expressions to expose blocked redexes.

$$\text{case}(u; \lambda\vec{a} : \vec{\alpha}.\lambda x : I\vec{q}\vec{a}.\forall z : \sigma.\tau; \vec{x_1} : \vec{\tau_1} \Rightarrow t_1 \mid \ldots \mid \vec{x_k} : \vec{\tau_k} \Rightarrow t_k)w \to_{\rho_1}$$
$$\text{case}(u; \lambda\vec{a} : \vec{\alpha}.\lambda x : I\vec{q}\vec{a}.\tau[w/z]; \vec{x_1} : \vec{\tau_1} \Rightarrow t_1 w \mid \ldots \mid \vec{x_k} : \vec{\tau_k} \Rightarrow t_k w)$$

$$\text{case}(\text{case}(u; Q; \vec{x_1} : \vec{\tau_1} \Rightarrow t_1 \mid \ldots \mid \vec{x_k} : \vec{\tau_k} \Rightarrow t_k); R; P) \to_{\rho_2}$$
$$\text{case}(u; R'; \vec{x_1} : \vec{\tau_1} \Rightarrow \text{case}(t_1; R''; P) \mid \ldots \mid \vec{x_k} : \vec{\tau_k} \Rightarrow \text{case}(t_k; R''; P))$$

In the second reduction rule, $P$ stands for a list of case patterns $\vec{y_1} : \vec{\sigma_1} \Rightarrow w_1 \mid \ldots \mid \vec{y_m} : \vec{\sigma_m} \Rightarrow w_m$. We assume $\vec{x_i}$ do not occur in $P$. Similarly, $Q, R, R', R''$ stand for the specifications of the return types, where $Q = \lambda\vec{a} : \vec{\alpha}.\lambda x : I_1\vec{q_1}\vec{a}.I\vec{v}$, $R = \lambda\vec{b} : \vec{\beta}.\lambda x : I_2\vec{q_2}\vec{b}.\tau$, $R' = \lambda\vec{a} : \vec{\alpha}.\lambda x : I_1\vec{q_1}\vec{a}.\tau$, $R'' = \lambda\vec{a} : \vec{\alpha}.\lambda x : I_1\vec{q_1}\vec{a}.\tau$.

We write $\to_\rho$ for the union of $\to_{\rho_1}$ and $\to_{\rho_2}$. Note that $\rho_1$-reduction may create $\beta$-redexes, and $\rho_2$-reduction may create $\iota$-redexes.

The right-hand sides of the $\rho$-rules may be ill-typed if $\sigma, \tau$ above depend on any of $\vec{a}, \vec{b}, x$, i.e., if the return type varies across the case expression branches. Moreover, even if the type of a $\rho$-redex subterm is preserved by a $\rho$-contraction, the type of the entire term might not be. For example, assume the following are provable in context $\Gamma$: $P : A \to s$, $F : \forall x : A.Px$, $t : A$, $t' : A$ and $p : Pt$. Then $\Gamma \vdash Ft : Pt$ but in general $\Gamma \nvdash Ft' : Pt$ unless $t =_{\beta\iota} t'$.

An analogous problem occurs when attempting to define $\eta$-long normal forms – normal forms $\eta$-expanded as much as possible without creating $\beta$-redexes. The

$\eta$-expansion of a term $t$ of type $\forall x : \alpha.\beta$ is $\lambda x : \alpha.tx$ where $x \notin \mathrm{FV}(t)$. We do not consider $\eta$-expansions for inductive types. If the conversion rule does not include $\eta$ (Coq's does since v8.4), then $\eta$-expanding a subterm may change the type of the entire term. Even assuming the conversion rule does include $\eta$, defining $\eta$-long forms in the presence of dependent types is not trivial if we consider $\eta$-expansions inside variable type annotations [16]. However, for our purposes a simpler definition by mutual induction on term structure is sufficient.

A term $t$ is a *long normal form in $\Gamma$* ($\Gamma$-*lnf*) if:

- $t = \lambda x : \alpha.t'$, and $\Gamma \vdash t : \forall x : \alpha.\beta$, and $t'$ is a $\Gamma$-$(x : \alpha)$-lnf (defined below);
- $t = x\vec{u}$, and $\Gamma \vdash t : \tau$ with $\tau$ not a product, and each $u_i$ is a $\Gamma$-lnf and not a case expression;
- $t = c\vec{q}\vec{v}$, and $\Gamma \vdash t : I\vec{q}\vec{w}$ with $\vec{q}$ the parameters, and $c$ is a constructor of $I$, and each $v_i$ is a $\Gamma$-lnf and not a case expression;
- $t = \mathsf{case}(x\vec{u}; \lambda\vec{a} : \vec{\alpha}.\lambda x : I\vec{q}\vec{a}.\sigma; \vec{x_1} : \vec{\tau_1} \Rightarrow t_1 \mid \ldots \mid \vec{x_k} : \vec{\tau_k} \Rightarrow t_k)$, and $\Gamma \vdash t : \tau$ with $\tau$ not a product, and each $u_i$ is a $\Gamma$-lnf and not a case expression, and each $t_i$ is a $\Gamma$-$(\vec{x_i} : \vec{\tau_i})$-lnf.

A term $t$ is a $\Gamma$-$\Delta$-*lnf* if:

- $\Delta = \langle \rangle$ and $t$ is a $\Gamma$-lnf;
- $\Delta = x : \alpha, \Delta'$, and $\alpha$ is not $I\vec{q}\vec{u}$ for $I$ non-recursive with $I \succ \vec{q}$, and $t$ is a $\Gamma, x : \alpha$-$\Delta'$-lnf;
- $\Delta = x : I\vec{q}\vec{u}, \Delta'$, and $I$ is non-recursive with $I \succ \vec{q}$, and $\vec{q}$ are the parameter values, and $t = \mathsf{case}(x; \lambda\vec{a} : \vec{\alpha}.\lambda x : I\vec{q}\vec{a}.\tau; \vec{x_1} : \vec{\tau_1} \Rightarrow t_1 \mid \ldots \mid \vec{x_k} : \vec{\tau_k} \Rightarrow t_k)$, and $\Gamma, \Delta \vdash t : \tau[\vec{u}/\vec{a}]$, and each $t_i$ is a $\Gamma$-$\Delta'$, $\vec{x_i} : \vec{\tau_i}$-lnf (then $x \notin \mathrm{FV}(t_i)$).

For the supported term formers (variables, constructors, applications, lambda-abstractions, case expressions), this definition essentially describes $\eta$-long $\beta\iota\rho$-normal forms transformed to satisfy the additional restrictions corresponding to the elimination and the eager simple elimination restrictions from Sect. 3.2.

Given an inhabitation problem $\Gamma \vdash ? : \tau$, our procedure searches for a minimal solution in $\Gamma$-lnf. Solutions in $\Gamma$-lnf might not exist for some solvable problems. As outlined above, there are essentially two reasons: (1) with dependent elimination the return type may vary across case branches, which in particular makes permutative conversions unsound; (2) replacing a proof with a different proof of the same proposition is not sound if proofs occur in types. Point (1) may be dealt with by disallowing dependent elimination, and (2) by assuming definitional proof irrelevance. Hence, we conjecture completeness (of an appropriate variant of the procedure) for CIC with definitional proof irrelevance, with only non-dependent elimination, and without fixpoints.

Here we only prove that in a first-order fragment for every inhabited type there exists an inhabitant in $\Gamma$-lnf. The precise definition of the considered first-order fragment may be found in the appendix. It is essentially intuitionistic first-order logic with two predicative sorts Prop and Set, non-dependent inductive types in Prop, non-dependent pattern-matching, and terms of a type in Set restricted to applicative form. We use $\vdash_{\mathrm{fo}}$ for the typing judgement of the first-order fragment.

For the theorem below, we consider a basic variant of our procedure (Sects. 3.1 and 3.2) which does not perform the looping check for conjectures of sort `Set`.

**Theorem 1 (Completeness for a first-order fragment).** *If the inhabitation problem $\Gamma \vdash_{fo}$? : $\tau$ has a solution, then the inhabitation procedure will find one.*

*Proof (sketch).* Assume $\Gamma \vdash_{fo} t : \tau$. It suffices to show that $t$ may be converted into a $\Gamma$-lnf with the same type.

First, one shows that $\beta\iota\rho$-reduction enjoys subject reduction and weak normalisation. The weak normalisation proof is analogous to [41, Theorem 6.1.8].

Next, one shows that $\beta\iota\rho$-normal forms may be expanded to $\eta$-long $\beta\iota\rho$-normal forms. Some care needs to be taken to avoid creating a $\rho$-redex when expanding a case expression.

Finally, one performs transformations corresponding to the elimination and the eager simple elimination restrictions. See the appendix for details.

# 5    Evaluation

We performed several empirical evaluations of our proof search procedure. First, on a collection of a few Coq libraries and separately on CompCert [27], we measured the effectiveness of the procedure as a standalone proof search tactic, as well as its effectiveness as a reconstruction backend for CoqHammer. We also measured the effectiveness of our procedure on pure intuitionistic first-order logic by evaluating it on the ILTP library [36] of first-order intuitionistic problems.

Our proof search procedure intends to provide general push-button automation for CIC without fixpoints, based on sound theoretical foundations. As such, it is in a category of its own, as far we know. Our evaluations in several different scenarios indicate the practical viability of our approach despite its generality. It should be noted that the tactics we compare against are not intended for full automation, but target specific small fragments of CIC or require hand-crafted hints for effective automation.

Detailed evaluation results, complete logs, Coq problem files and conversion programs are available in the online supplementary material [12]. The collection of Coq libraries and developments on which we evaluated our procedure includes: coq-ext-lib library, Hahn library, int-map (a library of maps indexed by binary integers), Coq files accompanying the first three volumes of the Software Foundations book series [1, 34, 35], a general topology library, several other projects from coq-contribs. The full list is available at [12].

The results of the standalone (left) and CoqHammer backend (right) evaluation on 4494 problems from a collection of Coq developments, and seperately the results on CompCert are presented below.

Coq libraries collection
standalone+i (4494 problems, 30s)

| tactic | proved | proved % |
|---|---|---|
| sauto+i | 1840 | 40.9% |
| yelles+i | 1552 | 34.5% |
| coq+i | 1229 | 27.3% |
| crush+i | 1134 | 25.2% |

Coq libraries collection
CoqHammer (4494 problems, 30s+30s)

| tactic | proved | proved % | re-proved % |
|---|---|---|---|
| sauto-12 | 1756 | 39.1% | 93.9-96.7% |
| coq-4 | 1243 | 27.7% | 79.1-87.5% |

CompCert
standalone+i (5495 problems, 30s)

| tactic | proved | proved % |
|---|---|---|
| sauto+i | 941 | 17.1% |
| yelles+i | 875 | 15.9% |
| coq+i | 372 | 6.8% |
| crush+i | 355 | 6.5% |

CompCert
CoqHammer (5353 problems, 30s+30s)

| tactic | proved | proved % | re-proved % |
|---|---|---|---|
| sauto-12 | 1373 | 25.6% | 87.0-95.4% |
| coq-4 | 616 | 11.5% | 42.0-78.9% |

For the evaluation on CompCert, the number of problems for the CoqHammer backend evaluation is smaller because the CoqHammer translation cannot handle some Coq goals (e.g. with existential metavariables) and these were not included.

For the standalone evaluation, we first try to apply our procedure, and if it fails then we try heuristic unfolding and then try to do induction on each available hypothesis followed by the tactic. This gives a better idea of the usefulness of our procedure because the core tactic itself cannot succeed on any problems that require non-trivial induction. For comparison, an analogous combination of standard Coq tactics (or crush) with unfolding and induction is used.

For the standalone evaluation, by "sauto+i" we denote our proof search procedure, by "yelles+i" the preliminary procedure form [14], by "crush+i" a slightly improved version of the crush tactic from [9], by "coq+i" a mix of standard Coq automation tactics (including eauto, lia, congruence, firstorder). All these are combined with induction, etc., as described above.

We also performed a standalone evaluation without combining the tactics with induction or unfolding. The results are presented below. For the standalone evaluation without induction, by "coq-no-fo" we denote the same mix of standard Coq automation tactics as "coq" but not including the firstorder tactic.

Coq libraries collection
standalone (4494 problems, 5s)

| tactic | proved | proved % |
|---|---|---|
| coq | 978 | 21.8% |
| sauto | 888 | 19.6% |
| crush | 663 | 14.8% |
| coq-no-fo | 607 | 13.5% |
| yelles | 602 | 13.4% |

CompCert
standalone (5495 problems, 5s)

| tactic | proved | proved % |
|---|---|---|
| sauto | 420 | 7.6% |
| yelles | 407 | 7.4% |
| coq | 286 | 5.2% |
| coq-no-fo | 237 | 4.3% |
| crush | 210 | 3.8% |

The results of the standalone evaluations indicate that our procedure is useful as a standalone Coq tactic in a push-button automated proof search scenario, performing comparably or better than other tactics available for Coq.

For the evaluation of our procedure as a CoqHammer backend, we use 12 variants of our tactics (including 3 variants based on the incomplete preliminary procedure from [14]) run in parallel (i.e. a separate core assigned to each variant) for 30s ("sauto-12" row). We included the variants of the preliminary procedure form [14] to increase the diversity of the solved problems. The procedure from [14], while much more ad-hoc and heuristic, is essentially a less general version of the present one. The point of this evaluation is to show that our approach may be engineered into an effective CoqHammer reconstruction backend, and not to compare the present procedure with its limited ad-hoc variant. For comparison, we used 4 variants of combinations of standard Coq automation tactics ("coq-4" row). We show the total number and the percentage of problems solved with any of the external provers and premise selection methods employed by CoqHammer. The external provers were run for 30s each. The reconstruction success rates ("re-proved" column) are calculated separately for each prover and a range is presented.

A common property of the chosen libraries is that they use the advanced features of Coq sparingly and are written in a style where proofs are broken up into many small lemmas. Some do not use much automation, and the Software Foundations files contain many exercises with relatively simple proofs. Moreover, some of the developments modify the core hints database which is then used by the tactics. The resulting problem set is suitable for comparing proof search procedures on a restricted subset of Coq logic, but does not necessarily reflect Coq usage in modern developments. This explains the high success rate compared to CompCert. Also, CompCert uses classical logic, while our procedure tries to find only intuitionistic proofs. Hence, a lower success rate is to be expected since it is harder or impossible to re-prove some of the lemmas constructively.

The results of the evaluation on the ILTP library v1.1.2 [36] follow.

ILTP (2574 problems, 600s)

| tactic | proved | proved % | 0.00 | 0.25 | 0.50 | 0.75 | 1.00 |
|---|---|---|---|---|---|---|---|
| hprover | 662 | 25.7% | 96.3% | 52.1% | 72.7% | 43.9% | 12.9% |
| tprover | 636 | 24.7% | 96.3% | 52.1% | 52.7% | 44.6% | 11.8% |
| yelles | 602 | 23.4% | 79.4% | 40.1% | 52.7% | 47% | 12.2% |
| sauto-3 | 760 | 29.5% | | | | | |
| hprove10 | 565 | 22.0% | 90.8% | 40.8% | 50.9% | 38.7% | 10.3% |
| firstorder | 467 | 18.1% | 95.4% | 56.3% | 58.2% | 20% | 6.7% |
| ileanCoP | 934 | 36.3% | 97.7% | 95.8% | 96.4% | 95.8% | 16.8% |

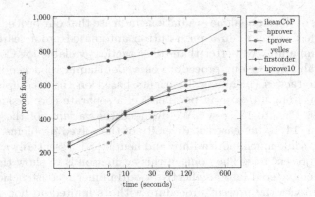

We compared our procedure with `firstorder` [11] and with ileanCoP 1.2 [31] – a leading connection-based first-order intuitionistic theorem prover. We converted the library files to appropriate formats (Coq or ileanCoP). For ileanCoP and `firstorder`, the converted problem files include equality axioms (reflexivity, symmetry, transitivity, congruence). These axioms were not added for our procedure because it can already perform limited equality reasoning. We used exhaustive variants of our tactics which perform backtracking on instantiations of existential metavariables and do not perform eager simple elimination, eager or unordered rewriting. The proof search procedure is run in an iterative deepening fashion, increasing the depth or cost bound on failure. The "hprover" row shows the result for the depth cost model, "tprover" for the tree cost model, "yelles" for the preliminary procedure from [14], and "sauto-3" the combination of the results for the above three. The columns labeled with a number R show the percentage of problems with difficulty rating R for which proofs were found. The graph below the table shows how many problems were solved within a given time limit.

The `firstorder` tactic is generally faster than our procedure, but it finds much fewer proofs for the problems with high difficulty rating. For `firstorder` we did not implement iterative deepening, because the depth limit is a global parameter not changeable at the tactic language level. We set the limit to 10. To provide a fair comparison, we also evaluated our proof search procedure with the depth cost model and the depth bound fixed at 10 ("hprove10").

In combination, the three iterative deepening variants of our procedure managed to find proofs for 80 theorems that were not proven by ileanCoP. Overall, the performance of ileanCoP is much better, but it does not produce proof terms and is restricted to pure intuitionistic first-order logic.

## 6  Examples

In this section, we give some examples of the use of our proof search procedure as a standalone Coq tactic. The core inhabitation procedure is implemented in the `sauto` tactic which uses the tree cost model and bounds the proof search by default. There are several other tactics which invoke different variants of the proof search procedure. The `ssimpl` tactic performs the simplifications, forward reasoning and eager actions described in Sects. 3.2 and 3.3. The implementation is available as part of a recent version of the CoqHammer tool [13,14], and it is used as the basis of its reconstruction tactics.

Our first example is a statement about natural numbers. It can be proven by `sauto` without any lemmas because the natural numbers, disjunction, existential quantification and equality are all inductive types.

```
Lemma lem_simple_nat : forall n, n = 0 \/ exists m, n = S m.
```

Note that because the proof requires inversion on `nat`, it cannot possibly be created by any of the standard Coq automation tactics.

Because $<$ is defined in terms of $\leq$ which is an inductive type, `sauto` can prove the following lemma about lists.

```
Lemma lem {A} (l : list A) : l <> nil -> length (tl l) < length l.
```

The next example concerns big-step operational semantics of simple imperative programs. The commands of an imperative program are defined with an inductive type `cmd`. The big-step operational semantics is represented with a dependent inductive type `==>` : `cmd * state -> state -> Prop` , and command equivalence `~~` : `cmd -> cmd -> Prop` is defined in terms of `==>` . We skip the details of these definitions.

Then `sauto` can fully automatically prove the following two lemmas. The first one states the associativity of command sequencing. The second establishes the equivalence of the while command with its one-step unfolding. On a computer with a 2.5 GHz processor, in both cases `sauto` finds a proof in less than 0.5 s.

```
Lemma lem_seq_assoc : forall c1 c2 c3 s s',
(Seq c1 (Seq c2 c3), s) ==> s' <-> (Seq (Seq c1 c2) c3, s) ==> s'.
```

```
Lemma lem_unfold_loop : forall b c,
   While b c ~~ If b (Seq c (While b c)) Skip.
```

Note again that both proofs require multiple inversions, and thus it is not possible to obtain them with standard Coq automation tactics.

According to Kunze [25], the following set-theoretical statement cannot be proven in reasonable time by either `firstorder` or JProver. The `sauto` tactic finds a proof in less than 0.3s. Below Seteq, Subset and In are variables of type $U \rightarrow U \rightarrow$ Prop; Sum of type $U \rightarrow U \rightarrow U$; and $U$ : Type.

```
(forall A B X, In X (Sum A B) <-> In X A \/ In X B) /\
(forall A B, Seteq A B <-> Subset A B /\ Subset B A) /\
(forall A B, Subset A B <-> forall X, In X A -> In X B) ->
(forall A, Seteq (Sum A A) A).
```

# 7    Conclusions and Future Work

We presented a practical general proof search procedure for Coq based on type inhabitation. This increases the power of out-of-the-box automation available for Coq and provides an effective reconstruction backend for CoqHammer. The empirical evaluations indicate that our approach to fully automated proof search in the Calculus of Inductive Constructions is practically viable.

For efficiency reasons, the inhabitation procedure is not complete in general, but it is complete for a first-order fragment of the Calculus of Inductive Constructions. We conjecture that a variant of our procedure could be shown complete for the Calculus of Inductive Constructions with definitional proof irrelevance, with only non-dependent elimination, and without fixpoints.

We implemented the proof search procedure in OCaml and Ltac as a Coq plugin. The plugin generates values in the Coq's tactic monad which contain callbacks to the plugin. Better efficiency would probably be achieved by directly generating Coq proof terms or sequences of basic tactics. This would, however, require much engineering work. Another disadvantage of the monadic implementation is that it limits the proof search strategy to depth-first order and precludes global caching. In the artificial intelligence literature, there are many approaches to graph and tree search [20] which might turn out to be better suited for an inhabitation procedure than the depth-first tree search.

**Acknowledgements.** The author thanks Ping Hou for feedback on the operation of the tactics and for pointing out some bugs.

# A    Completeness Proof for the First-Order Fragment

In this appendix we prove completeness of our proof search procedure for a first-order fragment of the Calculus of Inductive Constructions. First, we precisely define the first-order fragment.

## A.1    The First-Order Fragment

The system is essentially an extension of $\lambda$PRED from [4, Definition 5.4.5] with inductive types and higher-order functions.

A *preterm* is a sort $s \in \mathcal{S} = \{*^s, *^p, \Box^s, \Box^p\}$, a variable $x$, a constructor $c$, an inductive type $I$, an application $t_1 t_2$, an abstraction $\lambda x : \tau.t$, a dependent product $\forall x : \alpha.\beta$, or a case expression $\mathsf{case}_\tau^{I\vec{q}}(u; \vec{x_1} : \vec{\sigma_1} \Rightarrow t_1 \mid \ldots \mid \vec{x_k} : \vec{\sigma_k} \Rightarrow t_k)$. In a case expression, $u$ is the term matched on, the type of $u$ is $I\vec{q}$ where $\vec{q}$ are the values of the parameters, $\tau$ is the type of the case expression and of each of the branches $t_i$, and $t_i[\vec{v}/\vec{x}]$ is the value of the case expression if the value of $u$ is $c_i\vec{q}\vec{v}$. In comparison to the full CIC, we allow only non-dependent case expressions, i.e., the return type $\tau$ does not vary across branches. We omit the sub- and/or the superscript when clear or irrelevant.

The intuitive interpretation of the sorts is as follows. The sort $*^p$ (also written as Prop) is for propositions. First-order formulas are elements of $*^p$. The sort $*^s$ (also written as Set) is for sets – these form a simple type structure over a collection of first-order universes. For example, when $U : *^s$ then also $(U \to U) : *^s$. The sort $\Box^s$ is the sort of $*^s$. The presence of $\Box^s$ allows to declare set variables (i.e. of sort $*^s$) in the context. The sort $\Box^p$ is the sort of predicate types. We have $(\tau_1 \to \ldots \to \tau_n \to *^p) : \Box^p$ when $\tau_i : *^s$ for $i = 1, \ldots, n$.

An *inductive declaration*

$$I(\vec{p} : \vec{\rho}) : *^p := c_1 : \sigma_1 \mid \ldots \mid c_n : \sigma_n$$

declares an inductive type $I$ with *parameters* $\vec{p}$ and *arity* $\forall \vec{p} : \vec{\rho}.*^p$, with $n$ constructors $c_1, \ldots, c_n$ having types $\sigma_1, \ldots, \sigma_n$ respectively (in the context extended with $\vec{p} : \vec{\rho}$). We require:

- $\sigma_i = \forall x_i^1 : \tau_i^1. \ldots . \forall x_i^{k_i} : \tau_i^{k_i}.I\vec{p}$,
- $I$ does not occur in any $\tau_i^j$.

We could allow strictly positive occurrences of $I$ in $\sigma_i$, non-parameter arguments to $I$ or inductive types in $*^s$ as well as $*^p$. These modifications, however, would introduce some tedious technical complications. With the above definition, all inductive types are non-recursive.

The *arity* of a constructor $c_i$ is $\forall \vec{p} : \vec{\rho}.\sigma_i$, denoted $c_i(\vec{p} : \vec{\rho}) : \sigma_i$. We assume the well-foundedness of the relation $\succ$ defined by: $I_1 \succ I_2$ iff $I_2$ occurs in the arity of $I_1$ or the arity of a constructor of $I_1$.

An *environment* is a list of inductive declarations. We write $I \in E$ if a declaration of an inductive type $I$ occurs in the environment $E$. Analogously, we write $(I(\vec{p} : \vec{\rho}) : *^p) \in E$ and $(c(\vec{p} : \vec{\rho}) : \tau) \in E$, if a declaration of $I$ with arity $\forall \vec{p} : \vec{\rho}.*^p$ occurs in $E$, or a constructor $c(\vec{p} : \vec{\rho}) : \tau$ with arity $\forall \vec{p} : \vec{\rho}.\tau$ occurs in a declaration in $E$, respectively. A *context* $\Gamma$ is a list of pairs $x : \tau$ with $x$ a variable and $\tau$ a term. A *typing judgement* has the form $E; \Gamma \vdash t : \tau$ with $t, \tau$ preterms. A term $t$ is well-typed and has type $\tau$ in the context $\Gamma$ and environment $E$ if $E; \Gamma \vdash t : \tau$ may be derived using the rules from Fig. 1. We denote the empty list by $\langle \rangle$.

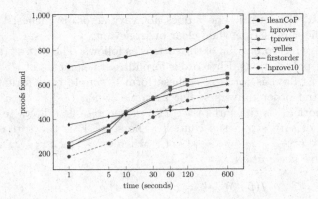

**Fig. 1.** Typing rules

The set $\mathcal{R} = \{(*^p, *^p), (*^s, *^p), (*^s, \square^p), (*^s, *^s)\}$ in Fig. 1 is the set of rules which determine the allowed dependent products.

The rule $(*^p, *^p)$ allows the formation of implication of two formulas:

$$\phi : *^p, \psi : *^p \vdash (\phi \to \psi) : *^p.$$

The rule $(*^s, *^p)$ allows quantification over sets:

$$A : *^s, \phi : *^p \vdash (\forall x : A.\phi) : *^p.$$

The rule $(*^s, \square^p)$ allows the formation of predicates:

$$A : *^s \vdash (A \to *^p) : \square^p,$$

hence

$$A : *^s, P : A \to *^s, x : A \vdash Px : *^p,$$

so $P$ is a predicate on $A$.

The rule $(*^s, *^s)$ allows the formation of function spaces between sets:

$$A : *^s, B : *^s \vdash (A \to B) : *^s.$$

Note that we permit quantification over higher-order functions but the formation of lambda-abstractions is allowed only for proofs (i.e. elements of propositions) and for predicates. Elements of sets are effectively restricted to applicative form.

Note that case expressions can occur only in proofs. Hence, including $\iota$ in the conversion rule is in fact superfluous.

In Fig. 1 we assume that the environment $E$ is *well-formed*, which is defined inductively: an empty environment is well-formed, and an environment $E, I(\vec{p} : \vec{\rho}) : *^p := c_1 : \tau_1 \mid \ldots \mid c_n : \tau_n$ (denoted $E, I$) is well-formed if $E$ is and:

– the constructors $c_1, \ldots, c_n$ are pairwise distinct and distinct from any constructors occurring in the declarations in $E$;

- $E; p_1 : \rho_1, \ldots, p_{j-1} : \rho_{j-1} \vdash \rho_j : \square$ with $\square \in \{\square^s, \square^p\}$, for each $j$;
- $E; \vec{p} : \vec{\rho}, i : *^p \vdash \tau'_j : *^p$ for $j = 1, \ldots, n$, where $\tau'_j$ is $\tau_j$ with all occurrences of $I\vec{p}$ replaced by $i$.

When $E, \Gamma$ are clear or irrelevant, we write $\Gamma \vdash t : \tau$ or $t : \tau$ instead of $E; \Gamma \vdash t : \tau$. In what follows, we assume a fixed a well-formed environment $E$ and omit it from the notation. We write $\Gamma \vdash t : \tau : s$ if $\Gamma \vdash t : \tau$ and $\Gamma \vdash \tau : s$.

Standard meta-theoretical properties hold for our system, including the substitution, thinning and generation lemmas, subject reduction for $\beta\iota$-reduction and uniqueness of types. We will use these properties implicitly. The proofs are analogous to [4, Section 5.2] and we omit them.

The available forms of inductive types and case expressions suffice to define all intuitionistic logical connectives with their introduction and elimination rules (see Sect. 2). They do not allow for an inductive definition of equality, however.

**Definition 1.** 1. A $\Gamma$-*proposition* is a term $\tau$ with $\Gamma \vdash \tau : *^p$.
2. A $\Gamma$-*proof* is a term $t$ such that $\Gamma \vdash t : \tau : *^p$ for some $\tau$.
3. A $\Gamma$-*set* is a term $\tau$ with $\Gamma \vdash \tau : *^s$.
4. A $\Gamma$-*element* is a term $t$ such that $\Gamma \vdash t : \tau : *^s$ for some $\tau$.

We often omit $\Gamma$ and talk about *propositions*, *proofs*, *sets* and *set elements*.

## A.2   Completeness Proof

The $\beta$-, $\iota$-, and $\rho$-reductions for our first-order system are:

$$(\lambda x : \tau.t_1)t_2 \to_\beta t_1[t_2/x]$$

$$\mathbf{case}_\tau^{I\vec{q}}(c_i\vec{q}\vec{v}; \vec{x_1} : \vec{\tau_1} \Rightarrow t_1 \mid \ldots \mid \vec{x_k} : \vec{\tau_k} \Rightarrow t_k) \to_\iota t_i[\vec{v}/\vec{x_i}]$$

$$\mathbf{case}_{\forall x:\alpha.\tau}^{I\vec{q}}(u; \vec{x_1} : \vec{\tau_1} \Rightarrow t_1 \mid \ldots \mid \vec{x_k} : \vec{\tau_k} \Rightarrow t_k)w \to_{\rho_1}$$
$$\mathbf{case}_{\tau[w/x]}^{I\vec{q}}(u; \vec{x_1} : \vec{\tau_1} \Rightarrow t_1 w \mid \ldots \mid \vec{x_k} : \vec{\tau_k} \Rightarrow t_k w)$$

$$\mathbf{case}_\tau^{I\vec{q}}(\mathbf{case}_{I\vec{q}}^{J\vec{p}}(u; \vec{x_1} : \vec{\tau_1} \Rightarrow t_1 \mid \ldots \mid \vec{x_k} : \vec{\tau_k} \Rightarrow t_k); P) \to_{\rho_2}$$
$$\mathbf{case}_\tau^{J\vec{p}}(u; \vec{x_1} : \vec{\tau_1} \Rightarrow \mathbf{case}_\tau^{I\vec{q}}(t_1; P) \mid \ldots \mid \vec{x_k} : \vec{\tau_k} \Rightarrow \mathbf{case}_\tau^{I\vec{q}}(t_k; P))$$

In the $\rho_2$-reduction rule, $P$ stands for a list of case patterns $\vec{y_1} : \vec{\sigma_1} \Rightarrow w_1 \mid \ldots \mid \vec{y_m} : \vec{\sigma_m} \Rightarrow w_m$. We assume $\vec{x_i}$ do not occur in $P$.

Because case expressions can occur only in proofs, subject reduction holds for $\rho$-reduction.

**Lemma 1.** *If* $\Gamma \vdash t : \tau : *^s$ *then* $t$ *does not contain case expressions or lambda-abstractions.*

*Proof.* Induction on $t$, using the generation lemma.

**Lemma 2.** *If* $\Gamma \vdash \tau : \forall \vec{x} : \vec{\alpha}.s$ *with* $s \in \mathcal{S}$ *then:*

*1. no $\Gamma$-proofs occur in $\tau$,*

2. *no case expressions occur in $\tau$.*

*Proof.* By induction on $\tau$.

**Corollary 1.** *If $\Gamma \vdash t : \tau$ and $t$ contains a case expression, then $\Gamma \vdash \tau : *^p$.*

**Corollary 2.** *If $\Gamma \vdash t : \varphi : *^p$ and $\Gamma, x : \varphi \vdash u : \tau$ then $\Gamma \vdash u[t/x] : \tau$.*

**Lemma 3. (Subject reduction for $\rho$).** *If $\Gamma \vdash t : \tau$ and $t \rightarrow_\rho t'$ then $\Gamma \vdash t' : \tau$.*

*Proof.* By Corollary 1, $t$ must be a $\Gamma$-proof if it contains a $\rho$-redex.

The lemma is shown by induction on the typing derivation, analogously to [4, Theorem 5.2.15] except that where the conversion rule is used we instead appeal to Corollary 2.

Implicitly, the following theorems, lemmas and definitions depend on the typing context $\Gamma$, which changes in the expected way when going under binders. We also implicitly consider types up to $\beta\iota$-equality.

**Theorem 2.** *The $\beta\iota\rho$-reduction is weakly normalising on typable terms.*

*Proof.* The proof is an adaptation of the proof of an analogous result for first-order intuitionistic natural deduction. See [41, Theorem 6.1.8].

Note that when the context is fixed, the type of each subterm is uniquely determined up to $\beta$-equality (the type does not contain proofs, so $\iota$-equality is redundant).

Set elements are in $\beta\iota\rho$-normal form, because they don't contain case expressions or lambda-abstractions. Hence, the only redexes which are not proofs must occur in types and have the form $(\lambda x : \alpha.\tau)t$ where $\alpha : *^s$. Since each contraction of a $\beta$-redex of this form strictly decreases the number of lambda-abstractions ($t$ is a set element not containing lambda-abstractions), $\beta$-reduction in types terminates. Moreover, redexes in types cannot be created by reducing $\beta\iota\rho$-redexes which are proofs, because abstraction over predicates (i.e. elements of $\square^p$) is not allowed. We may thus consider only the case when all types are in $\beta\iota\rho$-normal form and all redexes are proofs.

The *degree* $\delta(\alpha)$ of a set $\alpha$ is 0. The *degree* $\delta(\tau)$ of a proposition $\tau$ in $\beta$-normal form is defined inductively.

- If $\tau = \forall x : \tau_1.\tau_2$ then $\delta(\tau) = \delta(\tau_1) + \delta(\tau_2) + 1$.
- If $\tau = I\vec{q}$ with $I(\vec{p} : \vec{\rho}) : *^p := c_1 : \forall \vec{x_1} : \vec{\tau_1}.I\vec{p} \mid \ldots \mid c_k : \forall \vec{x_k} : \vec{\tau_k}.I\vec{p}$ then $\delta(\tau) = \delta(\vec{\sigma_1}) + \ldots + \delta(\vec{\sigma_k}) + 1$ where $\vec{\sigma_i} = \vec{\tau_i}[\vec{q}/\vec{p}]$ and $\delta(\vec{\sigma_i}) = \delta(\sigma_i^1, \ldots, \sigma_i^{m_i}) = \delta(\sigma_i^1) + \ldots + \delta(\sigma_i^{m_i})$.
- Otherwise $\delta(\tau) = 0$.

Formally, this definition is by induction on the lexicographic product of the multiset extension of the well-founded order $\succ$ on inductive types (we compare the multisets of inductive types occurring in $\tau$) and the size of $\tau$.

Note that $\delta(\tau[t/x]) = \delta(\tau)$ if $t : *^s$ or $t : *^p$. This is because set elements are not counted towards the degree and proofs do not occur in types.

The *degree* $\delta(t)$ of a redex $t$ is defined as follows.

- $\beta$- or $\rho_1$-redex: $t = t_1 t_2$ with $t_1 : \forall x : \alpha.\tau$. Then $\delta(t) = \delta(\forall x : \alpha.\tau)$.
- $\iota$- or $\rho_2$-redex: $t = \mathsf{case}_\tau^{I\vec{q}}(u; \ldots)$. Then $\delta(t) = \delta(I\vec{q})$.

Note that if $t : *^p$ is a redex and $u : *^s$ or $u : *^p$ then $\delta(t[u/x]) = \delta(t)$.

The *case-size* $\mathrm{cs}(t)$ of a term $t$ is defined inductively.

- If $t$ is not a case expression then $\mathrm{cs}(t) = 1$.
- If $t = \mathsf{case}(u; \vec{x_1} : \vec{\tau_1} \Rightarrow t_1 \mid \ldots \mid \vec{x_k} : \vec{\tau_k} \Rightarrow t_k)$ then $\mathrm{cs}(t) = \mathrm{cs}(t_1) + \ldots + \mathrm{cs}(t_k) + 1$.

The *redex-size* $\mathrm{rs}(t)$ of a redex $t$ is defined as follows.

- If $t = t_1 t_2$ (a $\beta$- or $\rho_1$-redex) then $\mathrm{rs}(t) = \mathrm{cs}(t_1)$.
- If $t = \mathsf{case}(u; \ldots)$ (a $\iota$- or $\rho_2$-redex) then $\mathrm{rs}(t) = \mathrm{cs}(u)$.

The *argument* and the *targets* of an application or a case expression $t$ are defined as follows.

- If $t = t_1 t_2$ then $t_1$ is the target and $t_2$ the argument.
- If $t = \mathsf{case}(a\vec{u}; \vec{x_1} : \vec{\tau_1} \Rightarrow t_1 \mid \ldots \mid \vec{x_k} : \vec{\tau_k} \Rightarrow t_k)$ with $a$ a constructor or a variable, then $a\vec{u}$ is the argument and $t_1, \ldots, t_k$ are the targets.
- If $t = \mathsf{case}(u; \vec{x_1} : \vec{\tau_1} \Rightarrow t_1 \mid \ldots \mid \vec{x_k} : \vec{\tau_k} \Rightarrow t_k)$ with $u$ a case expression, then $u$ is the target and $t_1, \ldots, t_k$ are the arguments.

A subterm occurrence $r$ is *to the right* of a subterm occurrence $r'$ (and $r'$ *to the left* of $r$) if $r'$ occurs to the right of $r$ in the in-order traversal of the term tree where we first traverse the targets of a subterm, then visit the subterm, and then traverse its arguments. With this definition, a subterm is to the right of its targets and to the left of its arguments.

Note that the rightmost redex $r$ of maximum degree does not occur in a target of a redex $r'$ of maximum degree and no redex $r''$ of maximum degree occurs in its argument (otherwise $r', r''$ would be to the right of $r$).

For a $\beta$- or $\rho$-redex, the redex-size is the case-size of the target. For a $\iota$-redex, the redex-size is 1. Let $r$ be the rightmost redex of maximum degree in $t$. Note that changing $r$ to a case expression $r'$ cannot increase the redex-size of a redex of maximum degree $r''$ in $t$ containing $r$. Indeed, otherwise $r''$ would be a $\beta$- or $\rho$-redex and $r$ would occur in its target, so $r''$ would be to the right of $r$. This implies that contracting the rightmost redex $r$ of maximum degree to $r'$ cannot increase the redex-size of another redex of maximum degree by exposing a case expression in $r'$. Note we have not (yet) ruled out the possibility of the contraction increasing the redex-size of a redex of maximum degree occurring inside $r$.

Let $n$ be the maximum degree of a redex in $t$ and $m$ the sum of redex-sizes of redexes of maximum degree in $t$. By induction on pairs $(n, m)$ ordered lexicographically, we show that $t$ is weakly $\beta\iota\rho$-normalising.

Choose the rightmost redex $r$ of maximum degree and contract it. This either decreases $n$ or leaves $n$ unchanged and decreases $m$.

- If $r = (\lambda x : \alpha.r_1)r_2 \to_\beta r_1[r_2/x]$ then no redexes of maximum degree occur in $r_2$ (because $r_2$ is the argument of a rightmost redex $r$ of maximum degree). So no redexes of maximum degree get duplicated.

  All redexes created by this contraction (either by exposing a possible $\lambda$-abstraction or case expression in $r_1$, or by substituting $r_2$ for $x$) are of smaller degree. Indeed, if e.g. $r_2 = \mathsf{case}_{I\vec{q}}(u; \ldots)$ is substituted for $x$ in $\mathsf{case}^{I\vec{q}}(x; \ldots)$, then $\alpha = I\vec{q}$ and the degree of the created $\rho_2$-redex is $\delta(I\vec{q}) = \delta(\alpha) < \delta(r)$. Other cases are similar: one notices that the degree of each redex created by substituting $r_2$ for $x$ is $\delta(\alpha) < \delta(r)$, and the degree of a redex created by exposing $r_1[r_2/x]$ is $\delta(\tau) < \delta(r)$ where $r_1 : \tau$.

  We also need to show that the $\beta$-contraction does not increase the redex-size of another redex of maximum degree. The contraction may increase the redex-size of a redex $r'$ in $r_1$ by substituting $r_2$ for $x$. But then $\delta(r') = \delta(\alpha) < \delta(r)$. As discussed before, the contraction cannot increase the redex-size of another redex of maximum degree by exposing $r_1[r_2/x]$.

  Therefore, either $n$ decreases if $r$ was the only redex of maximum degree, or $m$ decreases and $n$ does not change.

- If $r = \mathsf{case}_\tau^{I\vec{q}}(c_i\vec{q}\vec{v}; \vec{x_1} : \vec{\tau_1} \Rightarrow t_1 \mid \ldots \mid \vec{x_k} : \vec{\tau_k} \Rightarrow t_k) \to_\iota t_i[\vec{v}/\vec{x_i}]$ then no redexes of maximum degree occur in $\vec{v}$.

  Let $v_j : \sigma_j$. Because $\delta(\sigma_j) < \delta(I\vec{q})$, all redexes created by substituting $\vec{v}$ for $\vec{x_i}$ have smaller degree. If a redex is created by exposing $t_i[\vec{v}/\vec{x_i}]$ then $r$ occurs in $t$ as $ru$ or $\mathsf{case}(r; \ldots)$. The created redex is then $t_i[\vec{v}/\vec{x_i}]u$ or $\mathsf{case}(t_i[\vec{v}/\vec{x_i}]; \ldots)$ and has degree $\delta(\tau)$. But $ru$ or $\mathsf{case}(r; \ldots)$ was a redex of degree $\delta(\tau)$ which occurred to the right of $r$. This is only possible when $\delta(\tau) < \delta(I\vec{q})$.

  By a similar argument, the $\iota$-contraction may increase the redex-size only of redexes of smaller degree.

- If

$$r = \mathsf{case}_{\forall x:\alpha.\tau}^{I\vec{q}}(u; \vec{x_1} : \vec{\tau_1} \Rightarrow t_1 \mid \ldots \mid \vec{x_k} : \vec{\tau_k} \Rightarrow t_k)w \to_{\rho_1}$$
$$\mathsf{case}_{\tau[w/x]}^{I\vec{q}}(u; \vec{x_1} : \vec{\tau_1} \Rightarrow t_1 w \mid \ldots \mid \vec{x_k} : \vec{\tau_k} \Rightarrow t_k w) = r'$$

then no redexes of maximum degree occur in $w$.

New redexes of maximum degree may be created: $t_1 w, \ldots, t_k w$. However, the sum of the redex-sizes of these redexes is smaller than the redex-size of the contracted redex $r$, so $m$ decreases.

A redex may be created by exposing $r'$, but then the degree of this redex is $\delta(\tau) < \delta(r)$.

- If

$$r = \mathsf{case}_\tau^{I\vec{q}}(\mathsf{case}_{I\vec{q}}^{J\vec{p}}(u; \vec{x_1} : \vec{\tau_1} \Rightarrow t_1 \mid \ldots \mid \vec{x_k} : \vec{\tau_k} \Rightarrow t_k); P) \to_{\rho_2}$$
$$\mathsf{case}_\tau^{J\vec{p}}(u; \vec{x_1} : \vec{\tau_1} \Rightarrow \mathsf{case}_\tau^{I\vec{q}}(t_1; P) \mid \ldots \mid \vec{x_k} : \vec{\tau_k} \Rightarrow \mathsf{case}_\tau^{I\vec{q}}(t_k; P)) = r'$$

then no redexes of maximum degree occur in $P$.

New redexes of maximum degree may be created: $\mathsf{case}_\tau^{I\vec{q}}(t_i; P)$. The sum of the redex-sizes of these redexes is at most $\mathsf{cs}(t_1) + \ldots + \mathsf{cs}(t_k) < \mathsf{cs}(t_1) + \ldots + \mathsf{cs}(t_k) + 1 = \mathsf{rs}(r)$.

A redex may be created by exposing $r'$ if $r$ occurs in $ru$ or $\mathsf{case}(r; \ldots)$. But then $ru$ or $\mathsf{case}(r; \ldots)$ was a redex to the right of $r$ with the same degree $\delta(\tau)$

as the created redex $r'u$ or $\mathtt{case}(r'; \ldots)$, so the degree of the new redex must be smaller than $\delta(I\vec{q})$.

As discussed before, the redex-size of another redex of maximum degree cannot be increased by exposing $r'$.

**Lemma 4.** *If $t$ is a proof in $\beta\iota\rho$-normal form then one of the following holds.*

- $t = \lambda x : \alpha.t'$ and $t'$ is a proof in $\beta\iota\rho$-normal form.
- $t = xu_1 \ldots u_n$ and each $u_i$ is a proof or a set element in $\beta\iota\rho$-normal form.
- $t = c\vec{q}u_1 \ldots u_n$ with $\vec{q}$ the parameter values, and each $u_i$ is a proof or a set element in $\beta\iota\rho$-normal form.
- $t = \mathtt{case}(xu_1 \ldots u_n; \vec{y_1} : \vec{\sigma_1} \Rightarrow w_1 \mid \ldots \mid \vec{y_m} : \vec{\sigma_m} \Rightarrow w_m)$, and each $u_i$ is a proof or a set element in $\beta\iota\rho$-normal form, and each $w_i$ is a proof in $\beta\iota\rho$-normal form.

Note that well-typed constructor applications $c\vec{q}t_1 \ldots t_n$ must by definition (Fig. 1) always include all parameter values $\vec{q}$. In other words, partial application of a constructor to only some of the parameter values is not allowed.

A set element cannot contain lambda-abstractions or case-expressions, so it is always in $\beta\iota\rho$-normal form.

**Definition 2.** The *$\eta$-long form* of a set element $t$ is $t$.

The *$\eta$-long form* of a proof variable $x$ is defined by induction on the type $\tau : *^p$ of $x$ in normal form: if $\tau = \forall \vec{y} : \vec{\alpha}.\beta$ with $\beta$ not a product, then $\lambda \vec{y} : \vec{\alpha}.xy_1' \ldots y_n'$ where $y_i'$ is the $\eta$-long form of $y_i$. Note that the $\eta$-long form of $x$ is well-defined and its type is still $\tau$ because each $y_i$ is either a set element (then $y_i' = y_i$) or a proof variable (then $y_i$ does not occur in $\vec{\alpha}, \beta$).

The *$\eta$-long form* of a proof $t$ in $\beta\iota\rho$-normal form is defined by induction on $t$.

- If $t = \lambda x : \alpha.u$ and $u'$ is the $\eta$-long form of $u$, then $\lambda x : \alpha.u'$ is the $\eta$-long form of $t$.
- If $t = xt_1 \ldots t_k$, and $\forall \vec{y} : \alpha.\tau$ is the type of $t$ with $\tau$ not a product, and $t_i'$ is the $\eta$-long form of $t_i$, and $y_i'$ is the $\eta$-long form of (a proof variable or a set element) $y_i$, then $\lambda \vec{y} : \vec{\alpha}.xt_1' \ldots t_k'y_1' \ldots y_n'$ is the $\eta$-long form of $t$. For $k = 0$ this definition coincides with the definition of the $\eta$-long form of a proof variable.
- If $t = c\vec{q}t_1 \ldots t_k$, and $\forall \vec{y} : \alpha.\tau$ is the type of $t$ with $\tau$ not a product, and $\vec{q}$ are the parameters, and $t_i'$ is the $\eta$-long form of $t_i$, and $y_i'$ is the $\eta$-long form of (a proof variable or a set element) $y_i$, then $\lambda \vec{y} : \vec{\alpha}.c\vec{q}t_1' \ldots t_k'y_1' \ldots y_n'$ is the $\eta$-long form of $t$.
- If $t = \mathtt{case}(u; \vec{y_1} : \vec{\sigma_1} \Rightarrow w_1 \mid \ldots \mid \vec{y_m} : \vec{\sigma_m} \Rightarrow w_m)$ then let $u'$ be the $\eta$-long form of $u$ and $w_i'$ of $w_i$. Let $\tau = \forall \vec{z} : \vec{\alpha}.\beta$ be the type of $t$ in normal form, with $\beta$ not a product. Then also $w_i' : \tau$. Thus $w_i' = \lambda \vec{z} : \vec{\alpha}.w_i''$ because $w_i'$ is $\eta$-long. We may assume none of $\vec{z}$ occur in $u'$. We take $\lambda \vec{z} : \vec{\alpha}.\mathtt{case}(u'; \vec{y_1} : \vec{\sigma_1} \Rightarrow w_1'' \mid \ldots \mid \vec{y_m} : \vec{\sigma_m} \Rightarrow w_m'')$ as the $\eta$-long form of $t$.

A proof or set element $t$ in $\beta\iota\rho$-normal form is *$\eta$-long* if the $\eta$-long form of $t$ is $t$.

Note that we do not $\eta$-expand inductive type parameters or variable type annotations. A subterm of $t$ is *genuine* if it does not occur inside an inductive type parameter or a variable type annotation in $t$.

**Lemma 5.** *A proof $t$ in $\beta\iota\rho$-normal form is $\eta$-long iff for every genuine subterm $u$ of $t$ such that $u : \forall \vec{x} : \vec{\alpha}.\tau$ with $\tau$ not a product we have $u = \lambda \vec{x} : \vec{\alpha}.u'$.*

*Proof.* Induction on $t$.

**Lemma 6.** *The $\eta$-long form of a $\beta\iota\rho$-normal proof is $\beta\iota\rho$-normal and has the same type.*

*Proof.* Induction on the definition of $\eta$-long form.

**Definition 3.** The $\Delta$-*case-expansion* of a proof $t$ is defined inductively.

- The $\langle\rangle$-case-expansion of $t$ is $t$.
- If $\Delta = x : \alpha, \Delta'$ with $\alpha$ not an inductive type $I\vec{q}$ with $I \succ \vec{q}$, then the $\Delta'$-case-expansion of $t$ is the $\Delta$-case-expansion of $t$.
- If $\Delta = x : I\vec{q}, \Delta'$ with $I \succ \vec{q}$, and $c_1, \ldots, c_n$ are all constructors of $I$, and $c_i\vec{q} : \forall \vec{x_i} : \vec{\tau_i}.I\vec{q}$, and $t'$ is the $\Delta', \vec{x_i} : \vec{\tau_i}$-case-expansion of $t$, and $t_i$ is the $\iota$-normal form of $t'[c_i\vec{q}\vec{x_i}/x]$, then $\mathtt{case}(x; \vec{x_1} : \vec{\tau_1} \Rightarrow t_1 \mid \ldots \mid \vec{x_k} : \vec{\tau_k} \Rightarrow t_k)$ is the $\Delta$-case-expansion of $t$.

The induction is on the multiset extension $\succ_{\mathrm{mul}}$ of the bottom-extension $\succ_\perp$ of the well-founded order $\succ$ on inductive types. We compare the multisets $\mathcal{M}(\Delta)$ of inductive types of variables from $\Delta$, using $\perp$ for non-inductive types, e.g.,

$$\mathcal{M}(x : I_1, y : I_2\vec{q}, z : I_1, x' : \alpha, y' : \beta) = \{\perp, \perp, I_1, I_1, I_2\}$$

if $\alpha, \beta$ are not inductive types. In the last point of the definition, the requirement $I \succ \vec{q}$ guarantees $I \succ \vec{\tau_i}$, and thus $\mathcal{M}(\Delta) \succ_{\mathrm{mul}} \mathcal{M}(\Delta', \vec{x_i} : \vec{\tau_i})$.

**Lemma 7.** *The $\Delta$-case-expansion of a proof in $\beta\iota\rho$-normal form is $\beta\iota\rho$-normal and has the same type.*

*Proof.* By induction on the definition of $\Delta$-case-expansion, using Corollary 2. Note that in the last point of the definition, taking the $\iota$-normal form requires reducing only the redexes created directly by substituting $x$ with $c_i\vec{q}\vec{x_i}$. But then because the constructor arguments are variables $\vec{x_i}$, the $\iota$-contraction will result in one of the branches with some variables renamed, which cannot create new redexes if the original term was $\beta\iota\rho$-normal.

**Definition 4.** A *case-context* $C[t_1, \ldots, t_n]$ with branches $t_1, \ldots, t_n$ is defined inductively.

- If $t$ is not a case expression then it is an empty case-context with a single branch $t$.
- If $C_1[\vec{t_1}], \ldots, C_n[\vec{t_n}]$ are case-contexts with branches $\vec{t_1}, \ldots, \vec{t_n}$ respectively, then $\mathtt{case}(t; \vec{x_1} \Rightarrow C_1[\vec{t_1}] \mid \ldots \mid C_n[\vec{t_n}])$ is a case-context with branches $\vec{t_1}, \ldots, \vec{t_n}$ (i.e. the concatenation of the branch lists for $C_1, \ldots, C_n$).

We write $C[t_i]_i$ for a case-context with branches $t_1, \ldots, t_n$ with $n$ unspecified.

Note that for every term there exists a unique decomposition into a case-context $C[t_i]_i$ with branches $t_i$. We write $t = C[t_i]_i$ if $C[t_i]_i$ is the case-context decomposition of $t$. By definition, the branches of a case-context are not case expressions.

If $t = C[t_i]_i$ is a case-context, then by $C[t_i']_i$ we denote the term $t$ with branch $t_i$ replaced by $t_i'$ (which may now have a different case-context decomposition if $t_i'$ is a case expression). If $C_1[t_i]_i$ and $C_2[u_j]_j$ are case-contexts, then $C[w_{i,j}]_{i,j} = C_1[C_2[w_{i,j}]_j]_i$ is a case-context with branches $w_{i,j}$ if $w_{i,j}$ are not case expressions (and the result is well-typed).

**Definition 5.** The $\epsilon$-*normal form* of a set element $t$ is $t$.

The $\epsilon$- *normal form* of a $\beta\iota\rho$-normal proof $t$ is defined by induction on $t$.

- If $t = \lambda x : \alpha.u$, and $u'$ is the $(x : \alpha)$-case-expansion of the $\epsilon$-normal form of $u$, then $\lambda x : \alpha.u'$ is the $\epsilon$-normal form of $t$.
- If $t = x\vec{u}$, and $u_i'$ is the $\epsilon$-normal form of $u_i$, and $u_i' = C_i[w_{j_i}]_{j_i}$, then $C_1[C_2[\ldots C_n[xw_{j_1} \ldots w_{j_n}]_{j_n}]_{j_2}]_{j_1}$ is the $\epsilon$-normal form of $t$.
- If $t = c\vec{q}\vec{u}$ then the $\epsilon$-normal form of $t$ is defined analogously to the previous point (not modifying $\vec{q}$).
- If $t = \mathsf{case}(x\vec{u}; \vec{x_1} : \vec{\tau_1} \Rightarrow t_1 \mid \ldots \mid \vec{x_k} : \vec{\tau_k} \Rightarrow t_k)$, and $u_i'$ is the $\epsilon$-normal form of $u_i$, and $t_i'$ is the $(\vec{x_i} : \vec{\tau_i})$-case-expansion of the $\epsilon$-normal form of $t_i$, and $u_i' = C_i[w_{j_i}]_{j_i}$, then

$$C_1[C_2[\ldots C_n[\mathsf{case}(xw_{j_1} \ldots w_{j_n}; \vec{x_1} : \vec{\tau_1} \Rightarrow t_1' \mid \ldots \mid \vec{x_k} : \vec{\tau_k} \Rightarrow t_k']_{j_n} \ldots]_{j_2}]_{j_1}$$

is the $\epsilon$-normal form of $t$.

**Lemma 8.** *The $\epsilon$-normal form of a $\beta\iota\rho$-normal proof is $\beta\iota\rho$-normal and has the same type.*

*Proof.* Induction on the definition of $\epsilon$-normal form. We use Lemma 7 to handle case-expansions of $\epsilon$-normal forms in the first and the last point of the definition.

For the second point, note that if $u_i' = C_i[w_{j_i}]_{j_i}$ with the case-context $C_i$ nonempty, then each $w_{j_i}$ must be a proof because it is a branch of a case expression. Hence, $w_{j_i}$ cannot occur in the type of $xw_{j_1} \ldots w_{j_i}$. If the case-context $C_i$ is empty, then $w_{j_i} = u_i'$ and either $u_i'$ is a proof and it does not occur in the type of $xw_{j_1} \ldots w_{j_i}$, or it is a set element and $u_i' = u_i$. Thus each $xw_{j_1} \ldots w_{j_n}$ has the same type as $x\vec{u}$. It follows that $C_1[C_2[\ldots C_n[xw_{j_1} \ldots w_{j_n}]_{j_n} \ldots]_{j_2}]_{j_1}$ has the same type as $x\vec{u}$. An analogous observation applies to the case-context manipulations in the last point.

We restate the definition of long normal forms from Sect. 4 specialised to the first-order fragment.

**Definition 6.** Any set element is in *long normal form*. A proof $t$ is in *long normal form* (*lnf*) if:

- $t = \lambda x : \alpha.t'$, and $t'$ is in $(x : \alpha)$-ce-lnf (see below);

- $t = x\vec{u}$, and $t : \tau$ with $\tau$ not a product, and each $u_i$ is in lnf and not a case expression;
- $t = c\vec{q}\vec{w}$, and $t : I\vec{q}$, and each $w_i$ is in lnf and not a case expression;
- $t = \mathsf{case}(x\vec{u}; \vec{x_1} : \vec{\tau_1} \Rightarrow t_1 \mid \ldots \mid \vec{x_k} : \vec{\tau_k} \Rightarrow t_k)$, and $t : \tau$ with $\tau$ not a product, and each $u_i$ is in lnf and not a case expression, and each $t_i$ is in $(\vec{x_i} : \vec{\tau_i})$-ce-lnf.

A proof $t$ is in $\Delta$-*case-expanded long normal form* ($\Delta$-*ce-lnf*) if:

- $\Delta = \langle\rangle$ and $t$ is in lnf;
- $\Delta = x : \alpha, \Delta'$, and $\alpha$ is not an inductive type $I\vec{q}$ with $I \succ \vec{q}$, and $t$ is in $\Delta'$-ce-lnf;
- $\Delta = x : I\vec{q}, \Delta'$ with $I \succ \vec{q}$, and $t = \mathsf{case}(x; \vec{x_1} : \vec{\tau_1} \Rightarrow t_1 \mid \ldots \mid \vec{x_k} : \vec{\tau_k} \Rightarrow t_k)$, and each $t_i$ is in $\Delta', \vec{x_i} : \vec{\tau_i}$-ce-lnf and $x \notin \mathrm{FV}(t_i)$.

Formally, the definition is by mutual induction on pairs (size of $t$, length of $\Delta$) ordered lexicographically (with $\Delta = \langle\rangle$ for lnf).

**Lemma 9.** *If $t$ is in $\Delta$-ce-lnf and the type of $t$ is not a product, then $t$ is in lnf.*

*Proof.* Induction on the definition of $\Delta$-ce-lnf.

**Lemma 10.** *If $t$ is in $\Delta$-ce-lnf, $x \notin \Delta$ and $x : I\vec{q}$, then the $\iota$-normal form of $t[c_i\vec{q}\vec{y}/x]$ is in $\Delta$-ce-lnf.*

*Proof.* Induction on the definition of lnf and $\Delta$-ce-lnf.

**Lemma 11.** *If $t$ is a proof in lnf and the type of $t$ is not a product, then the $\Delta$-case-expansion of $t$ is in $\Delta$-ce-lnf.*

*Proof.* By induction on the definition of $\Delta$-case-expansion, using Lemma 10.

**Lemma 12.** *If $t = C[t_i]_i$ is in lnf, then so is each $t_i$.*

*Proof.* Induction on the case-context $C$, using Lemma 9.

**Lemma 13.** *If $u = C[u_i]_i$ is in $\Delta$-ce-lnf and all $u_i'$ are in lnf, then $C[u_i']_i$ is in $\Delta$-ce-lnf (assuming it is well-typed).*

*Proof.* Induction on the case-context $C$.

**Lemma 14.** *The $\epsilon$-normal form of an $\eta$-long $\beta\iota\rho$-normal proof is in long normal form.*

*Proof.* Induction on the definition of $\epsilon$-normal form, using the previous three lemmas.

Finally, we are ready to prove the completeness theorem. We consider a basic variant of our procedure (Sects. 3.1 and 3.2) which does not perform the looping check for conjectures of sort $*^s$. With first-order restrictions on term formation, it is to be understood that the procedure performs corresponding actions only when the resulting term is well-typed, e.g., the introduction and elimination actions are not performed for conjectures of sort $*^s$.

**Theorem 3 (Completeness for the first-order fragment).** *If the conjecture is a proposition or a set and the inhabitation problem has a solution, then our procedure will find one.*

*Proof.* One checks that the procedure with the restrictions outlined above performs an exhaustive search for (minimal) inhabitants in long normal form. Note that if the conjecture is a proposition or a set, then in any subgoal the conjecture is still a proposition or a set.

By Theorem 2, Lemma 6, Lemma 8 and Lemma 14, for any solution $t : \tau$ there exists a solution $t' : \tau$ in long normal form. This implies completeness of our procedure without the looping check. By Corollary 2, loop checking for propositional conjectures does not compromise completeness.

# References

1. Appel, A.: Verified Functional Algorithms. Software Foundations series, volume 3. Electronic textbook (2018)
2. Armand, M., Faure, G., Grégoire, B., Keller, C., Théry, L., Werner, B.: A modular integration of SAT/SMT solvers to Coq through proof witnesses. In: Jouannaud, J.-P., Shao, Z. (eds.) CPP 2011. LNCS, vol. 7086, pp. 135–150. Springer, Heidelberg (2011). https://doi.org/10.1007/978-3-642-25379-9_12
3. Baader, F., Nipkow, T.: Term Rewriting and All That. Cambridge University Press, Cambridge (1999)
4. Barendregt, H.: Lambda calculi with types. In: Handbook of Logic in Computer Science, vol. 2, pp. 118–310. Oxford University Press (1992)
5. Ben-Yelles, C.: Type-assignment in the lambda-calculus. Ph.D. thesis, University College Swansea (1979)
6. Bibel, W.: Automated Theorem Proving. Artificial Intelligence, 2nd edn. Vieweg, Wiesbaden (1987). https://doi.org/10.1007/978-3-322-90102-6
7. Blanchette, J., Kaliszyk, C., Paulson, L., Urban, J.: Hammering towards QED. J. Formaliz. Reason. **9**(1), 101–148 (2016)
8. Brown, C.E.: Satallax: an automatic higher-order prover. In: Gramlich, B., Miller, D., Sattler, U. (eds.) IJCAR 2012. LNCS (LNAI), vol. 7364, pp. 111–117. Springer, Heidelberg (2012). https://doi.org/10.1007/978-3-642-31365-3_11
9. Chlipala, A.: Certified Programming with Dependent Types. MIT Press, Cambridge (2013)
10. Coq Development Team: The Coq proof assistant, version 8.10.0 (2019). https://doi.org/10.5281/zenodo.3476303
11. Corbineau, P.: First-order reasoning in the calculus of inductive constructions. In: Berardi, S., Coppo, M., Damiani, F. (eds.) TYPES 2003. LNCS, vol. 3085, pp. 162–177. Springer, Heidelberg (2004). https://doi.org/10.1007/978-3-540-24849-1_11
12. Czajka, Ł.: Practical proof search for Coq by type inhabitation: supplementary evaluation material. http://www.mimuw.edu.pl/~lukaszcz/sauto/index.html
13. Czajka, Ł., Kaliszyk, C.: CoqHammer. https://github.com/lukaszcz/coqhammer
14. Czajka, Ł., Kaliszyk, C.: Hammer for Coq: automation for dependent type theory. J. Autom. Reason. **61**(1–4), 423–453 (2018)
15. Dowek, G.: A complete proof synthesis method for the cube of type systems. J. Logic Comput. **3**(3), 287–315 (1993)

16. Dowek, G., Huet, G., Werner, B.: On the definition of the eta-long normal form in type systems of the cube. In: Informal Proceedings of the Workshop on Types for Proofs and Programs (1993)

17. Dyckhoff, R.: Contraction-free sequent calculi for intuitionistic logic. J. Symb. Logic **57**(3), 795–807 (1992)

18. Dyckhoff, R.: Contraction-free sequent calculi for intuitionistic logic: a correction. J. Symb. Logic **83**(4), 1680–1682 (2018)

19. Dyckhoff, R., Negri, S.: Admissibility of structural rules for contraction-free systems of intuitionistic logic. J. Symb. Logic **65**(4), 1499–1518 (2000)

20. Edelkamp, S., Schrödl, S.: Heuristic Search - Theory and Applications. Academic Press, Cambridge (2012)

21. Ekici, B., et al.: SMTCoq: a plug-in for integrating SMT solvers into Coq. In: Majumdar, R., Kunčak, V. (eds.) CAV 2017. LNCS, vol. 10427, pp. 126–133. Springer, Cham (2017). https://doi.org/10.1007/978-3-319-63390-9_7

22. Gutiérrez, F., Ruiz, B.: Cut elimination in a class of sequent calculi for pure type systems. Electron. Notes Theor. Comput. Sci. **84**, 105–116 (2003)

23. Hindley, J.R.: Basic Simple Type Theory, Cambridge Tracts in Theoretical Computer Science, vol. 42. Cambridge University Press, Cambridge (1997)

24. Kreitz, C., Otten, J.: Connection-based theorem proving in classical and nonclassical logics. J. UCS **5**(3), 88–112 (1999)

25. Kunze, F.: Towards the integration of an intuitionistic first-order prover into Coq. HaTT **2016**, 30–35 (2016)

26. Lengrand, S., Dyckhoff, R., McKinna, J.: A focused sequent calculus framework for proof search in pure type systems. Logic. Methods Comput. Sci. **7**(1) (2011)

27. Leroy, X.: Formal verification of a realistic compiler. Commun. ACM **52**(7), 107–115 (2009)

28. Lindblad, F., Benke, M.: A tool for automated theorem proving in Agda. In: Filliâtre, J.-C., Paulin-Mohring, C., Werner, B. (eds.) TYPES 2004. LNCS, vol. 3839, pp. 154–169. Springer, Heidelberg (2006). https://doi.org/10.1007/11617990_10

29. Nipkow, T., Wenzel, M., Paulson, L.C. (eds.): Isabelle/HOL - A Proof Assistant for Higher-Order Logic. LNCS, vol. 2283. Springer, Heidelberg (2002). https://doi.org/10.1007/3-540-45949-9

30. Norell, U.: Towards a practical programming language based on dependent type theory. Ph.D. thesis, Chalmers University of Technology, September 2007

31. Otten, J.: leanCoP 2.0 and ileanCoP 1.2: high performance lean theorem proving in classical and intuitionistic logic (System Descriptions). In: Armando, A., Baumgartner, P., Dowek, G. (eds.) IJCAR 2008. LNCS (LNAI), vol. 5195, pp. 283–291. Springer, Heidelberg (2008). https://doi.org/10.1007/978-3-540-71070-7_23

32. Otten, J., Bibel, W.: Advances in connection-based automated theorem proving. In: Provably Correct Systems, pp. 211–241 (2017)

33. Paulin-Mohring, C.: Inductive definitions in the system Coq - rules and properties. TLCA **1993**, 328–345 (1993)

34. Pierce, B., et al.: Programming language foundations. Software Foundations series, volume 2. Electronic textbook, May 2018

35. Pierce, B., et al.: Logical Foundations. Software Foundations series, volume 1, Electronic textbook, May 2018

36. Raths, T., Otten, J., Kreitz, C.: The ILTP problem library for intuitionistic logic. J. Autom. Reason. **38**(1–3), 261–271 (2007)

37. Schmitt, S., Lorigo, L., Kreitz, C., Nogin, A.: JProver : integrating connection-based theorem proving into interactive proof assistants. IJCAR **2001**, 421–426 (2001)
38. Schubert, A., Urzyczyn, P., Zdanowski, K.: On the Mints hierarchy in first-order intuitionistic logic. Logic. Methods Comput. Sci. **12**(4), (2016)
39. Sørensen, M., Urzyczyn, P.: Lectures on the Curry-Howard Isomorphism. Elsevier, Amsterdam (2006)
40. Steen, A., Benzmüller, C.: The higher-order prover Leo-III. In: Galmiche, D., Schulz, S., Sebastiani, R. (eds.) IJCAR 2018. LNCS (LNAI), vol. 10900, pp. 108–116. Springer, Cham (2018). https://doi.org/10.1007/978-3-319-94205-6_8
41. Troelstra, A., Schwichtenberg, H.: Basic Proof Theory. Cambridge University Press, Cambridge (1996)
42. Urzyczyn, P.: Inhabitation in typed lambda-calculi (a syntactic approach). In: de Groote, P., Roger Hindley, J. (eds.) TLCA 1997. LNCS, vol. 1210, pp. 373–389. Springer, Heidelberg (1997). https://doi.org/10.1007/3-540-62688-3_47
43. Wajsberg, M.: Untersuchungen über den Aussagenkalkül von A. Heyting. Wiadomości Matematyczne **46**, 45–101 (1938)
44. Wajsberg, M.: On A. Heyting's propositional calculus. In: Surma, S. (ed.) Logical Works, pp. 132–172. PAN, Warsaw (1977)
45. Wallen, L.: Automated Proof Search in Non-classical Logics - Efficient Matrix Proof Methods for Modal and Intuitionistic Logics. MIT Press, Cambridge (1990)
46. Werner, B.: Une Théorie des Constructions Inductives. Ph.D. thesis, Paris Diderot University, France (1994)

# Quotients of Bounded Natural Functors

Basil Fürer[1], Andreas Lochbihler[2]([⊠]) [iD], Joshua Schneider[1]([⊠]) [iD],
and Dmitriy Traytel[1]([⊠]) [iD]

[1] Institute of Information Security, Department of Computer Science,
ETH Zürich, Zurich, Switzerland
{joshua.schneider,traytel}@inf.ethz.ch
[2] Digital Asset (Switzerland) GmbH, Zurich, Switzerland
mail@andreas-lochbihler.de

**Abstract.** The functorial structure of type constructors is the foundation for many definition and proof principles in higher-order logic (HOL). For example, inductive and coinductive datatypes can be built modularly from bounded natural functors (BNFs), a class of well-behaved type constructors. Composition, fixpoints, and—under certain conditions—subtypes are known to preserve the BNF structure. In this paper, we tackle the preservation question for quotients, the last important principle for introducing new types in HOL. We identify sufficient conditions under which a quotient inherits the BNF structure from its underlying type. We extend the Isabelle proof assistant with a command that automates the registration of a quotient type as a BNF by lifting the underlying type's BNF structure. We demonstrate the command's usefulness through several case studies.

## 1 Introduction

The functorial structure of type constructors forms the basis for many definition and proof principles in proof assistants. Examples include datatype and codatatype definitions [3,9,37], program synthesis [13,19,24], generalized term rewriting [36], and reasoning based on representation independence [6,19,23] and about effects [26,27].

A type constructor becomes a functor through a mapper operation that lifts functions on the type arguments to the constructed type. The mapper must be *functorial*, i.e., preserve identity functions (id) and distribute over function composition ($\circ$). For example, the list type constructor $\_\ list$ [1] has the well-known mapper $\mathsf{map} :: (\alpha \to \beta) \to \alpha\ list \to \beta\ list$, which applies the given function to every element in the given list. It is functorial:

$$\mathsf{map}\ \mathsf{id} = \mathsf{id} \qquad \mathsf{map}\ g \circ \mathsf{map}\ f = \mathsf{map}\ (g \circ f)$$

Most applications of functors can benefit from even richer structures. In this paper, we focus on bounded natural functors (BNFs) [37] (Sect. 2.1). A BNF

---

[1] Type constructors are written postfix in this paper.

N. Peltier and V. Sofronie-Stokkermans (Eds.): IJCAR 2020, LNAI 12167, pp. 58–78, 2020.
https://doi.org/10.1007/978-3-030-51054-1_4

comes with additional setter operators that return sets of occurring elements, called atoms, for each type argument. The setters must be *natural* transformations, i.e., commute with the mapper, and *bounded*, i.e., have a fixed cardinality bound on the sets they return. For example, set :: $\alpha$ *list* → $\alpha$ *set* returns the set of elements in a list. It satisfies set ∘ map $f = f\langle\_\rangle$ ∘ set, where $f\langle\_\rangle$ denotes the function that maps a set $X$ to $f\langle X\rangle = \{f\ x \mid x \in X\}$, i.e., the image of $X$ under $f$. Moreover, since lists are finite sequences, set $xs$ is always a finite set.

Originally, BNFs were introduced for modularly constructing datatypes and codatatypes [9] in the Isabelle/HOL proof assistant. Although (co)datatypes are still the most important use case, the BNF structure is used nowadays in other contexts such as reasoning via free theorems [29] and transferring theorems between types [22,28].

Several type definition principles in HOL preserve the BNF structure: composition (e.g., ($\alpha$ *list*) *list*), datatypes and codatatypes [37], and—under certain conditions—subtypes [7,28]. Subtypes include records and type copies. Accordingly, when a new type constructor is defined via one of these principles from an existing BNF, then the new type automatically comes with a mapper and setters and with theorems for the BNF properties.

One important type definition principle is missing above: quotients [18,19, 21,34,35] (Sect. 2.2). A quotient type identifies elements of an underlying type according to a (partial) equivalence relation ∼. That is, the quotient type is isomorphic to the equivalence classes of ∼. For example, unordered pairs $\alpha$ *upair* are the quotient of ordered pairs $\alpha \times \alpha$ and the equivalence relation $\sim_{upair}$ generated by $(x, y) \sim_{upair} (y, x)$. Similarly, finite sets, bags, and cyclic lists are quotients of lists where the equivalence relation permutes or duplicates the list elements as needed.

In this paper, we answer the question when and how a quotient type inherits its underlying type's BNF structure. It is well known that a quotient preserves the functorial properties if the underlying type's mapper preserves ∼; then the quotient type's mapper is simply the lifting of the underlying type's mapper to equivalence classes [3]. For setters, the situation is more complicated. We discovered that if the setters are defined as one would expect, the resulting structure may not preserve empty intersections, i.e., it is *unsound* in Adámek et al.'s [2] terminology. All BNFs, however, are sound. To repair the situation, we characterize the setters in terms of the mapper and identify a definition scheme for the setters that results in sound functors. We then derive sufficient conditions on the equivalence relation ∼ for the BNF properties to be preserved for these definitions (Sect. 3). With few exceptions, we omit proofs and refer to our technical report [15], which contains them.

Moreover, we have implemented an Isabelle/HOL command that automates the registration of a quotient type as a BNF (Sect. 4); the user merely needs to discharge the conditions on ∼. One of the conditions, subdistributivity, often requires considerable proof effort, though. We therefore developed a novel sufficient criterion using confluent relations that simplifies the proofs in our case studies (Sect. 3.4). Our implementation is distributed with the Isabelle2020 release.

*Contributions.* The main contributions of this paper are the following:

1. We identify sufficient criteria for when a quotient type preserves the BNF properties of the underlying type. Registering a quotient as a BNFs allows (co)datatypes to nest recursion through it. Consider for example node-labeled unordered binary trees

$$\text{datatype } \textit{ubtree} = \mathsf{Leaf} \mid \mathsf{Node} \textit{ nat } (\textit{ubtree upair})$$

   BNF use cases beyond datatypes benefit equally.
2. In particular, we show that the straightforward definitions would cause the functor to be unsound, and find better definitions that avoid unsoundness. This problem is not limited to BNFs. The lifting operations for Lean's QPFs [3] also suffer from unsoundness and our repair applies to them as well (Sect. 5.2).
3. We propose a sufficient criterion on $\sim$ for subdistributivity, which is typically the most difficult BNF property to show. We show with several examples that the criterion is applicable in practice and yields relatively simple proofs.
4. We have implemented an Isabelle/HOL command to register the quotient as a BNF, once the user has discharged the conditions on $\sim$. The command also generates proof rules for transferring theorems about the BNF operations from the underlying type to the quotient (Sect. 4.2). Several case studies demonstrate the command's usefulness. Some examples reformulate well-known BNFs as quotients (e.g., unordered pairs, distinct lists, finite sets). Others formally prove the BNF properties for the first time, e.g., cyclic lists, the free idempotent monoid, and regular expressions modulo ACI. These examples become part of the collection of formalized BNFs and can thus be used in datatype definitions and other BNF applications.

*Example 1.* To illustrate our contribution's usefulness, we consider linear dynamic logic (LDL) [14], an extension of linear temporal logic with regular expressions. LDL's syntax is usually given as two mutually recursive datatypes of formulas and regular expressions [5,14]. Here, we opt for nested recursion, which has the modularity benefit of being able to formalize regular expressions separately. We define regular expressions $\alpha \textit{ re}$:

$$\text{datatype } \alpha \textit{ re} = \mathsf{Atom} \ \alpha \mid \mathsf{Alt} \ (\alpha \textit{ re}) \ (\alpha \textit{ re}) \mid \mathsf{Conc} \ (\alpha \textit{ re}) \ (\alpha \textit{ re}) \mid \mathsf{Star} \ (\alpha \textit{ re})$$

Often, it is useful to consider regular expressions modulo some syntactic equivalences. For example, identifying expressions modulo the associativity, commutativity, and idempotence (ACI) of the alternation constructor Alt results in a straightforward construction of deterministic finite automata from regular expressions via Brzozowski derivatives [32]. We define the ACI-equivalence $\sim_{aci}$ as the least congruence relation satisfying:

$$\mathsf{Alt} \ (\mathsf{Alt} \ r \ s) \ t \sim_{aci} \mathsf{Alt} \ r \ (\mathsf{Alt} \ s \ t) \qquad \mathsf{Alt} \ r \ s \sim_{aci} \mathsf{Alt} \ s \ r \qquad \mathsf{Alt} \ r \ r \sim_{aci} r$$

Next, we define the quotient type of regular expressions modulo ACI $\alpha \textit{ re}_{aci}$ and the datatype of LDL formulas *ldl*, which uses nested recursion through $\alpha \textit{ re}_{aci}$.

```
quotient_type α re_aci = α re/~_aci
datatype ldl = Prop string | Neg ldl | Conj ldl ldl | Match (ldl re_aci)
```

For the last declaration to succeed, Isabelle must know that $\alpha\ re_{aci}$ is a BNF. We will show in Sect. 3.4 how our work allows us to lift $\alpha\ re$'s BNF structure to $\alpha\ re_{aci}$. ◇

## 2  Background

We work in Isabelle/HOL, Isabelle's variant of classical higher-order logic—a simply typed theory with Hilbert choice and rank-1 polymorphism. We refer to a textbook for a detailed introduction to Isabelle/HOL [31] and only summarize relevant notation here.

Types are built from type variables $\alpha$, $\beta$, ... via type constructors. A type constructor can be nullary (*nat*) or have some type arguments ($\alpha$ *list*, $\alpha$ *set*, ($\alpha$, $\beta$) *upair*). Type constructor application is written postfix. Exceptions are the binary type constructors for sums (+), products (×), and functions (→), all written infix. Terms are built from variables $x$, $y$, ... and constants c, d, ... via lambda-abstractions $\lambda x.\ t$ and applications $t\ u$. The sum type's embeddings are Inl and Inr and the product type's projections are fst and snd.

The primitive way of introducing new types in HOL is to take a non-empty subset of an existing type. For example, the type of lists could be defined as the set of pairs ($n :: nat$, $f :: nat \rightarrow \alpha$) where $n$ is the list's length and $f\ i$ is the list's $i$th element for $i < n$ and some fixed unspecified element of type $\alpha$ for $i \geq n$. To spare the users from such low-level encodings, Isabelle/HOL offers more high-level mechanisms for introducing new types, which are internally reduced to primitive subtyping. In fact, lists are defined as an inductive **datatype** $\alpha$ *list* = [] | $\alpha$ # $\alpha$ *list*, where [] is the empty list and # is the infix list constructor. Recursion in datatypes and their coinductive counterparts may take place only under well-behaved type constructors, the bounded natural functors (Sect. 2.1). Quotient types (Sect. 2.2) are another high-level mechanism for introducing new types.

For $n$-ary definitions, we use the vector notation $\overline{x}$ that denotes $x_1, \ldots, x_n$ where $n$ is clear from the context. Vectors spanning several variables indicate repetition with synchronized indices. For example, $\mathsf{map}_F\ \overline{(g \circ f)}$ abbreviates $\mathsf{map}_F\ (g_1 \circ f_1)\ \cdots\ (g_n \circ f_n)$. Abusing notation slightly, we write $\overline{\alpha} \rightarrow \beta$ for the $n$-ary function type $\alpha_1 \rightarrow \cdots \rightarrow \alpha_n \rightarrow \beta$.

To simplify notation, we identify the type of binary predicates $\alpha \rightarrow \beta \rightarrow \mathbb{B}$ and sets of pairs $(\alpha \times \beta)\ set$, and write $\alpha \otimes \beta$ for both. These types are different in Isabelle/HOL and the BNF ecosystem works with binary predicates. The identification allows us to use set operations, e.g., the subset relation $\subseteq$ or relation composition • (both written infix).

## 2.1 Bounded Natural Functors

A bounded natural functor (BNF) [37] is an $n$-ary type constructor $\overline{\alpha}\ F$ equipped with the following polymorphic constants. Here and elsewhere, $i$ implicitly ranges over $\{1, \ldots, n\}$:

$$\mathsf{map}_F :: \overline{(\alpha \to \beta)} \to \overline{\alpha}\ F \to \overline{\beta}\ F \qquad \mathsf{bd}_F :: bd\_type \otimes bd\_type$$
$$\mathsf{set}_{F,i} :: \overline{\alpha}\ F \to \alpha_i\ set \quad \text{for all } i \qquad \mathsf{rel}_F :: \overline{(\alpha \otimes \beta)} \to \overline{\alpha}\ F \otimes \overline{\beta}\ F$$

The *shape and content* intuition [37] is a useful way of thinking about elements of $\overline{\alpha}\ F$. The mapper $\mathsf{map}_F$ leaves the shape unchanged but modifies the contents by applying its function arguments. The $n$ setters $\mathsf{set}_{F,i}$ extract the contents (and dispose of the shape). For example, the shape of a list is given by its length, which map preserves. The cardinal bound $\mathsf{bd}_F$ is a fixed bound on the number of elements returned by $\mathsf{set}_{F,i}$. Cardinal numbers are represented in HOL using particular well-ordered relations [10]. Finally, the relator $\mathsf{rel}_F$ lifts relations on the type arguments to a relation on $\overline{\alpha}\ F$ and $\overline{\beta}\ F$. Thereby, it only relates elements of $\overline{\alpha}\ F$ and $\overline{\beta}\ F$ that have the same shape.

The BNF constants must satisfy the following properties:

MAP_ID    $\mathsf{map}_F\ \overline{\mathsf{id}} = \mathsf{id}$

MAP_COMP    $\mathsf{map}_F\ \overline{g} \circ \mathsf{map}_F\ \overline{f} = \mathsf{map}_F\ \overline{(g \circ f)}$

SET_MAP    $\mathsf{set}_{F,i} \circ \mathsf{map}_F\ \overline{f} = f_i\langle\_\rangle \circ \mathsf{set}_{F,i}$

MAP_CONG    $(\forall i.\ \forall z \in \mathsf{set}_{F,i}\ x.\ f_i\ z = g_i\ z) \implies \mathsf{map}_F\ \overline{f}\ x = \mathsf{map}_F\ \overline{g}\ x$

SET_BD    $|\mathsf{set}_{F,i}\ x| \leq_o \mathsf{bd}_F$

BD    $\mathsf{infinite\_card}\ \mathsf{bd}_F$

IN_REL    $\mathsf{rel}_F\ \overline{R}\ x\ y = \exists z.\ (\forall i.\ \mathsf{set}_{F,i}\ z \subseteq R_i) \wedge \mathsf{map}\ \overline{\mathsf{fst}}\ z = x \wedge \mathsf{map}\ \overline{\mathsf{snd}}\ z = y$

REL_COMP    $\mathsf{rel}_F\ \overline{R} \bullet \mathsf{rel}_F\ \overline{S} \subseteq \mathsf{rel}_F\ \overline{(R \bullet S)}$

Properties MAP_ID and MAP_COMP capture the mapper's functoriality; SET_MAP the setters' naturality. Moreover, the mapper and the setters must agree on what they identify as content (MAP_CONG). Any set returned by $\mathsf{set}_{F,i}$ must be bounded (SET_BD); the operator $\leq_o$ compares cardinal numbers [10]. The bound is required to be infinite (BD), which simplifies arithmetics. The relator can be expressed in terms of the mapper and the setter (IN_REL) and must distribute over relation composition (REL_COMP). The other inclusion, namely $\mathsf{rel}_F\ \overline{(R \bullet S)} \subseteq \mathsf{rel}_F\ \overline{R} \bullet \mathsf{rel}_F\ \overline{S}$, follows from these properties. We refer to REL_COMP as *subdistributivity* because it only requires one inclusion.

A useful derived operator is the action on sets $\boxed{F} :: \overline{\alpha\ set} \to \overline{\alpha}\ F\ set$, which generalizes the type constructor's action on its type arguments. Formally, $\boxed{F}\ \overline{A} = \{x \mid \forall i.\ \mathsf{set}_{F,i}\ x \subseteq A_i\}$. Note that we can write $z \in \boxed{F}\ \overline{R}$ to replace the equivalent $\forall i.\ \mathsf{set}_{F,i}\ z \subseteq R_i$ in IN_REL.

Most basic types are BNFs, notably, sum and product types. BNFs are closed under composition, e.g., $1 + \alpha \times \beta$ is a BNF with the mapper $\lambda f\ g.\ \mathsf{map}_{1+}\ (\mathsf{map}_\times\ f\ g)$, where 1 is the unit type (consisting of the single element $\star$) and $\mathsf{map}_{1+}\ h = \mathsf{map}_+\ \mathsf{id}\ h$. Moreover, BNFs support fixpoint operations, which correspond to (co)datatypes, and are closed under them [37]. For instance, the `datatype` command internally computes a least solution for the fixpoint type equation $\beta = 1 + \alpha \times \beta$ to define the $\alpha$ *list* type. Closure means that the resulting datatype, here $\alpha$ *list*, is equipped with the BNF structure, e.g., the mapper `map`. Also subtypes inherit the BNF structure under certain conditions [7]. For example, the subtype $\alpha$ *nelist* of non-empty lists $\{xs :: \alpha\ list \mid xs \neq []\}$ is a BNF.

## 2.2 Quotient Types

An equivalence relation $\sim$ on a type $T$ partitions the type into equivalence classes. Isabelle/HOL supports the definition of the quotient type $Q = T/\sim$, which yields a new type $Q$ isomorphic to the set of equivalence classes [21]. For example, consider $\sim_{fset}$ that relates two lists if they have the same set of elements, i.e., $xs \sim_{fset} ys$ iff set $xs =$ set $ys$. The following command defines the type $\alpha$ *fset* of finite sets as a quotient of lists:

> `quotient_type` $\alpha\ fset = \alpha\ list/\sim_{fset}$

This command requires a proof that $\sim_{fset}$ is, in fact, an equivalence relation.

The Lifting and Transfer tools [19, 22] automate the lifting of definitions and theorems from the raw type $T$ to the quotient $Q$. For example, the image operation on finite sets can be obtained by lifting the list mapper `map` using the command

> `lift_definition` fimage $:: (\alpha \to \beta) \to \alpha\ fset \to \beta\ fset$ is `map`

Lifting is only possible for terms that respect the quotient. For fimage, respectfulness states that map $f\ xs \sim_{fset}$ map $f\ ys$ whenever $xs \sim_{fset} ys$.

Lifting and Transfer are based on *transfer rules* that relate two terms of possibly different types. The `lift_definition` command automatically proves the transfer rule

$$(\mathsf{map}, \mathsf{fimage}) \in ((=) \mapsto \mathsf{cr}_{fset} \mapsto \mathsf{cr}_{fset})$$

where $A \mapsto B$ (right-associative) relates two functions iff they map $A$-related arguments to $B$-related results. The correspondence relation $\mathsf{cr}_{fset}$ relates a list with the finite set that it represents, i.e., the set whose corresponding equivalence class contains the list. Every quotient is equipped with such a correspondence relation. The meaning of the above rule is that applying map $f$ to a list representing the finite set $X$ results in a list that represents fimage $f\ X$, for all $f$. The transfer rule's relation $(=) \mapsto \mathsf{cr}_{fset} \mapsto \mathsf{cr}_{fset}$ is constructed according to the types of the related terms. This enables the composition of transfer rules to relate larger terms. For instance, the Transfer tool automatically derives

$$(\forall x.\ \mathsf{set}\ (\mathsf{map}\ \mathsf{id}\ x) = \mathsf{set}\ x) \longleftrightarrow (\forall X.\ \mathsf{fimage}\ \mathsf{id}\ X = X)$$

such that the equation $\forall X.$ fimage id $X = X$ can be proved by reasoning about lists.

# 3   Quotients of Bounded Natural Functors

We develop the theory for when a quotient type inherits the underlying type's BNF structure. We consider the quotient $\overline{\alpha} \, Q = \overline{\alpha} \, F/{\sim}$ of an $n$-ary BNF $\overline{\alpha} \, F$ over an equivalence relation $\sim$ on $\overline{\alpha} \, F$. The first idea is to define $\mathsf{map}_Q$ and $\mathsf{set}_{Q,i}$ in terms of $F$'s operations:

```
quotient_type α̅ Q = α̅ F/∼
lift_definition map_Q :: (α → β) → α̅ Q → β̅ Q is map_F
lift_definition set_{Q,i} :: α̅ Q → α_i set is set_{F,i}
```

These three commands require the user to discharge the following proof obligations:

$$\mathsf{equivp} \sim \qquad (1) \qquad\qquad x \sim y \Longrightarrow \mathsf{map}_F \, \overline{f} \, x \sim \mathsf{map}_F \, \overline{f} \, y \qquad (2)$$

$$x \sim y \Longrightarrow \mathsf{set}_{F,i} \, x = \mathsf{set}_{F,i} \, y \qquad (3)$$

The first two conditions are as expected: $\sim$ must be an equivalence relation, by (1), and compatible with $F$'s mapper, by (2), i.e., $\mathsf{map}_F$ preserves $\sim$. The third condition, however, demands that equivalent values contain the same atoms. This rules out many practical examples including the following simplified (and therefore slightly artificial) one.

*Example 2.* Consider $\alpha \, F_P = \alpha + \alpha$ with the equivalence relation $\sim_P$ generated by $\mathsf{Inl} \, x \sim_P \mathsf{Inl} \, y$, where $\mathsf{Inl}$ is the sum type's left embedding. That is, $\sim_P$ identifies all values of the form $\mathsf{Inl} \, z$ and thus $\alpha \, Q_P = \alpha \, F_P/{\sim_P}$ is isomorphic to the type $1 + \alpha$. However, $\mathsf{Inl} \, x$ and $\mathsf{Inl} \, y$ have different sets of atoms $\{x\}$ and $\{y\}$, assuming $x \neq y$.                               $\diamond$

We derive better definitions for the setters and conditions under which they preserve the BNF properties. To that end, we characterize setters in terms of the mapper (Sect. 3.1). Using this characterization, we derive the relationship between $\mathsf{set}_{Q,i}$ and $\mathsf{set}_{F,i}$ and identify the conditions on $\sim$ (Sect. 3.2). Next, we do the same for the relator (Sect. 3.3). We thus obtain the conditions under which $\overline{\alpha} \, Q$ preserves $F$'s BNF properties. One of the conditions, the relator's subdistributivity over relation composition, is often difficult to show directly in practice. We therefore present an easier-to-establish criterion for the special case where a confluent rewrite relation $\rightsquigarrow$ over-approximates $\sim$ (Sect. 3.4).

## 3.1   Characterization of the BNF Setter

We now characterize $\mathsf{set}_{F,i}$ in terms of $\mathsf{map}_F$ for an arbitrary BNF $\overline{\alpha} \, F$. Observe that $F$'s action $\boxed{F} \, \overline{A}$ on sets contains all values that can be built with atoms from $\overline{A}$. Hence, $\mathsf{set}_{F,i} \, x$ is the smallest set $A_i$ such that $x$ can be built from atoms in $A_i$. Formally:

$$\mathsf{set}_{F,i} \, x = \bigcap \{ A_i \mid x \in \boxed{F} \, \overline{\mathsf{UNIV}} \, A_i \, \overline{\mathsf{UNIV}} \} \qquad (4)$$

Only atoms of type $\alpha_i$ are restricted; all other atoms $\alpha_j$ may come from UNIV, the set of all elements of type $\alpha_j$. Moreover, $\boxed{F}$ can be defined without $\text{set}_{F,i}$, namely by trying to distinguish values using the mapper. Informally, $x$ contains atoms not from $\overline{A}$ iff $\text{map}_F \, \overline{f} \, x$ differs from $\text{map}_F \, \overline{g} \, x$ for some functions $\overline{f}$ and $\overline{g}$ that agree on $\overline{A}$. Hence, we obtain:

$$\boxed{F} \, \overline{A} = \{x \mid \forall \overline{f} \, \overline{g}. \, (\forall i. \, \forall a \in A_i. \, f_i \, a = g_i \, a) \longrightarrow \text{map}_F \, \overline{f} \, x = \text{map}_F \, \overline{g} \, x\} \quad (5)$$

*Proof.* From left to right is trivial with MAP_CONG. So let $x$ be such that $\text{map}_F \, \overline{f} \, x = \text{map}_F \, \overline{g} \, x$ whenever $f_i \, a = g_i \, a$ for all $a \in A_i$ and all $i$. By the definition of $\boxed{F}$, it suffices to show that $\text{set}_{F,i} \, x \subseteq A_i$. Set $f_i \, a = (a \in A_i)$ and $g_i \, a = \text{True}$. Then,

$$
\begin{aligned}
f_i \langle \text{set}_{F,i} \, x \rangle &= \text{set}_{F,i} \, (\text{map}_F \, \overline{f} \, x) && \text{by SET\_MAP} \\
&= \text{set}_{F,i} \, (\text{map}_F \, \overline{g} \, x) && \text{by choice of } x \text{ as } \overline{f} \text{ and } \overline{g} \text{ agree on } \overline{A} \\
&= (\lambda\_.\text{True}) \langle \text{set}_{F,i} \, x \rangle && \text{by SET\_MAP} \\
&\subseteq \{\text{True}\}
\end{aligned}
$$

Therefore, $\forall a \in \text{set}_{F,i} \, x. \, f_i \, a$, i.e., $\text{set}_{F,i} \, x \subseteq A_i$. $\qquad\qquad\square$

Equations 4 and 5 reduce the setters $\text{set}_{F,i}$ of a BNF to its mapper $\text{map}_F$. In the next section, we will use this characterization to derive a definition of $\text{set}_{Q,i}$ in terms of $\text{set}_{F,i}$. Yet, this definition does not give us naturality out of the box.

*Example 3 ([2, Example 4.2, part iii])* Consider the functor $\alpha \, F_{ae} = nat \to \alpha$ of infinite sequences with $x \sim_{ae} y$ whenever $\{n \mid x \, n \neq y \, n\}$ is finite. That is, two sequences are equivalent iff they are equal almost everywhere. Conditions (1) and (2) hold, but not the naturality for the corresponding $\text{map}_Q$ and $\text{set}_Q$. $\qquad\qquad\diamond$

Gumm [16] showed that $\text{set}_F$ as defined in terms of (4) and (5) is a natural transformation iff $\boxed{F}$ preserves wide intersections and preimages, i.e.,

$$\boxed{F} \, \overline{\left(\bigcap \mathscr{A}\right)} = \bigcap \{\boxed{F} \, \overline{A} \mid \forall i. \, A_i \in \mathscr{A}_i\} \quad (6)$$

$$\boxed{F} \, \overline{(f^{-1}\langle A \rangle)} = (\text{map}_F \, \overline{f})^{-1} \langle \boxed{F} \, \overline{A} \rangle. \quad (7)$$

where $f^{-1}\langle A \rangle = \{x \mid f \, x \in A\}$ denotes the preimage of $A$ under $f$. Then, $\boxed{F} \, \overline{A} = \{x \mid \forall i. \, \text{set}_{F,i} \, x \subseteq A_i\}$ holds. The quotient in Example 3 does not preserve wide intersections.

In theory, we have now everything we need to define the BNF operations on the quotient $\overline{\alpha} \, Q = \overline{\alpha} \, F/\!\!\sim$: Define $\text{map}_Q$ as the lifting of $\text{map}_F$. Define $\boxed{Q}$ and $\text{set}_{Q,i}$ using (5) and (4) in terms of $\text{map}_Q$, and the relator via IN_REL. Prove that $\boxed{Q}$ preserves preimages and wide intersections. Prove that $\text{rel}_Q$ satisfies subdistributivity (REL_COMP).

Unfortunately, the definitions and the preservation conditions are phrased in terms of $Q$, not in terms of $F$ and $\sim$. It is therefore unclear how $\text{set}_{Q,i}$ and $\text{rel}_Q$ relate to $\text{set}_{F,i}$ and $\text{rel}_F$. In practice, understanding this relationship is important:

we want to express the BNF operations and discharge the proof obligations in terms of $F$'s operations and later use the connection to transfer properties from $\mathsf{set}_F$ and $\mathsf{rel}_F$ to $\mathsf{set}_Q$ and $\mathsf{rel}_Q$. We will work out the precise relationships for the setters in Sect. 3.2 and for the relator in Sect. 3.3.

## 3.2 The Quotient's Setter

We relate $Q$'s setters to $F$'s operations and $\sim$. We first look at $\boxed{Q}$, which characterizes $\mathsf{set}_{Q,i}$ via (4). Let $[x]_\sim = \{y \mid x \sim y\}$ denote the equivalence class that $x :: \overline{\alpha}\ F$ belongs to, and $[A]_\sim = \{[x]_\sim \mid x \in A\}$ denote the equivalence classes of elements in $A$. We identify the values of $\overline{\alpha}\ Q$ with $\overline{\alpha}\ F$'s equivalence classes. Then, it follows using (1), (2), and (5) that $\boxed{Q}\ A = [\boxed{F}\ A]_\sim$ where

$$\boxed{F}\ \overline{A} = \{x \mid \forall \overline{f}\ \overline{g}.\ (\forall i.\ \forall a \in A_i.\ f_i\ a = g_i\ a) \longrightarrow \mathsf{map}_F\ \overline{f}\ x \sim \mathsf{map}_F\ \overline{g}\ x\} \quad (8)$$

Equation 8 differs from (5) only in that the equality in $\mathsf{map}_F\ \overline{f}\ x = \mathsf{map}_F\ \overline{g}\ x$ is replaced by $\sim$. Clearly $[\boxed{F}\ \overline{A}]_\sim \subseteq [\boxed{F}\ \overline{A}]_\sim$. The converse does not hold in general, as shown next.

*Example 2 (continued).* For the example viewing $1 + \alpha$ as a quotient of $\alpha\ F_P = \alpha + \alpha$ via $\sim_P$, we have $[\mathsf{Inl}\ x]_\sim \in [\boxed{F_P}\ \{\}]_{\sim_P}$ because $\mathsf{map}_{F_P}\ f\ (\mathsf{Inl}\ x) = \mathsf{Inl}\ (f\ x) \sim_P \mathsf{Inl}\ (g\ x) = \mathsf{map}_{F_P}\ g\ (\mathsf{Inl}\ x)$ for all $f$ and $g$. Yet, $\boxed{F_P}\ \{\}$ is empty, and so is $[\boxed{F_P}\ \{\}]_{\sim_P}$. $\diamond$

This problematic behavior occurs only for empty sets $A_i$. To avoid it, we change types: Instead of $\overline{\alpha}\ F/\sim$, we consider the quotient $\overline{(1 + \alpha)}\ F/\sim$, where $1 + \alpha_i$ adds a new atom $\circledast = \mathsf{Inl}\ \star$ to the atoms of type $\alpha_i$. We write $\mathfrak{e} :: \alpha \to 1 + \alpha$ for the embedding of $\alpha$ into $1 + \alpha$ (i.e., $\mathfrak{e} = \mathsf{Inr}$). Then, we have the following equivalence:

**Lemma 1.** $\boxed{F}\ \overline{A} = \{x \mid [\mathsf{map}_F\ \overline{\mathfrak{e}}\ x]_\sim \in [\boxed{F}\ \overline{(\{\circledast\} \cup \mathfrak{e}\langle A \rangle)}]_\sim\}$.

*Proof.* From left to right: Let $x \in \boxed{F}\ \overline{A}$ and set $f_i\ y = \mathfrak{e}\ y$ for $y \in A_i$ and $f_i\ y = \circledast$ for $y \notin A_i$. Then, $\mathsf{set}_{F,i}\ (\mathsf{map}_F\ \overline{f}\ x) = f_i \langle \mathsf{set}_{F,i}\ x \rangle$ by the naturality of $\mathsf{set}_{F,i}$ and $f_i \langle B \rangle \subseteq \{\circledast\} \cup \mathfrak{e}\langle A_i \rangle$ by $f_i$'s definition for any $B$. Hence $\mathsf{map}\ \overline{f}\ x \in \boxed{F}\ \overline{(\{\circledast\} \cup \mathfrak{e}\langle A \rangle)}$ as $\boxed{F}\ \overline{A} = \{x \mid \forall i.\ \mathsf{set}_{F,i}\ x \subseteq A_i\}$ by the BNF properties. So, $[\mathsf{map}_F\ \overline{\mathfrak{e}}\ x]_\sim \in [\boxed{F}\ \overline{(\{\circledast\} \cup \mathfrak{e}\langle A \rangle)}]_\sim$ because $\mathsf{map}_F\ \overline{\mathfrak{e}}\ x \sim \mathsf{map}\ \overline{f}\ x$ by (8) and $x \in \boxed{F}\ \overline{A}$.

From right to left: Let $x$ such that $\mathsf{map}_F\ \overline{\mathfrak{e}}\ x \sim y$ for some $y \in \boxed{F}\ \overline{(\{\circledast\} \cup \mathfrak{e}\langle A \rangle)}$. Let $\overline{f}$ and $\overline{g}$ such that $f_i\ a = g_i\ a$ for all $a \in A_i$ and all $i$. Then, $\mathsf{map}_F\ \overline{f}\ x \sim \mathsf{map}_F\ \overline{g}\ x$ holds by the following reasoning, where $\mathfrak{e}^{-1}$ denotes the left-inverse of $\mathfrak{e}$ and $\mathsf{map}_{1+}\ h$ satisfies $\mathsf{map}_{1+}\ h\ (\mathfrak{e}\ a) = \mathfrak{e}\ (h\ a)$ and $\mathsf{map}_{1+}\ h\ \circledast = \circledast$:

$$
\begin{aligned}
\mathsf{map}_F\ \overline{f}\ x &= \mathsf{map}_F\ \overline{\mathfrak{e}^{-1}}\ (\mathsf{map}_F\ \overline{(\mathsf{map}_{1+}\ f)}\ (\mathsf{map}_F\ \overline{\mathfrak{e}}\ x)) && \text{as } f_i = \mathfrak{e}^{-1} \circ \mathsf{map}_{1+}\ f_i \circ \mathfrak{e} \\
&\sim \mathsf{map}_F\ \overline{\mathfrak{e}^{-1}}\ (\mathsf{map}_F\ \overline{(\mathsf{map}_{1+}\ f)}\ y) && \text{by } \mathsf{map}_F\ \overline{\mathfrak{e}}\ x \sim y \text{ and (2)} \\
&= \mathsf{map}_F\ \overline{\mathfrak{e}^{-1}}\ (\mathsf{map}_F\ \overline{(\mathsf{map}_{1+}\ g)}\ y) && \text{by choice of } y \text{ and (5)} \\
&\sim \mathsf{map}_F\ \overline{\mathfrak{e}^{-1}}\ (\mathsf{map}_F\ \overline{(\mathsf{map}_{1+}\ g)}\ (\mathsf{map}_F\ \overline{\mathfrak{e}}\ x)) && \text{by } y \sim \mathsf{map}_F\ \overline{\mathfrak{e}}\ x \text{ and (2)} \\
&= \mathsf{map}_F\ \overline{g}\ x && \text{as } \mathfrak{e}^{-1} \circ \mathsf{map}_{1+}\ g_i \circ \mathfrak{e} = g_i \quad \square
\end{aligned}
$$

Lemma 1 allows us to express the conditions (6) and (7) on $\boxed{Q}$ in terms of $\sim$ and $\boxed{F}$. For wide intersections, the condition is as follows (the other inclusion holds trivially):

$$\forall i. \, \mathscr{A}_i \neq \{\} \wedge (\bigcap \mathscr{A}_i \neq \{\}) \implies \bigcap \{ [\boxed{F}\,\overline{A}]_\sim \mid \forall i. \, A_i \in \mathscr{A}_i \} \subseteq \left[ \bigcap \{ \boxed{F}\,\overline{A} \mid \forall i. \, A_i \in \mathscr{A}_i \} \right]_{\widetilde{\sim}} \tag{9}$$

The conclusion is as expected: for sets of the form $\boxed{F}\,\overline{A}$, taking equivalence classes preserves wide intersections. The assumption is the interesting part: preservation is needed only for *non-empty* intersections. Non-emptiness suffices because Lemma 1 relates $\boxed{F}\,\overline{A}$ to $\boxed{F}\,(\{\circledast\} \cup \epsilon\langle A \rangle)$ and all intersections of interest therefore contain $\circledast$. (The condition does not explicitly mention $\circledast$ because Lemma 1 holds for any element that is not in $A$.)

Condition 9 is satisfied trivially for equivalence relations that preserve $\mathsf{set}_{F,i}$, i.e., satisfy (3). Examples include permutative structures like finite sets and cyclic lists.

**Lemma 2.** *Suppose that $\sim$ satisfies* (3). *Then,* $[\boxed{F}\,\overline{A}]_\sim = \boxed{F}\,\overline{A}$ *and condition* (9) *holds.*

In contrast, the non-emptiness assumption is crucial for quotients that identify values with different sets of atoms, such as Example 2. In general, such quotients do *not* preserve empty intersections (Sect. 5).

We can factor condition (9) into a separate property for each type argument $i$:

$$\mathscr{A}_i \neq \{\} \wedge (\bigcap \mathscr{A}_i \neq \{\}) \implies \bigcap_{A \in \mathscr{A}_i} [\{x \mid \mathsf{set}_{F,i}\, x \subseteq A\}]_\sim \subseteq \left[ \{ x \mid \mathsf{set}_{F,i}\, x \subseteq \bigcap \mathscr{A}_i \} \right]_\sim \tag{10}$$

This form is used in our implementation (Sect. 4). It is arguably more natural to prove for a concrete functor $F$ because each property focuses on a single setter.

**Lemma 3.** *Suppose that $\sim$ satisfies* (1) *and* (2). *Then,* (9) *holds iff* (10) *holds for all $i$.*

Preservation of preimages amounts to the following unsurprising condition:

$$\forall i. \, f_i^{-1}\langle A_i \rangle \neq \{\} \implies (\mathsf{map}_F\, \overline{f})^{-1} \left\langle \bigcup [\boxed{F}\,\overline{A}]_\sim \right\rangle \subseteq \bigcup \left[ (\mathsf{map}_F\, \overline{f})^{-1} \langle \boxed{F}\,\overline{A} \rangle \right]_\sim \tag{11}$$

As for wide intersections, taking equivalence classes must preserve *non-empty* preimages (the inclusion from right to left holds trivially). Again, non-emptiness comes from $\circledast$ being contained in all sets of interest. We do not elaborate on preimage preservation any further as it follows from subdistributivity, which we will look at in the next subsection.

Under conditions (9) and (11), we obtain the following characterization for $\mathsf{set}_Q$:

**Theorem 1 (Setter characterization).** $\mathsf{set}_{Q,i}\, [x]_\sim = \bigcap_{y \in [\mathsf{map}_F\, \epsilon\, x]_\sim} \{ a \mid \epsilon\, a \in \mathsf{set}_{F,i}\, y \}.$

## 3.3   The Quotient's Relator

In the previous section, we have shown that it is not a good idea to naively lift the setter and a more general construction is needed. We now show that the same holds for the relator. The following straightforward definition

$$\texttt{lift\_definition } \mathsf{rel}_Q :: \overline{(\alpha \otimes \beta)} \to \overline{\alpha}\, Q \otimes \overline{\beta}\, Q \texttt{ is } \mathsf{rel}_F$$

relates two equivalence classes $[x]_\sim$ and $[y]_\sim$ iff there are representatives $x' \in [x]_\sim$ and $y' \sim [y]_\sim$ such that $(x', y') \in \mathsf{rel}_F\, \overline{R}$. This relator does not satisfy IN_REL.

*Example 2 (continued).* By the lifted definition, $([\mathsf{Inl}\; x]_{\sim_P}, [\mathsf{Inl}\; y]_{\sim_P}) \notin \mathsf{rel}_{Q_P}\, \{\}$ because there are no $(x', y')$ in the empty relation $\{\}$ that could be used to relate using $\mathsf{rel}_{F_P}$ the representatives $\mathsf{Inl}\; x'$ and $\mathsf{Inl}\; y'$. However, the witness $z = [\mathsf{Inl}\; (x, y)]_{\sim_P}$ satisfies the right-hand side of IN_REL as $\boxed{Q}\, \{\} = \{[\mathsf{Inl}\; \_]_{\sim_P}\}$. $\diamondsuit$

So what is the relationship between $\mathsf{rel}_Q$ and $\mathsf{rel}_F$ and under what conditions does the subdistributivity property REL_COMP hold? Like for the setter, we avoid the problematic case of empty relations by switching to $1 + \alpha$. The relator $\mathsf{rel}_{1+}$ adds the pair $(\otimes, \otimes)$ to every relation $R$ and thereby ensures that all relations and their compositions are non-empty. Accordingly, we obtain the following characterization:

**Theorem 2 (Relator characterization).**

$$([x]_\sim, [y]_\sim) \in \mathsf{rel}_Q\, \overline{R} \longleftrightarrow (\mathsf{map}_F\, \overline{\mathsf{e}}\; x, \mathsf{map}_F\, \overline{\mathsf{e}}\; y) \in (\sim \bullet\, \mathsf{rel}_F\, \overline{(\mathsf{rel}_{1+}\, R)} \bullet \sim)$$

Moreover, the following condition on $\sim$ characterizes when $\mathsf{rel}_Q$ satisfies REL_COMP. Again, the non-emptiness assumptions for $R_i \bullet S_i$ come from $\mathsf{rel}_{1+}$ extending any relation $R$ with the pair $(\otimes, \otimes)$.

$$(\forall i.\; R_i \bullet S_i \neq \{\}) \implies \mathsf{rel}_F\, \overline{R} \bullet \sim \bullet\, \mathsf{rel}_F\, \overline{S} \subseteq\, \sim \bullet\, \mathsf{rel}_F\, \overline{(R \bullet S)} \bullet \sim \qquad (12)$$

It turns out that this condition implies the respectfulness of the mapper (2). Intuitively, the relator is a generalization of the mapper. Furthermore, it is well known that subdistributivity implies preimage preservation [17]. Since our conditions on $\sim$ characterize these preservation properties, it is no surprise that the latter implication carries over.

**Lemma 4.** *Condition* (12) *implies respectfulness* (2) *and preimage preservation* (11).

In summary, we obtain the following main preservation theorem:

**Theorem 3.** *The quotient* $\overline{\alpha}\, Q = \overline{\alpha}\, F/\sim$ *inherits the structure from the BNF* $\overline{\alpha}\, F$ *with the mapper* $\mathsf{map}_Q\, \overline{f}\, [x]_\sim = [\mathsf{map}_F\, \overline{f}\; x]_\sim$ *if* $\sim$ *satisfies the conditions* (1), (9), *and* (12). *The setters and relator are given by Theorems 1 and 2, respectively.*

*Example 4.* A terminated coinductive list $(\alpha, \beta)$ *tllist* is either a finite list of $\alpha$ values terminated by a single $\beta$ value, or an infinite list of $\alpha$ values. They can be seen as a quotient of pairs $\alpha$ *llist* $\times \beta$, where the first component stores the possibly infinite list given by a codatatype *llist* and the second component stores the terminator. The equivalence relation identifies all pairs with the same infinite list in the first component, effectively removing the terminator from infinite lists.[2] Let $(xs, b) \sim_{tllist} (ys, c)$ iff $xs = ys$ and, if $xs$ is finite, $b = c$. Like $\sim_P$ from Example 2, $\sim_{tllist}$ does not satisfy the naive condition (3).

> **codatatype** $\alpha$ *llist* = LNil | LCons $\alpha$ ($\alpha$ *llist*)
> **quotient_type** $(\alpha, \beta)$ *tllist* = $(\alpha$ *llist* $\times \beta)/\sim_{tllist}$

Our goal is the construction of (co)datatypes with recursion through quotients such as $(\alpha, \beta)$ *tllist*. As a realistic example, consider an inductive model of a finite interactive system that produces a possibly unbounded sequence of outputs *out* for every input *in*:

> **datatype** *system* = Step ($in \rightarrow (out, system)$ *tllist*)

This datatype declaration is only possible if *tllist* is a BNF in $\beta$. Previously, this had to be shown by manually defining the mapper and setters and proving the BNF properties. Theorem 3 identifies the conditions under which *tllist* inherits the BNF structure of its underlying type, and it allows us to automate these definitions and proofs. For *tlllist*, the conditions can be discharged easily using automatic proof methods and a simple lemma about *llist*'s relator (stating that related lists are either both finite or infinite).                                    ◇

## 3.4  Subdistributivity via Confluent Relations

Among the BNF properties, subdistributivity (REL_COMP) is typically the hardest to show. For example, distinct lists, type $\alpha$ *dlist*, have been shown to be a BNF. The manual proof requires 126 lines. Of these, the subdistributivity proof takes about 100 lines. Yet, with the theory developed so far, essentially the same argument is needed for the subdistributivity condition (12). We now present a sufficient criterion for subdistributivity that simplifies such proofs. For *dlist*, this shortens the subdistributivity proof to 58 lines. With our `lift_bnf` command (Sect. 4), the whole proof is now 64 lines, half of the manual proof.

Equivalence relations are often (or can be) expressed as the equivalence closure of a rewrite relation $\rightsquigarrow$. For example, the subdistributivity proof for distinct lists views $\alpha$ *dlist* as the quotient $\alpha$ *list*$/\sim_{dlist}$ with $xs \sim_{dlist} ys$ iff remdups $xs$ = remdups $ys$, where remdups $xs$ keeps only the last occurrence of every element in $xs$. So, $\sim_{dlist}$ is the equivalence closure of the following relation $\rightsquigarrow_{dlist}$, where · concatenates two lists:

$$xs \cdot [x] \cdot ys \rightsquigarrow_{dlist} xs \cdot ys \text{ if } x \in \text{set } ys$$

---

[2] Clearly, *tllist* could be defined directly as a codatatype. When Isabelle had no codatatype command, one of the authors formalized *tllist* via this quotient [25, version for Isabelle2013].

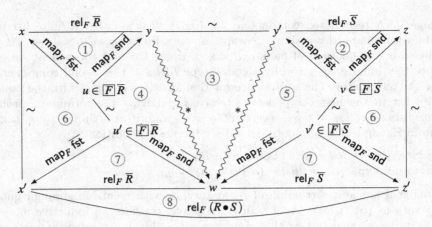

**Fig. 1.** Proof diagram for Theorem 4

We use the following notation: $\leftsquigarrow$ denotes the reverse relation, i.e., $x \leftsquigarrow y$ iff $y \rightsquigarrow x$. Further, $\overset{*}{\rightsquigarrow}$ denotes the reflexive and transitive closure, and $\overset{*}{\leftrightsquigarrow}$ the equivalence closure. A relation $\rightsquigarrow$ is confluent iff whenever $x \overset{*}{\rightsquigarrow} y$ and $x \overset{*}{\rightsquigarrow} z$, then there exists a $u$ such that $y \overset{*}{\rightsquigarrow} u$ and $z \overset{*}{\rightsquigarrow} u$—or, equivalently in pointfree style, if $(\overset{*}{\leftsquigarrow} \bullet \overset{*}{\rightsquigarrow}) \subseteq (\overset{*}{\rightsquigarrow} \bullet \overset{*}{\leftsquigarrow})$.

**Theorem 4 (Subdistributivity via confluent relations).** *Let an equivalence relation $\sim$ satisfy (2) and (3). Then, it also satisfies (9) and (12) if there is a confluent relation $\rightsquigarrow$ with the following properties:*

*(i) The equivalence relation is contained in $\rightsquigarrow$'s equivalence closure: $(\sim) \subseteq (\overset{*}{\leftrightsquigarrow})$.*

*(ii) The relation factors through projections: If $\mathsf{map}_F \, \overline{\mathsf{fst}} \, x \rightsquigarrow y$ then there exists a $y'$ such that $y = \mathsf{map}_F \, \overline{\mathsf{fst}} \, y'$ and $x \sim y'$, and similarly for $\mathsf{snd}$.*

*Proof.* The wide intersection condition (9) follows from (3) by Lemma 2. The proof for the subdistributivity condition (12) is illustrated in Fig. 1. The proof starts at the top with $(x, z) \in (\mathsf{rel}_F \, \overline{R} \bullet \sim \bullet \mathsf{rel}_F \, \overline{S})$, i.e., there are $y$ and $y'$ such that $(x, y) \in \mathsf{rel}_F \, \overline{R}$ and $y \sim y'$ and $(y', z) \in \mathsf{rel}_F \, \overline{S}$. We show $(x, z) \in (\sim \bullet \mathsf{rel}_F \, \overline{(R \bullet S)} \bullet \sim)$ by establishing the path from $x$ to $z$ via $x'$ and $z'$ along the three other borders of the diagram.

First ①, by IN_REL, there is a $u \in \boxed{F} \, \overline{R}$ such that $x = \mathsf{map}_F \, \overline{\mathsf{fst}} \, u$ and $y = \mathsf{map}_F \, \mathsf{snd} \, u$. Similarly, $\mathsf{rel}_F \, \overline{S} \, y' \, z$ yields a $v$ with the corresponding properties ②.

Second, by (i), $y \sim y'$ implies $y \overset{*}{\leftrightsquigarrow} y'$. Since $\rightsquigarrow$ is confluent, there exists a $w$ such that $y \overset{*}{\rightsquigarrow} w$ and $y' \overset{*}{\rightsquigarrow} w$ ③. By induction on $\overset{*}{\rightsquigarrow}$ using (ii), $y \overset{*}{\rightsquigarrow} w$ factors through the projection $y = \mathsf{map}_F \, \overline{\mathsf{snd}} \, u$ and we obtain a $u'$ such that $u \sim u'$ and $w = \mathsf{map}_F \, \overline{\mathsf{snd}} \, u'$ ④. Analogously, we obtain $v'$ corresponding to $y'$ and $v$ ⑤. Set $x' = \mathsf{map}_F \, \overline{\mathsf{fst}} \, u'$ and $z' = \mathsf{map}_F \, \overline{\mathsf{snd}} \, v'$. As $\mathsf{map}_F$ preserves $\sim$ by (2), we have $x \sim x'$ and $z \sim z'$ ⑥.

Next, we focus on the two triangles at the bottom ⑦. By Lemma 2 and (3), $u \sim u'$ and $u \in \boxed{F}\,\overline{R}$ imply $u' \in \boxed{F}\,\overline{R}$; similarly $v' \in \boxed{F}\,\overline{S}$. Now, $u'$ and $v'$ are the witnesses to the existential in IN_REL for $x'$ and $w$, and $w$ and $z'$, respectively. So $(x', w) \in \mathsf{rel}_F\,\overline{R}$ and $(w, z') \in \mathsf{rel}_F\,\overline{S}$, i.e., $(x', z') \in (\mathsf{rel}_F\,\overline{R} \bullet \mathsf{rel}_F\,\overline{S})$. Finally, as $F$ is a BNF, $(x', z') \in \mathsf{rel}_F\,\overline{(R \bullet S)}$ follows with subdistributivity REL_COMP ⑧.    □

*Example 5.* For distinct lists, we have $(\sim_{dlist}) = (\overset{*}{\leftrightsquigarrow}_{dlist})$ and $\rightsquigarrow_{dlist}$ is confluent. Yet, condition (ii) of Theorem 4 does not hold. For example, for $x = [(1, a), (1, b)]$, we have $\mathsf{map}_{list}$ fst $x = [1, 1] \rightsquigarrow_{dlist} [1]$. However, there is no $y$ such that $x \sim_{dlist} y$ and $\mathsf{map}_{list}$ fst $y = [1]$. The problem is that the projection $\mathsf{map}_{list}$ fst makes different atoms of $x$ equal and $\rightsquigarrow_{dlist}$ *removes* equal atoms, but the removal cannot be mimicked on $x$ itself. Fortunately, we can also *add* equal atoms instead of removing them. Define $\rightsquigarrow'_{dlist}$ by

$$xs \cdot ys \rightsquigarrow'_{dlist} xs \cdot [x] \cdot ys \text{ if } x \in \mathsf{set}\ ys$$

Then, $\rightsquigarrow'_{dlist}$ is confluent and factors through projections. So distinct lists inherit the BNF structure from lists by Theorem 4.    ◇

*Example 6.* The free monoid over atoms $\alpha$ consists of all finite lists $\alpha$ *list*. The free idempotent monoid $\alpha$ *fim* is then the quotient $\alpha$ *list*$/\sim_{fim}$ where $\sim_{fim}$ is the equivalence closure of the idempotence law for list concatenation

$$xs \cdot ys \cdot zs \rightsquigarrow_{fim} xs \cdot ys \cdot ys \cdot zs$$

We have oriented the rule such that it introduces rather than removes the duplication. In term rewriting, the rule is typically oriented in the other direction [20] such that the resulting rewriting system terminates; however, this classical relation $\leftsquigarrow_{fim}$ is not confluent: $abab\underline{cbabc}$ has two normal forms $\underline{ababc}babc \leftsquigarrow_{fim} \underline{ababc}\ abc$ and $ababc\underline{babc} \leftsquigarrow_{fim} abcbabc$ (redexes are underlined). In contrast, our orientation yields a confluent relation $\rightsquigarrow_{fim}$, although the formal proof requires some effort. The relation also factors through projections. So by Theorem 4, the free idempotent monoid $\alpha$ *fim* is also a BNF.    ◇

*Example 7.* A cyclic list is a finite list where the two ends are glued together. Abbot et al. [1] define the type of cyclic lists as the quotient that identifies lists whose elements have been shifted. Let $\rightsquigarrow_{rotate}$ denote the one-step rotation of a list, i.e.,

$$[] \rightsquigarrow_{rotate} []  \qquad\qquad [x] \cdot xs \rightsquigarrow_{rotate} xs \cdot [x]$$

The quotient $\alpha$ *cyclist* $= \alpha$ *list*$/\overset{*}{\leftrightsquigarrow}_{rotate}$ is a BNF as $\rightsquigarrow_{rotate}$ satisfies the conditions of Theorem 4.    ◇

*Example 1 (continued).* We prove the fact that $\alpha$ *re*$_{aci}$ is a BNF using Theorem 4. The confluent rewrite relation $\rightsquigarrow_{aci}$ that satisfies the conditions of Theorem 4 and whose equivalence closure is $\sim_{aci}$ is defined inductively as follows.

$$\text{Alt (Alt } r \text{ } s) \text{ } t \rightsquigarrow_{aci} \text{Alt } r \text{ (Alt } s \text{ } t) \qquad \text{Alt } r \text{ (Alt } s \text{ } t) \rightsquigarrow_{aci} \text{Alt (Alt } r \text{ } s) \text{ } t$$
$$\text{Alt } r \text{ } s \rightsquigarrow_{aci} \text{Alt } s \text{ } r \qquad\qquad\qquad r \rightsquigarrow_{aci} \text{Alt } r \text{ } r$$
$$r \rightsquigarrow_{aci} r$$
$$r \rightsquigarrow_{aci} r' \implies s \rightsquigarrow_{aci} s' \implies \text{Alt } r \text{ } s \rightsquigarrow_{aci} \text{Alt } r' \text{ } s'$$
$$r \rightsquigarrow_{aci} r' \implies s \rightsquigarrow_{aci} s' \implies \text{Conc } r \text{ } s \rightsquigarrow_{aci} \text{Conc } r' \text{ } s'$$
$$r \rightsquigarrow_{aci} r' \implies \text{Star } r \rightsquigarrow_{aci} \text{Star } r' \qquad\qquad\qquad\qquad\qquad \diamond$$

# 4   Implementation

We provide an Isabelle/HOL command that automatically lifts the BNF structure to quotient types. The command was implemented in 1590 lines of Isabelle/ML. It requires the user to discharge our conditions on the equivalence relation. Upon success, it defines the mapper, setters, and the relator, and proves the BNF axioms and transfer rules. All automated proofs are checked by Isabelle's kernel. Eventually, the command registers the quotient type with the BNF infrastructure for use in future (co)datatype definitions.

## 4.1   The lift_bnf command

Our implementation extends the interface of the existing lift_bnf command for subtypes [7]. Given a quotient type $\overline{\alpha} \, Q = \overline{\alpha} \, F/\sim$,

> lift_bnf $\overline{\alpha} \, Q$

asks the user to prove the conditions (9) and (12) of Theorem 3, where (9) is expressed in terms of (10) according to Lemma 3. Since the quotient construction already requires that $\sim$ be an equivalence relation, the remaining condition (1) holds trivially.

After the assumptions have been proved by the user, the command defines the BNF constants. Their definitions use an abstraction function $\mathsf{abs}_Q :: \overline{\alpha} \, F \to \overline{\alpha} \, Q$ and a representation function $\mathsf{rep}_Q :: \overline{\alpha} \, Q \to \overline{\alpha} \, F$, as in HOL $Q$ is distinct from (but isomorphic to) the set of equivalence classes. Concretely, we define the quotient's mapper by

$$\mathsf{map}_Q \, \overline{f} = \mathsf{abs}_Q \circ \mathsf{map}_F \, \overline{f} \circ \mathsf{rep}_Q$$

The quotient's setters use the function $\mathsf{set}_{1+}$, which maps $\mathfrak{e} \, a$ to $\{a\}$ and $\circledast$ to $\{\}$:

$$\mathsf{set}_{Q,i} = \left( \lambda x. \bigcap\nolimits_{y \in [\mathsf{map}_F \, \overline{\mathfrak{e}} \, x]_\sim} \bigcup \mathsf{set}_{1+} \langle \mathsf{set}_{F,i} \, y \rangle \right) \circ \mathsf{rep}_Q \qquad (13)$$

This definition is equivalent to the characterization in Theorem 1. The relator (Theorem 2) is lifted similarly using $\mathsf{rep}_Q$.

## 4.2   Transfer Rule Generation

The relationship of a quotient's BNF structure to its underlying type allows us to prove additional properties about the former. This is achieved by transfer rules, which drive Isabelle's Transfer tool [19] (Sect. 2.2). Our command automatically proves parametrized transfer rules for the lifted mapper, setters, and relator. Parametrized transfer rules are more powerful because they allow the refinement of nested types [22, Section 4.3]. They involve a parametrized correspondence relation $\mathsf{pcr}_Q \ \overline{A} = \mathsf{rel}_F \ \overline{A} \bullet \mathsf{cr}_Q$, where the parameters $\overline{A}$ relate the type arguments of $F$ and $Q$.

Since $\mathsf{map}_Q$ is lifted canonically, its transfer rule is unsurprising:

$$(\mathsf{map}_F, \mathsf{map}_Q) \in \left( \overline{(A \mapsto B)} \mapsto \mathsf{pcr}_Q \ \overline{A} \mapsto \mathsf{pcr}_Q \ \overline{B} \right)$$

Setters are not transferred to $\mathsf{set}_F$ but to the more complex function from (13):

$$\left( \lambda x. \ \bigcap_{y \in [\mathsf{map}_F \ \overline{e} \ x]_\sim} \bigcup \mathsf{set}_{1+} \langle \mathsf{set}_{F,i} \ y \rangle, \ \mathsf{set}_{Q,i} \right) \in (\mathsf{pcr}_Q \ \overline{A} \mapsto \mathsf{rel}_{\mathsf{set}} \ A_i)$$

where $(X, Y) \in \mathsf{rel}_{\mathsf{set}} \ A \longleftrightarrow (\forall x \in X. \ \exists y \in Y. \ (x, y) \in A) \wedge (\forall y \in Y. \ \exists x \in X. \ (x, y) \in A)$. Similarly, the rule for $Q$'s relator follows its definition in Theorem 2.

*Example 4 (continued).* Recall that terminated coinductive lists satisfy the conditions for lifting the BNF structure. Thus, we obtain the setter $\mathsf{set}_{tllist,2}$ :: $(\alpha, \beta) \ tllist \to \beta \ set$ among the other BNF operations. We want to prove that $\mathsf{set}_{tllist,2} \ x$ is empty for all infinite lists $x$. To make this precise, let the predicate lfinite :: $\alpha \ llist \to bool$ characterize finite coinductive lists. We lift it to $(\alpha, \beta) \ tllist$ by projecting away the terminator:

```
lift_definition tlfinite :: (α, β) tllist → bool is (λx. lfinite (fst x))
```

Therefore, we have to show that $\forall x. \ \neg \ \mathsf{tlfinite} \ x \implies \mathsf{set}_{tllist,2} \ x = \{\}$. Using the transfer rules for the setter and the lifted predicate tlfinite, the `transfer` proof method reduces the proof obligation to

$$\forall x'. \ \neg \ \mathsf{lfinite} \ (\mathsf{fst} \ x') \implies \bigcap_{y \in [\mathsf{map}_F \ \overline{e} \ x']_{\sim_{tllist}}} \bigcup \mathsf{set}_{1+} \langle \mathsf{set}_{F,2} \ y \rangle = \{\}$$

where $x' :: (\alpha, \beta) \ F$, and $(\alpha, \beta) \ F = (\alpha \ llist \times \beta)$ is the underlying functor of *tllist*. The rest of the proof, which need not refer to *tllist* anymore, is automatic.   $\diamond$

We have also extended `lift_bnf` to generate transfer rules for subtypes. There, the setters and the relator do not change: if $T$ is a subtype of $F$, e.g., then $\mathsf{set}_{T,i}$ is transferred to $\mathsf{set}_{F,i}$.

## 5   Related Work

Quotient constructions have been formalized and implemented, e.g., in Isabelle/HOL [19,21,34,35], HOL4 [18], Agda [40,41], Cedille [30], Coq [11,12],

Lean [3], and Nuprl [33]. None of these works look at the preservation of functor properties except for Avigad et al. [3] (discussed in Sect. 5.2) and Veltri [41]. Veltri studies the special case of when the delay monad is preserved by a quotient of weak bisimilarity, focusing on the challenges that quotients pose in intensional type theory.

Abbot et al. [1] introduce quotient containers as a model of datatypes with permutative structure, such as unordered pairs, cyclic lists, and multisets. The map function of quotient containers does not change the shape of the container. Quotient containers therefore cannot deal with quotients where the equivalence relation takes the identity of elements into account, such as distinct lists, finite sets, and the free idempotent monoid. Overall our construction strictly subsumes quotient containers.

## 5.1   Quotients in the Category of Sets

BNFs are accessible functors in the category of Sets. We therefore relate to the literature on when quotients preserve functors and their properties in Set.

Trnková [38] showed that all Set functors preserve non-empty intersections: in our notation $\boxed{F}\ A \cap \boxed{F}\ B = \boxed{F}\ (A \cap B)$ whenever $A \cap B \neq \{\}$. Empty intersections need not be preserved though. Functors that do are called regular [39] or sound [2]. All BNFs are sound as $\boxed{F}\ A = \{x \mid \mathsf{set}_F\ x \subseteq A\}$. As shown in Example 2, the naive quotient construction can lead to unsound functors.

Every unsound functor can be "repaired" by setting $\boxed{F}\ \{\}$ to the *distinguished points* $\mathsf{dp}_F$. We write $\boxed{F}'$ for the repaired action.

$$\boxed{F}'\ A = \begin{cases} \mathsf{dp}_F & \text{if } A = \{\} \\ \boxed{F}\ A & \text{otherwise} \end{cases} \tag{14}$$

Trnková characterizes the distinguished points $\mathsf{dp}_F$ as the natural transformations from $C_{1,0}$ to $F$ where $\boxed{C_{1,0}}\ \{\} = \{\}$ and $\boxed{C_{1,0}}\ A = \{\circledast\}$ for $A \neq \{\}$. Barr [4] and Gumm [16] use equalizers instead of natural transformations to define the distinguished points of univariate functors:

$$\mathsf{dp}_F = \{x \mid \mathsf{map}_F\ (\lambda\_.\ \mathsf{True})\ x = \mathsf{map}_F\ (\lambda\_.\ \mathsf{False})\ x\} \tag{15}$$

The case distinction in (14) makes it hard to work with repaired functors, especially as the case distinctions proliferate for multivariate functors. Instead, we repair the unsoundness by avoiding empty sets altogether: Our characterization $\boxed{F}\ A$ in Lemma 1 effectively derives the quotient from $(1 + \alpha)\ F$ instead of $\alpha\ F$. Moreover, our characterization of $\boxed{F}\ \overline{A}$ generalizes Barr and Gumm's definition of distinguished points: for $\overline{A} = \{\}$, (5) simplifies to (15). The resulting quotient is the same because $[\boxed{F}\ \overline{A}]_\sim = [\boxed{F}\ \overline{A}]_\sim$ if $A_i \neq \{\}$ for all $i$.

Given the other BNF properties, subdistributivity is equivalent to the functor preserving weak pullbacks. Adámek et al. [2] showed that an accessible Set functor preserves weak pullbacks iff it has a so-called dominated presentation in terms of flat equations $E$ over a signature $\Sigma$. This characterization does not immediately help with proving subdistributivity, though. For example, the finite set quotient $\alpha$ *fset* $= \alpha$ *list*$/\sim_{fset}$ comes with the signature $\Sigma = \{\sigma_n \mid n \in \mathbb{N}\}$ and the equations $\sigma_n(x_1, \ldots x_n) = \sigma_m(y_1, \ldots, y_m)$ whenever $\{x_1, \ldots, x_n\} = \{y_1, \ldots, y_m\}$. Proving domination for this presentation boils down to proving subdistributivity directly. Our criterion using a confluent relation (Theorem 4) is only sufficient, not necessary, but it greatly simplifies the actual proof effort.

## 5.2   Comparison with Lean's Quotients of Polynomial Functors

Avigad et al. [3] proposed quotients of polynomial functors (QPF) as a model for datatypes. QPFs generalize BNFs in that they require less structure: there is no setter and the relator need not satisfy subdistributivity. Nevertheless, the quotient construction is similar to ours. Without loss of generality, we consider in our comparison only the univariate case $\alpha\, Q = \alpha\, F/\sim$.

The main difference lies in the definition of the liftings $\mathsf{lift}_F$ of predicates $P :: \alpha \to \mathbb{B}$ and relations $R :: \alpha \otimes \beta$. In our notation, $\mathsf{lift}_F\, P$ corresponds to $\lambda x.\ x \in \boxed{F}\ \{a \mid P\, a\}$ and $\mathsf{lift}_F\, R$ to $\mathsf{rel}_F\, R$. QPF defines these liftings for the quotient $Q$ as follows:

$$\mathsf{lift}_Q\, P\, [x]_\sim = (\exists x' \in [x]_\sim.\, P\, x') \qquad \mathsf{lift}_Q\, R\, [x]_\sim\, [y]_\sim = (\exists x' \in [x]_\sim.\, \exists y' \in [y]_\sim.\, R\, x'\, y')$$

That is, these definitions correspond to the naive construction $\boxed{Q}\, A = [\boxed{F}\, A]_\sim$ and $\mathsf{rel}_Q\, R = [\mathsf{rel}_F\, R]_\sim$ where $[(x,y)]_\sim = ([x]_\sim, [y]_\sim)$. As discussed above, the resulting quotient may be an unsound functor. Consequently, lifting of predicates does not preserve empty intersections in general. This hinders modular proofs. For example, suppose that a user has already shown $\mathsf{lift}_Q\, P_1\, x$ and $\mathsf{lift}_Q\, P_2\, x$ for some value $x$ and two properties $P_1$ and $P_2$. Then, to deduce $\mathsf{lift}_F\, (\lambda a.\, P_1\, a \wedge P_2\, a)\, x$, they would have to prove that the two properties do not contradict each other, i.e., $\exists a.\, P_1\, a \wedge P_2\, a$. Obviously, this makes modular proofs harder as extra work is needed to combine properties.

QPF uses $\mathsf{lift}_F\, P$ in the induction theorem for datatypes. So when a datatype recurses through *tllist*, this spreads to proofs by induction: splitting a complicated inductive statement into smaller lemmas is not for free. Moreover, $\mathsf{lift}_Q$ holds for fewer values, as the next example shows. Analogous problems arise in QPF for relation lifting, which appears in the coinduction theorem.

*Example 4 (continued).* Consider the infinite repetition repeat $a :: (\alpha, \beta)$ *tllist* of the atom $a$ as a terminated lazy list. As repeat $a$ contains only $a$s, one would expect that $\mathsf{lift}_{tllist}\, (\lambda a'.\, a' = a)\, (\lambda\_.\, \mathsf{False})\, (\mathsf{repeat}\, a)$ holds. Yet, this property is provably false.                                                                                      ◇

These issues would go away if $\mathsf{lift}_Q$ was defined following our approach for $\boxed{Q}\, A = [\boxed{F}\, A]_\sim$ and $\mathsf{rel}_Q$ as in Theorem 2. These definitions do not rely on the additional BNF structure; only $\mathsf{map}_Q$ is needed and QPF defines $\mathsf{map}_Q$ like we do. The repair should therefore work for the general QPF case, as well.

# 6  Conclusion

We have described a sufficient criterion for quotient types to be able to inherit the BNF structure from the underlying type. We have demonstrated the effectiveness of the criterion by automating the BNF "inheritance" in the form of the `lift_bnf` command in Isabelle/HOL and used it (which amounts to proving the criterion) for several realistic quotient types. We have also argued that our treatment of the quotient's setter and relator to avoid unsoundness carries over to more general structures, such as Lean's QPFs.

As future work, we plan to investigate quotients of existing generalizations of BNFs to co- and contravariant functors [28] and functors operating on small-support endomorphisms and bijections [8]. Furthermore, we would like to provide better automation for proving subdistributivity via confluent rewrite systems as part of `lift_bnf`.

**Acknowledgment.** We thank David Basin for supporting this work, Ralf Sasse and Andrei Popescu for insightful discussions about confluent relations, BNFs, their preservation of wide intersections, and ways to express the setters in terms of the mapper, and Jasmin Blanchette and the anonymous IJCAR reviewers for numerous comments on earlier drafts of this paper, which helped to improve the presentation. Julian Biendarra developed the `lift_bnf` command for subtypes, which we extend to quotient types in this work.

# References

1. Abbott, M., Altenkirch, T., Ghani, N., McBride, C.: Constructing polymorphic programs with quotient types. In: Kozen, D. (ed.) MPC 2004. LNCS, vol. 3125, pp. 2–15. Springer, Heidelberg (2004). https://doi.org/10.1007/978-3-540-27764-4_2
2. Adámek, J., Gumm, H.P., Trnková, V.: Presentation of set functors: a coalgebraic perspective. J. Log. Comput. **20**(5), 991–1015 (2010)
3. Avigad, J., Carneiro, M., Hudon, S.: Data types as quotients of polynomial functors. In: Harrison, J., O'Leary, J., Tolmach, A. (eds.) ITP 2019 Leibniz International Proceedings in Informatics (LIPIcs), vol. 141, pp. 6:1–6:19. Schloss Dagstuhl–Leibniz-Zentrum für Informatik, Dagstuhl (2019)
4. Barr, M.: Terminal coalgebras in well-founded set theory. Theor. Comput. Sci. **114**(2), 299–315 (1993)
5. Basin, D., Krstić, S., Traytel, D.: Almost event-rate independent monitoring of metric dynamic logic. In: Lahiri, S., Reger, G. (eds.) RV 2017. LNCS, vol. 10548, pp. 85–102. Springer, Cham (2017). https://doi.org/10.1007/978-3-319-67531-2_6
6. Basin, D.A., Lochbihler, A., Sefidgar, S.R.: CryptHOL: game-based proofs in higher-order logic. J. Cryptol. **33**, 494–566 (2020)
7. Biendarra, J.: Functor-preserving type definitions in Isabelle/HOL. Bachelor thesis, Fakultät für Informatik, Technische Universität München (2015)
8. Blanchette, J.C., Gheri, L., Popescu, A., Traytel, D.: Bindings as bounded natural functors. PACMPL **3**(POPL), 22:1–22:34 (2019)

9. Blanchette, J.C., Hölzl, J., Lochbihler, A., Panny, L., Popescu, A., Traytel, D.: Truly modular (co)datatypes for Isabelle/HOL. In: Klein, G., Gamboa, R. (eds.) ITP 2014. LNCS, vol. 8558, pp. 93–110. Springer, Cham (2014). https://doi.org/10.1007/978-3-319-08970-6_7

10. Blanchette, J.C., Popescu, A., Traytel, D.: Cardinals in Isabelle/HOL. In: Klein, G., Gamboa, R. (eds.) ITP 2014. LNCS, vol. 8558, pp. 111–127. Springer, Cham (2014). https://doi.org/10.1007/978-3-319-08970-6_8

11. Chicli, L., Pottier, L., Simpson, C.: Mathematical quotients and quotient types in Coq. In: Geuvers, H., Wiedijk, F. (eds.) TYPES 2002. LNCS, vol. 2646, pp. 95–107. Springer, Heidelberg (2003). https://doi.org/10.1007/3-540-39185-1_6

12. Cohen, C.: Pragmatic quotient types in Coq. In: Blazy, S., Paulin-Mohring, C., Pichardie, D. (eds.) ITP 2013. LNCS, vol. 7998, pp. 213–228. Springer, Heidelberg (2013). https://doi.org/10.1007/978-3-642-39634-2_17

13. Cohen, C., Dénès, M., Mörtberg, A.: Refinements for free! In: Gonthier, G., Norrish, M. (eds.) CPP 2013. LNCS, vol. 8307, pp. 147–162. Springer, Cham (2013). https://doi.org/10.1007/978-3-319-03545-1_10

14. De Giacomo, G., Vardi, M.Y.: Linear temporal logic and linear dynamic logic on finite traces. In: Rossi, F. (eds.) IJCAI 2013, pp. 854–860. IJCAI/AAAI (2013)

15. Fürer, B., Lochbihler, A., Schneider, J., Traytel, D.: Quotients of bounded natural functors (extended report). Technical report (2020). https://people.inf.ethz.ch/trayteld/papers/ijcar20-qbnf/qbnf_report.pdf

16. Gumm, H.P.: From $T$-coalgebras to filter structures and transition systems. In: Fiadeiro, J.L., Harman, N., Roggenbach, M., Rutten, J. (eds.) CALCO 2005. LNCS, vol. 3629, pp. 194–212. Springer, Heidelberg (2005). https://doi.org/10.1007/11548133_13

17. Gumm, H.P., Schröder, T.: Types and coalgebraic structure. Algebra Univers. 53(2), 229–252 (2005)

18. Homeier, P.V.: A design structure for higher order quotients. In: Hurd, J., Melham, T. (eds.) TPHOLs 2005. LNCS, vol. 3603, pp. 130–146. Springer, Heidelberg (2005). https://doi.org/10.1007/11541868_9

19. Huffman, B., Kunčar, O.: Lifting and Transfer: a modular design for quotients in Isabelle/HOL. In: Gonthier, G., Norrish, M. (eds.) CPP 2013. LNCS, vol. 8307, pp. 131–146. Springer, Cham (2013). https://doi.org/10.1007/978-3-319-03545-1_9

20. Hullot, J.-M.: A catalogue of canonical term rewrite systems. Technical report CSL-113, SRI International (1980)

21. Kaliszyk, C., Urban, C.: Quotients revisited for Isabelle/HOL. In: Chu, W.C., Wong, W.E., Palakal, M.J., Hung, C. (eds.) SAC 2011, pp. 1639–1644. ACM (2011)

22. Kunčar, O.: Types, Abstraction and Parametric Polymorphism in Higher-Order Logic. Ph.D. thesis, Technical University Munich, Germany (2016)

23. Kunčar, O., Popescu, A.: From types to sets by local type definition in higher-order logic. J. Autom. Reasoning 62(2), 237–260 (2019)

24. Lammich, P., Lochbihler, A.: Automatic refinement to efficient data structures: a comparison of two approaches. J. Autom. Reasoning 63(1), 53–94 (2019)

25. Lochbihler, A.: Coinductive. Archive of Formal Proofs (2010). Formal proof development. http://isa-afp.org/entries/Coinductive.html

26. Lochbihler, A.: Effect polymorphism in higher-order logic (proof pearl). J. Autom. Reasoning 63(2), 439–462 (2019)

27. Lochbihler, A., Schneider, J.: Equational reasoning with applicative functors. In: Blanchette, J.C., Merz, S. (eds.) ITP 2016. LNCS, vol. 9807, pp. 252–273. Springer, Cham (2016). https://doi.org/10.1007/978-3-319-43144-4_16

28. Lochbihler, A., Schneider, J.: Relational parametricity and quotient preservation for modular (co)datatypes. In: Avigad, J., Mahboubi, A. (eds.) ITP 2018. LNCS, vol. 10895, pp. 411–431. Springer, Cham (2018). https://doi.org/10.1007/978-3-319-94821-8_24

29. Lochbihler, A., Sefidgar, S.R., Basin, D.A., Maurer, U.: Formalizing constructive cryptography using CryptHOL. In: CSF 2019, pp. 152–166. IEEE (2019)

30. Marmaduke, A., Jenkins, C., Stump, A.: Quotients by idempotent functions in Cedille. In: Bowman, W.J., Garcia, R. (eds.) TFP 2019. LNCS, vol. 12053, pp. 1–20. Springer, Cham (2020). https://doi.org/10.1007/978-3-030-47147-7_1

31. Nipkow, T., Klein, G.: Concrete Semantics - With Isabelle/HOL. Springer, Cham (2014). https://doi.org/10.1007/978-3-319-10542-0

32. Nipkow, T., Traytel, D.: Unified decision procedures for regular expression equivalence. In: Klein, G., Gamboa, R. (eds.) ITP 2014. LNCS, vol. 8558, pp. 450–466. Springer, Cham (2014). https://doi.org/10.1007/978-3-319-08970-6_29

33. Nogin, A.: Quotient types: a modular approach. In: Carreño, V.A., Muñoz, C.A., Tahar, S. (eds.) TPHOLs 2002. LNCS, vol. 2410, pp. 263–280. Springer, Heidelberg (2002). https://doi.org/10.1007/3-540-45685-6_18

34. Paulson, L.C.: Defining functions on equivalence classes. ACM Trans. Comput. Log. **7**(4), 658–675 (2006)

35. Slotosch, O.: Higher order quotients and their implementation in Isabelle/HOL. In: Gunter, E.L., Felty, A. (eds.) TPHOLs 1997. LNCS, vol. 1275, pp. 291–306. Springer, Berlin, Heidelberg (1997)

36. Sozeau, M.: A new look at generalized rewriting in type theory. J. Formalized Reasoning **2**(1), 41–62 (2010)

37. Traytel, D., Popescu, A., Blanchette, J.C.: Foundational, compositional (co)datatypes for higher-order logic: category theory applied to theorem proving. In: LICS 2012, pp. 596–605. IEEE Computer Society (2012)

38. Trnková, V.: Some properties of set functors. Commentationes Mathematicae Univ. Carol. **10**(2), 323–352 (1969)

39. Trnková, V.: On descriptive classification of set-functors I. Commentationes Mathematicae Univ. Carol. **12**(1), 143–174 (1971)

40. Veltri, N.: Two set-based implementations of quotients in type theory. In: Nummenmaa, J., Sievi-Korte, O., Mäkinen, E. (eds.) SPLST 2015. CEUR Workshop Proceedings, vol. 1525, pp. 194–205 (2015)

41. Veltri, N.: A Type-Theoretical Study of Nontermination. Ph.D. thesis, Tallinn University of Technology (2017)

# Trakhtenbrot's Theorem in Coq
## A Constructive Approach to Finite Model Theory

Dominik Kirst[1]([⊠])(iD) and Dominique Larchey-Wendling[2](iD)

[1] Saarland University, Saarland Informatics Campus, Saarbrücken, Germany
`kirst@ps.uni-saarland.de`
[2] Université de Lorraine, CNRS, LORIA, Vandœuvre-lès-Nancy, France
`dominique.larchey-wendling@loria.fr`

**Abstract.** We study finite first-order satisfiability (FSAT) in the constructive setting of dependent type theory. Employing synthetic accounts of enumerability and decidability, we give a full classification of FSAT depending on the first-order signature of non-logical symbols. On the one hand, our development focuses on Trakhtenbrot's theorem, stating that FSAT is undecidable as soon as the signature contains an at least binary relation symbol. Our proof proceeds by a many-one reduction chain starting from the Post correspondence problem. On the other hand, we establish the decidability of FSAT for monadic first-order logic, i.e. where the signature only contains at most unary function and relation symbols, as well as the enumerability of FSAT for arbitrary enumerable signatures. All our results are mechanised in the framework of a growing Coq library of synthetic undecidability proofs.

## 1 Introduction

In the wake of the seminal discoveries concerning the undecidability of first-order logic by Turing and Church in the 1930s, a broad line of work has been pursued to characterise the border between decidable and undecidable fragments of the original decision problem. These fragments can be grouped either by syntactic restrictions controlling the allowed function and relation symbols or the quantifier prefix, or by semantic restrictions on the admitted models (see [1] for a comprehensive description).

Concerning signature restrictions, already predating the undecidability results, Löwenheim had shown in 1915 that monadic first-order logic, admitting only signatures with at most unary symbols, is decidable [15]. Therefore, the successive negative results usually presuppose non-trivial signatures containing an at least binary symbol.

Turning to semantic restrictions, Trakhtenbrot proved in 1950 that, if only admitting finite models, the satisfiability problem over non-trivial signatures is still undecidable [21]. Moreover, the situation is somewhat dual to the unrestricted case, since finite satisfiability (FSAT) is still enumerable while, in the unrestricted case, validity is enumerable. As a consequence, finite validity cannot be characterised by a complete finitary deduction system and, resting on finite model theory, various natural problems in database theory are undecidable.

© Springer Nature Switzerland AG 2020
N. Peltier and V. Sofronie-Stokkermans (Eds.): IJCAR 2020, LNAI 12167, pp. 79–96, 2020.
https://doi.org/10.1007/978-3-030-51054-1_5

Conventionally, Trakhtenbrot's theorem is proved by (many-one) reduction from the halting problem for Turing machines (see e.g. [1,14]). An encoding of a given Turing machine $M$ can be given as a formula $\varphi_M$ such that the models of $\varphi_M$ correspond to the runs of $M$. Specifically, the finite models of $\varphi_M$ correspond to terminating runs of $M$ and so a decision procedure for finite satisfiability of $\varphi_M$ would be enough to decide whether $M$ terminates or not.

Although this proof strategy is in principle explainable on paper, already the formal definition of Turing machines, not to mention their encoding in first-order logic, is not ideal for mechanisation in a proof assistant. So for our Coq mechanisation of Trakhtenbrot's theorem, we follow a different strategy by starting from the Post correspondence problem (PCP), a simple matching problem on strings. Similar to the conventional proof, we proceed by encoding every instance $R$ of PCP as a formula $\varphi_R$ such that $R$ admits a solution iff $\varphi_R$ has a finite model. Employing the framework of synthetic undecidability [8,11], the computability of $\varphi_R$ from $R$ is guaranteed since all functions definable in constructive type theory are computable without reference to a concrete model of computation.

Both the conventional proof relying on Turing machines and our elaboration starting from PCP actually produce formulas in a custom signature well-suited for the encoding of the seed decision problems. The sharper version of Trakhtenbrot's theorem, stating that a signature with at least one binary relation (or one binary function and one unary relation) is enough to turn FSAT undecidable, is in fact left as an exercise in e.g. Libkin's book [14]. However, at least in a constructive setting, this generalisation is non-trivial and led us to mechanising a chain of signature transformations eliminating and compressing function and relation symbols step by step.

Complementing the undecidability result, we further formalise that FSAT is enumerable for enumerable signatures and decidable for monadic signatures. Again, both of these standard results come with their subtleties when explored in a constructive approach of finite model theory.

In summary, the main contributions of this paper are threefold:

- we provide an axiom-free Coq mechanisation comprising a full classification of finite satisfiability with regards to the signatures allowed;[1]
- we present a streamlined proof strategy for Trakhtenbrot's theorem well-suited for mechanisation and simple to explain informally, basing on PCP;
- we give a constructive account of signature transformations and the treatment of interpreted equality typically neglected in a classical development.

The rest of the paper is structured as follows. We first describe the type-theoretical framework for undecidability proofs and the representation of first-order logic in Sect. 2. We then outline our variant of Trakhtenbrot's theorem for a custom signature in Sect. 3. This is followed by a development of enough constructive finite model theory (Sect. 4) to conclude some decidability results (Sect. 5) as well as the final classification (Sect. 6). We end with a brief discussion of the Coq development and future work in Sect. 7.

---

[1] Downloadable from http://www.ps.uni-saarland.de/extras/fol-trakh/ and systematically hyperlinked with the definitions and theorems in this PDF.

## 2    First-Order Satisfiability in Constructive Type Theory

In order to make this paper accessible to readers unfamiliar with constructive type theory, we outline the required features of Coq's underlying type theory, the synthetic treatment of computability available in constructive mathematics, some properties of finite types, as well as our representation of first-order logic.

### 2.1    Basics of Constructive Type Theory

We work in the framework of a constructive type theory such as the one implemented in Coq, providing a predicative hierarchy of type universes $\mathbb{T}$ above a single impredicative universe $\mathbb{P}$ of propositions. On type level, we have the unit type $\mathbb{1}$ with a single element $* : \mathbb{1}$, the void type $\mathbb{0}$, function spaces $X \to Y$, products $X \times Y$, sums $X + Y$, dependent products $\forall x : X. F\,x$, and dependent sums $\{x : X \mid F\,x\}$. On propositional level, these types are denoted using the usual logical notation ($\top$, $\bot$, $\to$, $\wedge$, $\vee$, $\forall$, and $\exists$).

We employ the basic inductive types of Booleans ($\mathbb{B} ::= \mathsf{tt} \mid \mathsf{ff}$), of Peano natural numbers ($n : \mathbb{N} ::= 0 \mid \mathsf{S}\,n$), the option type ($\mathbb{O}\,X ::= \ulcorner x \urcorner \mid \emptyset$), and lists ($l : \mathbb{L}\,X ::= [] \mid x :: l$). We write $|l|$ for the *length* of a list, $l + m$ for the *concatenation* of $l$ and $m$, $x \in l$ for *membership*, and simply $f\,[x_1; \dots ; x_n] := [f\,x_1; \dots ; f\,x_n]$ for the *map* function. We denote by $X^n$ the type of vectors of length $n : \mathbb{N}$ and by $\mathbb{F}_n$ the finite types understood as indices $\{0, \dots, n-1\}$. The definitions/notations for lists are shared with vectors $v : X^n$. Moreover, when $i : \mathbb{F}_n$ and $x : X$, we denote by $v_i$ the $i$-th component of $v$ and by $v[x/i]$ the vector $v$ with $i$-th component updated to value $x$.

### 2.2    Synthetic (Un-)decidability

We review the main ingredients of our synthetic approach to decidability and undecidability [7,8,10,11,13,19], based on the computability of all functions definable in constructive type theory.[2] We first introduce standard notions of computability theory without referring to a formal model of computation, e.g. Turing machines.

**Definition 1.** A *problem* or predicate $p : X \to \mathbb{P}$ is

- *decidable* if there is $f : X \to \mathbb{B}$ with $\forall x.\, p\,x \leftrightarrow f\,x = \mathsf{tt}$.
- *enumerable* if there is $f : \mathbb{N} \to \mathbb{O}\,X$ with $\forall x.\, p\,x \leftrightarrow \exists n.\, f\,n = \ulcorner x \urcorner$.

These notions generalise to predicates of higher arity. Moreover, a type $X$ is

- *enumerable* if there is $f : \mathbb{N} \to \mathbb{O}\,X$ with $\forall x. \exists n.\, f\,n = \ulcorner x \urcorner$.
- *discrete* if equality on $X$ (i.e. $\lambda xy : X.\, x = y$) is decidable.
- a *data type* if it is both enumerable and discrete.

---

[2] A result shown and applied for many variants of constructive type theory and which Coq designers are committed to maintain as Coq evolves.

Using the expressiveness of dependent types, we equivalently tend to establish the decidability of a predicate $p : X \to \mathbb{P}$ by giving a function $\forall x : X. \, p\,x + \neg p\,x$. Note that it is common to mechanise decidability results in this synthetic sense (e.g. [2,16,17]). Next, decidability and enumerability transport along reductions:

**Definition 2.** A problem $p : X \to \mathbb{P}$ *(many-one) reduces* to $q : Y \to \mathbb{P}$, written $p \preceq q$, if there is a function $f : X \to Y$ such that $p\,x \leftrightarrow q\,(f\,x)$ for all $x : X$.[3]

**Fact 3.** *Assume $p : X \to \mathbb{P}$, $q : Y \to \mathbb{P}$ and $p \preceq q$: (1) if $q$ is decidable, then so is $p$ and (2) if $X$ and $Y$ are data types and $q$ is enumerable, then so is $p$.*

Item (1) implies that we can justify the undecidability of a target problem by reduction from a seed problem known to be undecidable, such as the halting problem for Turing machines. This is in fact the closest rendering of undecidability available in a synthetic setting, since the underlying type theory is consistent with the assumption that every problem is decidable.[4] Nevertheless, we believe that in the intended effective interpretation for synthetic computability, a typical seed problem is indeed undecidable and so are the problems reached by verified reductions.[5] More specifically, since the usual seed problems are not co-enumerable, (2) implies that the reached problems are not co-enumerable either.

Given its simple inductive characterisation involving only basic types of lists and Booleans, the (binary) Post correspondence problem (BPCP) is a well-suited seed problem for compact encoding into first-order logic.

**Definition 4.** Given a list $R : \mathbb{L}(\mathbb{L}\mathbb{B} \times \mathbb{L}\mathbb{B})$ of pairs $s/t$ of Boolean strings,[6] we define *derivability* of a pair $s/t$ from $R$ (denoted by $R \triangleright s/t$) and *solvability* (denoted by $\mathsf{BPCP}\,R$) by the following rules:

$$\frac{s/t \in R}{R \triangleright s/t} \qquad \frac{s/t \in R \quad R \triangleright u/v}{R \triangleright (s \mathbin{+\!\!+} u)/(t \mathbin{+\!\!+} v)} \qquad \frac{R \triangleright s/s}{\mathsf{BPCP}\,R}$$

**Fact 5.** *Given a list $R : \mathbb{L}(\mathbb{L}\mathbb{B} \times \mathbb{L}\mathbb{B})$, the derivability predicate $\lambda s\,t.R \triangleright s/t$ is decidable. However, the halting problem for Turing machines reduces to $\mathsf{BPCP}$.*

*Proof.* We give of proof of the decidability of $R \triangleright s/t$ by induction on $|s| + |t|$. We also provide a trivial proof of the equivalence of two definitions of BPCP. See [7,10] for details on the reduction from the halting problem to BPCP.  □

It might at first appear surprising that derivability $\lambda s\,t.R \triangleright s/t$ is decidable while BPCP is reducible from the halting problem (and hence undecidable). This simply illustrates that undecidability is caused by the unbounded existential quantifier in the equivalence $\mathsf{BPCP}\,R \leftrightarrow \exists s.\,R \triangleright s/s$.

---

[3] Or equivalently, the dependent characterisation. $\forall x : X. \, \{y : Y \mid p\,x \leftrightarrow q\,y\}$.

[4] As witnessed by classical set-theoretic models satisfying $\forall p : \mathbb{P}. \, p + \neg p$ (cf. [23]).

[5] This synthetic treatment of undecidability is discussed in more detail in [8] and [11].

[6] Notice that the list $R$ is viewed as a (finite) set of pairs $s/t \in R$ (hence ignoring the order or duplicates), while $s$ and $t$, which are also lists, are viewed a strings (hence repetitions and ordering matter for $s$ and $t$).

## 2.3    Finiteness

**Definition 6.** A *type $X$ is finite* if there is a list $l_X$ with $x \in l_X$ for all $x : X$ and a *predicate $p : X \to \mathbb{P}$ is finite* if there is a list $l_p$ with $\forall x.\, p\, x \leftrightarrow x \in l_p$.

Note that in constructive settings there are various alternative characterisations of finiteness[7] (bijection with $\mathbb{F}_n$ for some $n$; negated infinitude for some definition of infiniteness; etc.) and we opted for the above since it is easy to work with while transparently capturing the expected meaning. One can distinguish *strong* finiteness in $\mathbb{T}$ (i.e. $\{l_X : \mathbb{L}\, X \mid \forall x.\, x \in l_X\}$) from *weak* finiteness in $\mathbb{P}$ (i.e. $\exists l_X : \mathbb{L}\, X.\, \forall x.\, x \in l_X$), the list $l_X$ being required computable in the strong case.

We present three important tools for manipulating finite types: the *finite pigeon hole principle* (PHP) here established without assuming discreteness, the *well-foundedness* of strict orders over finite types, and *quotients* over strongly decidable equivalences that map onto $\mathbb{F}_n$. The proofs are given in Appendix A of the extended version of this paper [12].

For the finite PHP, the typical classical proof requires the discreteness of $X$ to design transpositions/permutations. Here we avoid discreteness completely, the existence of a duplicate being established without actually computing one.

**Theorem 7 (Finite PHP).** *Let $R : X \to Y \to \mathbb{P}$ be a binary relation and $l : \mathbb{L}\, X$ and $m : \mathbb{L}\, Y$ be two lists where $m$ is shorter than $l$ ($|m| < |l|$). If $R$ is total from $l$ to $m$ ($\forall x.\, x \in l \to \exists y.\, y \in m \land R\, x\, y$) then the values at two distinct positions in $l$ are related to the same $y$ in $m$, i.e. there exist $x_1, x_2 \in l$ and $y \in m$ such that $l$ has shape $l = \cdots +\!\!\!+ x_1 :: \cdots +\!\!\!+ x_2 :: \cdots$ and $R\, x_1\, y$ and $R\, x_2\, y$.*

Using the PHP, one can constructively show that, for a strict order over a finite type $X$, any descending chain has length bounded by the size of $X$.[8]

**Fact 8.** *Every strict order on a finite type is well-founded.*

Coq's type theory does not provide quotients in general (see e.g. [6]) but one can build computable quotients in certain conditions, here for a decidable equivalence relation of which representatives of equivalence classes are listable.

**Theorem 9 (Finite decidable quotient).** *Let $\sim\, : X \to X \to \mathbb{P}$ be a decidable equivalence with $\{l_r : \mathbb{L}\, X \mid \forall x \exists y.\, y \in l_r \land x \sim y\}$, i.e. finitely many equivalence classes.[9] Then one can compute the quotient $X/\!\sim$ onto $\mathbb{F}_n$ for some $n$, i.e. $n : \mathbb{N}$, $c : X \to \mathbb{F}_n$ and $r : \mathbb{F}_n \to X$ s.t. $\forall p.\, c\,(r\, p) = p$ and $\forall x\, y.\, x \sim y \leftrightarrow c\, x = c\, y$.*

Using Theorem 9 with identity over $X$ as equivalence, we get bijections between finite, discrete types and the type family $(\mathbb{F}_n)_{n:\mathbb{N}}$.[10]

**Corollary 10.** *If $X$ is a finite and discrete type then one can compute $n : \mathbb{N}$ and a bijection from $X$ to $\mathbb{F}_n$.*

---

[7] And these alternative characterisations are not necessarily constructively equivalent.

[8] i.e. the length of the enumerating list of $X$.

[9] Hence $l_r$ denotes a list of *representatives* of equivalence classes.

[10] For a given $X$, the value $n$ (usually called cardinal) is unique by the PHP.

## 2.4    Representing First-Order Logic

We briefly outline our representation of the syntax and semantics of first-order logic in constructive type theory (cf. [9]). Concerning the *syntax*, we describe terms and formulas as dependent inductive types over a signature $\Sigma = (\mathcal{F}_\Sigma; \mathcal{P}_\Sigma)$ of function symbols $f : \mathcal{F}_\Sigma$ and relation symbols $P : \mathcal{P}_\Sigma$ with arities $|f|$ and $|P|$, using binary connectives $\dot{\Box} \in \{\dot{\to}, \dot{\wedge}, \dot{\vee}\}$ and quantifiers $\dot{\nabla} \in \{\dot{\forall}, \dot{\exists}\}$:

$$t : \mathsf{Term}_\Sigma ::= x \mid f\,t \qquad\qquad (x : \mathbb{N},\ f : \mathcal{F}_\Sigma,\ t : \mathsf{Term}_\Sigma^{|f|})$$
$$\varphi, \psi : \mathsf{Form}_\Sigma ::= \dot{\perp} \mid P\,t \mid \varphi\,\dot{\Box}\,\psi \mid \dot{\nabla}\varphi \qquad (P : \mathcal{P}_\Sigma,\ t : \mathsf{Term}_\Sigma^{|P|})$$

Negation is defined as the abbreviation $\dot{\neg}\varphi := \varphi \dot{\to} \dot{\perp}$.

In the chosen de Bruijn representation [4], a bound variable is encoded as the number of quantifiers shadowing its binder, e.g. $\forall x.\,\exists y.\,P\,x\,u \to P\,y\,v$ may be represented by $\dot{\forall}\dot{\exists}P\,1\,4 \dot{\to} P\,0\,5$. The variables $2 = 4 - 2$ and $3 = 5 - 2$ in this example are the *free* variables, and variables that do not occur freely are called *fresh*, e.g. 0 and 1 are fresh. For the sake of legibility, we write concrete formulas with named binders and defer de Bruijn representations to the Coq development. For a formula $\varphi$ over a signature $\Sigma$, we define the list $\mathsf{FV}(\varphi) : \mathbb{L}\,\mathbb{N}$ of free variables, the list $\mathcal{F}_\varphi : \mathbb{L}\,\mathcal{F}_\Sigma$ of function symbols and the list $\mathcal{P}_\varphi : \mathbb{L}\,\mathcal{P}_\Sigma$ of relation symbols that actually occur in $\varphi$, all by recursion on $\varphi$.

Turning to *semantics*, we employ the standard (Tarski-style) model-theoretic semantics, evaluating terms in a given domain and embedding the logical connectives into the constructive meta-logic (cf. [22]):

**Definition 11.** A *model* $\mathcal{M}$ over a domain $D : \mathbb{T}$ is described by a pair of functions $\forall f.\, D^{|f|} \to D$ and $\forall P.\, D^{|P|} \to \mathbb{P}$ denoted by $f^\mathcal{M}$ and $P^\mathcal{M}$.

Given a *variable assignment* $\rho : \mathbb{N} \to D$, we recursively extend it to a *term evaluation* $\hat{\rho} : \mathsf{Term} \to D$ with $\hat{\rho}\,x := \rho x$ and $\hat{\rho}\,(f\,\boldsymbol{v}) := f^\mathcal{M}(\hat{\rho}\,\boldsymbol{v})$, and to the *satisfaction* relation $\mathcal{M} \vDash_\rho \varphi$ by

$$\mathcal{M} \vDash_\rho \dot{\perp} := \perp \qquad\qquad \mathcal{M} \vDash_\rho \varphi\,\dot{\Box}\,\psi := \mathcal{M} \vDash_\rho \varphi \,\Box\, \mathcal{M} \vDash_\rho \psi$$
$$\mathcal{M} \vDash_\rho P\,t := P^\mathcal{M}(\hat{\rho}\,t) \qquad\quad \mathcal{M} \vDash_\rho \dot{\nabla}\varphi := \nabla a : D.\,\mathcal{M} \vDash_{a \cdot \rho} \varphi$$

where each logical connective $\dot{\Box}/\dot{\nabla}$ is mapped to its meta-level counterpart $\Box/\nabla$ and where we denote by $a \cdot \rho$ the de Bruijn extension of $\rho$ by $a$, defined by $(a \cdot \rho)\,0 := a$ and $(a \cdot \rho)(1 + x) := \rho\,x$.[11]

A $\Sigma$-*model* is thus a dependent triple $(D, \mathcal{M}, \rho)$ composed of a domain $D$, a model $\mathcal{M}$ for $\Sigma$ over $D$ and an assignment $\rho : \mathbb{N} \to D$. It is *finite* if $D$ is finite, and *decidable* if $P^\mathcal{M} : D^{|P|} \to \mathbb{P}$ is decidable for all $P : \mathcal{P}_\Sigma$.

**Fact 12.** *Satisfaction* $\lambda\varphi.\,\mathcal{M} \vDash_\rho \varphi$ *is decidable for finite, decidable $\Sigma$-models.*

*Proof.* By induction on $\varphi$; finite quantification preserves decidability.      □

---

[11] The notation $a \cdot \rho$ illustrates that $a$ is pushed ahead of the sequence $\rho_0, \rho_1, \ldots$.

In this paper, we are mostly concerned with *finite satisfiability* of formulas. However, since some of the compound reductions hold for more general or more specific notions, we introduce the following variants:

**Definition 13 (Satisfiability).** For a formula $\varphi$ over a signature $\Sigma$, we write

- $\mathsf{SAT}(\Sigma)\,\varphi$ if there is a $\Sigma$-model $(D, \mathcal{M}, \rho)$ such that $\mathcal{M} \vDash_\rho \varphi$;
- $\mathsf{FSAT}(\Sigma)\,\varphi$ if additionally $D$ is finite and $\mathcal{M}$ is decidable;
- $\mathsf{FSATEQ}(\Sigma; \equiv)\,\varphi$ if the signature contains a distinguished binary relation symbol $\equiv$ interpreted as equality, i.e. $x \equiv^{\mathcal{M}} y \leftrightarrow x = y$ for all $x, y : D$.

Notice that in a classical treatment of finite model theory, models are supposed to be given *in extension*, i.e. understood as tables providing computational access to functions and relations values. To enable this view in our constructive setting, we restrict to decidable relations in the definition of FSAT, and from now on, *finite satisfiability is always meant to encompass a decidable model*. One could further require the domain $D$ to be discrete to conform more closely with the classical view; discreteness is in fact enforced by FSATEQ. However, we refrain from this requirement and instead show in Sect. 4.1 that FSAT and FSAT over discrete models are constructively equivalent.

# 3    Trakhtenbrot's Theorem for a Custom Signature

In this section, we show that BPCP reduces to $\mathsf{FSATEQ}(\Sigma_{\mathsf{BPCP}}; \equiv)$ for the special purpose signature $\Sigma_{\mathsf{BPCP}} := (\{\star^0, e^0, f_{\mathsf{tt}}^1, f_{\mathsf{ff}}^1\}; \{P^2, \prec^2, \equiv^2\})$. To this end, we fix an instance $R : \mathbb{L}(\mathbb{LB} \times \mathbb{LB})$ of BPCP (to be understood as a finite set of pairs of Boolean strings) and we construct a formula $\varphi_R$ such that $\varphi_R$ is finitely satisfiable if and only if $R$ has a solution.

Informally, we axiomatise a family $\mathcal{B}_n$ of models over the domain of Boolean strings of length bounded by $n$ and let $\varphi_R$ express that $R$ has a solution in $\mathcal{B}_n$. The axioms express enough equations and inversions of the constructions included in the definition of BPCP such that a solution for $R$ can be recovered.

Formally, the symbols in $\Sigma_{\mathsf{BPCP}}$ are used as follows: the functions $f_b$ and the constant $e$ represent $b :: (\cdot)$ and $[]$ for the encoding of strings $s$ as terms $\overline{s}$:

$$\overline{[]} \mathbin{+\!\!+\!\!+} \tau := \tau \qquad \overline{b :: s} \mathbin{+\!\!+\!\!+} \tau := f_b(\overline{s} \mathbin{+\!\!+\!\!+} \tau) \qquad \overline{s} := \overline{s} \mathbin{+\!\!+\!\!+} e$$

The constant $\star$ represents an undefined value for strings too long to be encoded in the finite model $\mathcal{B}_n$. The relation $P$ represents derivability from $R$ (denoted $R \triangleright \cdot / \cdot$ here) while $\prec$ and $\equiv$ represent strict suffixes and equality, respectively.

Expected properties of the intended interpretation can be captured formally as first-order formulas. First, we ensure that $P$ is proper (only subject to defined values) and that $\prec$ is a strict order (irreflexive and transitive):

$$\varphi_P := \dot{\forall} xy.\, P\,x\,y \dot{\rightarrow} x \not\equiv \star \dot{\wedge} y \not\equiv \star \qquad\qquad (P \text{ proper})$$
$$\varphi_\prec := (\dot{\forall} x.\, x \not\prec x) \dot{\wedge} (\dot{\forall} xyz.\, x \prec y \dot{\rightarrow} y \prec z \dot{\rightarrow} x \prec z) \qquad (\prec \text{ strict order})$$

Next, the image of $f_b$ is forced disjoint from $e$ and injective as long as $\star$ is not reached. We also ensure that the images of $f_{tt}$ and $f_{ff}$ intersect only at $\star$:

$$\varphi_f := \begin{pmatrix} f_{tt}\,\star \equiv \star \wedge f_{ff}\,\star \equiv \star \\ \dot\forall x.\, f_{tt}\,x \not\equiv e \\ \dot\forall x.\, f_{ff}\,x \not\equiv e \end{pmatrix} \wedge \begin{pmatrix} \dot\forall xy.\, f_{tt}\,x \not\equiv \star \dotrightarrow f_{tt}\,x \equiv f_{tt}\,y \dotrightarrow x \equiv y \\ \dot\forall xy.\, f_{ff}\,x \not\equiv \star \dotrightarrow f_{ff}\,x \equiv f_{ff}\,y \dotrightarrow x \equiv y \\ \dot\forall xy.\, f_{tt}\,x \equiv f_{ff}\,y \dotrightarrow f_{tt}\,x \equiv \star \wedge f_{ff}\,y \equiv \star \end{pmatrix}$$

Furthermore, we enforce that $P$ simulates $R\triangleright \cdot/\cdot$, encoding its inversion principle

$$\varphi_\triangleright := \dot\forall xy.\, P\,x\,y \dotrightarrow \bigvee_{s/t \in R} \dot\vee \begin{cases} x \equiv \overline{s} \wedge y \equiv \overline{t} \\ \dot\exists uv.\, P\,u\,v \wedge x \equiv \overline{s} \mathbin{+\!\!+} u \wedge y \equiv \overline{t} \mathbin{+\!\!+} v \wedge u/v \prec x/y \end{cases}$$

where $u/v \prec x/y$ denotes $(u \prec x \wedge v \equiv y)\dot\vee(v \prec y \wedge u \equiv x)\dot\vee(u \prec x \wedge v \prec y)$. Finally, $\varphi_R$ is the conjunction of all axioms plus the existence of a solution:

$$\varphi_R := \varphi_P \wedge \varphi_\prec \wedge \varphi_f \wedge \varphi_\triangleright \wedge \dot\exists x.\, P\,x\,x.$$

**Theorem 14.** $\mathsf{BPCP} \preceq \mathsf{FSATEQ}(\Sigma_{\mathsf{BPCP}}; \equiv)$.

*Proof.* The reduction $\lambda R.\,\varphi_R$ is proved correct by Lemmas 15 and 16. □

**Lemma 15.** $\mathsf{BPCP}\,R \to \mathsf{FSATEQ}(\Sigma_{\mathsf{BPCP}}; \equiv)\,\varphi_R$.

*Proof.* Assume $R \triangleright s/s$ holds for a string $s$ with $|s| = n$. We show that the model $\mathcal{B}_n$ over Boolean strings bounded by $n$ satisfies $\varphi_R$. To be more precise, we choose $D_n := \mathbb{O}\{s : \mathbb{LB} \mid |s| \leq n\}$ as domain, i.e. values in $D_n$ are either an (overflow) value $\emptyset$ or a (defined) dependent pair $\ulcorner(s, H_s)\urcorner$ where $H_s : |s| \leq n$. We interpret the function and relation symbols of the chosen signature by

$$e^{\mathcal{B}_n} := [] \quad f_b^{\mathcal{B}_n}\,\emptyset := \emptyset \qquad\qquad P^{\mathcal{B}_n}\,s\,t := R \triangleright s/t$$
$$\star^{\mathcal{B}_n} := \emptyset \quad f_b^{\mathcal{B}_n}\,s := \text{if } |s| < n \text{ then } b :: s \text{ else } \emptyset \quad s \prec^{\mathcal{B}_n} t := s \neq t \wedge \exists u.\, u \mathbin{+\!\!+} s = t$$

where we left out some explicit constructors and the excluded edge cases of the relations for better readability. As required, $\mathcal{B}_n$ interprets $\equiv$ by equality $=_{D_n}$.

Considering the desired properties of $\mathcal{B}_n$, first note that $D_n$ can be shown finite by induction on $n$. This however crucially relies on the proof irrelevance of the $\lambda x.\, x \leq n$ predicate.[12] The atoms $s \prec^{\mathcal{B}_n} t$ and $s \equiv^{\mathcal{B}_n} t$ are decidable by straightforward computations on Boolean strings. Decidability of $P^{\mathcal{B}_n}\,s\,t$ (i.e. $R \triangleright s/t$) was established in Fact 5. Finally, since $\varphi_R$ is a closed formula, any variable assignment $\rho$ can be chosen to establish that $\mathcal{B}_n$ satisfies $\varphi_R$, for instance $\rho := \lambda x.\emptyset$. Then showing $\mathcal{B}_n \vDash_\rho \varphi_R$ consists of verifying simple properties of the chosen functions and relations, with mostly straightforward proofs. □

**Lemma 16.** $\mathsf{FSATEQ}(\Sigma_{\mathsf{BPCP}}; \equiv)\,\varphi_R \to \mathsf{BPCP}\,R$.

---

[12] i.e. that for every $x : \mathbb{N}$ and $H, H' : x \leq n$ we have $H = H'$. In general, it is not always possible to establish finiteness of $\{x \mid P\,x\}$ if $P$ is not proof irrelevant.

*Proof.* Suppose that $\mathcal{M} \vDash_\rho \varphi_R$ holds for some finite $\Sigma_{\mathsf{BPCP}}$-model $(D, \mathcal{M}, \rho)$ interpreting $\equiv$ as equality and providing operations $f_b^\mathcal{M}$, $e^\mathcal{M}$, $\star^\mathcal{M}$, $P^\mathcal{M}$ and $\prec^\mathcal{M}$. Again, the concrete assignment $\rho$ is irrelevant and $\mathcal{M} \vDash_\rho \varphi_R$ ensures that the functions/relations behave as specified and that $P^\mathcal{M} x x$ holds for some $x : D$.

Instead of trying to show that $\mathcal{M}$ is isomorphic to some $\mathcal{B}_n$, we directly reconstruct a solution for $R$, i.e. we find some $s$ with $R \triangleright s/s$ from the assumption that $\mathcal{M} \vDash_\rho \varphi_R$ holds. To this end, we first observe that the relation $u/v \prec^\mathcal{M} x/y$ as defined above is a strict order and thus well-founded as an instance of Fact 8.

Now we can show that for all $x/y$ with $P^\mathcal{M} x y$ there are strings $s$ and $t$ with $x = \overline{s}$, $y = \overline{t}$ and $R \triangleright s/t$, by induction on the pair $x/y$ using the well-foundedness of $\prec^\mathcal{M}$. So let us assume $P^\mathcal{M} x y$. Since $\mathcal{M}$ satisfies $\varphi_\triangleright$ there are two cases:

- there is $s/t \in R$ such that $x = \overline{s}$ and $y = \overline{t}$. The claim follows by $R \triangleright s/t$;
- there are $u, v : D$ with $P^\mathcal{M} u v$ and $s/t \in R$ such that $x = \overline{s} \mathbin{+\!\!+} u$, $y = \overline{t} \mathbin{+\!\!+} v$, and $u/v \prec^\mathcal{M} x/y$. The latter makes the inductive hypothesis applicable for $P^\mathcal{M} u v$, hence yielding $R \triangleright s'/t'$ for some strings $s'$ and $t'$ corresponding to the encodings $u$ and $v$. This is enough to conclude $x = \overline{s \mathbin{+\!\!+} s'}$, $y = \overline{t \mathbin{+\!\!+} t'}$ and $R \triangleright (s \mathbin{+\!\!+} s')/(t \mathbin{+\!\!+} t')$ as wished.

Applying this fact to the assumed match $P^\mathcal{M} x x$ yields a solution $R \triangleright s/s$.    □

# 4   Constructive Finite Model Theory

Combined with Fact 5, Theorem 14 entails the undecidability (and non-co-enumerability) of FSATEQ over a custom (both finite and discrete) signature $\Sigma_{\mathsf{BPCP}}$. By a series of signature reductions, we generalise these results to any signature containing an at least binary relation symbol. In particular, we explain how to reduce $\mathsf{FSAT}(\Sigma)$ to $\mathsf{FSAT}(\mathbb{0}; \{\in^2\})$ for any discrete signature $\Sigma$, hence including $\Sigma_{\mathsf{BPCP}}$. We also provide a reduction from $\mathsf{FSAT}(\mathbb{0}; \{\in^2\})$ to $\mathsf{FSAT}(\{f^n\}; \{P^1\})$ for $n \geq 2$, which entails the undecidability of FSAT for signatures with one unary relation and an at least binary function. But first, let us show that FSAT is unaltered when further assuming discreteness of the domain.

## 4.1   Removing Model Discreteness and Interpreted Equality

We consider the case of models over a discrete domain $D$. Of course, in the case of $\mathsf{FSATEQ}(\Sigma; \equiv)$ the requirement that $\equiv$ is interpreted as a decidable binary relation which is equivalent to $=_D$ imposes the discreteness of $D$. But in the case of $\mathsf{FSAT}(\Sigma)$ nothing imposes such a restriction on $D$. However, as we argue here, we can always quotient $D$ using a suitable decidable congruence, making the quotient a discrete finite type while preserving first-order satisfaction.

**Definition 17.** We write $\mathsf{FSAT}'(\Sigma) \varphi$ if $\mathsf{FSAT}(\Sigma) \varphi$ on a discrete model.

Let us consider a fixed signature $\Sigma = (\mathcal{F}_\Sigma; \mathcal{P}_\Sigma)$. In addition, let us fix a finite type $D$ and a (decidable) model $\mathcal{M}$ of $\Sigma$ over $D$. We can conceive an equivalence over $D$ which is a congruence for all the interpretations of the symbols by $\mathcal{M}$,

namely *first-order indistinguishability* $x \doteq_\Sigma y := \forall \varphi \rho . \mathcal{M} \vDash_{x \cdot \rho} \varphi \leftrightarrow \mathcal{M} \vDash_{y \cdot \rho} \varphi$,
i.e. first-order semantics in $\mathcal{M}$ is not impacted when switching $x$ with $y$.

The facts that $\doteq_\Sigma$ is both an equivalence and a congruence are easy to prove
but, with this definition, there is little hope of establishing decidability of $\doteq_\Sigma$.
The main reason for this is that the signature may contain symbols of infinitely
many arities. So we fix two lists $l_{\mathcal{F}} : \mathbb{L}\,\mathcal{F}_\Sigma$ and $l_{\mathcal{P}} : \mathbb{L}\,\mathcal{P}_\Sigma$ of function and relation
symbols respectively and restrict the congruence requirement to these lists.

**Definition 18 (Bounded first-order indistinguishability).** We say that $x$
and $y$ are *first-order indistinguishable up to* $l_{\mathcal{F}}/l_{\mathcal{P}}$, and we write $x \doteq y$, if for
any $\rho : \mathbb{N} \to D$ and any first-order formula $\varphi$ built from the symbols in $l_{\mathcal{F}}$ and
$l_{\mathcal{P}}$ only, we have $\mathcal{M} \vDash_{x \cdot \rho} \varphi \leftrightarrow \mathcal{M} \vDash_{y \cdot \rho} \varphi$.

**Theorem 19.** *First-order indistinguishability $\doteq$ up to $l_{\mathcal{F}}/l_{\mathcal{P}}$ is a strongly decidable equivalence and a congruence for all the symbols in $l_{\mathcal{F}}/l_{\mathcal{P}}$.*

*Proof.* The proof is quite involved, we only give its sketch here; see Appendix B
of the extended version of this paper [12] for more details. The real difficulty is
to show the decidability of $\doteq$. To this end, we characterise $\doteq$ as a bisimulation,
i.e. we show that $\doteq$ is extensionally equivalent to Kleene's greatest fixpoint
$\mathrm{F}^\omega(\lambda uv.\top)$ of some $\omega$-continuous operator $\mathrm{F} : (D \to D \to \mathbb{P}) \to (D \to D \to \mathbb{P})$.
We then show that F preserves strong decidability. To be able to conclude, we
establish that F reaches its limit after $l := 2^{d \times d}$ iterations where $d := \operatorname{card} D$, the
length of a list enumerating the finite type $D$. To verify this upper bound, we
build the *weak powerset*, a list of length $l$ which contains all the weakly decidable
binary predicates of type $D \to D \to \mathbb{P}$, up to extensional equivalence. As all the
iterated values $\mathrm{F}^n(\lambda uv.\top)$ are strongly decidable, they all belong to the weak
powerset, so by Theorem 7, a duplicate is to be found in the first $l + 1$ steps,
ensuring that the sequence is stalled at $l$.                                    $\square$

We use the strongly decidable congruence $\doteq$ to quotient models onto discrete
ones (in fact $\mathbb{F}_n$ for some $n$) while preserving first-order satisfaction.

**Theorem 20.** *For every first-order signature $\Sigma$ and formula $\varphi$ over $\Sigma$, we have
$\mathsf{FSAT}(\Sigma)\,\varphi$ iff $\mathsf{FSAT}'(\Sigma)\,\varphi$, and as a consequence, both reductions $\mathsf{FSAT}(\Sigma) \preceq \mathsf{FSAT}'(\Sigma)$ and $\mathsf{FSAT}'(\Sigma) \preceq \mathsf{FSAT}(\Sigma)$ hold.*

*Proof.* $\mathsf{FSAT}(\Sigma)\,\varphi$ entails $\mathsf{FSAT}'(\Sigma)\,\varphi$ is the non-trivial implication. Hence we
consider a finite $\Sigma$-model $(D, \mathcal{M}, \rho)$ of $\varphi$ and we build a new finite $\Sigma$-model of
$\varphi$ which is furthermore discrete. We collect the symbols occurring in $\varphi$ as the
lists $l_{\mathcal{F}} := \mathcal{F}_\varphi$ (for functions) and $l_{\mathcal{P}} := \mathcal{P}_\varphi$ (for relations). By Theorem 19, first-
order indistinguishability $\doteq : D \to D \to \mathbb{P}$ up to $\mathcal{F}_\varphi/\mathcal{P}_\varphi$ is a strongly decidable
equivalence over $D$ and a congruence for the semantics of the symbols occurring
in $\varphi$. Using Theorem 9, we build the quotient $D/\doteq$ on a $\mathbb{F}_n$ for some $n : \mathbb{N}$. We
transport the model $\mathcal{M}$ along this quotient and because $\doteq$ is a congruence for
the symbols in $\varphi$, its semantics is preserved along the quotient. Hence, $\varphi$ has a
finite model over the domain $\mathbb{F}_n$ which is both finite and discrete.            $\square$

**Theorem 21.** *If $\equiv$ is a binary relation symbol in the signature $\Sigma$, one has a reduction* $\mathsf{FSATEQ}(\Sigma; \equiv) \preceq \mathsf{FSAT}(\Sigma)$.

*Proof.* Given a list $l_{\mathcal{F}}$ (resp. $l_{\mathcal{P}}$) of function (resp. relation) symbols, we construct a formula $\psi(l_{\mathcal{F}}, l_{\mathcal{P}}, \equiv)$ over the function symbols in $l_{\mathcal{F}}$ and relation symbols in $(\equiv :: l_{\mathcal{P}})$ expressing the requirement that $\equiv$ is an equivalence and a congruence for the symbols in $l_{\mathcal{F}}/l_{\mathcal{P}}$. Then we show that $\lambda\varphi.\,\varphi \wedge \psi(\mathcal{F}_\varphi, \equiv :: \mathcal{P}_\varphi, \equiv)$ is a correct reduction, where $\mathcal{F}_\varphi$ and $\mathcal{P}_\varphi$ list the symbols occurring in $\varphi$. □

## 4.2 From Discrete Signatures to Singleton Signatures

Let us start by converting a discrete signature to a finite and discrete signature.

**Lemma 22.** *For any formula $\varphi$ over a discrete signature $\Sigma$, one can compute a signature $\Sigma_{n,m} = (\mathbb{F}_n; \mathbb{F}_m)$, arity preserving maps $\mathbb{F}_n \to \mathcal{F}_\Sigma$ and $\mathbb{F}_m \to \mathcal{P}_\Sigma$ and an* equi-satisfiable *formula $\psi$ over $\Sigma_{n,m}$, i.e.* $\mathsf{FSAT}(\Sigma)\,\varphi \leftrightarrow \mathsf{FSAT}(\Sigma_{n,m})\,\psi$.

*Proof.* We use the discreteness of $\Sigma$ and bijectively map the lists of symbols $\mathcal{F}_\varphi$ and $\mathcal{P}_\varphi$ onto $\mathbb{F}_n$ and $\mathbb{F}_m$ respectively, using Corollary 10. We structurally map $\varphi$ to $\psi$ over $\Sigma_{n,m}$ along this bijection, which preserves finite satisfiability. □

Notice that $n$ and $m$ in the signature $\Sigma_{n,m}$ depend on $\varphi$, hence the above statement cannot be presented as a reduction between (fixed) signatures.

We now erase all function symbols by encoding them with relation symbols. To this end, let $\Sigma = (\mathcal{F}_\Sigma; \mathcal{P}_\Sigma)$ be a signature, we set $\Sigma' := (\mathbb{0}; \{\equiv^2\} + \mathcal{F}_\Sigma^{+1} + \mathcal{P}_\Sigma)$ where $\equiv$ is a new interpreted relation symbol of arity two and in the conversion, function symbols have arity lifted by one, hence the $\mathcal{F}_\Sigma^{+1}$ notation.

**Lemma 23.** *For any finite[13] type of function symbols $\mathcal{F}_\Sigma$, one has a reduction* $\mathsf{FSAT}'(\mathcal{F}_\Sigma; \mathcal{P}_\Sigma) \preceq \mathsf{FSATEQ}(\mathbb{0}; \{\equiv^2\} + \mathcal{F}_\Sigma^{+1} + \mathcal{P}_\Sigma; \equiv^2)$.

*Proof.* The idea is to recursively replace a term $t$ over $\Sigma$ by a formula which is "equivalent" to $x \equiv t$ (where $x$ is a fresh variable not occurring in $t$) and then an atomic formula like e.g. $P[t_1; t_2]$ by $\exists x_1\,x_2.\,x_1 \equiv t_1 \wedge x_2 \equiv t_2 \wedge P[x_1; x_2]$. We complete the encoding with a formula stating that every function symbol $f : \mathcal{F}_\Sigma$ is encoded into a total functional relation $P_f : \mathcal{F}_\Sigma^{+1}$ of arity augmented by 1. □

Next, assuming that the function symbols have already been erased, we explain how to merge the relation symbols in a signature $\Sigma = (\mathbb{0}; \mathcal{P}_\Sigma)$ into a single relation symbol, provided that there is an upper bound for the arities in $\mathcal{P}_\Sigma$.

**Lemma 24.** *The reduction* $\mathsf{FSAT}(\mathbb{0}; \mathcal{P}_\Sigma) \preceq \mathsf{FSAT}(\mathbb{0}; \{Q^{1+n}\})$ *holds when $\mathcal{P}_\Sigma$ is a finite and discrete type of relation symbols and $|P| \leq n$ holds for all $P : \mathcal{P}_\Sigma$.*

*Proof.* This comprises three independent reductions, see Fact 25 below. □

---

[13] In the Coq code, we prove the theorem for finite *or* discrete types of function symbols.

In the following, we denote by $\mathcal{F}_\Sigma^n$ (resp. $\mathcal{P}_\Sigma^n$) the same type of function (resp. relation) symbols but where the arity is uniformly converted to $n$.

**Fact 25.** *Let* $\Sigma = (\mathcal{F}_\Sigma; \mathcal{P}_\Sigma)$ *be a signature:*

1. $\mathsf{FSAT}(\mathcal{F}_\Sigma; \mathcal{P}_\Sigma) \preceq \mathsf{FSAT}(\mathcal{F}_\Sigma; \mathcal{P}_\Sigma^n)$ *if* $|P| \leq n$ *holds for all* $P : \mathcal{P}_\Sigma$;
2. $\mathsf{FSAT}(\mathbb{0}; \mathcal{P}_\Sigma^n) \preceq \mathsf{FSAT}(\mathcal{P}_\Sigma^0; \{Q^{1+n}\})$ *if* $\mathcal{P}_\Sigma$ *is finite;*
3. $\mathsf{FSAT}(\mathcal{F}_\Sigma^0; \mathcal{P}_\Sigma) \preceq \mathsf{FSAT}(\mathbb{0}; \mathcal{P}_\Sigma)$ *if* $\mathcal{F}_\Sigma$ *is discrete.*

*Proof.* For the first reduction, every atomic formula of the form $P\,\boldsymbol{v}$ with $|\boldsymbol{v}| = |P| \leq n$ is converted to $P\,\boldsymbol{v}'$ with $\boldsymbol{v}' := \boldsymbol{v} \mathbin{+\!\!+} [x_0; \ldots; x_0]$ and $|\boldsymbol{v}'| = n$ for an arbitrary term variable $x_0$. The rest of the structure of formulas is unchanged.

For the second reduction, we convert every atomic formula $P\,\boldsymbol{v}$ with $|\boldsymbol{v}| = n$ into $Q(P :: \boldsymbol{v})$ where $P$ now represents a constant symbol ($Q$ is fixed).

For the last reduction, we replace every constant symbol by a corresponding fresh variable chosen above all the free variables of the transformed formula. $\square$

## 4.3 Compressing $n$-ary Relations to Binary Membership

Let $\Sigma_n = (\mathbb{0}; \{P^n\})$ be a singleton signature where $P$ is of arity $n$. We now show that $P$ can be compressed to a binary relation modelling membership via a construction using hereditarily finite sets [18] (useful only when $n \geq 3$).

**Theorem 26.** $\mathsf{FSAT}'(\mathbb{0}; \{P^n\}) \preceq \mathsf{FSAT}(\mathbb{0}; \{\dot{\in}^2\})$.

Technically, this reduction is one of the most involved in this work, although in most presentations of Trakhtenbrot's theorem, this is left as an "easy exercise," see e.g. [14]. Maybe it is perceived so because it relies on the encoding of tuples in set theory, which is somehow natural for mathematicians,[14] but properly building the finite set model in constructive type theory was not that easy.

Here we only give an overview of the main tools. We encode an arbitrary $n$-ary relation $R : X^n \to \mathbb{P}$ over a finite type $X$ in the theory of *membership* over the signature $\Sigma_2 = (\mathbb{0}; \{\dot{\in}^2\})$. Membership is much weaker than set theory because the only required set-theoretic axiom is *extensionality*. Two sets are extensionally equal if their members are the same, and extensionality states that two extensionally equal sets belong to the same sets:

$$\dot{\forall} xy. (\dot{\forall} z.\, z \dot{\in} x \leftrightarrow z \dot{\in} y) \dot{\to} \dot{\forall} z.\, x \dot{\in} z \dot{\to} y \dot{\in} z \tag{1}$$

As a consequence, no first-order formula over $\Sigma_2$ can distinguish two extensionally equal sets. Notice that the language of membership theory (and set theory) does not contain any function symbol, hence, contrary to usual mathematical practices, there is no other way to handle a set than via its characterising formula which makes it a very cumbersome language to work with formally. However, this is how we have to proceed in the Coq development but here, we stick to meta-level "terms" in the prose for simplicity.

---

[14] In our case we use Kuratowski's encoding.

The ordered pair of two sets $p$ and $q$ is encoded as $(p,q) := \{\{p\}, \{p,q\}\}$ while the $n$-tuple $(t_1, \ldots, t_n)$ is encoded as $(t_1, (t_2, \ldots, t_n))$ recursively. The reduction function which maps formulas over $\Sigma_n$ to formulas over $\Sigma_2$ proceeds as follows. We reserve two first-order variables $d$ (for the domain $D$) and $r$ (for the relation $R$). We describe the recursive part of the reduction $\Sigma^r_{n \rightsquigarrow 2}$

$$\Sigma^r_{n \rightsquigarrow 2}(P\,\boldsymbol{v}) := \text{``tuple}\, \boldsymbol{v} \,\dot\in r\text{''} \qquad \Sigma^r_{n \rightsquigarrow 2}(\dot\forall z.\,\varphi) := \dot\forall z.\, z \,\dot\in d \,\dot\rightarrow\, \Sigma^r_{n \rightsquigarrow 2}(\varphi)$$
$$\Sigma^r_{n \rightsquigarrow 2}(\varphi \,\dot\Box\, \psi) := \Sigma^r_{n \rightsquigarrow 2}(\varphi) \,\dot\Box\, \Sigma^r_{n \rightsquigarrow 2}(\psi) \qquad \Sigma^r_{n \rightsquigarrow 2}(\dot\exists z.\,\varphi) := \dot\exists z.\, z \,\dot\in d \,\dot\wedge\, \Sigma^r_{n \rightsquigarrow 2}(\varphi)$$

ignoring the de Bruijn syntax (which would imply adding $d$ and $r$ as parameters). Notice that $d$ and $r$ should not occur freely in $\varphi$. In addition, we require that:

$$\begin{aligned}
\varphi_1 &:= \dot\in \text{ is extensional} & &\text{see Eq. 1;} \\
\varphi_2 &:= \dot\exists z.\, z \,\dot\in d & &\text{i.e. } d \text{ is non-empty;} \\
\varphi_3 &:= x_1 \,\dot\in d \,\dot\wedge\, \cdots \,\dot\wedge\, x_k \,\dot\in d & &\text{where } [x_1; \ldots; x_k] = \mathsf{FV}(\varphi).
\end{aligned}$$

This gives us the reduction function $\Sigma_{n \rightsquigarrow 2}(\varphi) := \varphi_1 \wedge \varphi_2 \wedge \varphi_3 \wedge \Sigma^r_{n \rightsquigarrow 2}(\varphi)$.

The *completeness* of the reduction $\Sigma_{n \rightsquigarrow 2}$ is the easy part. Given a finite model of $\Sigma_{n \rightsquigarrow 2}(\varphi)$ over $\Sigma_2$, we recover a model of $\varphi$ over $\Sigma_n$ by selecting as the new domain the members of $d$ and the interpretation of $P\,\boldsymbol{v}$ is given by testing whether the encoding of $\boldsymbol{v}$ as a $n$-tuple is a member of $r$.

The *soundness* of the reduction $\Sigma_{n \rightsquigarrow 2}$ is the formally involved part, with Theorem 27 below containing the key construction.

**Theorem 27.** *Given a decidable $n$-ary relation $R : X^n \to \mathbb{P}$ over a finite, discrete and inhabited type $X$, one can compute a finite and discrete type $Y$ equipped with a decidable relation $\in\, : Y \to Y \to \mathbb{P}$, two distinguished elements $d, r : Y$ and a pair of maps $i : X \to Y$ and $s : Y \to X$ s.t.*

1. *$\in$ is extensional;*
2. *extensionally equal elements of $Y$ are equal;*
3. *all $n$-tuples of members of $d$ exist in $Y$;*
4. *$\forall x : X.\, i\, x \in d$;*
5. *$\forall y : Y.\, y \in d \to \exists x.\, y = i\, x$;*
6. *$\forall x : X.\, s(i\, x) = x$;*
7. *$R\,\boldsymbol{v}$ iff $i(\boldsymbol{v})$ is a $n$-tuple member of $r$, for any $\boldsymbol{v} : X^n$.*

*Proof.* We give a brief outline of this quite involved proof, referring to the Coq code for details. The type $Y$ is built from the type of hereditarily finite sets based on [18], and when we use the word "set" below, it means hereditarily finite set. The idea is first to construct $d$ as a *transitive set* of which the elements are in bijection $i/s$ with the type $X$, hence $d$ is the cardinal of $X$ in the set-theoretic meaning. Then the iterated powersets $\mathcal{P}(d), \mathcal{P}^2(d), \ldots, \mathcal{P}^k(d)$ are all transitive as well and contain $d$ both as a member and as a subset. Considering $\mathcal{P}^{2n}(d)$ which contains all the $n$-tuples built from the members of $d$, we define $r$ as the set of $n$-tuples collecting the encodings $i(\boldsymbol{v})$ of vectors $\boldsymbol{v} : X^n$ such that $R\,\boldsymbol{v}$. We show $r \in p$ for $p$ defined as $p := \mathcal{P}^{2n+1}(d)$. Using the Boolean counterpart of $(\cdot) \in p$ for unicity of proofs, we then define $Y := \{z \mid z \in p\}$, restrict membership $\in$ to $Y$ and this gives the finite type equipped with all the required properties. Notice that the decidability requirement for $\in$ holds constructively because we work with hereditarily finite sets, and would not hold with arbitrary sets. $\qquad\square$

## 4.4   Summary: From Discrete Signatures to the Binary Signature

Combining all the previous results, we give a reduction from any discrete signature to the binary singleton signature.

**Theorem 28.** $\mathsf{FSAT}(\Sigma) \preceq \mathsf{FSAT}(\mathbb{0}; \{P^2\})$ *holds for any discrete signature* $\Sigma$.

*Proof.* Let us first consider the case of $\Sigma_{n,m} = (\mathbb{F}_n; \mathbb{F}_m)$, a signature over the finite and discrete types $\mathbb{F}_n$ and $\mathbb{F}_m$. Then we have a reduction $\mathsf{FSAT}(\mathbb{F}_n; \mathbb{F}_m) \preceq \mathsf{FSAT}(\mathbb{0}; \{P^2\})$ by combining Theorems 20, 21 and 26 and Lemmas 23 and 24.

Let us denote by $f_{n,m}$ the reduction $\mathsf{FSAT}(\mathbb{F}_n; \mathbb{F}_m) \preceq \mathsf{FSAT}(\mathbb{0}; \{P^2\})$. Let us now consider a fixed discrete signature $\Sigma$. For a formula $\varphi$ over $\Sigma$, using Lemma 22, we compute a signature $\Sigma_{n,m}$ and $\psi$ over $\Sigma_{n,m}$ s.t. $\mathsf{FSAT}(\Sigma)\,\varphi \leftrightarrow \mathsf{FSAT}(\mathbb{F}_n; \mathbb{F}_m)\,\psi$. The map $\lambda\varphi.f_{n,m}\,\psi$ is the required reduction.     □

**Lemma 29.** $\mathsf{FSAT}(\mathbb{0}; \{P^2\}) \preceq \mathsf{FSAT}(\{f^n\}; \{Q^1\})$ *when* $n \geq 2$.

*Proof.* We encode the binary relation $\lambda x\, y.\, P\,[x; y]$ with $\lambda x\, y.\, Q(f\,[x; y; \dots])$, using the first two parameters of $f$ to encode pairing. But since we need to change the domain of the model, we also use a fresh variable $d$ to encode the domain as $\lambda x.\, Q(f\,[d; x; \dots])$ and we restrict all quantifications to the domain similarly to the encoding $\Sigma_{n \rightsquigarrow 2}^r$ of Sect. 4.3.     □

We finish the reduction chains with the weakest possible signature constraints. The following reductions have straightforward proofs.

**Fact 30.** *One has reductions for the three statements below (for* $n \geq 2$*):*

1. $\mathsf{FSAT}(\mathbb{0}; \{P^2\}) \preceq \mathsf{FSAT}(\mathbb{0}; \{P^n\})$;
2. $\mathsf{FSAT}(\mathbb{0}; \{P^n\}) \preceq \mathsf{FSAT}(\Sigma)$ *if* $\Sigma$ *contains an* $n$-*ary relation symbol;*
3. $\mathsf{FSAT}(\{f^n\}; \{Q^1\}) \preceq \mathsf{FSAT}(\Sigma)$ *if* $\Sigma$ *contains an* $n$-*ary fun. and a unary rel.*

# 5   Decidability Results

Complementing the previously studied negative results, we now examine the conditions allowing for decidable satisfiability problems.

**Lemma 31 (FSAT over a fixed domain).** *Given a discrete signature* $\Sigma$ *and a discrete and finite type* $D$, *one can decide whether or not a formula over* $\Sigma$ *has a (finite) model over domain* $D$.

*Proof.* By Fact 12, satisfaction in a given finite model is decidable. It is also invariant under extensional equivalence, so we only need to show that there are finitely many (decidable) models over $D$ up to extensional equivalence.[15]     □

**Lemma 32.** *A formula over a signature* $\Sigma$ *has a finite and discrete model if and only if it has a (finite) model over* $\mathbb{F}_n$ *for some* $n : \mathbb{N}$.

---

[15] Without discreteness of $\Sigma$, it is impossible to build the list of models over $D = \mathbb{B}$.

*Proof.* If $\varphi$ has a model over a discrete and finite domain $D$, by Corollary 10, one can bijectively map $D$ to $\mathbb{F}_n$ and transport the model along this bijection. $\square$

**Lemma 33.** $\mathsf{FSAT}(\mathbb{O}; \mathcal{P}_\Sigma)$ *is decidable if* $\mathcal{P}_\Sigma$ *is discrete with uniform arity* 1.

*Proof.* By Lemma 22, we can assume $\mathcal{P}_\Sigma = \mathbb{F}_n$ w.l.o.g. We show that if $\varphi$ has a finite model then it must have a model over domain $\{v : \mathbb{B}^n \to \mathbb{B} \mid b\,v = \mathsf{tt}\}$ for some Boolean subset $b : (\mathbb{B}^n \to \mathbb{B}) \to \mathbb{B}$. Up to extensional equivalence, there are only finitely many such subsets $b$ and we conclude with Lemma 31. $\square$

**Lemma 34.** *For any finite type* $\mathcal{P}_\Sigma$ *of relation symbols and signatures of uniform arity* 1, *we have a reduction* $\mathsf{FSAT}(\mathbb{F}_n; \mathcal{P}_\Sigma) \preceq \mathsf{FSAT}(\mathbb{O}; \mathbb{L}\,\mathbb{F}_n \times \mathcal{P}_\Sigma + \mathcal{P}_\Sigma)$.

*Proof.* We implemented a proof somewhat inspired by that of Proposition 6.2.7 (Grädel) in [1, pp. 251] but the invariant suggested in the iterative process described there did not work out formally and we had to proceed in a single conversion step instead, switching from single symbols to lists of symbols. $\square$

If functions or relations have arity 0, one can always lift them to arity 1 using a fresh variable (of arbitrary value), like in Fact 25, item (1).

**Fact 35.** *The reduction* $\mathsf{FSAT}(\mathcal{F}_\Sigma; \mathcal{P}_\Sigma) \preceq \mathsf{FSAT}(\mathcal{F}_\Sigma^1; \mathcal{P}_\Sigma^1)$ *holds when all arities in* $\Sigma$ *are at most* 1, *where* $\mathcal{F}_\Sigma^1$ *and* $\mathcal{P}_\Sigma^1$ *denote arities uniformly updated to* 1.

# 6   Signature Classification

We conclude with the exact classification of FSAT regarding enumerability, decidability, and undecidability depending on the properties of the signature.

**Theorem 36.** *Given* $\Sigma = (\mathcal{F}_\Sigma; \mathcal{P}_\Sigma)$ *where both* $\mathcal{F}_\Sigma$ *and* $\mathcal{P}_\Sigma$ *are data types, the finite satisfiability problem for formulas over* $\Sigma$ *is enumerable.*

*Proof.* Using Theorem 20 and Lemmas 31 and 32, one constructs a predicate $Q : \mathbb{N} \to \mathsf{Form}_\Sigma \to \mathbb{B}$ s.t. $\mathsf{FSAT}(\Sigma)\,\varphi \leftrightarrow \exists n.\,Q\,n\,\varphi = \mathsf{tt}$. Then, it is easy to build a computable enumeration $e : \mathbb{N} \to \mathbb{O}\,\mathsf{Form}_\Sigma$ of $\mathsf{FSAT}(\Sigma) : \mathsf{Form}_\Sigma \to \mathbb{P}$. $\square$

**Theorem 37.** $\mathsf{FSAT}(\Sigma)$ *is decidable if* $\Sigma$ *is discrete with arities less or equal than* 1, *or if all relation symbols have arity* 0.

*Proof.* If all arities are at most 1, then by Fact 35, we can assume $\Sigma$ of uniform arity 1. Therefore, for a formula $\varphi$ over $\Sigma$ with uniform arity 1, we need to decide FSAT for $\varphi$. By Theorem 22, we can compute a signature $\Sigma_{n,m} = (\mathbb{F}_n; \mathbb{F}_m)$ and a formula $\psi$ over $\Sigma_{n,m}$ equi-satisfiable with $\varphi$. Using the reduction of Lemma 34, we compute a formula $\gamma$, equi-satisfiable with $\psi$, over a discrete signature of uniform arity 1, void of functions. We decide the satisfiability of $\gamma$ by Lemma 33.

If all relation symbols have arity 0, regardless of $\mathcal{F}_\Sigma$, no term can occur in formulas, hence neither can function symbols. Starting from $\varphi$ over $\Sigma = (\mathcal{F}_\Sigma; \mathcal{P}_\Sigma^0)$ where only $\mathcal{P}_\Sigma$ is assumed discrete, we compute an equi-satisfiable formula $\psi$ over $\Sigma' = (\mathbb{O}; \mathcal{P}_\Sigma^0)$ and we are back to the previous case. $\square$

**Theorem 38 (Full Trakhtenbrot).** *If $\Sigma$ contains either an at least binary relation symbol or a unary relation symbol together with an at least binary function symbol, then* BPCP *reduces to* FSAT$(\Sigma)$.

*Proof.* By Theorems 14, 21 and 28, Lemma 29, and Fact 30.    □

**Corollary 39.** *For an enumerable and discrete signature $\Sigma$ furthermore satisfying the conditions in Theorem 38,* FSAT$(\Sigma)$ *is both enumerable and undecidable, thus, more specifically, not co-enumerable.*

*Proof.* Follows by Facts 3 and 5.    □

Notice that even if the conditions on arities of Theorems 37 and 38 fully classify discrete signatures, it is not possible to decide which case holds unless the signature is furthermore finite. For a given formula $\varphi$ though, it is always possible to render it in the finite signature of used symbols.

## 7   Discussion

The main part of our Coq development directly concerned with the classification of finite satisfiability consists of 10k loc, in addition to 3k loc of (partly reused) utility libraries. Most of the code comprises the signature transformations with more than 4k loc for reducing discrete signatures to membership. Comparatively, the initial reduction from BPCP to FSATEQ$(\Sigma_{\mathsf{BPCP}})$ takes less than 500 loc.

Our mechanisation of first-order logic in principle follows previous developments [8,9] but also differs in a few aspects. Notably, we had to separate function from relation signatures to be able to express distinct signatures that agree on one sort of symbols computationally. Moreover, we found it favourable to abstract over the logical connectives in form of $\dot{\Box}$ and $\dot{\nabla}$ to shorten purely structural definitions and proofs. Finally, we did not use the Autosubst 2 [20] support for de Bruijn syntax to avoid its current dependency on the functional extensionality axiom.

We refrained from additional axioms since we included our development in the growing Coq library of synthetic undecidability proofs [11]. In this context, we plan to generalise some of the intermediate signature reductions so that they become reusable for other undecidability proofs concerning first-order logic over arbitrary models.

As further future directions, we want to explore and mechanise the direct consequences of Trakhtenbrot's theorem such as the undecidability of query containment and equivalence in data base theory or the undecidability of separation logic [3,5]. Also possible, though rather ambitious, would be to mechanise the classification of first-order satisfiability with regards to the quantifier prefix as comprehensively developed in [1]. Finally, we plan to mechanise the undecidability of semantic entailment and syntactic deduction in first-order axiom systems such as ZF set theory and Peano arithmetic.

**Funding.** The work of the second author was partially supported by the TICAMORE project (ANR grant 16-CE91-0002).

# References

1. Börger, E., Grädel, E., Gurevich, Y.: The Classical Decision Problem. Perspectives in Mathematical Logic. Springer, Heidelberg (1997)
2. Braibant, T., Pous, D.: An efficient Coq Tactic for deciding kleene algebras. In: Kaufmann, M., Paulson, L.C. (eds.) ITP 2010. LNCS, vol. 6172, pp. 163–178. Springer, Heidelberg (2010). https://doi.org/10.1007/978-3-642-14052-5_13
3. Brochenin, R., Demri, S., Lozes, E.: On the almighty wand. Inf. Comput. **211**, 106–137 (2012)
4. de Bruijn, N.G.: Lambda calculus notation with nameless dummies, a tool for automatic formula manipulation, with application to the Church-Rosser theorem. Indagationes Mathematicae (Proceedings) **75**(5), 381–392 (1972)
5. Calcagno, C., Yang, H., O'Hearn, P.W.: Computability and complexity results for a spatial assertion language for data structures. In: Hariharan, Ramesh, Vinay, V., Mukund, Madhavan (eds.) FSTTCS 2001. LNCS, vol. 2245, pp. 108–119. Springer, Heidelberg (2001). https://doi.org/10.1007/3-540-45294-X_10
6. Cohen, C.: Pragmatic quotient types in Coq. In: Blazy, S., Paulin-Mohring, C., Pichardie, D. (eds.) Interactive Theorem Proving, pp. 213–228. Springer, Berlin Heidelberg, Berlin, Heidelberg (2013)
7. Forster, Y., Heiter, E., Smolka, G.: Verification of PCP-related computational reductions in Coq. In: Avigad, J., Mahboubi, A. (eds.) ITP 2018. LNCS, vol. 10895, pp. 253–269. Springer, Cham (2018). https://doi.org/10.1007/978-3-319-94821-8_15
8. Forster, Y., Kirst, D., Smolka, G.: On synthetic undecidability in Coq, with an application to the Entscheidungs problem. In: International Conference on Certified Programs and Proofs, pp. 38–51. ACM (2019)
9. Forster, Y., Kirst, D., Wehr, D.: Completeness theorems for first-order logic analysed in constructive type theory. In: Symposium on Logical Foundations Of Computer Science, 2020, Deerfield Beach, Florida, U.S.A. January 2020
10. Forster, Y., Larchey-Wendling, D.: Certified undecidability of intuitionistic linear logic via binary stack machines and minsky machines. In: Proceedings of the 8th ACM SIGPLAN International Conference on Certified Programs and Proofs, pp. 104–117. ACM (2019)
11. Forster, Y., Larchey-Wendling, D., Dudenhefner, A., Heiter, E., Kirst, D., Kunze, F., Smolka, G., Spies, S., Wehr, D., Wuttke, M.: A Coq library of undecidable problems. In: CoqPL 2020. New Orleans, LA, United States (2020). https://github.com/uds-psl/coq-library-undecidability
12. Kirst, D., Larchey-Wendling, D.: Trakhtenbrot's Theorem in Coq, A Constructive Approach to Finite Model Theory (2020). https://arxiv.org/abs/2004.07390
13. Larchey-Wendling, D., Forster, Y.: Hilbert's tenth problem in Coq. In: 4th International Conference on Formal Structures for Computation and Deduction. LIPIcs, vol. 131, pp. 27:1–27:20, February 2019
14. Libkin, L.: Elements of Finite Model Theory, 1st edn. Springer, Heidelberg (2010)
15. Löwenheim, L.: Über Möglichkeiten im Relativkalkül. Mathematische Annalen **76**, 447–470 (1915). http://eudml.org/doc/158703
16. Maksimović, P., Schmitt, A.: HOCore in Coq. In: Urban, C., Zhang, X. (eds.) ITP 2015. LNCS, vol. 9236, pp. 278–293. Springer, Cham (2015). https://doi.org/10.1007/978-3-319-22102-1_19
17. Schäfer, S., Smolka, G., Tebbi, T.: Completeness and decidability of de Bruijn substitution algebra in Coq. In: Proceedings of the 2015 Conference on Certified Programs and Proofs, pp. 67–73. ACM (2015)

18. Smolka, G., Stark, K.: Hereditarily finite sets in constructive type theory. In: Blanchette, J.C., Merz, S. (eds.) ITP 2016. LNCS, vol. 9807, pp. 374–390. Springer, Cham (2016). https://doi.org/10.1007/978-3-319-43144-4_23

19. Spies, S., Forster, Y.: Undecidability of higher-order unification formalised in Coq. In: International Conference on Certified Programs and Proofs, CPP 2020, New Orleans, USA, January 2020

20. Stark, K., Schäfer, S., Kaiser, J.: Autosubst 2: reasoning with multi-sorted de Bruijn terms and vector substitutions. In: International Conference on Certified Programs and Proofs, pp. 166–180. ACM (2019)

21. Trakhtenbrot, B.A.: The impossibility of an algorithm for the decidability problem on finite classes. Dokl. Akad. Nok. SSSR **70**(4), 569–572 (1950)

22. Veldman, W., Waaldijk, F.: Some elementary results in intutionistic model theory. J. Symbol. Logic **61**(3), 745–767 (1996)

23. Werner, B.: Sets in types, types in sets. In: Abadi, M., Ito, T. (eds.) TACS 1997. LNCS, vol. 1281, pp. 530–546. Springer, Heidelberg (1997). https://doi.org/10.1007/BFb0014566

# Deep Generation of Coq Lemma Names
# Using Elaborated Terms

Pengyu Nie[1]([✉]), Karl Palmskog[2], Junyi Jessy Li[1], and Milos Gligoric[1]

[1] The University of Texas at Austin, Austin, TX, USA
{pynie,gligoric}@utexas.edu, jessy@austin.utexas.edu
[2] KTH Royal Institute of Technology, Stockholm, Sweden
palmskog@kth.se

**Abstract.** Coding conventions for naming, spacing, and other essentially stylistic properties are necessary for developers to effectively understand, review, and modify source code in large software projects. Consistent conventions in verification projects based on proof assistants, such as Coq, increase in importance as projects grow in size and scope. While conventions can be documented and enforced manually at high cost, emerging approaches automatically learn and suggest idiomatic names in Java-like languages by applying statistical language models on large code corpora. However, due to its powerful language extension facilities and fusion of type checking and computation, Coq is a challenging target for automated learning techniques. We present novel generation models for learning and suggesting lemma names for Coq projects. Our models, based on multi-input neural networks, are the first to leverage syntactic and semantic information from Coq's lexer (tokens in lemma statements), parser (syntax trees), and kernel (elaborated terms) for naming; the key insight is that learning from elaborated terms can substantially boost model performance. We implemented our models in a toolchain, dubbed ROOSTERIZE, and applied it on a large corpus of code derived from the Mathematical Components family of projects, known for its stringent coding conventions. Our results show that ROOSTERIZE substantially outperforms baselines for suggesting lemma names, highlighting the importance of using multi-input models and elaborated terms.

**Keywords:** Proof assistants · Coq · Lemma names · Neural networks

## 1 Introduction

Programming language source code with deficient coding conventions, such as misleading function and variable names and irregular spacing, is difficult for developers to effectively understand, review, and modify [8,52,67]. Code with haphazard adherence to conventions may also be more bug-prone [17]. The problem is exacerbated in large projects with many developers, where different source code files and components may have inconsistent and clashing conventions.

© Springer Nature Switzerland AG 2020
N. Peltier and V. Sofronie-Stokkermans (Eds.): IJCAR 2020, LNAI 12167, pp. 97–118, 2020.
https://doi.org/10.1007/978-3-030-51054-1_6

Many open source software projects manually document coding conventions that contributors are expected to follow, and maintainers willingly accept fixes of violations to such conventions [2]. Enforcement of conventions can be performed by static analysis tools [30, 59]. However, such tools require developers to write precise checks for conventions, which are tedious to define and often *incomplete*. To address this problem, researchers have proposed techniques for automatically learning coding conventions for Java-like languages from code corpora by applying statistical language models [4]. These models are applicable because code in these languages has high *naturalness* [35], i.e., statistical regularities and repetitiveness. Learned conventions can then be used to, e.g., suggest names in code.

Proof assistants, such as Coq [15], are increasingly used to formalize results in advanced mathematics [28, 29] and develop large trustworthy software systems, e.g., compilers, operating systems, file systems, and distributed systems [18, 44, 73]. Such projects typically involve contributions of many participants over several years, and require considerable effort to maintain over time. Coding conventions are essential for evolution of large verification projects, and are thus highly emphasized in the Coq libraries HoTT [37] and Iris [39], in Lean's Mathlib [9], and in particular in the influential Mathematical Components (MathComp) *family of Coq projects* [19]. Extensive changes to adhere to conventions, e.g., on naming, are regularly requested by MathComp maintainers for proposed external contributions [50], and its conventions have been adopted, to varying degrees, by a growing number of independent Coq projects [1, 13, 24, 66].

We believe these properties make Coq code related to MathComp an attractive target for automated learning and suggesting of coding conventions, in particular, for suggesting *lemma names* [7]. However, serious challenges are posed by, on the one hand, Coq's powerful language extension facilities and fusion of type checking and computation [12], and on the other hand, the idiosyncratic conventions used by Coq practitioners compared to software engineers. Hence, although suggesting lemma names is similar in spirit to suggesting method names in Java-like languages [74], the former task is more challenging in that lemma names are typically much shorter than method names and tend to include heavily abbreviated terminology from logic and advanced mathematics; a single character can carry significant information about a lemma's meaning. For example, the MathComp lemma names `card_support_normedTI` ("cardinality of support groups of a normed trivial intersection group") and `extprod_mulgA` ("associativity of multiplication operations in external product groups") concisely convey information on lemma statement structure and meaning through both abbreviations and suffixes, as when the suffix `A` indicates an associative property.

In this paper, we present novel generation models for learning and suggesting lemma names for Coq verification projects that address these challenges. Specifically, based on our knowledge of Coq and its implementation, we developed multi-input encoder-decoder neural networks for generating names that use information directly from Coq's internal data structures related to lexing, parsing, and type checking. In the context of naming, our models are the first to leverage the *lemma lemma statement* as well as the corresponding *syntax tree* and *elaborated term* (which we call *kernel tree*) processed by Coq's kernel [53].

We implemented our models in a toolchain, dubbed ROOSTERIZE, which we used to learn from a high-quality Coq corpus derived from the MathComp family. We then measured the performance of ROOSTERIZE using automatic metrics, finding that it significantly outperforms baselines. Using our best model, we also suggested lemma names for the PCM library [56, 66], which were manually reviewed by the project maintainer with encouraging results.

To allow ROOSTERIZE to use information directly from Coq's lexer, parser, and kernel, we extended the SerAPI library [26] to serialize Coq tokens, syntax trees, and kernel trees into a machine-readable format. This allowed us to achieve robustness against user-defined notations and other extensions to Coq syntax. Thanks to our integration with SerAPI and its use of metaprogramming, we expect our toolchain to only require modest maintenance as Coq evolves.

We make the following key contributions in this work:

- **Models**: We propose novel generation models based on multi-input neural networks to learn and suggest lemma names for Coq verification projects. A key property of our models is that they combine data from several Coq phases, including lexing, parsing, and term elaboration.

- **Corpus**: Advised by MathComp developers, we constructed a corpus of high-quality Coq code for learning coding conventions, totaling over 164k LOC taken from four core projects. We believe that our corpus can enable development of many novel techniques for Coq based on statistical language models.

- **Toolchain**: We implemented a toolchain, dubbed ROOSTERIZE, which suggests lemma names for a given Coq project. We envision ROOSTERIZE being useful during the review process of proposed contributions to a Coq project.

- **Evaluation**: We performed several experiments with ROOSTERIZE to evaluate our models using our corpus. Our results show that ROOSTERIZE performs significantly better than several strong baselines, as measured by standard automatic metrics [60]. The results also reveal that our novel multi-input models, as well as the incorporation of kernel trees, are important for suggestion quality. Finally, we performed a manual quality analysis by suggesting lemma names for a medium sized Coq project [56], evaluated by its maintainer, who found many of the suggestions useful for improving naming consistency.

The appendix of the extended version of the paper [57] describes more experiments, including an automatic evaluation on additional Coq projects. We provide artifacts related to our toolchain and corpus at: https://github.com/EngineeringSoftware/roosterize.

## 2  Background

This section gives brief background related to Coq and the Mathematical Components (MathComp) family of projects, as well as the SerAPI library.

```
1   Lemma mg_eq_proof L1 L2 (N1 : mgClassifier L1) : L1 =i L2 -> nerode L2 N1.
2   Proof. move => H0 u v. split => [/nerodeP H1 w|H1].
3     - by rewrite -!H0.
4     - apply/nerodeP => w. by rewrite !H0.
5   Qed.
```

**Fig. 1.** Coq lemma on the theory of regular languages, including proof script.

**Coq and Gallina:** Coq is a proof assistant based on dependent types, implemented in the OCaml language [15,20]. For our purposes, we view Coq as a programming language and type-checking toolchain. Specifically, Coq *files* are sequences of *sentences*, with each sentence ending with a period. Sentences are essentially either (a) commands for printing and other output, (b) definitions of functions, lemmas, and datatypes in the Gallina language [21], or (c) expressions in the Ltac tactic language [22]. We will refer to definitions of lemmas as in (b) as *lemma sentences*. Coq internally represents a lemma sentence both as a sequence of tokens (lexing phase) and as a syntax tree (parsing phase).

In the typical workflow for a Coq-based verification project, users write datatypes and functions and then interactively prove lemmas about them by executing different tactic expressions that may, e.g., discharge or split the current proof goal. Both statements to be proved and proofs are represented internally as *terms* produced during an *elaboration* phase [53]; we refer to elaborated terms as *kernel trees*. Hence, as tactics are successfully executed, they gradually build a kernel tree. The `Qed` command sends the kernel tree for a tentative proof to Coq's kernel for final certification. We call a collection of Ltac tactic sentences that build a kernel tree a *proof script*.

Figure 1 shows a Coq lemma and its proof script, taken verbatim from a development on the theory of regular languages [24]. Line 1 contains a lemma sentence with the lemma name `mg_eq_proof`, followed by a *lemma statement* (on the same line) involving the arbitrary languages L1 and L2, i.e., typed variables that are implicitly universally quantified. When Coq processes line 5, the kernel certifies that the kernel tree generated by the proof script (lines 2 to 4) has the type (is a proof) of the kernel tree for the lemma statement on line 1.

**MathComp and Lemma Naming:** The MathComp family of Coq projects, including in particular the MathComp library of general mathematical definitions and results [49], grew out of Gonthier's proof of the four-color theorem [28], with substantial developments in the context of the landmark proof of the odd order theorem in abstract algebra [29]. The MathComp library is now used in many projects outside of the MathComp family, such as in the project containing the lemma in Fig. 1 [23]. MathComp has documented naming conventions for two kinds of entities: (1) variables and (2) functions and lemmas [19]. Variable names tend to be short and simple, while function and lemma names can be long and consist of several *name components*, typically separated by an underscore, but sometimes using CamelCase. Examples of definition and lemma names in Fig. 1 include `mg_eq_proof`, `mgClassifier`, `nerode`, and `nerodeP`. Note that lemma

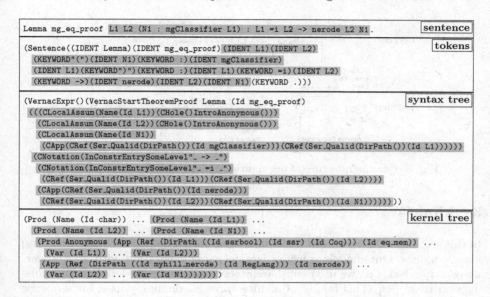

```
Lemma mg_eq_proof L1 L2 (N1 : mgClassifier L1) : L1 =i L2 -> nerode L2 N1.     sentence

(Sentence((IDENT Lemma)(IDENT mg_eq_proof)(IDENT L1)(IDENT L2)               tokens
  (KEYWORD"(")(IDENT N1)(KEYWORD :)(IDENT mgClassifier)
  (IDENT L1)(KEYWORD")")(KEYWORD :)(IDENT L1)(KEYWORD =i)(IDENT L2)
  (KEYWORD ->)(IDENT nerode)(IDENT L2)(IDENT N1)(KEYWORD .)))

(VernacExpr()(VernacStartTheoremProof Lemma (Id mg_eq_proof)               syntax tree
  ((((CLocalAssum(Name(Id L1))(CHole()IntroAnonymous()))
    (CLocalAssum(Name(Id L2))(CHole()IntroAnonymous()))
    (CLocalAssum(Name(Id N1))
      (CApp(CRef(Ser_Qualid(DirPath())(Id mgClassifier)))(CRef(Ser_Qualid(DirPath())(Id L1))))))
    (CNotation(InConstrEntrySomeLevel"_ -> _")
    (CNotation(InConstrEntrySomeLevel"_ =i _")
    (CRef(Ser_Qualid(DirPath())(Id L1)))(CRef(Ser_Qualid(DirPath())(Id L2))))
    (CApp(CRef(Ser_Qualid(DirPath())(Id nerode)))
    (CRef(Ser_Qualid(DirPath())(Id L2)))(CRef(Ser_Qualid(DirPath())(Id N1)))))))))

(Prod (Name (Id char)) ... (Prod (Name (Id L1)) ...                        kernel tree
  (Prod (Name (Id L2)) ... (Prod (Name (Id N1)) ...
    (Prod Anonymous (App (Ref (DirPath ((Id ssrbool) (Id ssr) (Id Coq))) (Id eq_mem)) ...
      (Var (Id L1)) ... (Var (Id L2)))
    (App (Ref (DirPath ((Id myhill_nerode) (Id RegLang))) (Id nerode)) ...
      (Var (Id L2)) ... (Var (Id N1)))))))))
```

**Fig. 2.** Coq lemma sentence at the top, with sexps for, from just below to bottom: tokens, syntax tree, and kernel tree; the lemma statement in each is highlighted.

names sometimes have *suffixes* to indicate their meaning, such as P in `nerodeP` which says that the lemma is a *characteristic property*. Coq functions tend to be named based on corresponding function definition bodies rather than just types (of the parameters and return value), analogously to methods in Java [47]. In contrast, MathComp lemma names tend to be based solely on the lemma statement. Hence, a more suitable name for the lemma in Fig. 1 is `mg_eq_nerode`.

**SerAPI and Coq Serialization:** SerAPI is an OCaml library and toolchain for machine interaction with Coq [26], which provides serialization and deserialization of Coq internal data structures to and from S-expressions (sexps) [51]. SerAPI is implemented using OCaml's PPX metaprogramming facilities [58], which enable modifying OCaml program syntax trees at compilation time. Figure 2 shows the lemma sentence on line 1 in Fig. 1, and below it, the corresponding (simplified) sexps for its tokens, syntax tree, and kernel tree, with the lemma statement highlighted in each representation. Note that the syntax tree omits the types of some quantified variables, e.g., for the types of L1 and L2, as indicated by the `CHole` constructor. Note also that during elaboration of the syntax tree into the kernel tree by Coq, an implicit variable `char` is added (all-quantified via `Prod`), and the extensional equality operator =i is translated to its globally unique *kernel name*, `Coq.ssr.ssrbool.eq_mem`. Hence, a kernel tree can be much larger and contain more information than the corresponding syntax tree.

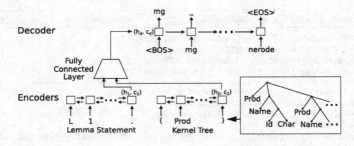

**Fig. 3.** Core architecture of our multi-input encoder-decoder models.

## 3    Models

In this section, we describe our multi-input generation models for suggesting Coq lemma names. Our models consider lemma name generation with an *encoder-decoder* mindset, i.e., we use neural architectures specifically designed for transduction tasks [68]. This family of architectures is commonly used for sequence generation, e.g., in machine translation [11] and code summarization [43], where it has been found to be much more effective than traditional probabilistic and retrieval-based approaches.

### 3.1    Core Architecture

Our encoders are Recurrent Neural Networks (RNNs) that learn a deep semantic representation of a given lemma statement from its tokens, syntax tree, and kernel tree. The decoder—another RNN—generates the descriptive lemma name as a sequence. The model is trained end-to-end, maximizing the probability of the generated lemma name given the input. In contrast to prior work in language-code tasks that uses a single encoder [27], we design multi-input models that leverage both syntactic and semantic information from Coq's lexer, parser, and kernel. A high-level visualization of our architecture is shown in Fig. 3.

**Encoding:** Our multi-input encoders combine different kinds of syntactic and semantic information in the encoding phase. We use a different encoder for each input, which are: lemma statement, syntax tree, and kernel tree.

Coq data structure instances can be large, with syntax trees having an average depth of 28.03 and kernel trees 46.51 in our corpus (we provide detailed statistics in Sect. 4). Therefore, we flatten the trees into sequences, which can be trained more efficiently than tree encoders without performance loss [38]. We flatten the trees with pre-order traversal, and we use "(" and ")" as the boundaries of the children of a node. In later parts of this paper, we use syntax and kernel trees to refer to their flattened versions. In Sect. 3.2, we introduce *tree chopping* to reduce the length of the resulting sequences.

To encode lemma statements and flattened tree sequences, we use bi-directional Long-Short Term Memory (LSTM) [36] networks. LSTMs are advanced

```
(Prod Anonymous (App (Ref (DirPath ((Id ssrbool) (Id ssr) (Id Coq))) (Id eq_mem)) ...
  ((App (Ref ... ))) ... ))
```

```
(Prod Anonymous (App eq_mem ... (App (Ref ... )) ... ))
```

**Fig. 4.** Kernel tree sexp before and after chopping; chopped parts are highlighted.

RNNs good at capturing long-range dependencies in a sequence, and are widely used in encoders [38]. A bi-directional LSTM learns stronger representations (than a uni-directional LSTM) by encoding a sequence from both left to right and right to left [75].

**Decoding:** We use an LSTM (left to right direction only) as our decoder. To obtain the initial hidden and cell states $(h_d, c_d)$ of the decoder, we learn a unified representation of these separate encoders by concatenating their final hidden and cell states $(h_i, c_i)$, and then applying a fully connected layer on the concatenated states: $h_d = W_h \cdot \texttt{concat}([h_i]) + b_h$ and $c_d = W_c \cdot \texttt{concat}([c_i]) + b_c$, where $W_h$, $W_c$, $b_h$, and $b_c$ are learnable parameters.

During training, we maximize the log likelihood of the reference lemma name given all input sequences. Standard beam search is used to reduce the search space for the optimal sequence of tokens. With regular decoding, at each time step the decoder generates a new token relying on the preceding *generated* token, which can be error-prone and leads to slow convergence and instability. We mitigate this problem by performing decoding with teacher forcing [72] such that the decoder relies on the preceding *reference* token. At test time, the decoder still uses the proceeding generated token as input.

**Attention:** With RNN encoders, the input sequence is compressed into the RNN's final hidden states, which results in a loss of information, especially for longer sequences. The attention mechanism [48] grants the decoder access to the encoder hidden and cell states for all previous tokens. At each decoder time step, an attention vector is calculated as a distribution over all encoded tokens, indicating which token the decoder should "pay attention to". To make the attention mechanism work with multiple encoders, we concatenate the hidden states of the $n$ encoders $[h_1, ..., h_n]$ and apply an attention layer on the result [70].

**Initialization:** Since there are no pre-trained token embeddings for Coq, we initialize each unique token in the vocabulary with a random vector sampled from the uniform distribution $U(-0.1, 0.1)$. These embeddings are trained together with the model. The hidden layer parameters of the encoders and decoders are also initialized with random vectors sampled from the same uniform distribution.

## 3.2 Tree Chopping

While syntax and kernel trees for lemma statements can be large, not all parts of the trees are relevant for naming. For instance, each constant reference is expanded to its fully qualified form in the kernel tree, but the added prefixes are usually related to directory paths and likely do not contain relevant information for generating the name of the lemma. Irrelevant information in long sequences can be detrimental to the model, since the model would have to reason about and encode all tokens in the sequence.

To this end, we implemented *chopping* heuristics for both syntax trees and kernel trees to remove irrelevant parts. The heuristics essentially: (1) replace the fully qualified name sub-trees with only the last component of the name; (2) remove the location information from sub-trees; (3) extract the singletons, i.e., non-leaf nodes that have only one child. Figure 4 illustrates the chopping of a kernel tree, with the upper box showing the tree before chopping with the parts to be removed highlighted, and the lower box showing the tree after chopping. In the example in the figure, we chopped a fully qualified name and extracted a singleton. These heuristics greatly reduce the size of the tree: for kernel trees, they reduce the average depth from 39.20 to 11.39.

Our models use chopped trees as the inputs to the encoders. As we discuss in more detail in Sect. 6, the chopped trees help the models to focus better on the relevant parts of the inputs. While the attention mechanism in principle could learn what the relevant parts of the trees are, our evaluation shows that it can easily be overwhelmed by large amounts of irrelevant information.

## 3.3 Copy Mechanism

We found it common for lemma name tokens to only occur in a single Coq file, whence they are unlikely to appear in the vocabulary learned from the training set, but can still appear in the respective lemma statement, syntax tree, or kernel tree. For example, mg occurs in both the lemma name and lemma statement in Fig. 1, but not outside the file the lemma is in. To account for this, we adopt the copy mechanism [64] which improves the generalizability of our model by allowing the decoder to *copy* from inputs rather than always choosing one word from the fixed vocabulary from the training set. To handle multiple encoders, similar to what we did with the attention layer, we concatenate the hidden states of each encoder and apply a copy layer on the concatenated hidden states.

## 3.4 Sub-tokenization

We sub-tokenize all inputs (lemma statements, syntax and kernel trees) and outputs (lemma names) in a pre-processing step. Previous work on learning from software projects has shown that sub-tokenization helps to reduce the sparsity of the vocabulary and improves the performance of the model [10]. However, unlike Java-like languages where the method names (almost) always follow the CamelCase convention, lemma names in Coq use a mix of snake_case, Camel-Case, prefixes, and suffixes, thus making sub-tokenization more complex. For example, extprod_mulgA should be sub-tokenized to extprod, _, mul, g, and A.

To perform sub-tokenization, we implemented a set of heuristics based on the conventions outlined by MathComp developers [19]. After sub-tokenization, the vocabulary size of lemma names in our corpus was reduced from 8,861 to

**Table 1.** Projects from the MathComp family used in our corpus.

| Project | | SHA | #Files | #Lemmas | #Toks | LOC | | LOC/file | |
|---|---|---|---|---|---|---|---|---|---|
| | | | | | | Spec. | Proof | Spec. | Proof |
| finmap | ↻ | 27642a8 | 4 | 940 | 78,449 | 4,260 | 2,191 | 1,065.00 | 547.75 |
| fourcolor | ↻ | 0851d49 | 60 | 1,157 | 560,682 | 9,175 | 27,963 | 152.92 | 466.05 |
| math-comp | ↻ | 748d716 | 89 | 8,802 | 1,076,096 | 38,243 | 46,470 | 429.70 | 522.13 |
| odd-order | ↻ | ca602a4 | 34 | 367 | 519,855 | 11,882 | 24,243 | 349.47 | 713.03 |
| Avg. | | N/A | 46.75 | 2,816.50 | 558,770.50 | 15,890.00 | 25,216.75 | 339.89 | 539.40 |
| Σ | | N/A | 187 | 11,266 | 2,235,082 | 63,560 | 100,867 | 63,560 | 100,867 |

**Table 2.** Statistics on the lemmas in the training, validation, and testing sets.

| | #Files | #Lemmas | Name | | Stmt | |
|---|---|---|---|---|---|---|
| | | | #Char | #SubToks | #Char | #SubToks |
| training | 152 | 8,861 | 10.14 | 4.22 | 44.16 | 19.59 |
| validation | 18 | 1,085 | 9.20 | 4.20 | 38.28 | 17.30 |
| testing | 17 | 1,320 | 9.76 | 4.34 | 48.49 | 23.20 |

2,328. When applying the sub-tokenizer on the lemma statements and syntax and kernel trees, we sub-tokenize the identifiers and not the keywords or operators.

### 3.5 Repetition Prevention

We observed that decoders often generated repeated tokens, e.g., mem_mem_mem. This issue also exists in natural language summarization [69]. We further observed that it is very unlikely to have repeated sub-tokens in lemma names used by proof engineers (only 1.37% of cases in our corpus). Hence, we simply forbid the decoder from repeating a sub-token (modulo "_") during beam search.

## 4 Corpus

We constructed a corpus of four large Coq projects from the MathComp family, totaling 164k lines of code (LOC). We selected these projects based on the recommendation of MathComp developers, who emphasized their high quality and stringent adherence to coding conventions. Our corpus is *self-contained*: there are inter-project dependencies within the corpus, but no project depends on a project outside the corpus (except Coq's standard library). All projects build with Coq version 8.10.2. Note that we need to be able to build projects to be able to extract tokens, syntax trees, and kernel trees.

**Constituent Projects:** Table 1 lists the projects in the corpus, along with basic information about each project. The table includes columns for the project identifier, revision SHA, number of files (#Files), number of lemmas (#Lemmas), number of tokens (#Toks), LOC for specifications (Spec.) and proof scripts

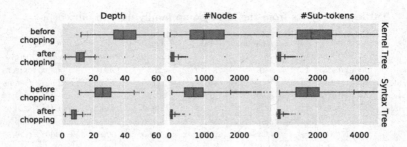

**Fig. 5.** Statistics on syntax and kernel trees.

(Proof), and average LOC per file for specifications and proof scripts. The math-comp SHA corresponds to version 1.9.0 of the library. The LOC numbers are computed with Coq's bundled `coqwc` tool. The last two rows of the table show the averages and sums across all projects.

**Corpus Statistics:** We extracted all lemmas from the corpus, and initially we obtained 15,005 lemmas in total. However, we found several outlier lemmas where the lemma statement, syntax tree and kernel tree were very large. To ensure stable training, and similar to prior work on generating method names for Java [47], we excluded the lemmas with the deepest 25% kernel trees. This left us with 11,266 lemmas. Column 4 of Table 1 shows the number of lemmas after filtering.

We randomly split corpus files into training, validation, and testing sets which contain 80%, 10%, 10% of the files, respectively. Table 2 shows statistics on the lemmas in each set, which includes columns for the number of files, the number of lemmas, the average number of characters and sub-tokens in lemma names, and the average number of characters and sub-tokens in lemma statements.

Figure 5 illustrates the changes of the depth, number of nodes and number of sub-tokens (after flattening) of the kernel trees (first row) and syntax trees (second row) before and after chopping. Our chopping process reduced tree depth by 70.9% for kernel trees and 70.7% for syntax trees, and reduced the number of nodes by 91.5% for kernel trees and 90.8% for syntax trees; after flattening, the resulting average sequence length is, for kernel trees 165 comparing to the original 2,056, and for syntax trees 144 comparing to the original 1,590. We provide additional statistics on lemmas before filtering in the appendix of the extended paper [57].

## 5    Implementation

In this section, we briefly describe our toolchain which implements the models in Sect. 3 and processes and learns from the corpus in Sect. 4; we dub this tool-chain ROOSTERIZE. The components of the toolchain can be divided into two categories: (1) components that interact with Coq or directly process information extracted from Coq, and (2) components concerned with machine learning and name generation.

The first category includes several OCaml-based tools integrated with SerAPI [26] (and thus Coq itself), and Python-based tools for processing of data obtained via SerAPI from Coq. All OCaml tools have either already been included in, or accepted for inclusion into, SerAPI itself. The tools are as follows:

**sercomp:** We integrated the existing program `sercomp` distributed with SerAPI into ROOSTERIZE to serialize Coq files to lists of sexps for syntax trees.

**sertok:** We developed an OCaml program dubbed `sertok` on top of SerAPI. The program takes a Coq file as input and produces sexps of all tokens found by Coq's lexer in the file, organized at the sentence level.

**sername:** We developed an OCaml program dubbed `sername` on top of SerAPI. The program takes a list of fully qualified (kernel) lemma names and produces sexps for the kernel trees of the corresponding lemma statements.

**postproc & subtokenizer:** We created two small independent tools in Python to post-process Coq sexps and perform sub-tokenization, respectively.

For the second category, we implemented our machine learning models in Python using two widely-used deep learning libraries: PyTorch [61] and Open-NMT [41]. More specifically, we extended the sequence-to-sequence models in OpenNMT to use multi-input encoders, and extended attention and copy layers to use multiple inputs. Source code for the components of ROOSTERIZE is available from: https://github.com/EngineeringSoftware/roosterize.

# 6   Evaluation

This section presents an extensive evaluation of our models as implemented in ROOSTERIZE. Our automatic evaluation (Sect. 6.2) compares ROOSTERIZE with a series of strong baselines and reports on ablation experiments; additional experiments, e.g., on chopping heuristics, are described in the appendix of the extended version of the paper [57]. Our manual quality assessment (Sect. 6.3) analyzes 150 comments we received from the maintainer of the PCM library on names suggested by ROOSTERIZE for that project using our best model.

## 6.1   Models and Baselines

We study the combinations of: (1) using individual input (lemma statement and trees) in a single encoder, or multi-input encoders with different mixture of these inputs; and (2) using the attention and copy mechanisms. Our inputs include: lemma statement (*Stmt*), syntax tree (*SynTree*), chopped syntax tree (*ChopSynTree*), kernel tree (*KnlTree*), and chopped kernel tree (*ChopKnlTree*). For multiple inputs, the models are named by concatenating inputs with "+"; a "+" is also used to denote the presence of attention (*attn*) or copy (*copy*). For example, Stmt+ChopKnlTree+attn+copy refers to a model that uses two encoders—one for lemma statement and one for chopped kernel tree—and uses attention and copy mechanisms.

**Table 3.** Results of Roosterize models.

| Group | Model | BLEU | Frag.Acc. | Top1 | Top5 |
|---|---|---|---|---|---|
| Multi-input +attn +copy | Stmt+ChopKnlTree+ChopSynTree+attn+copy | 45.4 | 22.2% | 7.5% | 16.5% |
| | Stmt+ChopKnlTree+attn+copy | **47.2** | **24.9%** | **9.6%** | **18.0%** |
| | Stmt+ChopSynTree+attn+copy | 37.7 | 18.1% | 6.1% | 10.6% |
| | ChopKnlTree+ChopSynTree+attn+copy | 45.4 | 22.9% | 7.6% | 15.3% |
| Single-input +attn +copy | ChopKnlTree+attn+copy | 42.9 | 19.8% | 5.0% | 11.7% |
| | ChopSynTree+attn+copy | 39.8 | 18.3% | 6.8% | 12.2% |
| | KnlTree+attn+copy | 37.0 | 14.2% | 2.2% | 8.4% |
| | SynTree+attn+copy | 31.0 | 10.8% | 2.8% | 6.1% |
| | Stmt+attn+copy | 38.9 | 19.4% | 6.9% | 11.6% |
| Multi-input +attn | Stmt+ChopKnlTree+ChopSynTree+attn | 24.5 | 8.6% | 0.4% | 0.9% |
| | Stmt+ChopKnlTree+attn | 25.6 | 8.5% | 0.9% | 1.7% |
| | Stmt+ChopSynTree+attn | 23.8 | 8.2% | 0.8% | 1.6% |
| | ChopKnlTree+ChopSynTree+attn | 28.4 | 10.9% | 1.8% | 3.4% |
| Single-input +attn | ChopKnlTree+attn | 19.5 | 4.9% | 0.6% | 1.3% |
| | ChopSynTree+attn | 28.9 | 12.1% | 1.5% | 2.9% |
| | KnlTree+attn | 14.1 | 1.6% | 0.0% | 0.0% |
| | SynTree+attn | 8.8 | 1.0% | 0.0% | 0.0% |
| | Stmt+attn | 26.9 | 11.1% | 1.1% | 2.5% |
| Multi-input | Stmt+ChopKnlTree+ChopSynTree | 17.7 | 3.5% | 0.1% | 0.2% |
| | Stmt+ChopKnlTree | 19.5 | 4.5% | 0.1% | 0.3% |
| | Stmt+ChopSynTree | 12.6 | 0.6% | 0.0% | 0.0% |
| | ChopKnlTree+ChopSynTree | 16.7 | 2.4% | 0.0% | 0.1% |
| Single-input | ChopKnlTree | 15.5 | 1.6% | 0.0% | 0.0% |
| | ChopSynTree | 14.5 | 0.8% | 0.1% | 0.1% |
| | KnlTree | 12.0 | 0.6% | 0.0% | 0.0% |
| | SynTree | 5.7 | 0.4% | 0.0% | 0.0% |
| | Stmt | 20.0 | 4.7% | 0.1% | 0.3% |
| - | Retrieval-based | 28.3 | 10.0% | 0.2% | 0.3% |

We consider the vanilla encoder-decoder models with only one input (lemma statement, kernel tree, or syntax tree) as baseline models. We also compare with a retrieval-based baseline model implemented using Lucene [6]: a k-nearest neighbors classifier using the tf-idf of the tokens in lemma statement as features.

Hyperparameters are tuned on the validation set within the following options: embedding dimensions from {200, 500, 1000}, number of hidden units in each LSTM from {200, 500, 1000}, number of stacked LSTM layers from {1, 2, 3}. We set the dropout rate between LSTM layers to 0.5. We set the output dimension of the fully connected layer for combining encoders to the same number as the number of hidden units in each LSTM. We checked the validation loss every 200 training steps (as defined in OpenNMT [41], which is similar to one training epoch on our dataset), and set an early stopping threshold of 3. We used the Adam [40] optimizer with a learning rate of 0.001. We used a beam size of 5 in beam search. All the experiments were run with one NVIDIA 1080-TI GPU and Intel Xeon E5-2620 v4 CPU.

## 6.2 Automatic Evaluation

**Metrics:** We use four automatic metrics which evaluate generated lemma names against the reference lemma name (as written by developers) in the testing set. Each metric captures a different level of granularity of the generation quality. *BLEU* [60] is a standard metric used in transduction tasks including language $\leftrightarrow$ code transduction. It calculates the number of n-grams in a generated sequence that also appear in the reference sequence, where one "n-gram" is n consecutive items in a sequence (in our case, one "n-gram" is n consecutive characters in the sequence of characters of the lemma name). We use it to compute the $1 \sim 4$-grams overlap between the characters in generated name and characters in the reference name, averaged between $1 \sim 4$-grams with smoothing method proposed by Lin and Och [46]. *Fragment accuracy* computes the accuracy of generated names on the fragment level, which is defined by splitting the name by underscores ("_"). For example, `map_determinant_mx` has a fragment accuracy of 66.7% when evaluated against `det_map_mx`. Unlike BLEU, fragment accuracy ignores the ordering of the fragments. Finally, *top-1 accuracy* and *top-5 accuracy* compute how often the true name fully matches the generated name or is one of the top-5 generated names.

**Results:** Table 3 shows the performance of the models. Similar models are grouped together. The first column shows the names of the model groups and the second column shows the names of the models. For each model, we show values for the four automatic metrics, BLEU, fragment accuracy (Frag.Acc.), top-1 accuracy (Top1), and top-5 accuracy (Top5). We repeated each experiment 3 times, with different random initialization each time, and computed the averages of each automated metric. We performed statistical significance tests—under significance level $p < 0.05$ using the bootstrap method [14]—to compare each pair of models. We use bold text to highlight the best value for each automatic metric, and gray background for baseline models. We make several observations:

**Finding #1:** The best overall performance (BLEU = 47.2) is obtained using the multi-input model with lemma statement and chopped kernel tree as inputs, which also includes copy and attention mechanisms (Stmt+ChopKnlTree+attn+copy). The improvements over all other models are statistically significant and all automatic metrics are consistent in identifying the best model. This shows the importance of using Coq's internal structures and focusing only on certain parts of those structures.

**Finding #2:** The copy mechanism brings statistically significant improvements to all models. This can be clearly observed by comparing groups 1 and 3 in the table, as well as groups 2 and 4. For example, BLEU for Stmt+attn and Stmt+attn+copy are 26.9 and 38.9, respectively. We believe that the copy mechanism plays an important role because many sub-tokens are specific to the file context and do not appear in the fixed vocabulary learned on the files in training set.

**Finding #3:** Using chopped trees greatly improves performance of models and the improvements brought by upgrading KnlTree to ChopKnlTree or SynTree to

**Table 4.** Manual quality analysis representative examples.

| |
|---|
| **Lemma statement:** `p s : supp (kfilter p s) = filter p (supp s)` |
| **Hand-written:** `supp_kfilt` **Roosterize:** `supp_kfilter` |
| **Comment:** ✓ Using only `kfilt` has cognitive overhead |
| **Lemma statement:** `g e k v f : path ord k (supp f) ->` `foldfmap g e (ins k v f) = g (k, v) (foldfmap g e f)` |
| **Hand-written:** `foldf_ins` **Roosterize:** `foldfmap_ins` |
| **Comment:** ✓ The whole function name is used in the suggested name |
| **Lemma statement:** `: transitive (@ord T)` |
| **Hand-written:** `trans` **Roosterize:** `ord_trans` |
| **Comment:** ✓ Useful to add the `ord` prefix to the name |
| **Lemma statement:** `s : sorted (@ord T) s -> sorted (@oleq T) s` |
| **Hand-written:** `sorted_oleq` **Roosterize:** `ord_sorted` |
| **Comment:** ✗ The conclusion content should have greater priority |
| **Lemma statement:** `x y : total_spec x y (ord x y) (x == y) (ord y x)` |
| **Hand-written:** `totalP` **Roosterize:** `ordP` |
| **Comment:** ✗ Maybe this lemma should be named `ord_totalP`? |
| **Lemma statement:** `p1 p2 s : kfilter (predI p1 p2) s =` `kfilter p1 (kfilter p2 s)` |
| **Hand-written:** `kfilter_predI` **Roosterize:** `eq_kfilter` |
| **Comment:** ✗ The suggested name is too generic |

ChopSynTree are statistically significant. For example, this can be clearly seen in the second group: BLEU for KnlTree+attn+copy and ChopKnlTree+attn+copy are 37.0 and 42.9, respectively. We believe that the size of the original trees, and a lot of irrelevant data in those trees, hurt the performance. The fact that ChopKnlTree and ChopSynTree both perform much better than using KnlTree or SynTree across all groups indicate that the chopped trees could be viewed as a form of supervised attention with flat values that helps later attention layers to focus better.

**Finding #4:** Although chopped syntax tree with attention outperforms (statistically significant) chopped kernel tree with attention (BLEU 28.9 vs. 19.5), chopped kernel tree with attention and copy by far outperforms (statistically significant) chopped syntax tree with attention and copy (BLEU 42.9 vs. 39.8). The copy mechanism helps kernel trees much more than the syntax trees, because the mathematical notations and symbols in the syntax trees get expanded to their names in the kernel trees, and some of them are needed as a part of the lemma names.

**Finding #5:** Lemma statement and syntax tree do not work well together, primarily because the two representations contain mostly the same information.

In which case, a model taking both as inputs may not work as well as using only one of the inputs, because more parameters need to be trained.

**Finding #6:** The retrieval-based baseline, which is the strongest among baselines, outperforms several encoder-decoder models without attention and copy or with only attention, but is worse than (statistically significant) all models with both attention and copy mechanisms enabled.

### 6.3   Manual Quality Analysis

While generated lemma names may not always match the manually written ones in the training set, they can still be semantically valid and conform to prevailing conventions. However, such name properties are not reflected in our automatic evaluation metrics, since these metrics only consider exactly matched tokens as correct. To obtain a more complete evaluation, we therefore performed a manual quality analysis of generated lemma names from ROOSTERIZE by applying it to a Coq project outside of our corpus, the PCM library [56]. This project depends on MathComp, and follows, to a degree, many of the MathComp coding conventions. The PCM library consists of 12 Coq files, and contains 690 lemmas.

We ran ROOSTERIZE with the best model (Stmt+ChopKnlTree+attn+copy) on the PCM library to get the top-1 suggestions for all lemma names. Overall, the ROOSTERIZE suggestions achieved a BLEU score of 36.3 and a fragment accuracy of 17%, and 36 suggestions (5%) exactly match the existing lemma names. Next, we asked the maintainer of the PCM library to evaluate the remaining 654 lemma names (those that do not match exactly) and send us feedback.

The maintainer spent one day on the task and provided comments on 150 suggested names. We analyzed these comments to identify patterns and trends. He found that 20% of the suggested names he inspected were of good quality, out of which more than half were of high quality. Considering that the analysis was of top-1 suggestions excluding exact matches, we find these figures encouraging. For low-quality names, a clear trend was that they were often "too generic". Similar observations have been made about the results from encoder-decoder models in dialog generation [45,65]. In contrast, useful suggestions were typically able to expand or elaborate on name components that are intuitively too concise, e.g., replacing `kfilt` with `kfilter`. Table 4 lists examples that are representative of these trends; checkmarks indicate useful suggestions, while crosses indicate unsuitability. We also include comments from the maintainer. As illustrated by the comments, even suggestions considered unsuitable may contain useful parts.

## 7   Discussion

Our toolchain builds on Coq 8.10.2, and thus we only used projects that support this version. However, we do not expect any fundamental obstacles in supporting future Coq releases. Thanks to the use of OCaml metaprogramming via PPX, which allowed eliding explicit references to the internal structure of Coq datatypes, SerAPI itself and our extensions to it are expected to require only modest effort to maintain as Coq evolves.

Our models and toolchain may not be applicable to Coq projects unrelated to the MathComp family of projects, i.e., projects which do not follow any MathComp conventions. To the best of our knowledge, MathComp's coding conventions are the most recognizable and well-documented in the Coq community; suggesting coding conventions based on learning from projects unrelated to MathComp are likely to give more ambiguous results that are difficult to validate manually. Our case study also included generating suggestions for a project outside the MathComp family, the PCM library, with encouraging results.

Our models are in principle applicable to proof assistants with similar foundations, such as Lean [54]. However, the current version of Lean, Lean 3, does not provide serialization of internal data structures as SerAPI does for Coq, which prevents direct application of our toolchain. Application of our models to proof assistants with different foundations and proof-checking toolchains, such as Isabelle/HOL, is even less straightforward, although the Archive of Formal Proofs (AFP) contains many projects with high-quality lemma names [25].

## 8    Related Work

**Naturalness and Coding Conventions:** Hindle et al. [35] first applied the concept of naturalness to Java-like languages, noting that program statement regularities and repetitiveness make statistical language models applicable for performing software engineering tasks [4]. Rahman et al. [62] validated the naturalness of other similar programming languages, and Hellendoorn et al. [31] found high naturalness in Coq code, providing motivation for our application of statistical language models to Coq. Allamanis et al. [2] used the concept of naturalness and statistical language models to learn and suggest coding conventions, including names, for Java, and Raychev et al. [63] used conditional random fields to learn and suggest coding conventions for JavaScript. To our knowledge, no previous work has developed *applications* of naturalness for proof assistants; Hellendorn et al. [31] only measured naturalness for their Coq corpus.

**Suggesting Names:** Prior work on suggesting names mostly concerns Java method names. Liu et al. [47] used a similarity matching algorithm, based on deep representations of Java method names and bodies learned with Paragraph Vector and convolutional neural networks, to detect and fix inconsistent Java method names. Allamanis et al. [3] used logbilinear neural language models supplemented by additional manual features to predict Java method and class names. Java method names have also been treated as short, descriptive "summaries" of its body; in this view, prior work has augmented attention mechanisms in convolutional networks [5], used sequence-to-sequence models to learn from descriptions (e.g., Javadoc comments) [27], and utilized the tree-structure of the code in a hierarchical attention network [74]. Unlike Java syntax trees, Coq syntax and kernel trees contain considerable semantic information useful for naming. In the work closest to our domain, Aspinall and Kaliszyk used a k-nearest neighbors multi-label classifier on a corpus for the HOL Light proof assistant to suggest names of lemmas [7]. However, their technique only suggests names that exist in the training data and therefore does not generalize. To our knowledge, ours is the first neural generation model for suggesting names in a proof assistant context.

**Mining and Learning for Proof Assistants:** Müller et al. [55] exported Coq kernel trees as XML strings to translate 49 Coq projects to the OMDoc theory graph format. Rather than translating documents to an independently specified format, we produce lightweight machine-readable representations of Coq's internal data structures. Wiedijk [71] collected early basic statistics on the core libraries of several proof assistants, including Coq and Isabelle/HOL. Blanchette et al. [16] mined the AFP to gather statistics such as the average number of lines of Isabelle/HOL specifications and proof scripts. However, these corpora were not used to perform learning. Komendantskaya et al. [32–34,42] used machine learning without neural networks to identify patterns in Coq tactic sequences and proof kernel trees, e.g., to find structural similarities between lemmas and simplify proof development. In contrast, our models capture similarity among several different representations of lemma *statements* to generate lemma names.

## 9 Conclusion

We presented novel techniques, based on neural networks, for learning and suggesting lemma names in Coq verification projects. We designed and implemented multi-input encoder-decoder models that use Coq's internal data structures, including (chopped) syntax trees and kernel trees. Additionally, we constructed a large corpus of high quality Coq code that will enable development and evaluation of future techniques for Coq. We performed an extensive evaluation of our models using the corpus. Our results show that the multi-input models, which use internal data structures, substantially outperform several baselines; the model that uses the lemma statement tokens and the chopped kernel tree with attention and copy mechanism performs the best. Based on our findings, we believe that multi-input models leveraging key parts of internal data structures can play a critical role in producing high-quality lemma name suggestions.

**Acknowledgments.** The authors thank Yves Bertot, Cyril Cohen, Emilio Jesús Gallego Arias, Gaëtan Gilbert, Hugo Herbelin, Anton Trunov, Théo Zimmermann, and the anonymous reviewers for their comments and feedback. This work was partially supported by the US National Science Foundation under Grant Nos. CCF-1652517 and IIS-1850153, and by the Swedish Foundation for Strategic Research under the TrustFull project.

## References

1. Affeldt, R., Garrigue, J.: Formalization of error-correcting codes: from Hamming to modern coding theory. In: Urban, C., Zhang, X. (eds.) ITP 2015. LNCS, vol. 9236, pp. 17–33. Springer, Cham (2015). https://doi.org/10.1007/978-3-319-22102-1_2

2. Allamanis, M., Barr, E.T., Bird, C., Sutton, C.: Learning natural coding conventions. In: International Symposium on the Foundations of Software Engineering, pp. 281–293. ACM, New York (2014). https://doi.org/10.1145/2635868.2635883
3. Allamanis, M., Barr, E.T., Bird, C., Sutton, C.: Suggesting accurate method and class names. In: Joint Meeting on Foundations of Software Engineering, pp. 38–49. ACM, New York (2015). https://doi.org/10.1145/2786805.2786849
4. Allamanis, M., Barr, E.T., Devanbu, P., Sutton, C.: A survey of machine learning for big code and naturalness. ACM Comput. Surv. **51**(4), 81:3–81:37 (2018). https://doi.org/10.1145/3212695
5. Allamanis, M., Peng, H., Sutton, C.: A convolutional attention network for extreme summarization of source code. In: International Conference on Machine Learning, pp. 2091–2100 (2016)
6. Apache Software Foundation: Apache Lucene (2020). https://lucene.apache.org. Accessed 23 Jan 2020
7. Aspinall, D., Kaliszyk, C.: What's in a theorem name? In: Blanchette, J.C., Merz, S. (eds.) ITP 2016. LNCS, vol. 9807, pp. 459–465. Springer, Cham (2016). https://doi.org/10.1007/978-3-319-43144-4_28
8. Avidan, E., Feitelson, D.G.: Effects of variable names on comprehension: an empirical study. In: International Conference on Program Comprehension, pp. 55–65. IEEE Computer Society, Washington (2017). https://doi.org/10.1109/ICPC.2017.27
9. Avigad, J.: Mathlib naming conventions (2016). https://github.com/leanprover-community/mathlib/blob/snapshot-2019-10/docs/contribute/naming.md. Accessed 23 Jan 2020
10. Babii, H., Janes, A., Robbes, R.: Modeling vocabulary for big code machine learning. CoRR abs/1904.01873 (2019). https://arxiv.org/abs/1904.01873
11. Bahdanau, D., Cho, K., Bengio, Y.: Neural machine translation by jointly learning to align and translate. In: International Conference on Learning Representations (2015). https://arxiv.org/abs/1409.0473
12. Barendregt, H., Barendsen, E.: Autarkic computations in formal proofs. J. Autom. Reason. **28**(3), 321–336 (2002). https://doi.org/10.1023/A:1015761529444
13. Bartzia, E.-I., Strub, P.-Y.: A formal library for elliptic curves in the Coq proof assistant. In: Klein, G., Gamboa, R. (eds.) ITP 2014. LNCS, vol. 8558, pp. 77–92. Springer, Cham (2014). https://doi.org/10.1007/978-3-319-08970-6_6
14. Berg-Kirkpatrick, T., Burkett, D., Klein, D.: An empirical investigation of statistical significance in NLP. In: Joint Conference on Empirical Methods in Natural Language Processing and Computational Natural Language Learning, pp. 995–1005. Association for Computational Linguistics, Stroudsburg (2012)
15. Bertot, Y., Castéran, P.: Interactive Theorem Proving and Program Development: Coq'Art: The Calculus of Inductive Constructions. Springer, Heidelberg (2004). https://doi.org/10.1007/978-3-662-07964-5
16. Blanchette, J.C., Haslbeck, M., Matichuk, D., Nipkow, T.: Mining the archive of formal proofs. In: Kerber, M., Carette, J., Kaliszyk, C., Rabe, F., Sorge, V. (eds.) CICM 2015. LNCS (LNAI), vol. 9150, pp. 3–17. Springer, Cham (2015). https://doi.org/10.1007/978-3-319-20615-8_1
17. Boogerd, C., Moonen, L.: Evaluating the relation between coding standard violations and faults within and across software versions. In: International Working Conference on Mining Software Repositories, pp. 41–50. IEEE Computer Society, Washington (2009). https://doi.org/10.1109/MSR.2009.5069479

18. Chen, H., Ziegler, D., Chajed, T., Chlipala, A., Kaashoek, M.F., Zeldovich, N.: Using crash Hoare logic for certifying the FSCQ file system. In: Symposium on Operating Systems Principles, pp. 18–37. ACM, New York (2015). https://doi.org/10.1145/2815400.2815402

19. Cohen, C.: Contribution guide for the Mathematical Components library (2018). https://github.com/math-comp/math-comp/blob/mathcomp-1.9.0/CONTRIBUTING.md. Accessed 14 Apr 2020

20. Coq Development Team: The Coq proof assistant, version 8.10.0, October 2019. https://doi.org/10.5281/zenodo.3476303

21. Coq Development Team: The Gallina specification language (2019). https://coq.inria.fr/distrib/V8.10.2/refman/language/gallina-specification-language.html. Accessed 17 Apr 2020

22. Delahaye, D.: A tactic language for the system Coq. In: Parigot, M., Voronkov, A. (eds.) LPAR 2000. LNAI, vol. 1955, pp. 85–95. Springer, Heidelberg (2000). https://doi.org/10.1007/3-540-44404-1_7

23. Doczkal, C., Kaiser, J.O., Smolka, G.: Regular language representations in Coq (2020). https://github.com/coq-community/reglang. Accessed 09 Apr 2020

24. Doczkal, C., Smolka, G.: Regular language representations in the constructive type theory of Coq. J. Autom. Reason. **61**(1), 521–553 (2018)

25. Eberl, M., Klein, G., Nipkow, T., Paulson, L., Thiemann, R.: Archive of Formal Proofs (2020). https://www.isa-afp.org. Accessed 23 Jan 2020

26. Gallego Arias, E.J.: SerAPI: machine-friendly, data-centric serialization for Coq. Technical report, MINES ParisTech (2016). https://hal-mines-paristech.archives-ouvertes.fr/hal-01384408

27. Gao, S., Chen, C., Xing, Z., Ma, Y., Song, W., Lin, S.: A neural model for method name generation from functional description. In: International Conference on Software Analysis, Evolution and Reengineering, pp. 414–421. IEEE Computer Society, Washington (2019). https://doi.org/10.1109/SANER.2019.8667994

28. Gonthier, G.: Formal proof–the four-color theorem. Not. Am. Math. Soc. **55**(11), 1382–1393 (2008)

29. Gonthier, G., et al.: A machine-checked proof of the odd order theorem. In: Blazy, S., Paulin-Mohring, C., Pichardie, D. (eds.) ITP 2013. LNCS, vol. 7998, pp. 163–179. Springer, Heidelberg (2013). https://doi.org/10.1007/978-3-642-39634-2_14

30. Google: Google-Java-Format (2020). https://github.com/google/google-java-format. Accessed 23 Jan 2020

31. Hellendoorn, V.J., Devanbu, P.T., Alipour, M.A.: On the naturalness of proofs. In: International Symposium on the Foundations of Software Engineering, New Ideas and Emerging Results, pp. 724–728. ACM, New York (2018). https://doi.org/10.1145/3236024.3264832

32. Heras, J., Komendantskaya, E.: ML4PG in computer algebra verification. In: Carette, J., Aspinall, D., Lange, C., Sojka, P., Windsteiger, W. (eds.) CICM 2013. LNCS (LNAI), vol. 7961, pp. 354–358. Springer, Heidelberg (2013). https://doi.org/10.1007/978-3-642-39320-4_28

33. Heras, J., Komendantskaya, E.: Proof pattern search in Coq/SSReflect. CoRR abs/1402.0081 (2014). https://arxiv.org/abs/1402.0081

34. Heras, J., Komendantskaya, E.: Recycling proof patterns in Coq: case studies. Math. Comput. Sci. **8**(1), 99–116 (2014). https://doi.org/10.1007/s11786-014-0173-1

35. Hindle, A., Barr, E.T., Su, Z., Gabel, M., Devanbu, P.: On the naturalness of software. In: International Conference on Software Engineering, pp. 837–847. IEEE Computer Society, Washington (2012). https://doi.org/10.1109/ICSE.2012.6227135
36. Hochreiter, S., Schmidhuber, J.: Long short-term memory. Neural Comput. **9**(8), 1735–1780 (1997). https://doi.org/10.1162/neco.1997.9.8.1735
37. HoTT authors: HoTT Conventions and Style Guide (2019). https://github.com/HoTT/HoTT/blob/V8.10/STYLE.md. Accessed 23 Jan 2020
38. Hu, X., Li, G., Xia, X., Lo, D., Jin, Z.: Deep code comment generation. In: International Conference on Program Comprehension, pp. 200–210. ACM, New York (2018). https://doi.org/10.1145/3196321.3196334
39. Iris authors: Iris Style Guide (2019). https://gitlab.mpi-sws.org/iris/iris/blob/iris-3.2.0/StyleGuide.md. Accessed 17 Apr 2020
40. Kingma, D.P., Ba, J.: Adam: a method for stochastic optimization. In: International Conference on Learning Representations (2015). https://arxiv.org/abs/1412.6980
41. Klein, G., Kim, Y., Deng, Y., Senellart, J., Rush, A.M.: OpenNMT: open-source toolkit for neural machine translation. In: Annual Meeting of the Association for Computational Linguistics, System Demonstrations, pp. 67–72. Association for Computational Linguistics, Stroudsburg (2017). https://doi.org/10.18653/v1/P17-4012
42. Komendantskaya, E., Heras, J., Grov, G.: Machine learning in Proof General: interfacing interfaces. In: Kaliszyk, C., Lüth, C. (eds.) International Workshop on User Interfaces for Theorem Provers. EPTCS, vol. 118, pp. 15–41. Open Publishing Association, Sydney (2013). https://doi.org/10.4204/EPTCS.118.2
43. LeClair, A., Jiang, S., McMillan, C.: A neural model for generating natural language summaries of program subroutines. In: International Conference on Software Engineering, pp. 795–806. IEEE Computer Society, Washington (2019). https://doi.org/10.1109/ICSE.2019.00087
44. Leroy, X.: Formal verification of a realistic compiler. Commun. ACM **52**(7), 107–115 (2009). https://doi.org/10.1145/1538788.1538814
45. Li, J., Galley, M., Brockett, C., Gao, J., Dolan, B.: A diversity-promoting objective function for neural conversation models. In: Conference of the North American Chapter of the Association for Computational Linguistics: Human Language Technologies, pp. 110–119. Association for Computational Linguistics, Stroudsburg (2016). https://doi.org/10.18653/v1/n16-1014
46. Lin, C., Och, F.J.: ORANGE: a method for evaluating automatic evaluation metrics for machine translation. In: International Conference on Computational Linguistics, pp. 501–507. Association for Computational Linguistics, Stroudsburg (2004)
47. Liu, K., et al.: Learning to spot and refactor inconsistent method names. In: International Conference on Software Engineering, pp. 1–12. IEEE Computer Society, Washington (2019). https://doi.org/10.1109/ICSE.2019.00019
48. Luong, T., Pham, H., Manning, C.D.: Effective approaches to attention-based neural machine translation. In: Empirical Methods in Natural Language Processing, pp. 1412–1421. Association for Computational Linguistics, Stroudsburg (2015). https://doi.org/10.18653/v1/d15-1166
49. Mahboubi, A., Tassi, E.: Mathematical Components Book (2017). https://math-comp.github.io/mcb/. Accessed 17 Apr 2020
50. Mathematical Components Team: Missing lemmas in Seq (2016). https://github.com/math-comp/math-comp/pull/41. Accessed 18 Apr 2020

51. McCarthy, J.: Recursive functions of symbolic expressions and their computation by machine, part I. Commun. ACM **3**(4), 184–195 (1960). https://doi.org/10.1145/367177.367199

52. Miara, R.J., Musselman, J.A., Navarro, J.A., Shneiderman, B.: Program indentation and comprehensibility. Commun. ACM **26**(11), 861–867 (1983). https://doi.org/10.1145/182.358437

53. de Moura, L., Avigad, J., Kong, S., Roux, C.: Elaboration in dependent type theory. CoRR abs/1505.04324 (2015). https://arxiv.org/abs/1505.04324

54. de Moura, L., Kong, S., Avigad, J., van Doorn, F., von Raumer, J.: The Lean theorem prover (system description). In: Felty, A.P., Middeldorp, A. (eds.) CADE 2015. LNCS (LNAI), vol. 9195, pp. 378–388. Springer, Cham (2015). https://doi.org/10.1007/978-3-319-21401-6_26

55. Müller, D., Rabe, F., Sacerdoti Coen, C.: The Coq library as a theory graph. In: Kaliszyk, C., Brady, E., Kohlhase, A., Sacerdoti Coen, C. (eds.) CICM 2019. LNCS (LNAI), vol. 11617, pp. 171–186. Springer, Cham (2019). https://doi.org/10.1007/978-3-030-23250-4_12

56. Nanevski, A., Ley-Wild, R., Sergey, I., Delbianco, G., Trunov, A.: The PCM library (2020). https://github.com/imdea-software/fcsl-pcm. Accessed 24 Jan 2020

57. Nie, P., Palmskog, K., Li, J.J., Gligoric, M.: Deep generation of Coq lemma names using elaborated terms. CoRR abs/2004.07761 (2020). https://arxiv.org/abs/2004.07761

58. OCaml Labs: PPX (2017). http://ocamllabs.io/doc/ppx.html. Accessed 23 Jan 2020

59. Ogura, N., Matsumoto, S., Hata, H., Kusumoto, S.: Bring your own coding style. In: International Conference on Software Analysis, Evolution and Reengineering, pp. 527–531. IEEE Computer Society, Washington (2018). https://doi.org/10.1109/SANER.2018.8330253

60. Papineni, K., Roukos, S., Ward, T., Zhu, W.: BLEU: a method for automatic evaluation of machine translation. In: Annual Meeting of the Association for Computational Linguistics, pp. 311–318. Association for Computational Linguistics, Stroudsburg (2002)

61. Paszke, A., et al.: Automatic differentiation in PyTorch. In: Autodiff Workshop (2017). https://openreview.net/forum?id=BJJsrmfCZ

62. Rahman, M., Palani, D., Rigby, P.C.: Natural software revisited. In: International Conference on Software Engineering, pp. 37–48. IEEE Computer Society, Washington (2019). https://doi.org/10.1109/ICSE.2019.00022

63. Raychev, V., Vechev, M., Krause, A.: Predicting program properties from "big code". In: Symposium on Principles of Programming Languages, pp. 111–124. ACM, New York (2015). https://doi.org/10.1145/2676726.2677009

64. See, A., Liu, P.J., Manning, C.D.: Get to the point: summarization with pointer-generator networks. In: Annual Meeting of the Association for Computational Linguistics, pp. 1073–1083. Association for Computational Linguistics, Stroudsburg (2017). https://doi.org/10.18653/v1/P17-1099

65. Serban, I.V., Sordoni, A., Bengio, Y., Courville, A., Pineau, J.: Building end-to-end dialogue systems using generative hierarchical neural network models. In: AAAI Conference on Artificial Intelligence, pp. 3776–3783. AAAI Press, Palo Alto (2016)

66. Sergey, I., Nanevski, A., Banerjee, A.: Mechanized verification of fine-grained concurrent programs. In: Conference on Programming Language Design and Implementation, pp. 77–87. ACM, New York (2015). https://doi.org/10.1145/2737924.2737964

67. Shneiderman, B., McKay, D.: Experimental investigations of computer program debugging and modification. Hum. Factors Soc. Ann. Meet. **20**(24), 557–563 (1976). https://doi.org/10.1177/154193127602002401
68. Sutskever, I., Vinyals, O., Le, Q.V.: Sequence to sequence learning with neural networks. In: Advances in Neural Information Processing Systems 27, pp. 3104–3112. MIT Press, Cambridge (2014)
69. Suzuki, J., Nagata, M.: Cutting-off redundant repeating generations for neural abstractive summarization. In: Conference of the European Chapter of the Association for Computational Linguistics, pp. 291–297. Association for Computational Linguistics, Stroudsburg (2017). https://doi.org/10.18653/v1/e17-2047
70. Unanue, I.J., Borzeshi, E.Z., Piccardi, M.: A shared attention mechanism for interpretation of neural automatic post-editing systems. In: Workshop on Neural Machine Translation and Generation, pp. 11–17. Association for Computational Linguistics, Stroudsburg (2018). https://doi.org/10.18653/v1/w18-2702
71. Wiedijk, F.: Statistics on digital libraries of mathematics. Stud. Logic Gramm. Rhetor. **18**(31), 137–151 (2009)
72. Williams, R.J., Zipser, D.: A learning algorithm for continually running fully recurrent neural networks. Neural Comput. **1**(2), 270–280 (1989). https://doi.org/10.1162/neco.1989.1.2.270
73. Woos, D., Wilcox, J.R., Anton, S., Tatlock, Z., Ernst, M.D., Anderson, T.: Planning for change in a formal verification of the Raft consensus protocol. In: Certified Programs and Proofs, pp. 154–165. ACM, New York (2016). https://doi.org/10.1145/2854065.2854081
74. Xu, S., Zhang, S., Wang, W., Cao, X., Guo, C., Xu, J.: Method name suggestion with hierarchical attention networks. In: Workshop on Partial Evaluation and Program Manipulation, pp. 10–21. ACM, New York (2019). https://doi.org/10.1145/3294032.3294079
75. Zhang, S., Zheng, D., Hu, X., Yang, M.: Bidirectional long short-term memory networks for relation classification. In: Pacific Asia Conference on Language, Information and Computation, pp. 207–212. Association for Computational Linguistics, Stroudsburg (2015). https://doi.org/10.18653/v1/p16-2034

# Extensible Extraction of Efficient Imperative Programs with Foreign Functions, Manually Managed Memory, and Proofs

Clément Pit-Claudel[1]([⊠]) , Peng Wang[2], Benjamin Delaware[3], Jason Gross[1] , and Adam Chlipala[1]

[1] MIT CSAIL, Cambridge, MA 02139, USA
{cpitcla,jgross,adamc}@csail.mit.edu
[2] Google, Mountain View, CA 94043, USA
wangpeng@google.com
[3] Purdue University, West Lafayette, IN 47907, USA
bendy@purdue.edu

**Abstract.** We present an original approach to sound program extraction in a proof assistant, using syntax-driven automation to derive correct-by-construction imperative programs from nondeterministic functional source code. Our approach does not require committing to a single inflexible compilation strategy and instead makes it straightforward to create domain-specific code translators. In addition to a small set of core definitions, our framework is a large, user-extensible collection of compilation rules each phrased to handle specific language constructs, code patterns, or data manipulations. By mixing and matching these pieces of logic, users can easily tailor extraction to their own domains and programs, getting maximum performance and ensuring correctness of the resulting assembly code.

Using this approach, we complete the first proof-generating pipeline that goes automatically from high-level specifications to assembly code. In our main case study, the original specifications are phrased to resemble SQL-style queries, while the final assembly code does manual memory management, calls out to foreign data structures and functions, and is suitable to deploy on resource-constrained platforms. The pipeline runs entirely within the Coq proof assistant, leading to final, linked assembly code with overall full-functional-correctness proofs in separation logic.

## 1 Introduction

The general area of correct-by-construction code generation is venerable, going back at least to Dijkstra's work in the 1960s [5]. Oftentimes, solutions offer a strict subset of the desiderata of generality, automation, and performance of synthesized code. This paper presents the final piece of a pipeline that sits at the sweet spot of all three, enabling semiautomatic refinement of high-level specifications into efficient low-level code in a proof-generating manner. Our initial

© Springer Nature Switzerland AG 2020
N. Peltier and V. Sofronie-Stokkermans (Eds.): IJCAR 2020, LNAI 12167, pp. 119–137, 2020.
https://doi.org/10.1007/978-3-030-51054-1_7

specification language is the rich, higher-order logic of Coq, and we support a high degree of automation through domain-specific refinement strategies, which in turn enable targeted optimization strategies for extracting efficient low-level code. In order to take advantage of these opportunities, we have built an extensible compilation framework that can be updated to handle new compilation strategies without sacrificing soundness. Our pipeline is *foundational*: it produces a fully linked assembly program represented as a Coq term with a proof that it meets the original high-level specification.

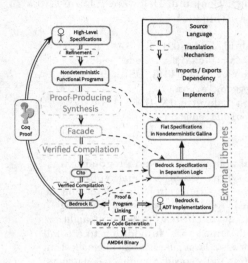

Our complete toolchain uses Fiat [4] to refine high-level specifications of abstract data types (ADTs) into nondeterministic functional programs depending on external data structures (expressed in a shallowly embedded Gallina DSL), then soundly extracts these programs to an imperative intermediate language (Facade) using a novel proof-generating extraction procedure. The resulting programs are then translated into the Cito [29] language by a newly written compiler, backed by a nontrivial soundness argument bridging two styles of operational semantics. A traditional verified compiler produces efficient Bedrock assembly [3] from the Cito level, which we soundly link against hand-verified

**Fig. 1.** The full pipeline, with this work's contributions in blue. Stick figures indicate user-supplied components. (Color figure online)

implementations of the required data structures. Beyond exploring a new technique for sound extraction of shallowly embedded DSLs (EDSLs), this work bridges the last remaining gap (extraction) to present the first mechanically certified automatic translation pipeline from declarative specifications to efficient assembly programs, as shown in Fig. 1.

In the original Fiat system, specifications were highly nondeterministic programs, and final implementations were fully deterministic programs obtained by repeatedly *refining* the specification, eventually committing to a single possible result. As a consequence, the generated code committed to a particular deterministic (and pure) implementation of external ADTs and functions that it relied on, reducing flexibility, optimization opportunities, and overall performance. Additionally, the final step in previous work using Fiat was to *extract* this code directly to OCaml, using Coq's popular but unverified extraction mechanism. Unfortunately, this meant that correctness of the compiled executable depended not only on the correctness of Coq's kernel but also on that of the extraction mechanism and of the OCaml compiler and runtime system. These two dependencies significantly decreased the confidence that users can place in programs synthesized by Fiat, and more generally in all programs extracted from Gallina code.

Our work overcomes these issues via a novel approach to extraction that is both *extensible* and *correct* and produces *efficient, stateful low-level code* from nondeterministic functional sources. The process runs within Coq, produces assembly code instead of OCaml code, and supports linking with handwritten or separately compiled verified assembly code.

Instead of refining specifications down to a fully deterministic Gallina program, as the original Fiat system did, we allow Fiat's final output to incorporate nondeterminism. These choices are resolved at a later stage by interpreting the nondeterminism as a postcondition specification in Hoare logic and linking against assembly code proven to meet that specification. Nondeterminism at runtime, which is not normally present in Gallina programs, is essential to support code derivation with flexible use of efficient low-level data structures. For example, if we represent a database with a type of binary trees that does not enjoy canonical representations, the same data may admit multiple concrete representations, each corresponding to a different ordering of results for an operation enumerating all database records.

Unlike certified compilers like CompCert [13] or CakeML [9], we do not implement our translator in the proof assistant's logic and prove it sound once and for all. Instead, we use proof-generating extraction: we phrase the translation problem in a novel sequent-calculus-style formulation that allows us to apply all of Coq's usual scriptable proof automation. The primary reason is that we want to make our compiler *extensible* by not committing to a specific compilation strategy: in our system, programmers can teach the compiler about new verified low-level data structures and code-generation strategies by introducing new lemmas explaining how to map a Gallina term to a particular imperative program[1]. Our automation then builds a (deeply embedded) syntax tree by repeatedly applying lemmas until the nondeterministic functional program is fully compiled. The many advantages of this approach (extensibility, ease of development, flexibility, performance, and ease of verification) do come at a cost, however: compilation is slower, care is needed to make the compiler robust to small variations in input syntax, and the extensible nature of the compiler makes it hard to characterize the supported source language precisely.

To summarize the benefits of our approach:

- It is lightweight: it does not require reifying the entirety of Gallina into a deeply embedded language before compiling. Instead, we use Coq's tactic language to drive compilation.
- It is extensible: each part of the compilation logic is expressed as a derivation rule, proved as an arbitrarily complex Coq theorem. Users can assemble a

---

[1] In fact, nondeterministic choices *cannot* be compiled systematically, as they can represent arbitrary complexity. Additionally, a proof-producing approach lets us elegantly bypass the issue of self-reference, since our original programs are shallowly embedded.

customized compiler by supplying their own compilation lemmas to extend the source language or improve the generated code.

- It is well-suited to compiling EDSLs: we support nondeterminism in input programs (standard extraction requires deterministic code).
- It allows us to link against axiomatically specified foreign functions and data structures, implemented and verified separately.
- It compiles to a relatively bare language with explicit memory management.

To demonstrate the applicability of this approach, Sect. 6 presents a set of microbenchmarks of Fiat programs manipulating variables, conditions, and nested lists of machine words, as well as a more realistic example of SQL-like programs similar to those of the original Fiat paper. These benchmarks start from high-level specifications of database queries and pass automatically through our pipeline to closed assembly programs, complete with full-functional-correctness specifications and proofs in separation logic. Source code and compilation instructions for the framework and benchmarks are available online at https://github.com/mit-plv/fiat/tree/ijcar2020.

## 2   A Brief Outline of Our Approach

We begin with an example of the pipeline in action. Below are an SQL-style query finding all titles by an author and a Fiat-generated implementation (right):

```
SELECT title FROM Books        rows ← IndexedByAuthor.bfind $books $author;
WHERE Books.by = $author       ret (map (λ row ⇒ row.Title) rows)
```

The generated code relies on a Fiat module  IndexedByAuthor , which is not an executable implementation of the required functionality; rather, it specifies certain methods nondeterministically, implying that bfind returns the expected rows in some *undetermined* order. The order may even be different for every call, as might arise, for instance, with data structures like splay trees that adjust their layout even during logically read-only operations.

Such nondeterministic programs are the starting point for our new refinement phases. The ultimate output of the pipeline is a library of assembly code in the Bedrock framework [3], obtained by extracting to a new language, *Facade*, built as a layer on top of the Cito C-like language [29], and then compiling to Bedrock.

The output for our running example might look like the code on the right. Note that this code works directly with pointers to heap-allocated mutable objects, handling all memory management by itself, including for intermediate values. The general  IndexedByAuthor  interface has been replaced with calls to a

```
rows = BTree.find($books, $author);
out  = StringList.new();
While (not TupleList.empty?(rows))
    row = TupleList.pop!(rows);
    title = Tuple.get(row, 0);
    StringList.push!(out, title);
EndWhile;
TupleList.delete!(rows);
StringList.reverse!(out);
```

concrete module  BTree  providing binary search trees of tuples, and the call to
map  became an imperative loop. We implement and verify  BTree  in Bedrock
assembly, and then we link code and proofs to obtain a binary and an end-to-end
theorem guaranteeing full functional correctness of assembly libraries, for code
generated automatically from high-level specifications.

The heart of our contribution is spanning the gap from *nondeterministic
functional programs* (written in Gallina) to *imperative low-level programs* (written in Facade) using an extensible, proof-generating framework. We phrase this
derivation problem as one of finding a proof of a Hoare triple, where the precondition and postcondition are known, but the Facade program itself must be
derived during the proof. The central goal from our running example looks as follows, where  ?1  stands for the overall Facade program that we seek, and where
we unfold IndexedByAuthor.bfind (Subsect. 4.3 defines these triples precisely).

$$[\![\text{"books"} \mapsto \text{ret } \$books]\!] :: \underset{\varnothing}{\overset{?1}{\rightsquigarrow}} [\![\text{"out"} \mapsto \begin{array}{l} r \leftarrow \underline{\text{shuffle}}(\$books \cap \{b \,|\, b.\text{by} = \$author\}); \\ \text{ret } (\text{map } (\lambda \; row \Rightarrow row.\text{Title}) \; r) \end{array}]\!]$$
$$[\![\text{"author"} \mapsto \text{ret } \$author]\!]$$

The actual implementation of ?1 is found by applying lemmas to decompose this goal into smaller, similar goals representing subexpressions of the final
program. These lemmas form a tree of deduction steps, produced automatically
by a syntax-directed compilation script written in Coq's *Ltac* tactic language.
Crucially, the derivation implemented by this script can include *any* adequately
phrased lemma, allowing new implementation strategies. Composed with the
automation that comes before and after this stage, we have a fully automated,
proof-generating pipeline from specifications to libraries of assembly code.

# 3  An Example of Proof-Producing Extraction

We begin by illustrating the compilation process on the example Fiat program
from Sect. 2. We synthesize a Facade program p according to the following spec-
ification[2], which we summarize as $\boxed{args} \underset{\varnothing}{\overset{p}{\rightsquigarrow}} [\![\text{"out"} \mapsto p]\!] :: \boxed{args}$:

- p, when started in an initial state containing the arguments $author and
  $books, must be safe (it must not violate function preconditions, access unde-
  fined variables, leak memory, etc.).
- p, when started in a proper initial state, must reach (if it terminates) a state
  where the variable "out" has one of the values allowed by the nondeterministic
  program p shown above.

Replacing p with our example, we need to find a program p such that

$$[\![\text{"books"} \mapsto \text{ret } \$books]\!] :: \underset{\varnothing}{\overset{p}{\rightsquigarrow}} [\![\text{"out"} \mapsto \begin{array}{l} rows \leftarrow \underline{\text{shuffle}}(\$books \cap \ldots); \\ \text{ret } (\text{map } (\lambda \; row \Rightarrow row.\text{Title}) \; rows) \end{array}]\!] ::$$
$$[\![\text{"author"} \mapsto \text{ret } \$author]\!] \quad\quad\quad [\![\text{"books"} \mapsto \text{ret } \$books]\!] :: [\![\text{"author"} \mapsto \text{ret } \$author]\!]$$

We use our first *compilation lemma* (with a few examples shown in Fig. 2) to
connect the semantics of Fiat's bind operation (the $\leftarrow$ operator of monads [27])
to the meaning of $\rightsquigarrow$, which yields the following synthesis goal:

---

[2] In the following, underlined variables such as comp are Fiat computations, and ital-
icized variables such as $r$ are Gallina variables.

$$\dfrac{\forall\, v_0.\ v_0 \in \underline{v} \implies t\ v_0 \underset{[k\,\mapsto\,v_0]\,::\,ext}{\overset{p}{\rightsquigarrow}} t'\ v_0}{[\![k \mapsto \underline{v}\ \text{as}\ v_0]\!] :: t\ v_0 \underset{ext}{\overset{p}{\rightsquigarrow}} [\![k \mapsto \underline{v}\ \text{as}\ v_0]\!] :: t'\ v_0}$$

(a) The CHOMP rule: to synthesize a program whose pre- and postconditions share the same prefix $[\![k \mapsto \underline{v}]\!]$, it is enough to synthesize a program that works for any constant values permitted by the Fiat computation $\underline{v}$.

$$\dfrac{st \underset{ext}{\overset{p}{\rightsquigarrow}} [\![\_ \mapsto \underline{\text{comp}}\ \text{as}\ x]\!] :: [\![k \mapsto \underline{f}\ x]\!] :: st'}{st \underset{ext}{\overset{p}{\rightsquigarrow}} \left[\!\left[k \mapsto \boxed{\begin{array}{l} x \leftarrow \underline{\text{comp}} \\ \underline{f}\ x \end{array}}\right]\!\right] :: st'}$$

(b) The BIND rule: dependencies between consecutive bindings in Fiat states accurately model the semantics of Fiat's bind operation.

$$[\![ls \mapsto \text{ret}\ \ell]\!] :: t \underset{ext}{\overset{\text{Pinit}}{\rightsquigarrow}} [\![\text{out} \mapsto \text{ret}\ a_0]\!] :: [\![ls \mapsto \text{ret}\ \ell]\!] :: t$$

$$\dfrac{\forall\, h\ \underline{a}\ \ell.\ [\![\text{hd} \mapsto \text{ret}\ h]\!] :: [\![\text{out} \mapsto \underline{a}]\!] :: t \underset{[ls\,\mapsto\,\ell]\,::\,ext}{\overset{\text{Pbody}}{\longrightarrow}} [\![\text{out} \mapsto \underline{f}\ \underline{a}\ h]\!] :: t}{[\![ls \mapsto \text{ret}\ \ell]\!] :: t \underset{ext}{\overset{\text{LOOP}(\text{Pinit},\ \text{Pbody},\ ls)}{\longrightarrow}} [\![\text{out} \mapsto \text{fold}_L\ \underline{f}\ (\text{ret}\ a_0)\ \ell]\!] :: t}\quad\text{FOLDL}$$

(c) The FOLDL rule, connecting a fold of $\underline{f}$ on $\ell$ with an initial value $a_0$ and the imperative computation of the same value using destructive iteration on a mutable list.

```
Pinit ;
end = List.empty?(ls);
While (not end)
   hd = List.pop!(ls); Pbody ; end = List.empty?(ls);
EndWhile;
List.delete!(ls);
```

(d) The Facade LOOP($p_{init}$, $p_{body}$, ls) macro.

Fig. 2. A few rules of our synthesizing compiler.

$$[\![\text{"books"} \mapsto \text{ret}\ \$books]\!] :: [\![\text{"author"} \mapsto \text{ret}\ \$author]\!] \underset{\varnothing}{\overset{p}{\rightsquigarrow}} [\![\text{"tmp"} \mapsto \underline{\text{shuffle}}(\$books \cap \{b\ |\ b.by = \$author\})\ \text{as}\ r]\!] :: [\![\text{"out"} \mapsto \text{ret}\ (\text{map}\ (\lambda\ r \Rightarrow r.\text{Title})\ r)]\!] :: [\![\text{"books"} \mapsto \text{ret}\ \$books]\!] :: [\![\text{"author"} \mapsto \text{ret}\ \$author]\!]$$

In this step, we have broken down the assignment to "out" of a Fiat-level bind (rows ← ...; ...) into the assignment of two variables: "tmp" to the intermediate list of authors, and "out" to the final result. The :: operator separates entries in a list of bindings of Facade variables to nondeterministic Fiat terms. The ordering of the individual bindings matters: the Fiat term that we assign to "out" depends on the particular value chosen for "tmp" bound locally as $r$.

We then break down the search for p into the search for two smaller programs: the first ($p_1$) starts in the initial state (abbreviated to $\boxed{args}$) and is only concerned with the assignment to "tmp"; the second ($p_2$) starts in a state where "tmp" is already assigned and uses that value to construct the final result.

$$\boxed{args} \underset{\varnothing}{\overset{p_1}{\rightsquigarrow}} [\![\text{"tmp"} \mapsto \underline{\text{shuffle}}(\$books \cap \ldots)\ \text{as}\ r]\!] :: \boxed{args}$$

$$[\![\text{"tmp"} \mapsto \underline{\text{shuffle}}(\$books \cap \ldots)\ \text{as}\ r]\!] :: \boxed{args} \underset{\varnothing}{\overset{p_2}{\rightsquigarrow}} [\![\text{"tmp"} \mapsto \underline{\text{shuffle}}(\$books \cap \ldots)\ \text{as}\ r]\!] :: [\![\text{"out"} \mapsto \text{ret}\ (\text{map}\ (\lambda\ r \Rightarrow r.\text{Title})\ r)]\!] :: \boxed{args}$$

At this point, a lemma about connecting the meaning of the nondeterministic selection of authors and the Facade-level `BTree.find` function tells us that `tmp= BTree.find($books, $author)` is a good choice for $p_1$ (this is the *call rule* for for `BTree.find`). We are therefore only left with $p_2$ to synthesize: noticing the common prefix of the starting and ending states, we apply a rule (called *chomp* in our development) allowing us to set aside the common prefix and focus on the tail of the pre- and post-states, transforming the problem into

$$\forall\ r.\ r \in \underline{\text{shuffle}}(\$books \cap \{b \mid b.by = \$author\}) \implies$$

$$\boxed{args} \xrightarrow[["tmp" \mapsto r]]{p_2} [\![\,"out" \mapsto \text{ret}\ (\text{map}\ (\lambda\ r \Rightarrow r.\text{Title})\ r)]\!] :: \boxed{args}$$

The additional mapping pictured under the $\rightsquigarrow$ arrow indicates that the initial and final states must both map `"tmp"` to the same value $r$. In this form, we can first rewrite `map` to $\text{fold}_L$, at which point the synthesis goal matches the conclusion of the $fold_L$ rule shown in Fig. 2c: given a program $p_{init}$ to initialize the accumulator and a program $p_{body}$ to implement the body of the fold, the Facade program defined by the macro $\text{LOOP}(p_{init}, p_{body}, \text{rows})$ obeys the specification above. This gives us two new synthesis goals, which we can handle recursively, in a fashion similar to the one described above. Once these obligations have been resolved, we arrive at the desired Facade program.

# 4 Proof-Generating Extraction of Nondeterministic Functional Programs: From Fiat to Facade

## 4.1 The Facade Language

We start with a brief description of our newly designed target language, Facade. Facade is an Algol-like untyped imperative language operating on Facade states, which are finite maps from variable names to Facade values (either scalars, or nonnull values of heap-allocated ADTs). Syntactically, Facade includes standard programming constructs like assignments, conditionals, loops, function calls, and recursion. What distinguishes the language is its operational semantics, pictured partially in Fig. 3. First, that semantics follows that of Cito in supporting modularity by modeling calls to externally defined functions via preconditions and postconditions. Second, *linearity* is baked into Facade's operational semantics, which enforce that every ADT value on the heap will be referred to by exactly one live variable (no aliasing and no leakage) to simplify reasoning about the formal connection to functional programs: if every object has at most one referent, then we can almost pretend that variables hold abstract values instead of pointers to mutable objects. In practice, we have not found this requirement overly constraining for our applications: one can automatically introduce copying when needed, or one can require the external ADTs to provide nondestructive iteration.

The program semantics manipulates local-variable environments where ADTs are associated with high-level models. For instance, a finite set is modeled as a mathematical set, not as e.g. a hash table. A key parameter to the compiler soundness theorem is *a separation-logic abstraction relation, connecting the*

Statement $s ::=$

Skip $| s ; s | x = e$

If $e$ Then $s$ Else $s$ EndIf

While $e$ $s$ EndWhile

$x = $ Call $l(\overline{x})$

$$\frac{[\![e]\!]_{\mathsf{st_i}} = \mathtt{Scalar}(\_) \qquad st(x) \neq \mathtt{ADT}(\_)}{\varPsi \vdash (\mathsf{st}, \mathsf{x} = \mathsf{e}) \Downarrow [\mathsf{x} \mapsto [\![e]\!]_{\mathsf{st}}] :: \mathsf{st}} \text{ Assign}$$

$$\frac{\varPsi(1) = \mathtt{AX}(\mathsf{pre}, \mathsf{post}) \qquad st(x) \neq \mathtt{ADT}(\_)}{\mathsf{pre}(\mathsf{st}(\overline{\mathsf{y}})) \qquad |\overline{v}| = |\overline{\mathsf{y}}| \qquad \mathsf{post}(\mathsf{st}(\overline{\mathsf{y}}) \rhd \overline{v}, r)}{\varPsi \vdash (\mathsf{st}, \mathsf{x} = \mathtt{Call}\, l(\overline{\mathsf{y}})) \Downarrow [\mathsf{x} \mapsto r] :: [\overline{\mathsf{y}} \mapsto \overline{v}] :: \mathsf{st}} \text{ CallAx}$$

**Fig. 3.** Selected syntax & operational semantics of Facade [28].

*domain of high-level ADT models to mutable memories of bytes.* By picking different relations at the appropriate point in our pipeline, we can justify linking with different low-level implementations of high-level concepts. No part of our automated translation from Fiat to Facade need be aware of which relation is chosen, and the same result of that process can be reused for different later choices. This general approach to stateful encapsulation is largely inherited from Cito, though with Facade we have made it even easier to use.

Facade's operational semantics are defined by two predicates, $\varPsi \vdash (\mathsf{p}, \mathsf{st})\!\downarrow$ and $\varPsi \vdash (\mathsf{p}, \mathsf{st}) \Downarrow \mathsf{st}'$, expressing respectively that the Facade program p will run safely when started in Facade state st, and that p may reach state st' when started from st (this latter predicate essentially acts as a big-step semantics of Facade). Both predicates are parameterized over a context $\varPsi$ mapping function names to their axiomatic specifications. The semantics is nondeterministic in the sense that there can be more than one possible st'.

Modularity is achieved through the CALLAX rule, allowing a Facade program to call a function via its specification in $\varPsi$. A function call produces a return value $r$ and a list of output values $\overline{v}$ representing the result of in-place modification of input ADT arguments $\overline{y}$. A precondition is a predicate pre on the values assigned to the input arguments of the callee by the map st. A postcondition is a predicate post on these input values, output values $\overline{v}$, and return value $r$. The semantics prescribes that a function call will nondeterministically pick a list of output values and a return value satisfying post and use them to update the relevant variables in the caller's postcall state (possibly deallocating them).

Linearity is achieved by a set of syntactic and semantic provisions. For instance, variables currently holding ADT values cannot appear on the righthand sides of assignments, to avoid aliasing. They also cannot appear on the lefthand sides of assignments, to avoid losing their current payloads and causing memory leaks.

We have implemented a verified translation from Facade to Cito, and from there we reuse established infrastructure to connect into the Bedrock framework for verified assembly code. Its soundness proof has the flavor of justifying a new type system for an existing language, since Facade's syntax matches that of Cito rather closely.

$$\frac{\forall\, \mathtt{k},\, \mathtt{v}.\ \mathtt{st(k)} = \mathtt{Scalar}(v) \rightarrow \mathit{ext}(\mathtt{k}) = \mathtt{Scalar}(v) \qquad \forall\, \mathtt{k},\, \mathtt{v}.\ \mathtt{st(k)} = \mathtt{ADT}(v) \leftrightarrow \mathit{ext}(\mathtt{k}) = \mathtt{ADT}(v)}{\mathtt{st} \lesssim \varnothing \uplus \mathit{ext}}\ \text{EqvStNil}$$

$$\frac{\mathtt{st(k)} = \mathtt{wrap}(v') \qquad v' \in \underline{v} \qquad \mathtt{st\text{-}\{k\}} \lesssim (st\ v') \uplus \mathit{ext}}{\mathtt{st} \lesssim [\![\mathtt{k} \mapsto \underline{v}\ \mathtt{as}\ v]\!] :: (st\ v) \uplus \mathit{ext}}\ \text{EqvStCons}$$

**Fig. 4.** Equivalence relation on Fiat and Facade states. Because Facade does not allow us to leak ADT values, we require that all bindings pointing to ADT values in $\mathtt{st}$ be reflected in $st \uplus \mathit{ext}$, and vice versa. For scalars, we only require that bindings in $st \uplus \mathit{ext}$ be present in $\mathtt{st}$.

### 4.2 Fiat and Facade States

We connect Fiat's semantics to those of Facade by introducing a notion of *Fiat states*, which allow us to express pre and post-conditions in a concise and homogeneous way, facilitating syntax-driven compilation. Each Fiat state (denoted as $st$) describes a set of Facade states (denoted as $\mathtt{st}$): in Facade, machine states are unordered collections of names and values. Fiat states, on the other hand, are ordered collections of bindings (sometimes called *telescopes*), each containing a variable name and a set of permissible values for that variable.

Forexample,thetelescope$[\![\mathtt{"x"} \mapsto \{x \mid x > 0\}\ \mathtt{as}\ x]\!] :: [\![\mathtt{"y"} \mapsto \mathtt{ret}\ (x,\ x + 1)]\!]$ describes all machine states in which $\mathtt{"x"}$ maps to a positive value $x$ and $\mathtt{"y"}$ maps to the pair $(\mathtt{x},\ \mathtt{x} + 1)$. Each variable in a Fiat state is annotated with a function $\mathtt{wrap}$ describing how to inject values of its type in and out of the concrete type used at the Facade level (e.g. a linked list may be extracted to a vector, as in our example).

Finally, to be able to implement the aforementioned *chomp* rule, Fiat states are extended with an unordered map ($\mathit{ext}$) from names to concrete values. A full Fiat state is thus composed of a telescope $st$ and an extra collection of bindings $\mathit{ext}$, written $st \uplus \mathit{ext}$. We relate Fiat states to Facade states using the ternary predicate $\mathtt{st} \lesssim st \uplus \mathit{ext}$ defined in Fig. 4, which ensures that the values assigned to variables in the Facade state $\mathtt{st}$ are compatible with the bindings described in the Fiat state $st \uplus \mathit{ext}$.

### 4.3 Proof-Generating Extraction by Synthesis

Armed with this predicate, we are ready for the full definition of $st \overset{\mathtt{p}}{\underset{\mathit{ext}}{\rightsquigarrow}} st'$:

- $\forall\, \mathtt{st}.\ \mathtt{st} \lesssim st \uplus \mathit{ext} \implies (\mathtt{p}, \mathtt{st})\!\downarrow$
  For any initial Facade state $\mathtt{st}$, if $\mathtt{st}$ is in relation with the Fiat state $st$ extended by $\mathit{ext}$, then it is safe to run the Facade program $\mathtt{p}$ from state $\mathtt{st}$.
- $\forall\, \mathtt{st}, \mathtt{st'}.\ \mathtt{st} \lesssim st \uplus \mathit{ext} \wedge (\mathtt{p}, \mathtt{st}) \Downarrow \mathtt{st'} \implies \mathtt{st'} \lesssim st' \uplus \mathit{ext}$
  For all initial and final Facade states $\mathtt{st}$ and $\mathtt{st'}$, if $\mathtt{st}$ is in relation with the Fiat state $st$ extended by $\mathit{ext}$, and if running the Facade program $\mathtt{p}$ starting from $\mathtt{st}$ may produce the Facade state $\mathtt{st'}$, then $\mathtt{st'}$ is in relation with the Fiat state $st'$ extended by $\mathit{ext}$.

This definition is enough to concisely and precisely phrase the three types of lemmas required to synthesize Facade programs: properties of the $\leadsto$ relation used to drive the proof search and provide the extraction architecture; connections between the $\leadsto$ relation and Fiat's semantics, used to reduce extraction of Fiat programs to that of Gallina programs; and connections between Fiat and Facade, such as the FoldL rule of Fig. 2c (users provide additional lemmas of the latter kind to extend the scope of the compiler and broaden the range of source programs that the compiler is able to handle).

With these lemmas, we can phrase certified extraction as a proof-search problem that can be automated effectively. Starting from a Fiat computation $\underline{f}\ x_1 \ldots x_n$ mixing Gallina code with calls to external ADTs, we generate a specification $\uparrow\!\underline{f}\!\uparrow$ based on the $\leadsto$ predicate (which itself is defined in terms of Facade's operational semantics):

$$\uparrow\!\underline{f}\!\uparrow \triangleq \exists\,p.\ \forall\,x_1 \ldots x_n.\, [\![\texttt{"x}_1\texttt{"} \mapsto \texttt{ret}\ x_1]\!] :: \ldots :: [\![\texttt{"x}_n\texttt{"} \mapsto \texttt{ret}\ x_n]\!] \overset{p}{\underset{\varnothing}{\leadsto}} [\![\texttt{"out"} \mapsto \underline{f}\ x_1 \ldots x_n]\!] \quad (1)$$

From this starting point, extraction proceeds by analyzing the shapes of the pre- and post-states to determine applicable compilation rules, which are then used to build a Facade program progressively. This stage explains why we chose strongly constrained representations for pre and post-states: where typical verification tasks compute verification conditions from the program's source, we compute the program from carefully formulated pre- and postconditions (proper care in designing the compilation rules and their preconditions obviates the need for backtracking).

In practice, this syntax-driven process is implemented by a collection of matching functions written in Ltac. These may either fail, or solve the current goal by applying a lemma, or produce a new goal by applying a compilation lemma of the form shown in Fig. 2. Our extraction architecture is extensible: the main loop exposes hooks that users can rebind to call their own matching rules. Examples of such rules are provided in Sect. 6.1. Our focus is on extracting efficient code from Gallina EDSLs, so the set of rules is tailored to each domain and does not cover all possible programs (in particular, we do not have support for arbitrary fixpoints or pattern-matching constructs; we use custom lemmas mapping specific matches to specific code snippets or external functions). When the compiler encounters an unsupported construct $\underline{C}$, it stops and presents the user with a goal of the form $pre \overset{?}{\underset{ext}{\leadsto}} [\![k \mapsto \underline{C}]\!] :: post$, indicating which piece is missing so the user can provide the missing lemmas and tactics.

In our experience, debugging proof search and adding support for new constructs is relatively easy, though it does require sufficient familiarity with Coq. Typically, our compiler would have two classes of users: *library developers*, who interactively implement support for new DSLs (developing compilation tactics requires manual labor similar to writing a domain-specific compiler); and *final users*, who write programs within supported DSLs and use fully automated compilation tactics.

# 5  The Complete Proof-Generating Pipeline

The components presented in the previous section form the final links in an automated pipeline lowering high-level specifications to certified Bedrock modules, whose correctness is guaranteed by Theorem 1.

Starting from a Fiat ADT specification $\underline{\text{ADT}}_{\text{spec}}$ (a collection of high-level method specifications $\underline{\text{m}}_{\text{spec}}$, as shown in Fig. 5a), we obtain by refinement under a relation $\approx$ a Fiat ADT implementation $\underline{\text{ADT}}_{\text{impl}}$ (a collection of nondeterministic functional programs $\underline{\text{m}}_{\text{impl}}$, as shown in Fig. 5b). Each method of this implementation is assigned an operational specification $\lceil\underline{\text{m}}_{\text{impl}}\rceil$ (Eq. 1), from which we extract (using proof-producing synthesis, optionally augmented with user-specified lemmas and tactics) a verified Facade implementation $\text{m}_{\text{impl}}$ (Sect. 4.3) that calls into a number of external functions ($\Psi$, Fig. 3), as shown in Fig. 5c.

Finally, we package the resulting Facade methods into a Facade *module*. This module imports $\Psi$ (i.e. it must be linked against implementations of the functions in $\Psi$) and exports axiomatic specifications straightforwardly lifted *from the original high-level specifications* into Facade-style axiomatic specifications (of the style demonstrated in the call rule of Fig. 3): for each high-level specification $\underline{\text{meth}}_{\text{spec}}$, we export the following (written $\lceil\underline{\text{meth}}_{\text{spec}}\rceil$):

```
Pre ≜ λ args ⇒ ∃ rₛ r_I xs. rₛ ≈ r_I ∧ args = [r_I] ⧺ xs
Post ≜ λ args ⇒ ∃ rₛ r_I r'ₛ r'_I v xs.
    rₛ ≈ r_I ∧ (r'ₛ, v) ∈ meth_spec rₛ xs ∧
    r'ₛ ≈ r'_I ∧ args = [(r_I, r'_I)] ⧺ (zip xs xs)
```

Since we are working in an object-oriented style at the high level, our low-level code follows a convention of an extra "self" argument added to each method, written in this logical formulation as $r_S$ for spec-level "self" values and $r_I$ for implementation-level "self" values.

A *generic* proof guarantees that the operational specifications $\lceil\underline{\text{meth}}_{\text{impl}}\rceil$ used to synthesize Facade code indeed refine the axiomatic specifications $\lceil\underline{\text{meth}}_{\text{spec}}\rceil$ exported by our Facade module. Compiling this Facade module via our new formally verified Facade compiler produces a correct Bedrock module, completing Theorem 1:

**Theorem 1.** *Starting from a valid refinement $\underline{\text{ADT}}_{\text{impl}}$ of a Fiat ADT specification $\underline{\text{ADT}}_{\text{spec}}$ with methods $\underline{\text{meth}}_{\text{impl}}$ and $\underline{\text{meth}}_{\text{spec}}$ and a set of Facade programs synthesized from each $\lceil\underline{\text{meth}}_{\text{impl}}\rceil$, we can build a certified Bedrock module whose methods satisfy the axiomatic specifications $\lceil\underline{\text{meth}}_{\text{spec}}\rceil$.*

The final Bedrock module satisfies the *original, high-level* Fiat specifications. It specifies its external dependencies $\Psi$, for which verified assembly implementations must be provided as part of the final *linking* phase, which happens entirely inside of Coq. After linking, we obtain a closed, executable Bedrock module, exposing an axiomatic specification directly derived from the original, high-level ADT specification. Our implementation links against verified hand-written implementations of low-level indexing structures, though it would be possible to use the output of any compiler emitting Bedrock assembly code.

# 6  Evaluation

## 6.1  Microbenchmarks

We first evaluated our pipeline by extracting a collection of twenty six Gallina programs manipulating machine words, lists, and nested lists, with optional nondeterministic choices. Extraction takes a few seconds for each program, ranging from simple operations such as performing basic arithmetic, allocating data structures, calling compiler intrinsics, or sampling arbitrary numbers to more

```
Definition SchedulerSpec: ADT ≜ QueryADTRep SchedulerSchema {
  Def Init: rep ≜ empty,

  Insert a new process, failing if newpid already exists.
  Def Spawn (r: rep) (newpid cpu st: W32): B ≜
  Insert <pid:: newpid, state:: st, cpu:: cpu> into r.Procs,

  Find all processes in a state st and return their PIDs.
  Def Enumerate (r: rep) (st: W32): list W32 ≜
  pids ← For p in r.Procs Where p.state = st Return p.pid; ret (r, pids),

  Find a process by PID and return its CPU time.
  Def GetCPUTime (r: rep) (id: W32): list W32 ≜
  cpu ← For p in r.Procs Where p.pid = id Return p.cpu; ret (r, cpu) }.
```

(a) The original Fiat specification of a process scheduler. The refinement process derives an efficient functional implementation of this specification by implementing it using nested trees keyed on the process ID, followed by the process state.

```
Spawn
a ← bfind r (_, $newpid, _);
if length (snd a) == 0 then
   u ← binsert (fst a)
        [$newpid, $state, $cpu];
   ret (fst u, true)
else ret (fst a, false)
```

```
Enumerate
a ← bfind r ($state, _, _);
ret (fst a, revmap (Get 0) (snd a))

GetCPUTime
a ← bfind r (_, $id, _);
ret (fst a, revmap (Get 2) (snd a))
```

(b) The output of Fiat, after refining the specifications presented in Fig. 5a. Notice the use of bfind and binsert, two nondeterministic methods of a bag ADT. In this example, the bag data structure that bfind depends on is expected to provide fast lookups by state and then by ID.

```
Enumerate
procs = BTree.findFirst(rep, $state);
return = List[W32].new();
test = List[Tuple].empty?(procs);
While (not test)
  head = List[Tuple].pop!(procs);
  head' = Tuple.get(head, 0);
  Call Tuple.delete!(head, 3);
  Call List[W32].push(ret, head');
  test = List[Tuple].empty?(procs)
EndWhile;
Call List[Tuple].delete!(procs);
```

(c) Facade output for the **Enumerate** method. The low-level data structure that we will link against is a nested tree, indexed by state and then by process ID. The call to findFirst returns a list of all processes in a particular state. **GetCPUTime** has a similar shape but is implemented using a skip-scan (or "loose index scan") on the first level of the nested tree (process states), followed by a search on the second level of the tree (process IDs).

**Fig. 5.** Different stages of a process-scheduler compilation example (see also the annotated 'ProcessScheduler.v' file).

complex operations involving sequence manipulations, like reversing, filtering, reducing (e.g. reading in a number written as a list of digits in a given base), flattening, and duplicating or replacing elements. All examples, and the corresponding outputs, are included in a literate Coq file available online. These examples illustrate that our extraction engine supports a fluid, extensible source language, including subsets of Gallina and many nondeterministic Fiat programs.

## 6.2   Relational Queries

To evaluate our full pipeline in realistic conditions, we targeted the query-structure ADT library of the Fiat paper [4] as well as an ADT modeling process scheduling inspired by Hawkins et al. [7]. This benchmark starts from high-level Fiat specifications (as shown in Fig. 5a) and outputs a closed Bedrock module, linked against a hand-verified nested-binary-tree implementation.

From Fiat specifications we derive a collection of nondeterministic Fiat programs (one per ADT method, as demonstrated in Fig. 5b), then extract each method to Facade Fig. 5c) and compile to Bedrock. Extraction is fully automatic; it draws from the default pool of extraction lemmas (about conditionals, constants, arithmetic operations, etc.) and from bag-specific lemmas that we added to the compiler (these manually verified *call rules* connect the pure bag specifications used in Fiat sources to Bedrock-style specifications of mutable binary search trees using the $\leadsto$ relation).

Figure 6 presents the results of our experimental validation. We compare our own verified implementation ("Fiat") against the corresponding SQL queries executed by SQLite 3.8.2 (using an in-memory database) and PostgreSQL 9.3.11 ("PG"). For increasingly large collections of processes, we run 20,000 Enumerate queries to locate the 10 active processes, followed by 10,000 GetCPUTime queries for arbitrary process IDs. In all cases, the data is indexed by (state, PID) to allow for constant-time Enumerate queries (the number of active

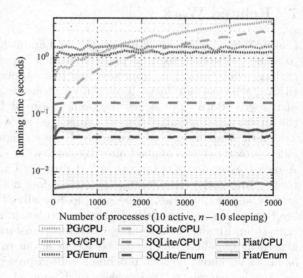

**Fig. 6.** Process scheduler benchmarks.

processes is kept constant) and logarithmic-time GetCPUTime queries (assuming a B-tree-style index and skip-scans).

Our implementation behaves as expected: it beats SQLite and PostgreSQL by 1.5 and 2.5 orders of magnitude respectively on GetCPUTime, and competes honorably with SQLite (while beating PostgreSQL by one order of magnitude) on Enumerate. Notice the red curves on the graph: without an explicit "$state IN (0, 1)" clause, both database management systems missed the skipscan opportunity and exhibited asymptotically suboptimal linear-time behavior, so we had to tweak the queries fed to PostgreSQL and SQLite to obtain good GetCPUTime performance (in contrast, the optimizer in our system can be guided explicitly by adding compiler hints in the form of extra tactics, without modifying the specifications).

Of course, our implementation does much less work than a database engine; the strength of our approach is to expose an SQL-style interface while enabling generation of specialized data-structure-manipulation code, allowing programmers to benefit from the conciseness and clarity of high-level specifications without incurring the overheads of a full-fledged DBMS.

*Trusted Base.* Our derivation assumes ensemble extensionality and Axiom K. Our trusted base comprises the Coq 8.4 checker [25] ($\sim$10 000 lines of OCaml code), the semantics of the Bedrock IL and the translator from it to x86 assembly ($\sim$1200 lines of Gallina code), an assembler, and wrappers for extracted methods ($\sim$50 lines of x86 assembly). We used Proof General [2] for development.

# 7    Related Work

Closely related to our work is a project by Lammich [10] that uses Isabelle/HOL to refine functional programs to an embedded imperative language that requires garbage collection. This approach has been applied to various complex algorithms, whereas our focus is on fully automatic derivation from highly regular specs. Both approaches use some form of linearity checking to bridge the functional-imperative gap (Lammich et al. use separation logic [20] and axiomatic semantics, while we apply Facade's lighter-weight approach: decidable syntactic checks applied after-the-fact, with no explicit pointer reasoning). A recent extension [11] targets LLVM directly. Crucially, the initial work only targets Imperative/HOL and its extension does not support linking against separately verified libraries, while our pipeline allows linking, inside of Coq, low-level programs against verified libraries written in any language of the Bedrock ecosystem. Finally, we have integrated our translation into an automated proof-generating pipeline from relational specifications to executable assembly code—as far as we know, no such pipeline has been presented before.

Another closely related project by Kumar et al. [8,17] focuses on extracting terms written in a purely functional subset of HOL4's logic into the CakeML dialect of ML. The main differences with our pipeline are optimization opportunities, extensibility, and external linking. Indeed, while the compiler to CakeML bridges a relatively narrow gap (between two functional languages with expressive type systems and automatic memory management), our extraction procedure connects two very different languages, opening up many more opportunities for optimizations (including some related to memory management). We expose these opportunities to our users by letting them freely extend the compiler based on their domain-specific optimization knowledge.

Recent work by Protzenko et al. [19] achieves one of our stated goals (efficient extraction to low-level code, here from F* to C) but does not provide formal correctness guarantees for the extracted code (the tool, KreMLin, consists of over 15,000 lines of unverified OCaml code). Additionally, KreMLin requires source programs to be written in a style matching that of the extracted code: instead of extending the compiler with domain-specific representation choices and optimizations, users must restrict their programs to the Low* subset of F*.

One last related project is the compiler of the COGENT language [18]. Its sources are very close to Facade's (it allows for foreign calls to axiomatically specified functions, but it does not permit iteration or recursion except through foreign function calls), and its compiler also produces low-level code without a garbage collector. Our projects differ in architecture and in spirit: COGENT is closer to a traditional verified compiler, producing consecutive embeddings of a source program (from C to a shallow embedding in Isabelle/HOL) and generating equivalence proofs connecting each of them. COGENT uses a linear type system to establish memory safety, while we favor extensibility over completeness, relying on lemmas to justify the compilation of arbitrary Gallina constructs.

We draw further inspiration from a number of other efforts:

*Program Extraction.* Program extraction (a facility offered by Coq and other proof assistants) is a popular way of producing executable binaries from verified code. Extractors are rather complex programs, subjected to varying degrees of scrutiny: for example, the theory behind Coq's extraction was mechanically formalized and verified [14], but the corresponding concrete implementation itself is unverified. The recent development of CertiCoq [1], a verified compiler for Gallina, has significantly improved matters over unverified extraction, but it only supports pure Gallina programs, and it uses a fixed compilation strategy. In contrast, our pipeline ensures that nondeterministic specifications are preserved down to the generated Bedrock code and grants user fine control over the compilation process.

*Compiler Verification.* Our compilation strategy allows Fiat programs to depend on separately compiled libraries. This contrasts with verified compilers like CakeML [9] or CompCert [13]: in the latter, correctness guarantees only extend to linking with modules written in CompCert C and compiled with the same version of the compiler. Recent work [23] generalized these guarantees to cover cross-language compilation, but these developments have not yet been used to perform functional verification of low-level programs assembled from separately verified components.

An alternative approach, recently used to verify an operating-system kernel [21], is to validate individual compiler outputs. This is particularly attractive as an extension of existing compilers, but it generally falls short when trying to verify complex optimizations, such as our high-level selection of algorithms and data structures. In the same vein, verified compilers often rely on unverified programs to solve complex problems such as register allocation, paired with verified checkers to validate solutions. In our context, the solver is the proof-producing extraction logic, and the verifier is Coq's kernel: our pipeline produces proofs that witness the correctness of the resulting Facade code.

*Extensible Compilation.* Multiple research projects let users add optimizations to existing compilers. Some, like Racket [26], do not focus on verification. Others, like Rhodium [12], let users phrase and verify transformations using DSLs. Unfortunately, most of these tools are unverified and do not provide end-to-end guarantees. One recent exception is XCert [24], which lets CompCert users soundly describe program transformations using an EDSL. Our approach is similar insofar as we assemble DSL compilers from collections of verified rewritings.

*Program Synthesis.* Our approach of program generation via proofs follows in the deductive-synthesis tradition started in the 1980s [15]. We use the syntactic structure of our specialized pre- and postconditions to drive synthesis: the idea of strongly constraining the search space is inherited from the syntax-guided approach pioneered in the *Sketch* language [22]. That family of work uses SMT solvers where we use a proof assistant, offering more baseline automation with less fundamental flexibility.

*Formal Decompilation.* Instead of deriving low-level code from high-level specifications, some authors have used HOL-family proof assistants to translate unverified low-level programs (in assembly [16] or C [6]) into high-level code suitable for verification. Decompilation is an attractive approach for existing low-level code, or when compiler verification is impractical.

## 8    Conclusion

The extraction technique presented in this paper is a convenient and lightweight approach for generating certified extracted programs, reducing the trusted base of verified programs to little beyond a proof assistant's kernel. We have shown our approach to be suitable for the extraction of DSLs embedded in proof assistants, using it to compile a series of microbenchmarks and to do end-to-end proof-generating derivation of assembly code from SQL-style specifications. Crucially, the latter derivations work via linking with verified implementations of assembly code that our derivation pipeline could never produce directly. To ease this transition, we developed Facade, a new language designed to facilitate reasoning about memory allocation in synthesized extracted programs. In the process, we have closed the last gap in the first automatic and mechanically certified translation pipeline from declarative specifications to assembly-language libraries, supporting user-guided optimizations and parameterization over abstract data types implemented, compiled, and verified using arbitrary languages and tools.

**Acknowledgments.** This work has been supported in part by NSF grants CCF-1512611 and CCF-1521584, and by DARPA under agreement number FA8750-16-C-0007. The U.S. Government is authorized to reproduce and distribute reprints for Governmental purposes notwithstanding any copyright notation thereon. The views and conclusions contained herein are those of the authors and should not be interpreted as necessarily representing the official policies or endorsements, either expressed or implied, of DARPA or the U.S. Government.

# References

1. Anand, A., et al.: CertiCoq: a verified compiler for Coq. In: The Third International Workshop on Coq for PL, CoqPL 2017, January 2017
2. Aspinall, D.: Proof general: a generic tool for proof development. In: Graf, S., Schwartzbach, M. (eds.) TACAS 2000. LNCS, vol. 1785, pp. 38–43. Springer, Heidelberg (2000). https://doi.org/10.1007/3-540-46419-0_3
3. Chlipala, A.: The Bedrock structured programming system: combining generative metaprogramming and Hoare logic in an extensible program verifier. In: ACM SIGPLAN International Conference on Functional Programming, ICFP 2013, Boston, MA, USA, 25–27 September 2013, pp. 391–402 (2013). https://doi.org/10.1145/2500365.2500592
4. Delaware, B., Pit-Claudel, C., Gross, J., Chlipala, A.: Fiat: deductive synthesis of abstract data types in a proof assistant. In: ACM SIGPLAN-SIGACT Symposium on Principles of Programming Languages, POPL 2015, Mumbai, India, 15–17 January 2015, pp. 689–700 (2015). https://doi.org/10.1145/2676726.2677006
5. Dijkstra, E.W.: A constructive approach to the problem of program correctness, August 1967. https://www.cs.utexas.edu/users/EWD/ewd02xx/EWD209.PDF, circulated privately
6. Greenaway, D., Andronick, J., Klein, G.: Bridging the gap: automatic verified abstraction of C. In: Beringer, L., Felty, A. (eds.) ITP 2012. LNCS, vol. 7406, pp. 99–115. Springer, Heidelberg (2012). https://doi.org/10.1007/978-3-642-32347-8_8
7. Hawkins, P., Aiken, A., Fisher, K., Rinard, M.C., Sagiv, M.: Data representation synthesis. In: ACM SIGPLAN Conference on Programming Language Design and Implementation, PLDI 2011, San Jose, CA, USA, 4–8 June 2011, pp. 38–49 (2011). https://doi.org/10.1145/1993498.1993504
8. Ho, S., Abrahamsson, O., Kumar, R., Myreen, M.O., Tan, Y.K., Norrish, M.: Proof-producing synthesis of CakeML with I/O and local state from monadic HOL functions. In: Galmiche, D., Schulz, S., Sebastiani, R. (eds.) IJCAR 2018. LNCS (LNAI), vol. 10900, pp. 646–662. Springer, Cham (2018). https://doi.org/10.1007/978-3-319-94205-6_42
9. Kumar, R., Myreen, M.O., Norrish, M., Owens, S.: CakeML: a verified implementation of ML. In: ACM SIGPLAN-SIGACT Symposium on Principles of Programming Languages, POPL 2014, San Diego, CA, USA, 20–21 January 2014, pp. 179–192 (2014). https://doi.org/10.1145/2535838.2535841

10. Lammich, P.: Refinement to imperative/HOL. In: Urban, C., Zhang, X. (eds.) ITP 2015. LNCS, vol. 9236, pp. 253–269. Springer, Cham (2015). https://doi.org/10.1007/978-3-319-22102-1_17

11. Lammich, P.: Generating verified LLVM from Isabelle/HOL. In: International Conference on Interactive Theorem Proving, ITP 2019, Portland, OR, USA, 9–12 September 2019, pp. 22:1–22:19 (2019). https://doi.org/10.4230/LIPIcs.ITP.2019.22

12. Lerner, S., Millstein, T.D., Rice, E., Chambers, C.: Automated soundness proofs for dataflow analyses and transformations via local rules. In: ACM SIGPLAN-SIGACT Symposium on Principles of Programming Languages, POPL 2005, Long Beach, California, USA, 12–14 January 2005, pp. 364–377 (2005). https://doi.org/10.1145/1040305.1040335

13. Leroy, X.: Formal certification of a compiler back-end or: programming a compiler with a proof assistant. In: ACM SIGPLAN-SIGACT Symposium on Principles of Programming Languages, POPL 2006, Charleston, South Carolina, USA, 11–13 January 2006, pp. 42–54 (2006). https://doi.org/10.1145/1111037.1111042

14. Letouzey, P.: A new extraction for coq. In: Geuvers, H., Wiedijk, F. (eds.) TYPES 2002. LNCS, vol. 2646, pp. 200–219. Springer, Heidelberg (2003). https://doi.org/10.1007/3-540-39185-1_12

15. Manna, Z., Waldinger, R.J.: A deductive approach to program synthesis. ACM Trans. Program. Lang. Syst. $2$(1), 90–121 (1980). https://doi.org/10.1145/357084.357090

16. Myreen, M.O., Gordon, M.J.C., Slind, K.: Decompilation into logic - improved. In: Formal Methods in Computer-Aided Design, FMCAD 2012, Cambridge, UK, 22–25 October 2012, pp. 78–81 (2012). https://ieeexplore.ieee.org/document/6462558/

17. Myreen, M.O., Owens, S.: Proof-producing synthesis of ML from higher-order logic. In: ACM SIGPLAN International Conference on Functional Programming, ICFP 2012, Copenhagen, Denmark, 9–15 September 2012, pp. 115–126 (2012). https://doi.org/10.1145/2364527.2364545

18. O'Connor, L., et al.: COGENT: certified compilation for a functional systems language. CoRR abs/1601.05520 (2016). https://arxiv.org/abs/1601.05520

19. Protzenko, J., et al.: Verified low-level programming embedded in F*. In: Proceedings of the ACM on Programming Languages 1(ICFP), pp. 17:1–17:29 (2017). https://doi.org/10.1145/3110261

20. Reynolds, J.C.: Separation logic: a logic for shared mutable data structures. In: IEEE Symposium on Logic in Computer Science, LICS 2002, Copenhagen, Denmark, 22–25 July 2002, pp. 55–74 (2002). https://doi.org/10.1109/LICS.2002.1029817

21. Sewell, T.A.L., Myreen, M.O., Klein, G.: Translation validation for a verified OS kernel. In: ACM SIGPLAN Conference on Programming Language Design and Implementation, PLDI 2013, Seattle, WA, USA, 16–19 June 2013, pp. 471–482 (2013). https://doi.org/10.1145/2491956.2462183

22. Solar-Lezama, A.: The sketching approach to program synthesis. In: Hu, Z. (ed.) APLAS 2009. LNCS, vol. 5904, pp. 4–13. Springer, Heidelberg (2009). https://doi.org/10.1007/978-3-642-10672-9_3

23. Stewart, G., Beringer, L., Cuellar, S., Appel, A.W.: Compositional CompCert. In: ACM SIGPLAN-SIGACT Symposium on Principles of Programming Languages, POPL 2015, Mumbai, India, 15–17 January 2015, pp. 275–287 (2015). https://doi.org/10.1145/2676726.2676985

24. Tatlock, Z., Lerner, S.: Bringing extensibility to verified compilers. In: ACM SIG-PLAN Conference on Programming Language Design and Implementation, PLDI 2010, Toronto, Ontario, Canada, 5–10 June 2010, pp. 111–121 (2010). https://doi.org/10.1145/1806596.1806611
25. The Coq Development Team: The Coq Proof Assistant Reference Manual (2012). https://coq.inria.fr, version 8.4
26. Tobin-Hochstadt, S., St-Amour, V., Culpepper, R., Flatt, M., Felleisen, M.: Languages as libraries. In: ACM SIGPLAN Conference on Programming Language Design and Implementation, PLDI 2011, San Jose, CA, USA, 4–8 June 2011, pp. 132–141 (2011). https://doi.org/10.1145/1993498.1993514
27. Wadler, P.: Comprehending monads. Math. Struct. Comput. Sci. **2**(4), 461–493 (1992). https://doi.org/10.1017/S0960129500001560
28. Wang, P.: The Facade language. Technical report, MIT CSAIL (2016). https://people.csail.mit.edu/wangpeng/facade-tr.pdf
29. Wang, P., Cuellar, S., Chlipala, A.: Compiler verification meets cross-language linking via data abstraction. In: ACM International Conference on Object Oriented Programming Systems Languages & Applications, OOPSLA 2014, part of SPLASH 2014, Portland, OR, USA, 20–24 October 2014, pp. 675–690 (2014). https://doi.org/10.1145/2660193.2660201

# Validating Mathematical Structures

Kazuhiko Sakaguchi(✉)

University of Tsukuba, Tsukuba, Japan
sakaguchi@logic.cs.tsukuba.ac.jp

**Abstract.** Sharing of notations and theories across an inheritance hierarchy of mathematical structures, e.g., groups and rings, is important for productivity when formalizing mathematics in proof assistants. The packed classes methodology is a generic design pattern to define and combine mathematical structures in a dependent type theory with records. When combined with mechanisms for implicit coercions and unification hints, packed classes enable automated structure inference and subtyping in hierarchies, e.g., that a ring can be used in place of a group. However, large hierarchies based on packed classes are challenging to implement and maintain. We identify two hierarchy invariants that ensure modularity of reasoning and predictability of inference with packed classes, and propose algorithms to check these invariants. We implement our algorithms as tools for the Coq proof assistant, and show that they significantly improve the development process of Mathematical Components, a library for formalized mathematics.

## 1 Introduction

Mathematical structures are a key ingredient of modern formalized mathematics in proof assistants, e.g., [1,18,25,41] [10, Chap. 2 and Chap. 4] [20, Sect. 3] [30, Chap. 5] [46, Sect. 4]. Since mathematical structures have an inheritance/subtyping hierarchy such that "a ring is a group and a group is a monoid", it is usual practice in mathematics to reuse notations and theories of superclasses implicitly to reason about a subclass. Similarly, the sharing of notations and theories across the hierarchy is important for productivity when formalizing mathematics.

The packed classes methodology [16,17] is a generic design pattern to define and combine mathematical structures in a dependent type theory with records. Hierarchies using packed classes support multiple inheritance, and maximal sharing notations and theories. When combined with mechanisms for implicit coercions [32,33] and for extending unification procedure, such as the canonical structures [26,33] of the Coq proof assistant [42], and the unification hints [4] of the Lean theorem prover [6,27] and the Matita interactive theorem prover [5], packed classes enable subtyping and automated inference of structures in hierarchies. Compared to approaches based on type classes [22,40], packed classes are more robust, and their inference approach is efficient and predictable [1]. The success of the packed classes methodology in formalized mathematics can be seen in the Mathematical Components library [45] (hereafter MathComp), the Coquelicot

© Springer Nature Switzerland AG 2020
N. Peltier and V. Sofronie-Stokkermans (Eds.): IJCAR 2020, LNAI 12167, pp. 138–157, 2020.
https://doi.org/10.1007/978-3-030-51054-1_8

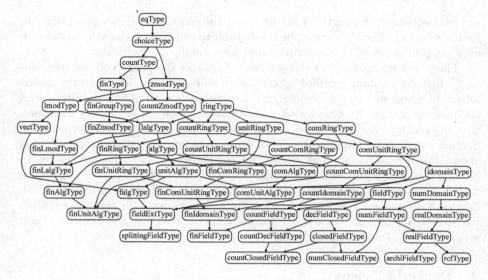

**Fig. 1.** The hierarchy of structures in the MathComp library 1.10.0

library [8], and especially the formal proof of the Odd Order Theorem [20]. It has also been successfully applied for program verification tasks, e.g., a hierarchy of monadic effects [2] and a hierarchy of partial commutative monoids [28] for Fine-grained Concurrent Separation Logic [39].

In spite of its success, the packed classes methodology is hard to master for library designers and requires a substantial amount of work to maintain as libraries evolve. For instance, the strict application of packed classes requires defining quadratically many implicit coercions and unification hints in the number of structures. To give some figures, the MathComp library 1.10.0 uses this methodology ubiquitously to define the 51 mathematical structures depicted in Fig. 1, and declares 554 implicit coercions and 746 unification hints to implement their inheritance. Moreover, defining new intermediate structures between existing ones requires fixing their subclasses and their inheritance accordingly; thus, it can be a challenging task.

In this paper, we indentify two hierarchy invariants concerning implicit coercions and unification hints in packed classes, and propose algorithms to check these invariants. We implement our algorithms as tools for the Coq system, evaluate our tools on a large-scale development, the MathComp library 1.7.0, and then successfully detect and fix several inheritance bugs with the help of our tools. The invariant concerning implicit coercions ensures the modularity of reasoning with packed classes and is also useful in other approaches, such as type classes and telescopes [26, Sect. 2.3], in a dependent type theory. This invariant was proposed before as a *coherence* of inheritance graphs [7]. The invariant concerning unification hints, that we call *well-formedness*, ensures the predictability of structure inference. Our tool not only checks well-formedness, but also generates an exhaustive set of assertions for structure inference, and these assertions can be tested inside Coq. We state the predictability of inference as a metatheorem on a simplified model of hierarchies, that we formally prove in Coq.

The paper is organized as follows: Sect. 2 reviews the packed classes methodology using a running example. Section 3 studies the implicit coercion mechanism of Coq, and then presents the new coherence checking algorithm and its implementation. Section 4 reviews the use of canonical structures for structure inference in packed classes, and introduces the notion of well-formedness. Section 5 defines a simplified model of hierarchies and structure inference, and shows the metatheorem that states the predictability of structure inference. Section 6 presents the well-formedness checking algorithm and its implementation. Section 7 evaluates our checking tools on the MathComp library 1.7.0. Section 8 discusses related work and concludes the paper. Our running example for Sect. 2, Sect. 4, and Sect. 6, the formalization for Sect. 5, and the evaluation script for Sect. 7 are available at [37].

## 2    Packed Classes

This section reviews the packed classes methodology [16,17] through an example, but elides canonical structures. Our example is a minimal hierarchy with multiple inheritance, consisting of the following four algebraic structures (Fig. 2):

**Additive monoids** $(A, +, 0)$: Monoids have an associative binary operation $+$ on the set $A$ and an identity element $0 \in A$.

**Semirings** $(A, +, 0, \times, 1)$: Semirings have the monoid axioms, commutativity of addition, multiplication, and an element $1 \in A$. Multiplication $\times$ is an associative binary operation on $A$ that is left and right distributive over addition. 0 and 1 are absorbing and identity elements with respect to multiplication, respectively.

**Additive groups** $(A, +, 0, -)$: Groups have the monoid axioms and a unary operation $-$ on $A$. $-x$ is the additive inverse of $x$ for any $x \in A$.

**Rings** $(A, +, 0, -, \times, 1)$: Rings have all the semiring and group axioms, but no additional axioms.

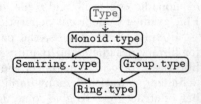

**Fig. 2.** Hierarchy diagram for monoids, semirings, groups, and rings, where an arrow from X.type to Y.type means that Y directly inherits from X. The monoid structure is the superclass of all other structures. Semirings and groups directly inherit from monoids. Rings directly inherit from semirings and groups, and indirectly inherit from monoids.

We start by defining the base class, namely, the Monoid structure.

```
1   Module Monoid.
2
3   Record mixin_of (A : Type) := Mixin {
4      zero : A;
5      add : A → A → A;
```

```
6    addA : associative add;              (* 'add' is associative.                *)
7    addOx : left_id zero add;            (* 'zero' is the left and right         *)
8    addx0 : right_id zero add;           (*    identity element w.r.t. 'add'.    *)
9    }.
10
11   Record class_of (A : Type) := Class { mixin : mixin_of A }.
12
13   Structure type := Pack { sort : Type; class : class_of sort }.
14
15   End Monoid.
```

The above definitions are enclosed by the `Monoid` module, which forces users to write qualified names such as `Monoid.type`. Thus, we can reuse the same name of record types (`mixin_of`, `class_of`, and `type`), their constructors (`Mixin`, `Class`, and `Pack`), and constants (e.g., `sort` and `class`) for other structures to indicate their roles. Structures are written as records that have three different roles: *mixins*, *classes*, and *structures*. The mixin record (line 3) gathers operators and axioms newly introduced by the `Monoid` structure. Since monoids do not inherit any other structure in Fig. 2, those are all the monoid operators, namely 0 and +, and their axioms. The class record (line 11) assembles all the mixins of the superclasses of the `Monoid` structure (including itself), which is the singleton record consisting of the monoid mixin. The structure record (line 13) is the actual interface of the structure that bundles a carrier of type `Type` and its class instance. `Record` and `Structure` are synonyms in Coq, but we reserve the latter for actual interfaces of structures. In a hierarchy of algebraic structures, a carrier set `A` has type `Type`; hence, for each structure, the first field of the `type` record should have type `Type`, and the `class_of` record should be parameterized by that carrier. In general, it can be other types, e.g., `Type` $\rightarrow$ `Type` in the hierarchy of functors and monads [2], but should be fixed in each hierarchy of structures.

Mixin and monoid records are internal definitions of mathematical structures; in contrast, the structure record is a part of the interface of the monoid structure when reasoning about monoids. For this reason, we lift the projections for `Monoid.mixin_of` to definitions and lemmas for `Monoid.type` as follows.

```
Definition zero {A : Monoid.type} : Monoid.sort A :=
  Monoid.zero _ (Monoid.mixin _ (Monoid.class A)).
Definition add {A : Monoid.type} :
  Monoid.sort A → Monoid.sort A → Monoid.sort A :=
  Monoid.add _ (Monoid.mixin _ (Monoid.class A)).
Lemma addA {A : Monoid.type} : associative (@add A).
Lemma addOx {A : Monoid.type} : left_id (@zero A) (@add A).
Lemma addx0 {A : Monoid.type} : right_id (@zero A) (@add A).
```

The curly brackets enclosing `A` mark it as an implicit argument; in contrast, `@` is the explicit application symbol that deactivates the hiding of implicit arguments.

Since a monoid instance `A : Monoid.type` can be seen as a type equipped with monoid axioms, it is natural to declare `Monoid.sort` as an implicit coercion. The types of `zero` can be written and shown as $\forall A$ : `Monoid.type`, `A` rather than $\forall A$ : `Monoid.type`, `Monoid.sort A` thanks to this implicit coercion.

```
Coercion Monoid.sort : Monoid.type >-> Sortclass.
```

Next, we define the `Semiring` structure. Since semirings inherit from monoids and the semiring axioms interact with the monoid operators, e.g., distributivity of multiplication over addition, the semiring mixin should take `Monoid.type` rather than `Type` as its argument.

```
Module Semiring.

Record mixin_of (A : Monoid.type) := Mixin {
  one : A;
  mul : A → A → A;
  addC : commutative (@add A);           (* 'add' is commutative.                *)
  mulA : associative mul;                (* 'mul' is associative.                *)
  mul1x : left_id one mul;               (* 'one' is the left and right          *)
  mulx1 : right_id one mul;              (*    identity element w.r.t. 'mul'.    *)
  mulDl : left_distributive mul add;     (* 'mul' is left and right              *)
  mulDr : right_distributive mul add;    (*    distributive over 'add'.          *)
  mul0x : left_zero zero mul;            (* 'zero' is the left and right         *)
  mulx0 : right_zero zero mul;           (*    absorbing element w.r.t. 'mul'.   *)
}.
```

The `Semiring` class packs the `Semiring` mixin together with the `Monoid` class to assemble the mixin records of monoids and semirings. We may also assemble all the required mixins as record fields directly rather than nesting class records, yielding what is called the *flat* variant of packed classes [14, Sect. 4]. Since the semiring mixin requires `Monoid.type` as its type argument, we have to bundle the monoid class with the carrier to provide that `Monoid.type` instance, as follows.

```
Record class_of (A : Type) :=
  Class { base : Monoid.class_of A; mixin : mixin_of (Monoid.Pack A base) }.

Structure type := Pack { sort : Type; class : class_of sort }.
```

The inheritance from monoids to semirings can then be expressed as a canonical way to construct a monoid from a semiring as below.

```
Local Definition monoidType (cT : type) : Monoid.type :=
  Monoid.Pack (sort cT) (base _ (class cT)).

End Semiring.
```

Following the above method, we declare `Semiring.sort` as an implicit coercion, and then lift `mul`, `one`, and the semiring axioms from projections for the mixin to definitions for `Semiring.type`.

```
1  Coercion Semiring.sort : Semiring.type >-> Sortclass.
2  Definition one {A : Semiring.type} : A :=
3    Semiring.one _ (Semiring.mixin _ (Semiring.class A)).
4  Definition mul {A : Semiring.type} : A → A → A :=
5    Semiring.mul _ (Semiring.mixin _ (Semiring.class A)).
6  Lemma addC {A : Semiring.type} : commutative (@add (Semiring.monoidType A)).
7  ...
```

In the statement of the `addC` axiom (line 6 just above), we need to explicitly write `Semiring.monoidType A` to get the canonical `Monoid.type` instance for `A : Semiring.type`. We omit this subtyping function `Semiring.monoidType` by declaring it as an implicit coercion. In general, for a structure S inheriting from other structures, we define implicit coercions from S to all its superclasses.

```
Coercion Semiring.monoidType : Semiring.type >-> Monoid.type.
```

The `Group` structure is monoids extended with an additive inverse. Following the above method, it can be defined as follows.

```
Module Group.

Record mixin_of (A : Monoid.type) := Mixin {
  opp : A → A;
  addNx : left_inverse zero opp add;    (* 'opp x' is the left and right   *)
  addxN : right_inverse zero opp add;   (*    additive inverse of 'x'.     *)
}.

Record class_of (A : Type) :=
  Class { base : Monoid.class_of A; mixin : mixin_of (Monoid.Pack A base) }.

Structure type := Pack { sort : Type; class : class_of sort }.

Local Definition monoidType (cT : type) : Monoid.type :=
  Monoid.Pack (sort cT) (base _ (class cT)).

End Group.

Coercion Group.sort : Group.type >-> Sortclass.
Coercion Group.monoidType : Group.type >-> Monoid.type.
Definition opp {A : Group.type} : A → A :=
  Group.opp _ (Group.mixin _ (Group.class A)).
...
```

The `Ring` structure can be seen both as groups extended by the semiring axioms and as semirings extended by the group axioms. Here, we define it in the first way, but one may also define it in the second way. Since rings have no other axioms than the group and semiring axioms, no additional `mixin_of` record is needed.[1]

```
Module Ring.

Record class_of (A : Type) := Class {
  base : Group.class_of A;
  mixin : Semiring.mixin_of (Monoid.Pack A (Group.base A base)) }.

Structure type := Pack { sort : Type; class : class_of sort }.
```

The ring structure inherits from monoids, groups, and semirings. Here, we define implicit coercions from the ring structure to those superclasses.

---

[1] One may also define a new structure that inherits from multiple existing classes and has an extra mixin, e.g., by defining commutative rings instead of rings in this example, and left algebras as `lalgType` in MathComp.

```
Local Definition monoidType (cT : type) : Monoid.type :=
  Monoid.Pack (sort cT) (Group.base _ (base _ (class cT))).
Local Definition groupType (cT : type) : Group.type :=
  Group.Pack (sort cT) (base _ (class cT)).
Local Definition semiringType (cT : type) : Semiring.type :=
  Semiring.Pack (sort cT) (Semiring.Class _ (Group.base _ (base _ (class cT)))
                          (mixin _ (class cT))).
End Ring.

Coercion Ring.sort : Ring.type >-> Sortclass.
Coercion Ring.monoidType : Ring.type >-> Monoid.type.
Coercion Ring.semiringType : Ring.type >-> Semiring.type.
Coercion Ring.groupType : Ring.type >-> Group.type.
```

# 3   Coherence of Implicit Coercions

This section describes the implicit coercion mechanism of Coq and the *coherence property* [7] of inheritance graphs that ensures modularity of reasoning with packed classes, and presents the coherence checking mechanism we implemented in Coq. More details on implicit coercions can be found in the Coq reference manual [44], and its typing algorithm is described in [32]. First, we define classes and implicit coercions.

**Definition 3.1. (Classes** [32, Sect. 3.1] [44]**).** *A class with n parameters is a defined name* $C$ *with a type* $\forall(x_1 : T_1) \dots (x_n : T_n)$, *sort where sort is* **SProp**, **Prop**, **Set**, *or* **Type**. *Thus, a class with parameters is considered a single class and not a family of classes. An object of class* $C$ *is any term of type* $C\, t_1 \dots t_n$.

**Definition 3.2. (Implicit coercions).** *A name* $f$ *can be declared as an implicit coercion from a source class* $C$ *to a target class* $D$ *with* $k$ *parameters if the type of* $f$ *has the form* $\forall x_1 \dots x_k (y : C\, t_1 \dots t_n), D\, u_1 \dots u_m$. *We then write* $f : C \rightarrowtail D$.[2]

An implicit coercion $f : C \rightarrowtail D$ can be seen as a subtyping $C \leq D$ and applied to fill type mismatches to a term of class $C$ placed in a context that expects to have a term of class $D$. Implicit coercions form an inheritance graph with classes as nodes and coercions as edges, whose path $[f_1; \dots; f_n]$ where $f_i : C_i \rightarrowtail C_{i+1}$ can also be seen as a subtyping $C_1 \leq C_{n+1}$; thus, we write $[f_1; \dots; f_n] : C_1 \rightarrowtail C_{n+1}$ to indicate $[f_1; \dots; f_n]$ is an inheritance path from $C_1$ to $C_{n+1}$. The Coq system pre-computes those inheritance paths for any pair of source and target classes, and updates to keep them closed under transitivity when a new implicit coercion is declared [32, Sect. 3.3] [44, Sect. 8.2.5 "Inheritance Graph"]. The coherence of inheritance graphs is defined as follows.

---

[2] In fact, the target classes can also be functions (**Funclass**) and sorts (**Sortclass**); that is to say, a function returning functions, types, or propositions can be declared as an implicit coercion. In this paper, we omit these cases to simplify the presentation, but our discussion can be generalized to these cases.

**Definition 3.3. (Definitional equality** [43] [15, Sect. 3.1]**).** *Two terms* $t_1$ *and* $t_2$ *are said to be definitionally equal, or convertible, if they are equivalent under* $\beta\delta\iota\zeta$-*reduction and* $\eta$-*expansion. This equality is denoted by the infix symbol* $\equiv$.

**Definition 3.4. (Coherence** [7, Sect. 3.2] [32, Sect. 7]**).** *An inheritance graph is coherent if and only if the following two conditions hold.*

1. *For any circular inheritance path* $p : C \rightarrowtail C$, $p\,x \equiv x$, *where* $x$ *is a fresh variable of class* $C$.
2. *For any two inheritance paths* $p, q : C \rightarrowtail D$, $p\,x \equiv q\,x$, *where* $x$ *is a fresh variable of class* $C$.

Before our work, if multiple inheritance paths existed between the same source and target class, only the oldest one was kept as a valid one in the inheritance graph in Coq, and all the others were reported as *ambiguous paths* and ignored. We improved this mechanism to report only paths that break the coherence conditions and also to minimize the number of reported ambiguous paths [34,35]. The second condition ensures the modularity of reasoning with packed classes. For example, proving $\forall(R : \text{Ring.type})\,(x, y : R), (-x) \times y = -(x \times y)$ requires using both Semiring.monoidType (Ring.semiringType R) and Group.monoidType (Ring.groupType R) implicitly. If those Monoid instances are not definitionally equal, it will prevent us from proving the lemma by reporting type mismatch between R and R.

Convertibility checking for inheritance paths consisting of implicit coercions as in Definition 3.2 requires constructing a composition of functions for a given inheritance path. One can reduce any unification problem to this well-typed term construction problem, that in the higher-order case is undecidable [19]. However, the inheritance paths that make the convertibility checking undecidable can never be applied as implicit coercions in type inference, because they do not respect the uniform inheritance condition.

**Definition 3.5. (Uniform inheritance condition** [44] [32, Sect. 3.2]**).** *An implicit coercion* $f$ *between classes* $C \rightarrowtail D$ *with* $n$ *and* $m$ *parameters, respectively, is uniform if and only if the type of* $f$ *has the form*

$$\forall(x_1 : A_1)\ldots(x_n : A_n)\,(x_{n+1} : C\,x_1\ldots x_n), D\,u_1\ldots u_m.$$

*Remark 3.1.* Names that can be declared as implicit coercions are defined as constants that respect the uniform inheritance condition in [32, Sect. 3.2]. However, the actual implementation in the modern Coq system accepts almost any function as in Definition 3.2 as a coercion.

Saïbi claimed that the uniform inheritance condition "ensures that any coercion can be applied to any object of its source class" [32, Sect. 3.2], but the actual condition ensures additional properties. The number and ordering of parameters of a uniform implicit coercion are the same as those of its source class; thus, convertibility checking of uniform implicit coercions $f, g : C \rightarrowtail D$ does not require any special treatment such as permuting parameters of $f$ and $g$. Moreover, function composition preserves this uniformity, that is, the following lemma holds.

**Lemma 3.1.** *For any uniform implicit coercions $f : C \rightarrowtail D$ and $g : D \rightarrowtail E$, the function composition of the inheritance path $[f; g] : C \rightarrowtail E$ is uniform.*

*Proof.* Let us assume that $C$, $D$, and $E$ are classes with $n$, $m$, and $k$ parameters respectively, and $f$ and $g$ have the following types:

$$f : \forall (x_1 : T_1) \ldots (x_n : T_n)\,(x_{n+1} : C\,x_1 \ldots x_n), D\,u_1 \ldots u_m,$$
$$g : \forall (y_1 : U_1) \ldots (y_m : U_m)\,(y_{m+1} : D\,y_1 \ldots y_m), E\,v_1 \ldots v_k.$$

Then, the function composition of $f$ and $g$ can be defined and typed as follows:

$$g \circ f := \lambda(x_1 : T_1) \ldots (x_n : T_n)\,(x_{n+1} : C\,x_1 \ldots x_n), g\,u_1 \ldots u_m\,(f\,x_1 \ldots x_n\,x_{n+1})$$
$$: \forall (x_1 : T_1) \ldots (x_n : T_n)\,(x_{n+1} : C\,x_1 \ldots x_n),$$
$$E\,(v_1\{y_1/u_1\} \ldots \{y_m/u_m\}\{y_{m+1}/f\,x_1 \ldots x_n\,x_{n+1}\})$$
$$\vdots$$
$$(v_k\{y_1/u_1\} \ldots \{y_m/u_m\}\{y_{m+1}/f\,x_1 \ldots x_n\,x_{n+1}\}).$$

The terms $u_1, \ldots, u_m$ contain the free variables $x_1, \ldots, x_{n+1}$ and we omitted substitutions for them by using the same names for the binders in the above definition. Nevertheless, $(g \circ f) : C \rightarrowtail E$ respects the uniform inheritance condition.    □

In the above definition of the function composition $g \circ f$ of implicit coercions, the types of $x_1, \ldots, x_n, x_{n+1}$ and the parameters of $g$ can be automatically inferred in Coq; thus, it can be abbreviated as follows:

$$g \circ f := \lambda(x_1 : \_) \ldots (x_n : \_)\,(x_{n+1} : \_), g \underbrace{\_ \ldots \_}_{m \text{ parameters}}(f\,x_1 \ldots x_n\,x_{n+1}).$$

For implicit coercions $f_1 : C_1 \rightarrowtail C_2, f_2 : C_2 \rightarrowtail C_3, \ldots, f_n : C_n \rightarrowtail C_{n+1}$ that have $m_1, m_2, \ldots, m_n$ parameters respectively, the function composition of the inheritance path $[f_1; f_2; \ldots; f_n]$ can be written as follows by repeatedly applying Lemma 3.1 and the above abbreviation.

$$f_n \circ \cdots \circ f_2 \circ f_1 := \lambda(x_1 : \_) \ldots (x_{m_1} : \_)\,(x_{m_1+1} : \_),$$
$$f_n \underbrace{\_ \ldots \_}_{m_n \text{ parameters}}(\ldots (f_2 \underbrace{\_ \ldots \_}_{m_2 \text{ parameters}}(f_1\,x_1 \ldots x_{m_1}\,x_{m_1+1})) \ldots).$$

If $f_1, \ldots, f_n$ are all uniform, the numbers of their parameters $m_1, \ldots, m_n$ are equal to the numbers of parameters of $C_1, \ldots, C_n$. Consequently, the type inference algorithm always produces the typed closed term of most general function composition of $f_1, \ldots, f_n$ from the above term. If not all of $f_1, \ldots, f_n$ are uniform, type inference may fail or produce an open term, but if this produces a typed closed term, it is the most general function composition of $f_1, \ldots, f_n$.

Our coherence checking mechanism constructs the function composition of $p : C \rightarrowtail C$ and compares it with the identity function of class $C$ to check the first condition, and also constructs the function compositions of $p, q : C \rightarrowtail D$ and performs the conversion test for them to check the second condition.

# 4    Automated Structure Inference

This section reviews how the automated structure inference mechanism [26] works on our example and in general. The first example is $0+1$, whose desugared form is @add _ (@zero _) (@one _), where holes _ stand for implicit pieces of information to be inferred. The left- and right-hand sides of the top application can be type-checked without any use of canonical structures, as follows:

$?_M$ : Monoid.type ⊢ @add $?_M$ (@zero $?_M$) : Monoid.sort $?_M$ → Monoid.sort $?_M$,
$?_{SR}$ : Semiring.type ⊢ @one $?_{SR}$ : Semiring.sort $?_{SR}$,

where $?_M$ and $?_{SR}$ represent unification variables. Type-checking the application requires solving a unification problem Monoid.sort $?_M$ ≜ Semiring.sort $?_{SR}$, which is not trivial and which Coq does not know how to solve without hints. By declaring Semiring.monoidType : Semiring.type → Monoid.type as a canonical instance, Coq can become aware of that instantiating $?_M$ with Semiring.monoidType $?_{SR}$ is the canonical solution to this unification problem.

Canonical Semiring.monoidType.

The Canonical command takes a definition with a body of the form $\lambda x_1 \ldots x_n$, $\{|p_1 := (f_1 \ldots); \ldots; p_m := (f_m \ldots)|\}$, and then synthesizes unification hints between the projections $p_1, \ldots, p_m$ and the head symbols $f_1, \ldots, f_m$, respectively, except for unnamed projections. Since Semiring.monoidType has the following body, the above Canonical declaration synthesizes the unification hint between Monoid.sort and Semiring.sort that we need:

```
fun cT : Semiring.type ⇒
  {| Monoid.sort := Semiring.sort cT;
     Monoid.class := Semiring.base cT (Semiring.class cT) |}.
```

In general, for any structures A and B such that B inherits from A with an implicit coercion B.aType : B.type >-> A.type, B.aType should be declared as a canonical instance to allow Coq to solve unification problems of the form A.sort $?_A$ ≜ B.sort $?_B$ by instantiating $?_A$ with B.aType $?_B$.

The second example is $-1$, whose desugared form is @opp _ (@one _). The left- and right-hand sides of the top application can be type-checked as follows:

$?_G$ : Group.type ⊢ @opp $?_G$ : Group.sort $?_G$ → Group.sort $?_G$,
$?_{SR}$ : Semiring.type ⊢ @one $?_{SR}$ : Semiring.sort $?_{SR}$.

In order to type check the application, Coq has to unify Group.sort $?_G$ with Semiring.sort $?_{SR}$, which, again, is not trivial. Moreover, groups and semirings do not inherit from each other; therefore, this case is not an instance of the above criteria to define canonical instances. Nevertheless, this unification problem means that $?_G$ : Group.type and $?_{SR}$ : Semiring.type are the same, and they are equipped with both group and semiring axioms, that is, the ring structure. Thus, its canonical solution should be introducing a fresh unification variable $?_R$ : Ring.type and instantiating $?_G$ and $?_{SR}$ with Ring.groupType $?_R$ and

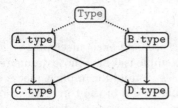
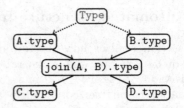

**Fig. 3.** A minimal hierarchy that has ambiguous joins. Both structure C and D directly inherit from the structures A and B; thus, A and B have two joins.

**Fig. 4.** A hierachy that disambiguates the join of A and B in Fig. 3 by redefining C and D to inherit from a new structure join(A, B) that inherits from A and B.

Ring.semiringType ?$_R$, respectively. Right after defining Ring.semiringType, this unification hint can be defined as follows.

```
Local Definition semiring_groupType (cT : type) : Group.type :=
  Group.Pack (Semiring.sort (semiringType cT)) (base _ (class cT)).
```

This definition is definitionally equal to Ring.groupType, but has a different head symbol, Semiring.sort instead of Ring.sort, in its first field Group.sort. Thus, the unification hint we need between Group.sort and Semiring.sort can be synthesized by the following declarations.

```
Canonical Ring.semiring_groupType.
```

This unification hint can also be defined conversely as follows. Whichever of those is acceptable, but at least one of them should be declared.

```
Local Definition group_semiringType (cT : type) : Semiring.type :=
  Semiring.Pack (Group.sort (groupType cT))
    (Semiring.Class _ (Group.base _ (base _ (class cT))) (mixin _ (class cT))).
```

For any structures $A$ and $B$ that have common (non-strict) subclasses $\mathcal{C}$, we say that $C \in \mathcal{C}$ is a *join* of $A$ and $B$ if $C$ does not inherit from any other structures in $\mathcal{C}$. For example, if we add the structure of commutative rings to the hierarchy of Sect. 2, the commutative ring structure is a common subclass of the group and semiring structures, but is not a join of them because the commutative ring structure inherits from the ring structure which is another common subclass of them. In general, the join of any two structures must be unique, and we should declare a canonical instance to infer the join $C$ as the canonical solution of unification problems of the form A.sort ?$_A$ ≜ B.sort ?$_B$. For any structures $A$ and $B$ such that $B$ inherits from $A$, $B$ is the join of $A$ and $B$; thus, the first criteria to define canonical instances is just an instance of the second criteria.

Figure 3 shows a minimal hierarchy that has ambiguous joins. If we declare that C (resp. D) is the canonical join of A and B in this hierarchy, it will also be accidentally inferred for a user who wants to reason about D (resp. C). Since C and D do not inherit from each other, inferred C (resp. D) can never be instantiated with D (resp. C); therefore, we have to disambiguate it as in Fig. 4, so that the join of A and B can be specialized to both C and D afterwards.

# 5    A Simplified Formal Model of Hierarchies

In this section, we define a simplified formal model of hierarchies and show a metatheorem that ensures the predictability of structure inference. First, we define the model of hierarchies and inheritance relations.

**Definition 5.1. (Hierarchy and inheritance relations).** *A hierarchy $\mathcal{H}$ is a finite set of structures partially ordered by a non-strict inheritance relation $\leadsto^*$; that is, $\leadsto^*$ is reflexive, antisymmetric, and transitive. We denote the corresponding strict (irreflexive) inheritance relation by $\leadsto^+$. $A \leadsto^* B$ and $A \leadsto^+ B$ respectively mean that $B$ non-strictly and strictly inherits from $A$.*

**Definition 5.2. (Common subclasses).** *The (non-strict) common subclasses of $A, B \in \mathcal{H}$ are $\mathcal{C} := \{C \in \mathcal{H} \mid A \leadsto^* C \wedge B \leadsto^* C\}$. The minimal common subclasses of $A$ and $B$ is $\mathrm{mcs}(A,B) := \mathcal{C} \setminus \{C \in \mathcal{H} \mid \exists C' \in \mathcal{C}, C' \leadsto^+ C\}$.*

**Definition 5.3. (Well-formed hierarchy).** *A hierarchy $\mathcal{H}$ is said to be well-formed if the minimal common subclasses of any two structures are unique; that is, $|\mathrm{mcs}(A,B)| \leq 1$ for any $A, B \in \mathcal{H}$.*

**Definition 5.4. (Extended hierarchy).** *An extended hierarchy $\bar{\mathcal{H}} := \mathcal{H} \dot{\cup} \{\top\}$ is a hierarchy $\mathcal{H}$ extended with $\top$ which means a structure that strictly inherits from all the structures in $\mathcal{H}$; thus, the inheritance relation is extended as follows:*

$$A \bar{\leadsto}^* \top \iff \text{true},$$
$$\top \bar{\leadsto}^* B \iff \text{false} \qquad\qquad \textit{(if } B \neq \top),$$
$$A \bar{\leadsto}^* B \iff A \leadsto^* B \qquad \textit{(if } A \neq \top \textit{ and } B \neq \top).$$

**Definition 5.5. (Join).** *The join is a binary operator on an extended well-formed hierarchy $\bar{\mathcal{H}}$, defined as follows:*

$$\mathrm{join}(A,B) = \begin{cases} C & \textit{(if } A, B \in \mathcal{H} \textit{ and } \mathrm{mcs}(A,B) = \{C\}), \\ \top & \textit{(otherwise).} \end{cases}$$

We encoded the above definitions on hierarchies in Coq by using the structure of partially ordered finite types `finPOrderType` of the `mathcomp-finmap` library [12] and proved the following theorem.

**Theorem 5.1.** *The join operator on an extended well-formed hierarchy is associative, commutative, and idempotent; that is, an extended well-formed hierarchy is a join-semilattice.*

If the unification algorithm of Coq respects our model of hierarchies and joins during structure inference, Theorem 5.1 implies that permuting, duplicating, and contracting unification problems do not change the result of inference; thus, it states the predictability of structure inference at a very abstract level.

# 6    Validating Well-Formedness of Hierarchies

This section presents a well-formedness checking algorithm that can also generate the exhaustive set of assertions for joins. We implemented this checking mechanism as a tool `hierarchy.ml` written in OCaml, which is available as a MathComp developer utility [45, /etc/utils/hierarchy.ml]. Our checking tool outputs the assertions as a Coq proof script which can detect missing and misimplemented unification hints for joins.

Since a hierarchy must be a finite set of structures (Definition 5.1), Definitions 5.2, 5.5, and 5.3 give us computable (yet inefficient) descriptions of joins and the well-formedness; in other words, for a given hierarchy $\mathcal{H}$ and any two structures $A, B \in \mathcal{H}$, one may enumerate their minimal common subclasses. Algorithm 1 is the checking algorithm we present, that takes in input an inheritance relation in the form of an indexed family of strict subclasses $\mathsf{subof}(A) := \{B \in \mathcal{H} \mid A \leadsto^+ B\}$. The `join` function in this algorithm takes two structures as arguments, checks the uniqueness of their join, and then returns the join if it uniquely exists. In this function, the enumeration of minimal common subclasses is done by constructing the set of common subclasses $\mathcal{C}$, and filtering out $\mathsf{subof}(C)$ from $\mathcal{C}$ for every $C \in \mathcal{C}$. In this filtering process, which is written as a **foreach** statement, we can skip elements already filtered out and do not need to care about ordering of picking up elements, thanks to transitivity.

The `hierarchy.ml` utility extracts the inheritance relation from a Coq library by interacting with `coqtop`, and then executes Algorithm 1 to check the well-formedness and to generate assertions. The assertions generated from our running example are shown below.

```
1   check_join Group.type Monoid.type Group.type.
2   check_join Group.type Ring.type Ring.type.
3   check_join Group.type Semiring.type Ring.type.
4   check_join Monoid.type Group.type Group.type.
5   check_join Monoid.type Ring.type Ring.type.
6   check_join Monoid.type Semiring.type Semiring.type.
7   check_join Ring.type Group.type Ring.type.
8   check_join Ring.type Monoid.type Ring.type.
9   check_join Ring.type Semiring.type Ring.type.
10  check_join Semiring.type Group.type Ring.type.
11  check_join Semiring.type Monoid.type Semiring.type.
12  check_join Semiring.type Ring.type Ring.type.
```

An assertion `check_join t1 t2 t3` asserts that the join of t1 and t2 is t3, and `check_join` is implemented as a tactic that fails if the assertion is false. For instance, if we do not declare `Ring.semiringType` as a canonical instance, the assertion of line 9 fails and reports the following error.

```
There is no join of Ring.type and Semiring.type but it is expected to be
Ring.type.
```

One may declare incorrect canonical instances that overwrite an existing join. For example, the join of groups and monoids must be groups; however, defining the following canonical instance in the `Ring` section overwrites this join.

```
Local Definition bad_monoid_groupType : Group.type :=
  Group.Pack (Monoid.sort monoidType) (base _ class).
```

**Parameters:** $\mathcal{H}$ is the set of all the structures. subof is the map from structures to their strict subclasses, which is required to be transitive: $\forall A\,B \in \mathcal{H}, A \in \text{subof}(B) \Rightarrow \text{subof}(A) \subset \text{subof}(B)$.

**Function** join$(A, B)$:
   $\mathcal{C} := (\text{subof}(A) \cup \{A\}) \cap (\text{subof}(B) \cup \{B\})$;
                     /* $\mathcal{C}$ is the set of all the common subclasses of $A$ and $B$. */
   **foreach** $C \in \mathcal{C}$ **do** $\mathcal{C} \leftarrow \mathcal{C} \setminus \text{subof}(C)$;
   /* Since subof is transitive, removed elements of $\mathcal{C}$ can be skipped in this loop, and the ordering of picking elements from $\mathcal{C}$ does not matter. */
   **if** $\mathcal{C} = \emptyset$ **then return** $\top$;          /* There is no join of $A$ and $B$. */
   **else if** $\mathcal{C}$ *is singleton* $\{C\}$ **then return** $C$;  /* $C$ is the join of $A$ and $B$. */
   **else fail**;                  /* The join of $A$ and $B$ is ambiguous. */

**foreach** $A \in \mathcal{H}, B \in \mathcal{H}$ **do**
   $C := \text{join}(A, B)$; **if** $A \neq B \wedge C \neq \top$ **then print** "check_join $A\ B\ C$.";
**end**

**Algorithm 1:** Well-formedness checking and assertions generation

By declaring Ring.bad_monoid_groupType as a canonical instance, the join of Monoid.type and Group.type is still Group.type, but the join of Group.type and Monoid.type becomes Ring.type, because of asymmetry of the unification mechanism. The assertion of line 1 fails and reports the following error.

```
The join of Group.type and Monoid.type is Ring.type but it is expected to be
Group.type.
```

# 7 Evaluation

This section reports the results of applying our tools to the MathComp library 1.7.0 and on recent development efforts to extend the hierarchy in MathComp using our tools. MathComp 1.7.0 provides the structures depicted in Fig. 1, except comAlgType and comUnitAlgType, and lacked a few of the edges; thus, its hierarchy is quite large. Our coherence checking mechanism found 11 inconvertible multiple inheritance paths in the ssralg library. Fortunately, those paths concern proof terms and are intended to be irrelevant; hence, we can ensure no implicit coercion breaks the modularity of reasoning. Our well-formedness checking tool discovered 7 ambiguous joins, 8 missing unification hints, and one overwritten join due to an incorrect declaration of a canonical instance. These inheritance bugs were found and fixed with the help of our tools; thus, similar issues cannot be found in the later versions of MathComp.

The first issue was that inheritance from the CountRing structures (countZmodType and its subclasses with the prefix count) to the FinRing structures (finZmodType and its subclasses) was not implemented, and consequently it introduced 7 ambiguous joins. For instance, finZmodType did not inherit from countZmodType as it should; consequently, they became ambiguous joins of countType and zmodType. 6 out of 8 missing unification hints should infer

CountRing or FinRing structures. The other 2 unification hints are in numeric field (numFieldType and realFieldType) structures. Fixing the issue of missing inheritance from CountRing to FinRing was a difficult task without tool support. The missing inheritance itself was a known issue from before our tooling work, but the sub-hierarchy consisting of the GRing, CountRing, and FinRing structures in Fig. 1 is quite dense; as a result, it prevents the library developers from enumerating joins correctly without automation [38].

The second issue was that the following canonical finType instance for extremal_group overwrote the join of finType and countType, which should be finType.

Canonical extremal_group_finType := FinType _ extremal_group_finMixin.

In this declaration, FinType is a packager [26, Sect. 7] that takes a type T and a Finite mixin of T as its arguments and construct a finType instance from the given mixin and the canonical countType instance for T. However, if one omits its first argument T with a placeholder as in the above, the packager may behave unpredictably as a unification hint. In the above case, the placeholder was instantiated with extremal_group_countType by type inference; as a result, it incorrectly overwrote the join of finType and countType.

Our tools can also help finding inheritance bugs when extending the hierarchy of MathComp, improve the development process by reducing the reviewing and maintenance burden, and allow developers and contributors to focus better on mathematical contents and other design issues. For instance, Hivert [24] added new structures of commutative algebras and redefined the field extension and splitting field structures to inherit from them. In this extension process, he fixed some inheritance issues with help from us and our tools; at the same time, we made sure there is no inheritance bug without reviewing the whole boilerplate code of structures. We ported the order sub-library of the mathcomp-finmap library [12] to MathComp, redefined numeric domain structures [10, Chap. 4] [11, Sect. 3.1] to inherit from ordered types, and factored out the notion of norms and absolute values as normed Abelian groups [1, Sect. 4.2] with the help of our tools [13,36]. This modification resulted in approximately 10,000 lines of changes; thus, reducing the reviewing burden was an even more critical issue. This work is motivated by an improvement of the MathComp Analysis library [3] [30, Part II], which extends the hierarchy of MathComp with some algebraic and topological structures [1, Sect. 4] [30, Chap. 5] and is another application of our tools.

# 8    Conclusion and Related Work

This paper has provided a thorough analysis of the packed classes methodology, introduced two invariants that ensure the modularity of reasoning and the predictability of structure inference, and presented systematic ways to check those invariants. We implemented our invariant checking mechanisms as a part of the Coq system and a tool bundled with MathComp. With the help of these tools, many inheritance bugs in MathComp have been found and fixed. The MathComp development process has also been improved significantly.

Coq had no coherence checking mechanism before our work. Saïbi [32, Sect. 7] claimed that the coherence property "is too restrictive in practice" and "it is better to replace conversion by Leibniz equality to compare path coercions because Leibniz equality is a bigger relation than conversion". However, most proof assistants based on dependent type theories including Coq still rely heavily on conversion, particularly in their type checking/inference mechanisms. Coherence should not be relaxed with Leibniz equality; otherwise, the type mismatch problems described in Sect. 3 will occur. With our coherence checking mechanism, users can still declare inconvertible multiple inheritance at their own risk and responsibility, because ambiguous paths messages are implemented as warnings rather than errors. The Lean system has an implicit coercion mechanism based on type class resolution, that allows users to define and use non-uniform implicit coercions; thus, coherence checking can be more difficult. Actually, Lean has no coherence checking mechanism; thus, users get more flexibility with this approach but need to be careful about being coherent.

There are three kinds of approaches to defining mathematical structures in dependent type theories: *unbundled, semi-bundled,* and *bundled* approaches [46, Sect. 4.1.1]. The unbundled approach uses an interface that is parameterized by carriers and operators, and gathers axioms as its fields, e.g., [41]; in contrast, the semi-bundled approach bundles operators together with axioms as in `class_of` records, but still places carriers as parameters, e.g., [46]. The bundled approach uses an interface that bundles carriers together with operators and axioms, e.g., packed classes and telescopes [26, Sect. 2.3] [9,18,29]. The above difference between definitions of interfaces, in particular, whether carriers are bundled or not, leads to the use of different instance resolution and inference mechanisms: type classes [22,40] for the unbundled and semi-bundled approaches, and canonical structures or other unification hint mechanisms for the bundled approach. Researchers have observed unpredictable behaviors [23] and efficiency issues [46, Sect. 4.3] [41, Sect. 11] in inference with type classes; in contrast, structure inference with packed classes is predictable, and Theorem 5.1 states this predictability more formally, except for concrete instance resolution. The resolution of canonical structures is carried out by consulting a table of unification hints indexed by pairs of two head symbols and optionally with its recursive application and backtracking [21, Sect. 2.3]. The packed classes methodology is designed to use this recursive resolution not for structure inference [17, Sect. 2.3] but only for parametric instances [26, Sect. 4] such as lists and products, and not to use backtracking. Thus, there is no efficiency issue in structure inference, except that nested class records and chains of their projections exponentially slow down the conversion which flat variant of packed classes [14, Sect. 4] can mitigate. In the unbundled and semi-bundled approaches, a carrier may be associated with multiple classes; thus, inference of join and our work on structure inference (Sect. 4, 5, and 6) are problems specific to the bundled approach. A detailed comparison of type classes and packed classes has also been provided in [1]. There are a few mechanisms to extend the unification engines of proof assistants other than canonical structures that can implement structure inference for packed classes: unification hints [4] and coercions pullback [31]. For any of those cases, our invariants are fundamental properties to implement packed classes and structure inference, but the invariant checking we propose has not been made yet at all.

Packed classes require the systematic use of records, implicit coercions, and canonical structures. This leads us to automated generation of structures from their higher-level descriptions [14], which is work in progress.

**Acknowledgements.** We appreciate the support from the STAMP (formerly MARELLE) project-team, INRIA Sophia Antipolis. In particular, we would like to thank Cyril Cohen and Enrico Tassi for their help in understanding packed classes and implementing our checking tools. We are grateful to Reynald Affeldt, Yoichi Hirai, Yukiyoshi Kameyama, Enrico Tassi, and the anonymous reviewers for helpful comments on early drafts. We are deeply grateful to Karl Palmskog for his careful proofreading and comments. We would like to thank the organizers and participants of The Coq Workshop 2019 where we presented preliminary work of this paper. In particular, an insightful question from Damien Pous was the starting point of our formal model of hierarchies presented in Sect. 5. This work was supported by JSPS Research Fellowships for Young Scientists and JSPS KAKENHI Grant Number 17J01683.

# References

1. Affeldt, R., Cohen, C., Kerjean, M., Mahboubi, A., Rouhling, D., Sakaguchi, K.: Competing inheritance paths in dependent type theory: a case study in functional analysis. In: Peltier, N., Sofronie-Stokkermans, V. (eds.) IJCAR 2020. LNCS, vol. 12167, pp. 3–20. Springer, Cham (2020)
2. Affeldt, R., Nowak, D., Saikawa, T.: A hierarchy of monadic effects for program verification using equational reasoning. In: Hutton, G. (ed.) MPC 2019. LNCS, vol. 11825, pp. 226–254. Springer, Cham (2019). https://doi.org/10.1007/978-3-030-33636-3_9
3. Analysis Team: Mathematical components compliant analysis library (2017). https://github.com/math-comp/analysis
4. Asperti, A., Ricciotti, W., Sacerdoti Coen, C., Tassi, E.: Hints in unification. In: Berghofer, S., Nipkow, T., Urban, C., Wenzel, M. (eds.) TPHOLs 2009. LNCS, vol. 5674, pp. 84–98. Springer, Heidelberg (2009). https://doi.org/10.1007/978-3-642-03359-9_8
5. Asperti, A., Ricciotti, W., Sacerdoti Coen, C., Tassi, E.: The matita interactive theorem prover. In: Bjørner, N., Sofronie-Stokkermans, V. (eds.) CADE 2011. LNCS (LNAI), vol. 6803, pp. 64–69. Springer, Heidelberg (2011). https://doi.org/10.1007/978-3-642-22438-6_7
6. Avigad, J., de Moura, L., Kong, S.: Theorem Proving in Lean (2019). https://leanprover.github.io/theorem_proving_in_lean/
7. Barthe, G.: Implicit coercions in type systems. In: Berardi, S., Coppo, M. (eds.) TYPES 1995. LNCS, vol. 1158, pp. 1–15. Springer, Heidelberg (1996). https://doi.org/10.1007/3-540-61780-9_58

8. Boldo, S., Lelay, C., Melquiond, G.: Coquelicot: a user-friendly library of real analysis for coq. Math. Comput. Sci. **9**(1), 41–62 (2014). https://doi.org/10.1007/s11786-014-0181-1

9. de Bruijn, N.G.: Telescopic mappings in typed lambda calculus. Inf. Comput. **91**(2), 189–204 (1991). https://doi.org/10.1016/0890-5401(91)90066-B

10. Cohen, C.: Formalized algebraic numbers: construction and first-order theory. (Formalisation des nombres algébriques : construction et théorie du premier ordre). Ph.D. thesis, Ecole Polytechnique X (2012). https://pastel.archives-ouvertes.fr/pastel-00780446

11. Cohen, C., Mahboubi, A.: Formal proofs in real algebraic geometry: from ordered fields to quantifier elimination. Logical Methods Comput. Sci. **8**(1) (2012). https://doi.org/10.2168/LMCS-8(1:2)2012

12. Cohen, C., Sakaguchi, K.: A finset and finmap library (2019). https://github.com/math-comp/finmap

13. Cohen, C., Sakaguchi, K., Affeldt, R.: Dispatching order and norm, and anticipating normed modules (2019). https://github.com/math-comp/math-comp/pull/270

14. Cohen, C., Sakaguchi, K., Tassi, E.: Hierarchy Builder: algebraic hierarchies made easy in Coq with Elpi (system description) (2020). Accepted in the Proceedings of FSCD 2020 https://hal.inria.fr/hal-02478907

15. Coquand, T., Huet, G.: The calculus of constructions. Inf. Comput. **76**(2), 95–120 (1988). https://doi.org/10.1016/0890-5401(88)90005-3

16. Garillot, F.: Generic proof tools and finite group theory. (Outils génériques de preuve et théorie des groupes finis). Ph.D. thesis, École Polytechnique, Palaiseau, France (2011). https://tel.archives-ouvertes.fr/pastel-00649586

17. Garillot, F., Gonthier, G., Mahboubi, A., Rideau, L.: Packaging mathematical structures. In: Berghofer, S., Nipkow, T., Urban, C., Wenzel, M. (eds.) TPHOLs 2009. LNCS, vol. 5674, pp. 327–342. Springer, Heidelberg (2009). https://doi.org/10.1007/978-3-642-03359-9_23

18. Geuvers, H., Pollack, R., Wiedijk, F., Zwanenburg, J.: A constructive algebraic hierarchy in Coq. J. Symb. Comput. **34**(4), 271–286 (2002). https://doi.org/10.1006/jsco.2002.0552

19. Goldfarb, W.D.: The undecidability of the second-order unification problem. Theor. Comput. Sci. **13**(2), 225–230 (1981). https://doi.org/10.1016/0304-3975(81)90040-2

20. Gonthier, G., et al.: A machine-checked proof of the odd order theorem. In: Blazy, S., Paulin-Mohring, C., Pichardie, D. (eds.) ITP 2013. LNCS, vol. 7998, pp. 163–179. Springer, Heidelberg (2013). https://doi.org/10.1007/978-3-642-39634-2_14

21. Gonthier, G., Ziliani, B., Nanevski, A., Dreyer, D.: How to make ad hoc proof automation less ad hoc. J. Funct. Program. **23**(4), 357–401 (2013). https://doi.org/10.1017/S0956796813000051

22. Haftmann, F., Wenzel, M.: Constructive type classes in isabelle. In: Altenkirch, T., McBride, C. (eds.) TYPES 2006. LNCS, vol. 4502, pp. 160–174. Springer, Heidelberg (2007). https://doi.org/10.1007/978-3-540-74464-1_11

23. Hales, T.: A review of the Lean theorem prover, September 2018. https://jiggerwit.wordpress.com/2018/09/18/a-review-of-the-lean-theorem-prover/, Blog entry of 18 September 2018

24. Hivert, F.: Commutative algebras (2019). https://github.com/math-comp/math-comp/pull/406

25. Hölzl, J., Immler, F., Huffman, B.: Type classes and filters for mathematical analysis in Isabelle/HOL. In: Blazy, S., Paulin-Mohring, C., Pichardie, D. (eds.) ITP 2013. LNCS, vol. 7998, pp. 279–294. Springer, Heidelberg (2013). https://doi.org/10.1007/978-3-642-39634-2_21

26. Mahboubi, A., Tassi, E.: Canonical structures for the working coq user. In: Blazy, S., Paulin-Mohring, C., Pichardie, D. (eds.) ITP 2013. LNCS, vol. 7998, pp. 19–34. Springer, Heidelberg (2013). https://doi.org/10.1007/978-3-642-39634-2_5

27. de Moura, L., Kong, S., Avigad, J., van Doorn, F., von Raumer, J.: The lean theorem prover (system description). In: Felty, A.P., Middeldorp, A. (eds.) CADE 2015. LNCS (LNAI), vol. 9195, pp. 378–388. Springer, Cham (2015). https://doi.org/10.1007/978-3-319-21401-6_26

28. Nanevski, A., Ley-Wild, R., Sergey, I., Delbianco, G., Trunov, A.: The PCM library (2019). https://github.com/imdea-software/fcsl-pcm

29. Pollack, R.: Dependently typed records in type theory. Formal Aspects Comput. **13**(3), 386–402 (2002). https://doi.org/10.1007/s001650200018

30. Rouhling, D.: Formalisation tools for classical analysis - a case study in control theory. (Outils pour la Formalisation en Analyse Classique - Une Étude de Cas en Théorie du Contrôle). Ph.D. thesis, Université Côte d'Azur (2019). https://hal.inria.fr/tel-02333396

31. Sacerdoti Coen, C., Tassi, E.: Working with mathematical structures in type theory. In: Miculan, M., Scagnetto, I., Honsell, F. (eds.) TYPES 2007. LNCS, vol. 4941, pp. 157–172. Springer, Heidelberg (2008). https://doi.org/10.1007/978-3-540-68103-8_11

32. Saïbi, A.: Typing algorithm in type theory with inheritance. In: POPL 1997, pp. 292–301. ACM (1997). https://doi.org/10.1145/263699.263742

33. Saïbi, A.: Outils Génériques de Modélisation et de Démonstration pour la Formalisation des Mathématiques en Théorie des Types. Application à la Théorie des Catégories. (Formalization of Mathematics in Type Theory. Generic tools of Modelisation and Demonstration. Application to Category Theory). Ph.D. thesis, Pierre and Marie Curie University, Paris, France (1999). https://tel.archives-ouvertes.fr/tel-00523810

34. Sakaguchi, K.: Coherence checking for coercions (2019). https://github.com/coq/coq/pull/11258

35. Sakaguchi, K.: Relax the ambiguous path condition of coercion (2019). https://github.com/coq/coq/pull/9743

36. Sakaguchi, K.: Non-distributive lattice structures (2020). https://github.com/math-comp/math-comp/pull/453

37. Sakaguchi, K.: Supplementary materials of this manuscript (2020). http://logic.cs.tsukuba.ac.jp/~sakaguchi/src/vms-ijcar2020.tar.gz

38. Sakaguchi, K., Cohen, C.: Fix inheritances from countalg to finalg (2019). https://github.com/math-comp/math-comp/pull/291

39. Sergey, I., Nanevski, A., Banerjee, A.: Mechanized verification of fine-grained concurrent programs. In: PLDI 2015, pp. 77–87. ACM (2015). https://doi.org/10.1145/2737924.2737964

40. Sozeau, M., Oury, N.: First-class type classes. In: Mohamed, O.A., Muñoz, C., Tahar, S. (eds.) TPHOLs 2008. LNCS, vol. 5170, pp. 278–293. Springer, Heidelberg (2008). https://doi.org/10.1007/978-3-540-71067-7_23

41. Spitters, B., van der Weegen, E.: Type classes for mathematics in type theory. Math. Struct. Comput. Sci. **21**(4), 795–825 (2011). https://doi.org/10.1017/S0960129511000119

42. The Coq Development Team: The Coq Proof Assistant Reference Manual (2020). https://coq.inria.fr/distrib/V8.11.0/refman/. The PDF version with numbered sections is available at https://zenodo.org/record/3744225
43. The Coq Development Team: Sect. 4.4.3 "Conversion rules". In: [42] (2020). https://coq.inria.fr/distrib/V8.11.0/refman/language/cic#conversion-rules
44. The Coq Development Team: Sect. 8.2 "Implicit Coercions". In: [42] (2020). https://coq.inria.fr/distrib/V8.11.0/refman/addendum/implicit-coercions
45. The Mathematical Components project: The mathematical components repository (2019). https://github.com/math-comp/math-comp
46. The mathlib Community: The Lean mathematical library. In: CPP 2020, pp. 367–381. ACM (2020). https://doi.org/10.1145/3372885.3373824

# Teaching Automated Theorem Proving by Example: PyRes 1.2
## (System Description)

Stephan Schulz[1]([✉]) and Adam Pease[2]

[1] DHBW Stuttgart, Stuttgart, Germany
schulz@eprover.org
[2] Articulate Software, New York, USA
apease@articulatesoftware.com

**Abstract.** PyRes is a complete theorem prover for classical first-order logic. It is not designed for high performance, but to clearly demonstrate the core concepts of a saturating theorem prover. The system is written in extensively commented Python, explaining data structures, algorithms, and many of the underlying theoretical concepts. The prover implements binary resolution with factoring and optional negative literal selection. Equality is handled by adding the basic axioms of equality. PyRes uses the given-clause algorithm, optionally controlled by weight- and age evaluations for clause selection. The prover can read TPTP CNF/FOF input files and produces TPTP/TSTP proof objects.

Evaluation shows, as expected, mediocre performance compared to modern high-performance systems, with relatively better performance for problems without equality. However, the implementation seems to be sound and complete.

## 1 Introduction

Modern automated theorem provers for first order logic such as E [7,8], Vampire [3], SPASS [12] or iProver [2] are powerful systems. They use optimised data structures, often very tight coding, and complex work flows and intricate algorithms in order to maximise performance. Moreover, most of these programs have evolved over years or even decades. As a result, they are quite daunting for even talented new developers to grasp, and present a very high barrier to entry. On the other hand, minimalist systems like leanCoP [5] do not represent typical current ATP systems, in calculus, structure, or implementation language.

Textbooks and scientific papers, on the other hand, often leave students without a clear understanding of how to translate theory into actual working code.

With PyRes, we try to fill the gap, by presenting a sound and complete theorem prover for first order logic based on widely used calculus and architecture, that is written in an accessible language with a particular focus on readability, and that explains the important concepts of each module with extensive high-level comments. We follow an object oriented design and explain data structures and algorithms as they are used.

© Springer Nature Switzerland AG 2020
N. Peltier and V. Sofronie-Stokkermans (Eds.): IJCAR 2020, LNAI 12167, pp. 158–166, 2020.
https://doi.org/10.1007/978-3-030-51054-1_9

PyRes consists of a series of provers, from a very basic system without any optimisations and with naive proof search to a prover for full first-order logic with some calculus refinements and simplification techniques. Each variant gracefully extends the previous one with new concepts. The final system is a saturation-style theorem prover based on Resolution and the given-clause algorithm, optionally with CNF transformation and subsumption.

The system is written in Python, a language widely used in education, scientific computing, data science and machine learning. While Python is quite slow, it supports coding in a very readable, explicit style, and its object-oriented features make it easy to go from more basic to more advanced implementations.

Students have found PyRes very useful in getting a basic understanding of the architecture and algorithms of an actual theorem prover, and have claimed that it enabled them to come to grips with the internals of E much faster than they could have done otherwise.

PyRes is available as open source/free software, and can be downloaded from https://github.com/eprover/PyRes.

## 2    Preliminaries

We assume the standard setting for first-order predicate logic. A signature consists of finite sets $P$ (of predicate symbols) and $F$ (of function symbols) with associated arities. We write e.g. $f/n \in F$ to indicate that $f$ is a function symbol of arity $n$. We also assume an enumerable set $V = \{X, Y, Z, \ldots\}$ of variables. Each variable is a *term*. Also, if $f/n \in F$ and $t_1, \ldots, t_n$ are terms, then so is $f(t_1, \ldots, t_n)$. This includes the special case of constants (function symbols with arity 0), for which we omit the parentheses. An *atom* is composed similarly from $p/n \in P$ and $n$ terms. A *literal* is either an atom, or a negated atom. A *clause* is a (multi-)set of literals, interpreted as the universal closure of the disjunction of its literals and written as such. As an example, $p(X, g(a))$ is an atom (and a literal), $\neg q(g(X), a)$ is a literal, and $p(X, g(a)) \vee \neg q(g(X), a) \vee p(X, Y)$ is a three-literal clause. A first-order formula is either an atom, or is composed of existing formulas $F, G$ by negation $\neg F$, quantification ($\forall X : F$ and $\exists X : F$), or any of the usual binary Boolean operators ($F \vee G$, $F \wedge G$, $F \rightarrow G$, $F \leftrightarrow G, \ldots$). We assume a reasonable precedence of operators and allow the use of parentheses where necessary or helpful. A substitution is a mapping from variables to terms, and is continued to terms, atoms, literals and clauses in the obvious way. A *match* from $s$ onto $t$ is a substitution $\sigma$ such that $\sigma(s) = t$ (where $s$ and $t$ can be terms, atoms, or literals). A *unifier* is similarly a substitution $\sigma$ such that $\sigma(s) = \sigma(t)$. Of particular importance are *most general unifiers*. If two terms are unifiable, a most general unifier is easy to compute, and, up to the renaming of variables, unique. We use $mgu(s, t)$ to denote the most general unifier of $s$ and $t$.

PyRes implements standard resolution as described in [6], but like most implementations, it separates resolution and factoring. It also optionally adds negative literal selection. This refinement of resolution allows the *selection* of an

$$(\text{BR})\frac{C \vee A \quad D \vee \neg B}{\sigma(C \vee D)} \text{ if } \sigma = mgu(A, B) \quad (\text{BF})\frac{C \vee L \vee M}{\sigma(C \vee L)} \text{ if } \sigma = mgu(L, M)$$

$$(\text{CS})\frac{C \quad \sigma(C) \vee D}{C}$$

$A, B$ stand for atoms, $L, M$ stand for literals, and $C, D$ are arbitrary clauses.
If negative literal selection is employed, additional constraints to (BR) are that $\neg B$ is *selected*, and that no literal in $C$ is selected. (CS) is a contraction rule, i.e. it *replaces* the premises by the conclusion, in effect removing the larger clause.

**Fig. 1.** Binary resolution with subsumption

arbitray negative literal in clauses that have at least one negative literal, and the restriction of resolution inferences involving this clause to those that resolve on the selected literal [1]. PyRes also supports subsumption, i.e. the discarding of clauses covered by a more general clause. The inference system, consisting of *binary resolution* (BR), *binary factoring* (BF) and *clause subsumption* (CS) is shown in Fig. 1.

## 3 System Design and Implementation

### 3.1 Architecture

The system is based on a layered software architecture. At the bottom is code for the lexical scanner. This is followed by the logical data types (terms, literals, clauses and formulas), with their associated input/output functions. Logical operations like unification and matching are implemented as separate modules, as are the generating inference rules and subsumption. On top of this, there are clause sets and formula sets, and the proof state of the given-clause algorithms, with two sets of clauses - one for those clauses that have been processed and one set that has not yet been processed.

From a logical perspective, the system is structured as a pipeline, starting with the parser, optionally followed by the clausifier and a module that adds equality axioms if equality is present, then followed by the core saturation algorithm, and finally, in the case of success, proof extraction and printing. To keep the learning curve simple, we have created 3 different provers: `pyres-simple` is a minimal system for clausal logic, `pyres-cnf` adds heuristics, indexing, and subsumption, and `pyres-fof` extends the pipeline to support full first-order logic with equality [11].

### 3.2 Implementation

Python is a high-level multi-paradigm programming language that combines both imperative and functional programming with an object-oriented inheritance system. It includes a variety of built-in data types, including lists, associative arrays/hashes and even sets. It shares dynamic typing/polymorphism, lambdas, and a built-in list datatype with LISP, one of the classical languages for symbolic AI and theorem proving. This enables us to implement both terms, the most frequent data type in a saturating prover, and atoms, as simple nested lists (s-expressions), using Python's built-in strings for function symbols, predicate symbols, and variables.

Literals are implemented as a class, with polarity, atom, and a flag to indicate literals selected for inference. Logical formulas are implemented as a class of recursive objects, with atoms as the base case and formulas being constructed with the usual operators and quantifiers. Top-level formulas are wrapped in a container object with meta-information. Both these formula containers and clauses are implemented as classes sharing a common super-class `Derivable` that provides for meta-information such as name and origin (read from input or derived via an inference record). The `Clause` class extends this with a list of literals, a TPTP style type, and an optional heuristic evaluation. The `WFormula` class extends it with a type and the recursive `Formula` object.

The `ClauseSet` class implements simple clause sets. In addition to methods for adding and removing clauses, it also has an interface to return potential inference partners for a literal, and to return a superset of possibly subsuming or subsumed clauses for a query clause. In the basic version, these simply return all clauses (clause/literal pairs for resolution) from the set. However, in the derived class `IndexedClauseSet`, simple indexing techniques (*top symbol hashing* for resolution and *predicate abstraction indexing*, a new technique for subsumption) return much smaller candidate sets. Resolution, factoring, and subsumption are implemented as plain functions.

The core of the provers is a given-clause saturation algorithm, based on two clause sets, the processed clauses and the unprocessed clauses. In the most basic case, clauses are processed first-in-first out. At each operation of the main loop, the oldest unprocessed clause is extracted from the unprocessed clauses. All its factors, and all resolvents between this *given clause* and all processed clauses, are computed and added to the unprocessed set. The clause itself is added to the processed set. The algorithm stops if the given clause is empty (i.e. an explicit contradiction), or if it runs out of unprocessed clauses. Figure 2 shows the substantial methods of `SimpleProofState`. In contrast to most pseudo-code versions, this actually working code shows e.g. that clauses have to be made variable-disjoint (here by creating a copy with fresh variables).

The more powerful variant `pyres-cnf` adds literal selection, heuristic clause selection with multiple evaluations in the style of E [9], and subsumption to this loop. For clause selection, each clause is assigned a list of heuristic evaluations (e.g. symbol counting and abstract creation time), and the prover selects the next clause in a fixed scheme according to this evaluation (e.g. 5 out of 6 times, it picks the smallest clause, once it picks the oldest). Subsumption checks are performed between the given clause and the processed clauses. Forward subsumption checks if the given clause is subsumed by any processed clause. If so, it is discarded.

```python
def processClause(self):
    """
    Pick a clause from unprocessed and process it. If the empty
    clause is found, return it. Otherwise return None.
    """
    given_clause = self.unprocessed.extractFirst()
    given_clause = given_clause.freshVarCopy()
    print("#", given_clause)
    if given_clause.isEmpty():
        # We have found an explicit contradiction
        return given_clause
    new = []
    factors     = computeAllFactors(given_clause)
    new.extend(factors)
    resolvents = computeAllResolvents(given_clause, self.processed)
    new.extend(resolvents)
    self.processed.addClause(given_clause)
    for c in new:
        self.unprocessed.addClause(c)
    return None

def saturate(self):
    """
    Main proof procedure. If the clause set is found
    unsatisfiable, return the empty clause as a witness. Otherwise
    return None.
    """
    while self.unprocessed:
        res = self.processClause()
        if res != None:
            return res
    else:
        return None
```

While most of the code should be self-explanatory, [] stands for the empty list, and extend() is a list method that adds the elements of another list at the end of a given list.

**Fig. 2.** Simple saturation

Backward subsumption removes processed clauses that are subsumed by the given clause.

For the full first-order pyres-fof, we first parse the input into a formula set, and use a naive clausifier to convert it to clause normal form.

The code base has a total of 8553 lines (including comments, docstrings, and unit tests), or 3681 lines of effective code. For comparison, our prover E has about 377000 lines of code (about 53000 actual C statements), or 170000 when excluding the automatically generated strategy code.

## 3.3   Experiences

We would like to share some experiences about coding a theorem prover in Python. First, the high level of abstraction makes many tasks very straightforward to code. Python's high-level data types are a good match to the theory of automated theorem proving, and the combination of object-orientation with inheritance and polymorphism is particularly powerful.

Python also has good development tools. In particularly, the built-in unit-test framework (and the coverage tool) are very helpful in testing partial products and gaining confidence in the quality of the code. The Python profiler (cProfile) is easy to use and produces useful results. On the negative side, the lack of a strict type system and the ad-hoc creation of variables has sometimes caused confusion. In particular, when processing command line options, the relevant function sometimes has to set global variables. If these are not explicitly declared as `global`, a new local variable will be created, shadowing the global variable. Also, a misspelled name of a class- or structure member will silently create that member, not throw an error. As an example, we only found out after extensive testing that the prover never applied backward subsumption, not because of some logic error or algorithmic problem, but because we set the value of `backward_subsuption` (notice the missing letter "m") in the parameter set to `True` trying to enable it.

Overall, however, programming a prover in Python proved to be a lot easier and faster than in e.g. programming in C, and resulted in more compact and easier to read code. This does come at the price of performance, of course. It might be an interesting project to develop datatype and algorithm libraries akin to NumPy, TensorFlow, or scikit-learn for ATP application, to bring together the best of both worlds.

## 4   Experimental Evaluation

We have evaluated PyRes (in the `pyres-fof` incarnation) with different parameter settings on all clausal (CNF) and unsorted first-order (FOF) problems from TPTP 7.2.0. Table 1 summarizes the results. We have also included some data from E 2.4, a state-of-the-art high-performance prover, Prover9 [4] (release 1109a), and leanCoP 2.2. Prover9 has been used as a standard reference in the CASC competition for several years. LeanCoP is a very compact prover written in Prolog. Experiments were run on StarExec Miami, a spin-off of the original StarExec project [10]. The machines were equipped with 256 GB of RAM and Intel Xeon CPUs running at 3.20 GHz. The per-problem time-limit was set to 300 s. For Prover9 and leanCoP, we used data included with the TPTP 7.2.0 distribution.

The *Best* configuration for PyRes enables forward and backward subsumption, negative literal selection (always select the largest literal by symbol count), uses indexing for subsumption and resolution, and processes given clauses interleaving smallest (by symbol count) and oldest clauses with a ratio of 5 to 1. The other configurations are modified from the *Best* configuration as described in the table. For the *Best* configuration (and the E results), we break the number of solutions into proofs and (counter-)saturations, for the other configurations we only include the total number of successes. All data (and the system and scripts used) is available at http://www.eprover.eu/E-eu/PyRes1.2.html.

It should be noted that Prover9, E, and leanCoP are all using an automatic mode to select different heuristics and strategies. leanCoP also uses strategy scheduling, i.e. it successively tries several different strategies.

We present the results for different problem classes: UEQ (unit problems with equality), CNE (clausal problems without equality), CEQ (clausal problem with equality, but excluding UEQ), FNE (FOF problems without equality) and FEQ (FOF problems with equality).

**Table 1.** PyRes performance (other systems for comparison)

| Strategy | UEQ | CNE | CEQ | FNE | FEQ | All |
|---|---|---|---|---|---|---|
| Class size | (1193) | (2383) | (4442) | (1771) | (6305) | (16094) |
| Best (all) | 116 | 1048 | 587 | 765 | 860 | 3376 |
| Best (proofs) | 113 | 946 | 499 | 632 | 725 | 2915 |
| Best (sat) | 3 | 102 | 88 | 133 | 135 | 461 |
| No indexing | 116 | 1042 | 567 | 736 | 829 | 3290 |
| No subsumption | 37 | 448 | 94 | 425 | 123 | 1127 |
| Forward sub. only | 115 | 1039 | 581 | 765 | 861 | 3361 |
| Backward sub. only | 40 | 541 | 106 | 479 | 143 | 1309 |
| No literal selection | 73 | 737 | 321 | 584 | 478 | 2193 |
| E 2.4 auto (all) | 813 | 1939 | 2648 | 1484 | 4054 | 10938 |
| E 2.4 auto (proofs) | 797 | 1621 | 2415 | 1171 | 3849 | 9853 |
| E 2.4 auto (sat) | 16 | 318 | 233 | 313 | 205 | 1085 |
| Prover9-1109a (all) | 728 | 1316 | 1678 | 709 | 2001 | 6432 |
| LeanCoP 2.2 (all) | 6 | 0 | 0 | 969 | 1826 | 2801 |

A note on the UEQ results: Most of the problems are specified as unit problems in CNF. A small number are expressed in first-order format. While the original specifications are unit equality, the added equality axioms are non-unit. This explains the rather large decrease in the number of successes if negative literal selection is disabled.

Overall, in the *Best* configuration, PyRes solves 3376 of the 16094 problems. Disabling indexing increases run time by a factor of around 3.7 (for problems with the same search behaviour), but this translates to only about 90 lost successes. Disabling subsumption, on the other hand, reduces the number of solutions found by 2/3rd. However, if we compare the effect of forward and backward subsumption, we can see that forward subsumption is crucial, while backward subsumption plays a very minor role. If we look at the detailed data, there are about 10 times more clauses removed by forward subsumption than by backward subsumption. This reflects the fact that usually smaller clauses are processed first, and a syntactically bigger clause cannot subsume a syntactically smaller clause. Finally, looking at negative literal selection, we can see that this extremely simple feature increases the number of solutions by over 1100.

Comparing PyRes and E, we can see the difference between a rather naive resolution prover and a high-performance superposition prover. Maybe not unexpectedly, the advantage of the more modern calculus is amplified for problems with equality. Overall, PyRes can solve about 30% of the problems E can solve. But there is a clear partition into problems with equality (14% in UEQ, 22% in CEQ, 21% in FEQ) and problems without equality (54% in CNE, 52% in FNE). PyRes does relatively much better with the latter classes. Prover9 falls in between E and PyRes. LeanCoP, for the categories it can handle, is similar to Prover9, but like PyRes is relatively stronger on problems without equality, and relatively weaker on problems with equality.

## 5   Conclusion

We have described PyRes, a theorem prover developed as a pedagogical example to demonstrate saturation-based theorem proving in an accessible, readable, well-documented way. The system's complexity is orders of magnitude lower than that of high-performance provers, and first exposure to students has been very successful. We hope that the lower barrier of entry will enable more students to enter the field.

Despite its relative simplicity, PyRes demonstrates many of the same properties as high-performance provers. Indexing speeds the system up significantly, but only leads to a moderate increase in the number of problems solved. Simple calculus refinements like literal selection and subsumption (the most basic simplification technique) have much more impact, as have search heuristics.

It is tempting to extend the system to e.g. the superposition calculus. However, implementing term orderings and rewriting would probably at least double the code base, something that is in conflict with the idea of a small, easily understood system. We are, however, working on a Java version, to see if the techniques demonstrated in Python can be easily transferred to a new language by developers not intimately familiar with automated theorem proving.

## References

1. Bachmair, L., Ganzinger, H.: Rewrite-based equational theorem proving with selection and simplification. J. Logic Comput. **3**(4), 217–247 (1994)
2. Korovin, K.: iProver – an instantiation-based theorem prover for first-order logic (system description). In: Armando, A., Baumgartner, P., Dowek, G. (eds.) IJCAR 2008. LNCS (LNAI), vol. 5195, pp. 292–298. Springer, Heidelberg (2008). https://doi.org/10.1007/978-3-540-71070-7_24

3. Kovács, L., Voronkov, A.: First-order theorem proving and VAMPIRE. In: Sharygina, N., Veith, H. (eds.) CAV 2013. LNCS, vol. 8044, pp. 1–35. Springer, Heidelberg (2013). https://doi.org/10.1007/978-3-642-39799-8_1

4. McCune, W.W.: Prover9 and Mace4 (2005–2010). http://www.cs.unm.edu/~mccune/prover9/. Accessed 29 Mar 2016

5. Otten, J.: leanCoP 2.0 and ileanCoP 1.2: high performance lean theorem proving in classical and intuitionistic logic (system descriptions). In: Armando, A., Baumgartner, P., Dowek, G. (eds.) IJCAR 2008. LNCS (LNAI), vol. 5195, pp. 283–291. Springer, Heidelberg (2008). https://doi.org/10.1007/978-3-540-71070-7_23

6. Robinson, J.A.: A machine-oriented logic based on the resolution principle. J. ACM 12(1), 23–41 (1965)

7. Schulz, S.: E – a brainiac theorem prover. J. AI Commun. 15(2/3), 111–126 (2002)

8. Schulz, S., Cruanes, S., Vukmirović, P.: Faster, higher, stronger: E 2.3. In: Fontaine, P. (ed.) CADE 2019. LNCS (LNAI), vol. 11716, pp. 495–507. Springer, Cham (2019). https://doi.org/10.1007/978-3-030-29436-6_29

9. Schulz, S., Möhrmann, M.: Performance of clause selection heuristics for saturation-based theorem proving. In: Olivetti, N., Tiwari, A. (eds.) IJCAR 2016. LNCS (LNAI), vol. 9706, pp. 330–345. Springer, Cham (2016). https://doi.org/10.1007/978-3-319-40229-1_23

10. Stump, A., Sutcliffe, G., Tinelli, C.: StarExec: a cross-community infrastructure for logic solving. In: Demri, S., Kapur, D., Weidenbach, C. (eds.) IJCAR 2014. LNCS (LNAI), vol. 8562, pp. 367–373. Springer, Cham (2014). https://doi.org/10.1007/978-3-319-08587-6_28

11. Sutcliffe, G., Schulz, S., Claessen, K., Van Gelder, A.: Using the TPTP language for writing derivations and finite interpretations. In: Furbach, U., Shankar, N. (eds.) IJCAR 2006. LNCS (LNAI), vol. 4130, pp. 67–81. Springer, Heidelberg (2006). https://doi.org/10.1007/11814771_7

12. Weidenbach, C., Dimova, D., Fietzke, A., Kumar, R., Suda, M., Wischnewski, P.: SPASS version 3.5. In: Schmidt, R.A. (ed.) CADE 2009. LNCS (LNAI), vol. 5663, pp. 140–145. Springer, Heidelberg (2009). https://doi.org/10.1007/978-3-642-02959-2_10

# Beyond Notations: Hygienic Macro Expansion for Theorem Proving Languages

Sebastian Ullrich[1]([✉]) and Leonardo de Moura[2]

[1] Karlsruhe Institute of Technology, Karlsruhe, Germany
sebastian.ullrich@kit.edu
[2] Microsoft Research, Redmond, USA
leonardo@microsoft.com

**Abstract.** In interactive theorem provers (ITPs), extensible syntax is not only crucial to lower the cognitive burden of manipulating complex mathematical objects, but plays a critical role in developing reusable abstractions in libraries. Most ITPs support such extensions in the form of restrictive "syntax sugar" substitutions and other ad hoc mechanisms, which are too rudimentary to support many desirable abstractions. As a result, libraries are littered with unnecessary redundancy. Tactic languages in these systems are plagued by a seemingly unrelated issue: accidental name capture, which often produces unexpected and counterintuitive behavior. We take ideas from the Scheme family of programming languages and solve these two problems simultaneously by proposing a novel *hygienic macro system* custom-built for ITPs. We further describe how our approach can be extended to cover type-directed macro expansion resulting in a single, uniform system offering multiple abstraction levels that range from supporting simplest syntax sugars to elaboration of formerly baked-in syntax. We have implemented our new macro system and integrated it into the upcoming version (v4) of the Lean theorem prover. Despite its expressivity, the macro system is simple enough that it can easily be integrated into other systems.

## 1 Introduction

*Mixfix* notation systems have become an established part of many modern ITPs for attaching terse and familiar syntax to functions and predicates of arbitrary arity.

| | |
|---|---|
| `_⊢_:_ = Typing` | *Agda* |
| `Notation "Ctx ⊢ E : T" := (Typing Ctx E T).` | *Coq* |
| `notation typing ("_ ⊢ _ : _")` | *Isabelle* |
| `notation Γ ` ⊢ ` e ` : ` τ := Typing Γ e τ` | *Lean 3* |

As a further extension, all shown systems also allow *binding names* inside mixfix notations.

**Electronic supplementary material** The online version of this chapter (https://doi.org/10.1007/978-3-030-51054-1_10) contains supplementary material, which is available to authorized users.

© Springer Nature Switzerland AG 2020
N. Peltier and V. Sofronie-Stokkermans (Eds.): IJCAR 2020, LNAI 12167, pp. 167–182, 2020.
https://doi.org/10.1007/978-3-030-51054-1_10

```
syntax ∃ A (λ x → P) = ∃[ x ∈ A ] P                        Agda
Notation "∃ x , P" := (exists (fun x => P)).                Coq
notation exists (binder "∃")                            Isabelle
notation `∃` binder `,` r:(scoped P, Exists P) := r       Lean 3
```

While these extensions differ in the exact syntax used, what is true about all of them is that at the time of the notation declaration, the system already, statically knows what parts of the term are bound by the newly introduced variable. This is in stark contrast to *macro systems* in Lisp and related languages where the expansion of a *macro* (a syntactic substitution) can be specified not only by a *template expression* with placeholders like above, but also by arbitrary *syntax transformers*, i.e. code evaluated at compile time that takes and returns a syntax tree.[1] As we move to more and more expressive notations and ideally remove the boundary between built-in and user-defined syntax, we argue that we should no more be limited by the static nature of existing notation systems and should instead introduce syntax transformers to the world of ITPs.

However, as usual, with greater power comes greater responsibility. By using arbitrary syntax transformers, we lose the ability to statically determine what parts of the macro template can be bound by the macro input (and vice versa). Thus it is no longer straightforward to avoid *hygiene* issues (i.e. accidental capturing of identifiers; [11]) by automatically renaming identifiers. We propose to learn from and adapt the macro hygiene systems implemented in the Scheme family of languages for interactive theorem provers in order to obtain more general but still well-behaved notation systems.

After giving a practical overview of the new, macro-based notation system we implemented in the upcoming version of Lean (Lean 4) in Sect. 2, we describe the issue of hygiene and our general hygiene algorithm, which should be just as applicable to other ITPs, in Sect. 3. Section 4 gives a detailed description of the implementation of this algorithm in Lean 4. In Sect. 5, we extend the use case of macros from mere syntax substitutions to type-aware elaboration. Finally, we have already encountered hygiene issues in the current version of Lean in a different part of the system: the tactic framework. We discuss how these issues are inevitable when implementing reusable tactic scripts and how our macro system can be applied to this hygiene problem as well in Sect. 6.

*Contributions.* We present a system for hygienic macros optimized for theorem proving languages as implemented[2] in the next version of the Lean theorem prover, Lean 4.

- We describe a novel, efficient hygiene algorithm to employ macros in ITP languages at large: a combination of a white-box, effect-based approach for detecting newly introduced identifiers and an efficient encoding of scope metadata.

---

[1] These two macro declaration styles are commonly referred to as *pattern-based* vs. *procedural*.

[2] https://github.com/leanprover/lean4/blob/IJCAR20/src/Init/Lean/Elab.

- We show how such a macro system can be seamlessly integrated into existing elaboration designs to support type-directed expansion even if they are not based on homogeneous source-to-source transformations.
- We show how hygiene issues also manifest in tactic languages and how they can be solved with the same macro system. To the best of our knowledge, the tactic language in Lean 4 is the first tactic language in an established theorem prover that is automatically hygienic in this regard.

## 2  The New Macro System

Lean's current notation system as shown in Sect. 1 is still supported in Lean 4, but based on a much more general macro system; in fact, the `notation` keyword itself has been reimplemented as a macro, more specifically as a *macro-generating macro* making use of our tower of abstraction levels. The corresponding Lean 4 command[3] for the example from the previous section `notation Γ "⊢" e ":" τ => Typing Γ e τ` expands to the macro declaration

```
macro Γ:term "⊢" e:term ":" τ:term : term => `(Typing $Γ $e $τ)
```

where the *syntactic category* (term) of placeholders and of the entire macro is now specified explicitly. The right-hand side uses an explicit *syntax quasiquotation* to construct the syntax tree, with syntax placeholders (*antiquotations*) prefixed with $. As suggested by the explicit use of quotations, the right-hand side may now be an arbitrary Lean term computing a syntax object; in other words, there is no distinction between pattern-based and procedural macros in our system. We can now use this abstraction level to implement simple command-level macros, for example.

```
macro "defthunk" id:ident ":=" e:term : command =>
`(def $id:ident := Thunk.mk (fun _ => $e))
defthunk big := mkArray 100000 true
```

Syntactic categories can be specified explicitly for antiquotations as in $id:ident where otherwise ambiguous. `macro` itself is another command-level macro that, for our `notation` example, expands to two commands

```
syntax term "⊢" term ":" term : term
macro_rules
| `($Γ ⊢ $e : $τ) => `(Typing $Γ $e $τ)
```

that is, a pair of parser extension (which we will not further discuss in this paper) and syntax transformer. Our reason for ultimately separating these two concerns is that we can now obtain a well-structured syntax tree pre-expansion, i.e. a *concrete* syntax tree, and use it to implement source code tooling such as auto-completion, go-to-definition, and refactorings. Implementing even just the most basic of these tools for the Lean 3 frontend that combined parsing and notation expansion meant that they had to be implemented right inside

---

[3] All examples including full context can be found in the supplemental material at https://github.com/Kha/macro-supplement.

the parser, which was not an extensible or even maintainable approach in our experience.

Both `syntax` and `macro_rules` are in fact further macros for regular Lean definitions encoding procedural metaprograms, though users should rarely need to make use of this lowest abstraction level explicitly. Both commands can only be used at the top level; we are not currently planning support for local macros.

There is no more need for the complicated scoped syntax since the desired translation can now be specified naturally, without any need for further annotations.

```
notation "∃" b "," P => Exists (fun b => P)
```

The lack of static restrictions on the right-hand side ensures that this works just as well with custom binding notations, even ones whose translation cannot statically be determined before substitution.

```
syntax "{" term "|" term "}" : term
macro_rules
| `({$x ∈ $s | $p}) => `(setOf (fun $x => $x ∈ $s ∧ $p))
| `({$b      | $p}) => `(setOf (fun $b => $p))

notation "⋃" b "," p => Union {b | p}
```

Here we explicitly make use of the and `macro_rules` abstraction level for its convenient syntactic pattern matching syntax. and `macro_rules` are "open" in the sense that multiple transformers for the same `syntax` declaration can be defined; they are tried in reverse declaration order by default up to the first match (though this can be customized using explicit priority annotations).

```
macro_rules
| `({$x ≤ $e | $p}) => `(setOf (fun $x => $x ≤ $e ∧ $p))
```

As a final example, we present a partial reimplementation of the arithmetic "bigop" notations found[4] in Coq's Mathematical Components library [12] such as

```
\sum_ (i <- [0, 2, 4] | i != 2) i
```

for summing over a filtered sequence of elements. The specific bigop notations are defined in terms of a single `\big_` fold operator; however, because Coq's notation system is unable to abstract over this new indexing syntax, every specific bigop notation has to redundantly repeat every specific index notation before delegating to `\big_` . In total, the 12 index notations for `\big_` are duplicated for 3 different bigops in the file.

```
Notation "\sum_ ( i <- r ) F"     := (\big[addn/0]_(i <- r) F).
Notation "\sum_ ( i <- r | P ) F" := (\big[addn/0]_(i <- r | P) F).
...
```

In contrast, using our system, we can introduce a new syntactic category for index notations, interpret it once in `\big_` , and define new bigops on top of it without any redundancy.

---

[4] https://github.com/math-comp/math-comp/blob/master/mathcomp/ssreflect/bigop.v.

```
declare_syntax_cat index
syntax ident "<-" term : index
syntax ident "<-" term "|" term : index
...

macro "Σ" "(" idx:index ")" F:term : term =>
`(\big_ [HasAdd.add, 0] ($idx:index) $F)
```

The full example is included in the supplement.

# 3  Hygiene Algorithm

In this section, we will give a mostly self-contained description of our algorithm for automatic hygiene applied to a simple recursive macro expander; we postpone comparisons to existing hygiene algorithms to Sect. 7.

Hygiene issues occur when transformations such as macro expansions lead to an unexpected capture (rebinding) of identifiers. For example, given the notation

```
notation "const" e => fun x => e
```

we would not expect the term const x to be closed because intuitively there is no x in scope at the argument position of const ; that the implementation of the macro makes use of the name internally should be of no concern to the macro user.

Thus hygiene issues can also be described as a *confusion of scopes* when syntax parts are removed from their original context and inserted into new contexts, which makes name resolution strictly after macro expansion (such as in a compiler preceded by a preprocessor) futile. Instead we need to track scopes *as metadata* before and during macro expansion so as not to lose information about the original context of identifiers. Specifically,

1. when an identifier captured in a syntax quotation matches one or more top-level symbols[5], the identifier is annotated with a list of these symbols as *top-level* scopes to preserve its *extra-macro* context (which, because of the lack of local macros, can only contain top-level bindings), and
2. when a macro is expanded, all identifiers freshly introduced by the expansion are annotated with a new *macro* scope to preserve the *intra-macro* context. Macro scopes are appended to a list, i.e. ordered by expansion time. This full "history of expansions" is necessary to treat macro-producing macros correctly, as we shall see in Sect. 3.2.

Thus, the expansion of the above term is (an equivalent of) fun x.1 => x where 1 is a fresh macro scope appended to the macro-introduced x , preventing it from capturing the x from the original input. In general, we will style hygienic identifiers in the following as $n.msc_1.msc_2.....msc_n\{tsc_1,...,tsc_n\}$ where $n$ is the original name, msc are macro scopes, and tsc top-level scopes, eliding

---

[5] Lean allows overloaded top-level bindings whereas local bindings are shadowing.

the braces if there are no top-level scopes as in the example above. We use the dot notation to suggest both the ordered nature of macro scopes and their eventual implementation in Sect. 4. We will now describe how to implement these operations in a standard macro expander.

## 3.1  Expansion Algorithm

The macro expander described in this section bundles the execution of macros and insertion of their results with interspersed name resolution to track scopes and ensure hygiene of identifiers. As we shall see below, top-level scopes on binding names are always discarded by it. Thus we will define a *symbol* more formally as an identifier together with a list of macro scopes, such as x.1 above.

Given a *global context* (a set of symbols), the expander does a conventional top-down expansion, keeping track of an initially-empty *local context* (another set of symbols). When a binding is encountered, the local context is extended with that symbol; top-level scopes on bindings are discarded since they are only meaningful on references. When a reference, i.e. an identifier not in binding position, is encountered, it is resolved according to the following rules:

1. If the local context has an entry for the same symbol, the reference binds to the corresponding local binding; any top-level scopes are ignored.
2. Otherwise, if the identifier is annotated with one or more top-level scopes or matches one or more symbols in the global context, it binds to all of these (to be disambiguated by the elaborator).
3. Otherwise, the identifier is unbound and an error is generated.

In the common incremental compilation mode of ITPs, every command is fully processed before subsequent commands. Thus, an expander for such a system will not extend the global context by itself, but pass the fully expanded command to the next compilation step before being called again with the next command's unexpanded syntax tree and a possibly extended global context.

Notably, our expander does not add macro scopes to identifiers by itself, either, much in contrast to other expansion algorithms. We instead delegate this task to the macro itself, though in a completely transparent way for all pattern-based and for many procedural macros. We claim that a macro should in fact be interpreted as an *effectful* computation since two expansions of the same identifier-introducing macro should not return the same syntax tree to avoid unhygienic interactions between them. Thus, as a *side effect*, it should apply a fresh macro scope to each captured identifier. In particular, a syntax quotation should not merely be seen as a datum, but implemented as an effectful value that obtains and applies this fresh scope to all the identifiers contained in it to immediately ensure hygiene for pattern-based macros. Procedural macros producing identifiers not originating from syntax quotations might need to obtain and make use of the fresh macro scope explicitly. We give a specific monad-based [14] implementation of effectful syntax quotations as a regular macro in Sect. 4.

## 3.2   Examples

Given the following input,

```
def x := 1
def e := fun y => x
notation "const" e => fun x => e
def y := const x
```

we incrementally parse, expand, and elaborate each declaration before advancing to the next one. For a first, trivial example, let us focus on the expansion of the second line. At this point, the global context contains the symbol x (plus any default imports that we will ignore here). Descending into the right-hand side of the definition, we first add y to the local context. The reference x does not match any local definitions, so it binds to the matching top-level definition.

In the next line, the built-in `notation` macro expands to the definitions

```
syntax "const" term : term
macro_rules
| `(const $e) => `(fun x => $e)
```

When a top-level macro application unfolds to multiple declarations, we expand and elaborate these incrementally as well to ensure that declarations are in the global context of subsequent declarations. When recursively expanding the `macro_rules` declaration (we will assume for this example that `macro_rules`itself is primitive) in the global context {x, e}, we first visit the syntax quotation on the left-hand side. The identifier e inside of it is in an antiquotation and thus not captured by the quotation. It is in binding position for the right-hand side, so we add e to the local context. Visiting the right-hand side, we find the quotation-captured identifier x and annotate it with the matching top-level definition of the same name; we do not yet know that it is in a binding position. When visiting the reference e, we see that it matches a local binding and do not add top-level scopes.

```
macro_rules
| `(const $e) => `(fun x{x} => $e)
```

Visiting the last line

```
def y := const x
```

with the global context {x, e}, we descend into the right-hand side. We expand the `const` macro given a fresh macro scope 2, which is applied to any captured identifiers.

```
def y := fun x.2{x} => x
```

We add the symbol x.2 (discarding the top-level scope x) to the local context and finally visit the reference x. The reference does not match the local binding x.2 but does match the top-level binding x, so it binds to the latter.

```
def y := fun x.2 => x
```

Now let us briefly look at a more complex macro-macro example demonstrating use of the macro scopes stack:

```
macro "m" n:ident : command => `(
  def f := 1
  macro "mm" : command => `(def $n:ident := f    def f := $n:ident))
```

If we call m f, we apply a macro scope 1 to all captured identifiers, then incrementally process the two new declarations.

```
  def f.1 := 1
  macro "mm" : command => `(def f := f.1{f.1}    def f.1{f.1} := f)
```

If we call the new macro mm, we apply one more macro scope 2.

```
        def f.2 := f.1.2{f.1}    def f.1.2{f.1} := f.2
```

When processing these new definitions, we see that the scopes ensure the expected name resolution. In particular, we now have global declarations f.1, f.2, and f.1.2 that show that storing only a single macro scope would have led to a collision.

# 4   Implementation

Syntax objects in Lean 4 are represented as an inductive type of *nodes* (or nonterminals), *atoms* (or terminals), and, as a special case of nonterminals, *identifiers*.

```
inductive Syntax
| node  (kind : Name) (args : Array Syntax)
| atom  (info : Option SourceInfo) (val : String)
| ident (info : Option SourceInfo) (rawVal : String) (val : Name)
    (preresolved : List (Nat × List String))
| missing
```

An additional constructor represents *missing* parts from syntax error recovery. Atoms and identifiers are annotated with source location metadata unless generated by a macro. Identifiers carry macro scopes inline in their Name while top-level scopes are held in a separate list. The additional Nat is an implementation detail of Lean's hierarchical name resolution.

The type Name of hierarchical names precedes the implementation of the macro system and is used throughout Lean's implementation for referring to (namespaced) symbols.

```
inductive Name
| anonymous : Name
| str : Name → String → Name
| num : Name → Nat    → Name
```

The syntax `a.b is a literal of type Name for use in meta-programs. The numeric part of Name is not accessible from the surface syntax and reserved for internal names; similar designs are found in other ITPs. By reusing Name for storing macro scopes, but not top-level scopes, we ensure that the new definition of *symbol* from Sect. 3.1 coincides with the existing Lean type and no changes to the implementation of the local or global context are necessary for adopting the macro system.

A Lean 4 implementation of the expansion algorithm described in the previous section is given in Fig. 1; the full implementation including examples is included in the supplement. As a generalization, syntax transformers have the type Syntax → `TransformerM Syntax` where the TransformerM monad gives access to the global context and a fresh macro scope per macro expansion. The expander itself uses an extended ExpanderM monad that also stores the local context and the set of registered macros. We use the Lean equivalent of Haskell's do notation [13] to program in these monads.

As usual, the expander has built-in knowledge of some "core forms" (lines 3–17) with special expansion behavior, while all other forms are assumed to be macros and expanded recursively (lines 20–22). Identifiers form one base case of the recursion. As described in the algorithm, they are first looked up in the local context (recall that the val of an identifier includes macro scopes), then as a fall back in the global context plus its own top-level scopes. mkTermId : Name → `Syntax` creates an identifier without source information or top-level scopes, which are not needed after expansion. mkOverloadedConstant implements the Lean special case of overloaded symbols to be disambiguated by elaboration; systems without overloading support should throw an ambiguity error instead in this case.

As an example of a core binding form, the expansion of a single-parameter fun is shown in lines 13–17 of Fig. 1. It recursively expands the given parameter type, then expands the body in a new local context extended with the value of id. Here `getIdentVal : Syntax` → `Name` in particular implements the discarding of top-level scopes from binders.

Finally, in the macro case, we fetch the syntax transformer for the given node kind, call it in a new context with a fresh current macro scope, and recurse.

Syntax quotations are given as one example of a macro: they do not have built-in semantics but transform into code that constructs the appropriate syntax tree (expandStxQuot in Fig. 2). More specifically, a syntax quotation will, at runtime, query the current macro scope msc from the surrounding TransformerM monad and apply it to all captured identifiers, which is done in quoteSyntax. quoteSyntax recurses through the quoted syntax tree, reflecting its constructors. Basic datatypes such as String and Name are turned into Syntax via the typeclass method quote. For antiquotations, we return their contents unreflected. In the case of identifiers, we resolve possible global references at compile time and reflect them, while msc is applied at runtime. Thus a quotation `(a + $b) inside a global context where the symbol a matches declarations a.a and b.a is transformed to the equivalent of

```
do msc ← getCurrMacroScope;
   pure (Syntax.node `plus
     [Syntax.ident none "a" (addMacroScope `a msc) [`a.a, `b.a],
      Syntax.atom none "+", b])
```

This implementation of syntax quotations itself makes use of syntax quotations for simplicity and thus is dependent on its own implementation in the previous stage of the compiler. Indeed, the helper variable msc must be renamed should the

```
1   partial def expand : Syntax → ExpanderM Syntax
2   | stx => match_syntax stx with
3     | `($id:ident) => do
4       let val := getIdentVal id;
5       gctx ← getGlobalContext;
6       lctx ← getLocalContext;
7       if lctx.contains val then
8         pure (mkTermId val)
9       else match resolve gctx val ++ getPreresolved id with
10        | []          => throw ("unknown identifier " ++ toString val)
11        | [(id, _)] => pure (mkTermId id)
12        | ids         => pure (mkOverloadedIds ids)
13    | `(fun ($id:ident : $ty) => $e) => do
14      let val := getIdentVal id;
15      ty ← expand ty;
16      e ← withLocal val (expand e);
17      `(fun ($(mkTermId val) : $ty) => $e)
18    | ...  -- other core forms
19    | _ => do
20      t ← getTransformerFor stx.getKind;
21      stx ← withFreshMacroScope (t stx);
22      expand stx
```

**Fig. 1.** Abbreviated implementation of a recursive expander for our macro system

name already be in scope and used inside an antiquotation. Note that quoteSyntax is allowed to reference the same msc as expandStxQuot because they are part of the same macro call and the current macro scope is unchanged between them.

## 5   Integrating Macros into Elaboration

The macro system as described so far can handle most syntax sugars of Lean 3 except for ones requiring type information. For example, the *anonymous constructor* ⟨e, ...⟩ is sugar for (c e ...) if the expected type of the expression is known and it is an inductive type with a single constructor c. While trivial to parse, there is no way to implement this syntax as a macro if expansion is done strictly prior to elaboration. To the best of our knowledge, none of the ITPs listed in the introduction support hygienic elaboration extensions of this kind, but we will show how to extend their common elaboration scheme in that way in this section.

Elaboration[6] can be thought of as a function elabTerm : Syntax → ElabM in an appropriate monad ElabM[7] from a (concrete or abstract) surface-level syntax tree type Syntax to a fully-specified core term type Expr [15]. We have presented the (concrete) definition of Syntax in Lean 4 in Sect. 4; the particular definition

---

[6] At the term level; other levels work analogously but with different output types.

[7] Or some other encoding of effects.

```
1   partial def quoteSyntax : Syntax → TransformerM Syntax
2   | Syntax.ident info rawVal val preresolved => do
3     gctx ← getGlobalContext;
4     let preresolved := resolve gctx val ++ preresolved;
5     `(Syntax.ident none $(quote rawVal) (addMacroScope $(quote val) msc)
          $(quote preresolved))
6   | stx@(Syntax.node k args) =>
7     if isAntiquot stx then pure (getAntiquotTerm stx)
8     else do
9       args ← args.mapM quoteSyntax;
10      `(Syntax.node $(quote k) $(quote args))
11  | Syntax.atom info val => `(Syntax.atom none $(quote val))
12  | Syntax.missing => pure Syntax.missing
13
14  def expandStxQuot (stx : Syntax) : TransformerM Syntax := do
15  stx ← quoteSyntax (stx.getArg 1);
16  `(do msc ← getCurrMacroScope; pure $stx)
```

**Fig. 2.** Simplified syntax transformer for syntax quotations

of Expr is not important here. While such an elaboration system could readily be composed with a type-insensitive macro expander such as the one presented in Sect. 3, we would rather like to *intertwine* the two to support type-sensitive but still hygienic-by-default macros (henceforth called *elaborators*) without having to reimplement macros of the kind discussed so far. Indeed, these can automatically be adapted to the new type given an adapter between the two monads, similarly to the adaption of macros to *expanders* in [6]:

```
def transformerToElaborator (m : Syntax → TransformerM Syntax) :
    Syntax → ElabM Expr :=
fun stx => do stx' ← (transformerMToElabM m) stx; elabTerm stx'
```

Because most parts of our hygiene system are implemented by the expander for syntax quotations, the only changes to an elaboration system necessary for supporting hygiene are storing the current macro scope in the elaboration monad (to be passed to the expansion monad in the adapter) and allocating a fresh macro scope in elabTerm and other recursion points, which morally now represent the starting point of a macro's expansion. Thus elaborators immediately benefit from hygiene as well whenever they use syntax quotations to construct unelaborated helper syntax objects to pass to elabTerm. In order to support syntax quotations in these two and other monads, we generalize their implementation to a new monad typeclass implemented by both monads.

```
class MonadQuotation (m : Type → Type) :=
(getCurrMacroScope : m MacroScope)
(withFreshMacroScope {α : Type} : m α → m α)
```

The second operation is not used by syntax quotations directly, but can be used by procedural macros to manually enter new macro call scopes.

As an example, the following is a simplified implementation of the anonymous constructor syntax mentioned above.

```
@[termElab anonymousCtor]
def elabAnonymousCtor (stx : Syntax) : ElabM Expr :=
match_syntax stx with
| `(($args*)) => do
  expectedType ← getExpectedType;
  match Expr.getAppFn expectedType with
  | Expr.const constName _ _ => do
    ctors ← getCtors constName;
    match ctors with
    | [ctor] => do
      stx ← `($(mkCTermId ctor) $(getSepElems args)*);
      elabTerm stx
  ... -- error handling
```

The [termElab] attribute registers this elaborator for the given syntax node kind. $args* is an antiquotation *splice* that extracts/injects a syntactic sequence of elements into/from an Array Syntax. The array by default includes separators such as "," as Syntax.atoms in order to be lossless, which we here filter out using getSepElems. The function mkCTermId : Name → Syntax synthesizes a hygienic reference to the given constant name by storing it as a top-level scope and applying a reserved macro scope to the constructed identifier.

This implementation fails if the expected type is not yet sufficiently known at this point. The actual implementation[8] of this elaborator extends the code by *postponing* elaboration in this case. When an elaborator requests postponement, the system returns a fresh metavariable as a placeholder and associates the input syntax tree with it. Before finishing elaboration, postponed elaborators associated with unsolved metavariables are retried until they all ultimately succeed, or else elaboration is stuck because of cyclic dependencies and an error is signed.

# 6   Tactic Hygiene

Lean 3 includes a tactic framework that, much like macros, allows users to write custom automation either procedurally inside a Tactic monad (renamed to TacticM in Lean 4) or "by example" using tactic language quotations, or in a mix of both [9]. For example, Lean 3 uses a short tactic block to prove injection lemmas for data constructors.

```
def mkInjEq : Tactic Unit :=
`[intros; apply propext; apply Iff.intro; ...]
```

Unfortunately, this code unexpectedly broke in Lean 3 when used from a library for homotopy type theory that defined its own propext and Iff.intro declarations;[9] in other words, Lean 3 tactic quotations are unhygienic and required

---

[8] https://github.com/leanprover/lean4/blob/IJCAR20/src/Init/Lean/Elab/
BuiltinNotation.lean#L43.

[9] https://github.com/leanprover/lean/pull/1913.

manual intervention in this case. Just like with macros, the issue with tactics is that binding structure in such embedded terms is not known at declaration time. Only at tactic run time do we know all local variables in the current context that preceding tactics may have added or removed, and therefore the scope of each captured identifier.

Arguably, the Lean 3 implementation also exhibited a lack of hygiene in the handling of tactic-introduced identifiers: it did not prevent users from referencing such an identifier outside of the scope it was declared in.

```
def myTac : Tactic Unit := `[intro h]
lemma triv (p : Prop) : p → p := begin myTac; exact h end
```

Coq's similar Ltac tactic language [5] exhibits the same issue and users are advised not to introduce fixed names in tactic scripts but to generate fresh names using the fresh tactic first,[10] which can be considered a manual hygiene solution.

Lean 4 instead extends its automatically hygienic macro implementation to tactic scripts by allowing regular macros in the place of tactic invocations.

```
macro "myTac" : tactic => `(intro h; exact h)
theorem triv (p : Prop) : p → p := begin myTac end
```

By the same hygiene mechanism described above, introduced identifiers such as h are renamed so as not to be accessible outside of their original scope, while references to global declarations are preserved as top-level scope annotations. Thus Lean 4's tactic framework resolves both hygiene issues discussed here without requiring manual intervention by the user. Expansion of tactic macros in fact does not precede but is integrated into the *tactic evaluator* evalTactic : Syntax → TacticM such that recursive macro calls are expanded lazily.

```
syntax "repeat" tactic : tactic
macro_rules
| `(tactic| repeat $t) => `(tactic| try ($t; repeat $t))
```

Here the *quotation kind* tactic followed by a pipe symbol specifies the parser to use for the quotation, since tactic syntax may otherwise overlap with term syntax. macro automatically infers it from the given syntax category, but cannot be used here because the parser for repeat would not yet be available in the right-hand side. When $t eventually fails, the recursion is broken without visiting and expanding the subsequent repeat macro call. The try tactical is used to ignore this eventual failure.

While we believe that macros will cover most use cases of tactic quotations in Lean 3, their use within larger TacticM metaprograms can be recovered by passing such a quotation to evalTactic:

$$Syntax → TacticM$$

TacticM implements the MonadQuotation typeclass for this purpose.

---

[10] https://github.com/coq/coq/issues/9474.

# 7    Related Work

The main inspiration behind our hygiene implementation was Racket's new *Sets of Scopes* [10] hygiene algorithm. Much like in our approach, Racket annotates identifiers both with scopes from their original context as well as with additional macro scopes when introduced by a macro expansion. However, there are some significant differences: Racket stores both types of scopes in a homogeneous, unordered set and does name resolution via a maximum-subset check. For both simplicity of implementation and performance, we have reduced scopes to the bare minimal representation using only strict equality checks, which we can easily encode in our existing Name implementation. In particular, we only apply scopes to matching identifiers and only inside syntax quotations. This optimization is of special importance because top-level declarations in Lean and other ITPs are not part of a single, mutually recursive scope as in Racket, but each open their own scope over all subsequent declarations, which would lead to a total number of scope annotations quadratic in the number of declarations using the Sets of Scopes algorithm. Finally, Racket detects macro-introduced identifiers using a "black-box" approach without the macro's cooperation following the marking approach of [11]: a fresh macro scope is applied to all identifiers in the macro input, then inverted on the macro output. While elegant, a naive implementation of this approach can result in quadratic runtime compared to unhygienic expansion and requires further optimizations in the form of lazy scope propagation [7], which is difficult to implement in a pure language such as Lean. Our "white-box" approach based on the single primitive of an effectful syntax quotation, while slightly easier to escape from in procedural syntax transformers, is simple to implement, incurs minimal overhead, and is equivalent for pattern-based macros.

The idea of automatically handling hygiene in the macro, and not in the expander, was introduced in [4], though only for pattern-based macros. MetaML [18] refined this idea by tying hygiene more specifically to syntax quotations that could be used in larger metaprogram contexts, which Template Haskell [17] interpreted as effectful (monadic) computations requiring access to a fresh-names generator, much like in our design. However, both of the latter systems should perhaps be characterized more as metaprogramming frameworks than Scheme-like macro systems: there are no "macro calls" but only explicit splices and so only built-in syntax with known binding semantics can be captured inside syntax quotations. Thus the question of which captured identifiers to rename becomes trivial again, just like in the basic notation systems discussed in Sect. 1.

While the vast majority of research on hygienic macro systems has focused on S-expression-based languages, there have been previous efforts on marrying that research with non-parenthetical syntax, with different solutions for combining syntax tree construction and macro expansion. The Dylan language requires macro syntax to use predefined terminators and eagerly scans for the end of a macro call using this knowledge [2], while in Honu [16] the syntactic structure of a macro call is discovered during expansion by a process called "enforestation". The Fortress [1] language strictly separates the two concerns into grammar extensions and transformer declarations, much like we do. Dylan and Fortress are restricted to pattern-based macro declarations and thus can make use of simple hygiene algorithms while Honu uses the full generality of the Racket macro expander. On the other hand, Honu's authors "explicitly trade expressiveness for syntactic simplicity" [16]. In order to express the full Lean language and desirable extensions in a macro system, we require both unrestricted syntax of macros and procedural transformers.

Many theorem provers such as Coq, Agda, Idris, and Isabelle not already based on a macro-powered language provide restricted syntax extension mechanisms, circumventing hygiene issues by statically determining binding as seen in Sect. 1. Extensions that go beyond that do not come with automatic hygiene guarantees. Agda's macros[11], for example, operate on the De Bruijn index-based core term level and are not hygienic.[12] The ACL2 prover in contrast uses a subset of Common Lisp as its input language and adapts the hygiene algorithm of [7] based on renaming [8]. The experimental Cur [3] theorem prover is a kind of dual to our approach: it takes an established language with hygienic macros, Racket, and extends it with a dependent type system and theorem proving tools. ACL2 does not support tactic scripts, while in Cur they can be defined via regular macros. However, this approach does not currently provide tactic hygiene as defined in Sect. 6.[13]

# 8    Conclusion

We have proposed a new macro system for interactive theorem provers that enables syntactic abstraction and reuse far beyond the usual support of mixfix notations. Our system is based on a novel hygiene algorithm designed with a focus on minimal runtime overhead as well as ease of integration into pre-existing codebases, including integration into standard elaboration designs to support type-directed macro expansion. Despite that, the algorithm is general enough to provide a complete hygiene solution for pattern-based macros and provides flexible hygiene for procedural macros. We have also demonstrated how our macro system can address unexpected name capture issues that haunt existing tactic frameworks. We have implemented our method in the upcoming version (v4) of the Lean theorem prover; it should be sufficiently attractive and straightforward to implement to be adopted by other interactive theorem proving systems as well.

**Acknowledgments.** We are very grateful to the anonymous reviewers, David Thrane Christiansen, Gabriel Ebner, Matthew Flatt, Sebastian Graf, Alexis King, Daniel Selsam, and Max Wagner for extensive comments, corrections, and advice.

---

[11] https://agda.readthedocs.io/en/v2.6.0.1/language/reflection.html#macros
[12] https://github.com/agda/agda/issues/3819.
[13] https://github.com/wilbowma/cur/issues/104.

# References

1. Allen, E., et al.: The Fortress language specification. Sun Microsyst. **139**(140), 116 (2005)
2. Bachrach, J., Playford, K., Street, C.: D-expressions: Lisp power. Dylan style, Style DeKalb IL (1999)
3. Chang, S., Ballantyne, M., Turner, M., Bowman, W.J.: Dependent type systems as macros. Proc. ACM Program. Lang. **4**(POPL), 1–29 (2019)
4. Clinger, W., Rees, J.: Macros that work. In: Proceedings of the 18th ACM SIGPLAN-SIGACT Symposium on Principles of Programming Languages, pp. 155–162 (1991)
5. Delahaye, D.: A tactic language for the system Coq. In: Logic for Programming and Automated Reasoning, 7th International Conference, LPAR 2000, Proceedings, pp. 85–95 (2000)
6. Dybvig, R.K., Friedman, D.P., Haynes, C.T.: Expansion-passing style: beyond conventional macros. In: Proceedings of the 1986 ACM Conference on LISP and Functional Programming, pp. 143–150 (1986)
7. Dybvig, R.K., Hieb, R., Bruggemån, C.: Syntactic abstraction in scheme. Lisp Symbolic Comput. **5**(4), 295–326 (1993)
8. Eastlund, C., Felleisen, M.: Hygienic macros for ACL2. In: Page, R., Horváth, Z., Zsók, V. (eds.) TFP 2010. LNCS, vol. 6546, pp. 84–101. Springer, Heidelberg (2011). https://doi.org/10.1007/978-3-642-22941-1_6
9. Ebner, G., Ullrich, S., Roesch, J., Avigad, J., de Moura, L.: A metaprogramming framework for formal verification. Proc. ACM Program. Lang. **1**(ICFP) (2017).https://doi.org/10.1145/3110278
10. Flatt, M.: Binding as sets of scopes. In: Proceedings of the 43rd Annual ACM SIGPLAN-SIGACT Symposium on Principles of Programming Languages, POPL 2016, pp. 705–717, ACM, New York (2016). https://doi.org/10.1145/2837614.2837620, http://doi.acm.org/10.1145/2837614.2837620
11. Kohlbecker, E., Friedman, D.P., Felleisen, M., Duba, B.: Hygienic macro expansion. In: Proceedings of the 1986 ACM Conference on LISP and Functional Programming, pp. 151–161 (1986)
12. Mahboubi, A., Tassi, E.: Mathematical components. https://math-comp.github.io/mcb/
13. Marlow, S., et al.: Haskell 2010 language report (2010). https://www.haskell.org/onlinereport/haskell2010
14. Moggi, E.: Notions of computation and monads. Inf. Comput. **93**(1), 55–92 (1991)
15. de Moura, L., Avigad, J., Kong, S., Roux, C.: Elaboration in dependent type theory (2015)
16. Rafkind, J., Flatt, M.: Honu: syntactic extension for algebraic notation through enforestation. In: ACM SIGPLAN Notices, vol. 48, pp. 122–131. ACM (2012)
17. Sheard, T., Jones, S.P.: Template meta-programming for Haskell. In: Proceedings of the 2002 ACM SIGPLAN Workshop on Haskell, pp. 1–16. ACM (2002)
18. Taha, W., Sheard, T.: MetaML and multi-stage programming with explicit annotations. Theor. Comput. Sci. **248**(1–2), 211–242 (2000)

# Formalizations

# Formalizing the Face Lattice of Polyhedra

Xavier Allamigeon[1,2]([⊠]), Ricardo D. Katz[3], and Pierre-Yves Strub[4]

[1] Inria, Palaiseau, France
xavier.allamigeon@inria.fr
[2] CMAP, CNRS, Ecole Polytechnique, Institut Polytechnique de Paris,
Palaiseau, France
[3] CIFASIS–CONICET, Rosario, Argentina
katz@cifasis-conicet.gov.ar
[4] LIX, CNRS, École Polytechnique, Institut Polytechnique de Paris,
Palaiseau, France
pierre-yves@strub.nu

**Abstract.** Faces play a central role in the combinatorial and computational aspects of polyhedra. In this paper, we present the first formalization of faces of polyhedra in the proof assistant COQ. This builds on the formalization of a library providing the basic constructions and operations over polyhedra, including projections, convex hulls and images under linear maps. Moreover, we design a special mechanism which automatically introduces an appropriate representation of a polyhedron or a face, depending on the context of the proof. We demonstrate the usability of this approach by establishing some of the most important combinatorial properties of faces, namely that they constitute a family of graded atomistic and coatomistic lattices closed under sublattices.

## 1 Introduction

A face of a polyhedron is defined as the set of points reaching the maximum (or minimum) of a linear function over the polyhedron. Faces are ubiquitous in the theory of polyhedra, and especially in the complexity analysis of optimization algorithms. As an illustration, the simplex method, one of the most widely used algorithms for solving linear programming, finds an optimal solution by iterating over the *graph* of the polyhedron, i.e. the adjacency graph of vertices and edges, which respectively constitute the 0- and 1-dimensional faces. The problem of finding a pivoting rule, i.e. a way to iterate over the graph, which ensures to reach an optimal vertex in a polynomial number of steps, is a central problem in computational optimization, related with Smale's ninth problem for the twenty-first century [25]. Faces of polyhedra are also involved in the worst-case complexity analysis of other

The first author was partially supported by the ANR CAPPS project (ANR-17-CE40-0018), and by a public grant as part of the Investissement d'avenir project, reference ANR-11-LABX-0056-LMH, LabEx LMH, in a joint call with Gaspard Monge Program for optimization, operations research and their interactions with data sciences. The third author was partially supported by the ANR SCRYPT project (ANR-18-CE25-0014).

N. Peltier and V. Sofronie-Stokkermans (Eds.): IJCAR 2020, LNAI 12167, pp. 185–203, 2020.
https://doi.org/10.1007/978-3-030-51054-1_11

optimization methods, such as interior point methods; see [2,13]. This has motivated several mathematical problems on the combinatorics of faces, which are of independent interest. For example, the question of finding a polynomial bound on the diameter of the graphs of polyhedra (in the dimension and the number of defining inequalities) is still unresolved, despite recent progress [6,7,23]. We refer to [12] for a recent account on the subject.

Other applications of polyhedra and their faces arise in formal verification, in which passing from a representation by inequalities to a representation as the convex hull of finitely many points and vice versa, is a critical computational step. The correctness analysis of the algorithms solving this problem, extensively relies on the understanding of the mathematical structure of faces, in particular of vertices, edges and facets (i.e. 1-co-dimensional faces).

In this paper, we formalize a significant part of the properties of faces in the proof assistant COQ. As usually happens in the formalization of mathematics, one of the key difficulties is to find the right representation for objects in the proof assistant. For polyhedra and their faces, the choice of the representation depends on the context. In more detail, every polyhedron admits infinitely many descriptions by linear inequality systems. In mathematics textbooks, proofs are carried out by choosing one (often arbitrary) inequality system for a polyhedron $\mathcal{P}$, and then manipulating the faces of $\mathcal{P}$ or other subsequent polyhedra through inequality systems which derive from the one chosen for $\mathcal{P}$. Proving that these are valid inequality systems is usually trivial for the reader, but not for the proof assistant. We exploit the so-called *canonical structures* of COQ in order to achieve this step automatically. This allows us to obtain proof scripts which only focus on the relevant mathematical content, and which are closer to what mathematicians write.

Thanks to this approach, we show that the faces of a polyhedron $\mathcal{P}$ form a finite lattice, in which the order is the set inclusion, the bottom and top elements are respectively the empty set and $\mathcal{P}$, and the meet operation is the set intersection. We establish that the face lattice is both atomistic and coatomistic, meaning that every element is the join (resp. the meet) of a finite set of atoms (resp. coatoms). Atoms and coatoms respectively correspond to minimal and maximal elements distinct from the top and bottom elements. Moreover, we prove that the face lattice is graded, i.e. every maximal chain has the same length. Finally, we show that the family of face lattices of polytopes (convex hulls of finitely many points) is closed under taking sublattices, i.e. any sublattice of the face lattice of a polytope is isomorphic to the face lattice of another polytope. As a consequence of that, we prove that any sublattice of height two is isomorphic to a diamond.

Formalizing these results requires the introduction of several important and non-trivial notions. First of all, our work relies on the construction of a library manipulating polyhedra, which provides all the basic operations over them, including intersections, projections, convex hulls, as well as special classes of polyhedra such as affine subspaces. Dealing with faces also requires to formalize the dimension of a polyhedron, and its relation with the dimension of its affine hull, i.e. the smallest affine subspace containing it. Some classes of faces also retain a particular attention, such as vertices, edges and facets. For instance, we formalize the vertex figure, which is a geometric construction to manipulate the faces containing a fixed vertex.

Throughout this work, we have drawn much inspiration from the textbooks of Schrijver [24] and Ziegler [28] to guide us in our approach. The source code of our formalization is done within the Coq-Polyhedra project, and is available at https://github.com/Coq-Polyhedra/Coq-Polyhedra/tree/IJCAR-2020, in the directory `theories`. We rely on the Mathematical Components library [18] (abridged MathComp thereafter) for basic data structures such as finite sets, ordered fields, and vector spaces.

The paper is organized as follows. In Sect. 2, we present how we define the basic operations and constructions over polyhedra. Section 3 deals with the central problem of finding an appropriate representation of faces, and explains how this leads to a seamless formalization of important properties like the dimension. Section 4 demonstrates the practical usability of our approach, by presenting the formalization of the face lattice and its main characteristics. Finally, we discuss related work in Sect. 5. A link to the relevant source files is given beside section titles in order to help the reader finding the results in the source code of the formalization.

## 2   Constructing a Library Manipulating Polyhedra

### 2.1   The Quotient Type of Polyhedra[1,2]

We recall that a *(convex) polyhedron* of $\mathbb{R}^n$ is defined as the intersection of finitely many *halfspaces* $\{x \in \mathbb{R}^n \colon \langle \alpha, x \rangle \geq \beta\}$, where $\alpha \in \mathbb{R}^n$, $\beta \in \mathbb{R}$, and $\langle \cdot, \cdot \rangle$ is the Euclidean scalar product, i.e. $\langle y, z \rangle := \sum_{1 \leq i \leq n} y_i z_i$. Equivalently, a polyhedron can be represented as the solution set of a linear affine system $Ax \geq b$, where $A \in \mathbb{R}^{m \times n}$ and $b \in \mathbb{R}^m$, in which case each inequality $A_i x \geq b_i$ corresponds to a halfspace.

Throughout the paper, we use the variable `n : nat` to represent the dimension of the ambient space. Instead of dealing with polyhedra over the reals, we introduce a variable `R : realFieldType` which represents an abstract ordered field with decidable ordering. In this setting, `'cV[R]_n` (or `'cV_n` for short) stands for the type of column vectors of size `n` over the field `R`.

As we mentioned earlier, the representation by inequalities (or halfspaces) of a convex polyhedron $\mathcal{P}$ is not unique. The first step in our work is to introduce a quotient structure, in order to define the basic operations (membership of a point, inclusion, etc.) regardless of the exact representation of the polyhedron. The quotient structure is based on a concrete type denoted by `'hpoly[R]_n` (or simply `'hpoly_n`, when `R` is clear from the context). The prefix letter "h" is taken from the terminology *H-polyhedron* or *H-representation* which is used to refer to

---

[1]  https://github.com/Coq-Polyhedra/Coq-Polyhedra/tree/IJCAR-2020/theories/hpolyhedron.v.

[2]  https://github.com/Coq-Polyhedra/Coq-Polyhedra/tree/IJCAR-2020/theories/polyhedron.v.

representations by halfspaces. The elements of 'hpoly_n are records consisting of a matrix $A \in \mathbb{R}^{m \times n}$ and a vector $b \in \mathbb{R}^m$ representing the system $Ax \geq b$:

```
Record hpoly := HPoly { m : nat; A : 'M_(m,n); b : 'cV_m }.
```

We equip 'hpoly_n with a membership predicate stating that, given P : 'hpoly_n and x : 'cV_n, we have x \in P if and only if x satisfies the system of inequalities represented by P. Two H-polyhedra are *equivalent* when they correspond to the same solution set, i.e. their membership predicate agree. We prove that this equivalence relation is decidable, by exploiting the implementation of the simplex method of [3]. The latter allows us to check that an inequality $\langle \alpha, x \rangle \geq \beta$ is valid over an H-polyhedron P : 'hpoly_n by minimizing the linear function $x \mapsto \langle \alpha, x \rangle$ over P, and checking that the optimal value is greater than or equal to $\beta$. Then, deciding whether P Q : 'hpoly_n are equivalent amounts to checking that each inequality in the system defining Q is valid over P, and vice versa.

The quotient structure is built following the approach of [10]. This introduces a quotient type, denoted here by 'poly[R]_n (or simply 'poly_n). Its elements are referred to as *polyhedra* and represent equivalence classes of H-polyhedra. In practice, each polyhedron is a record formed by a canonical representative of the class, and the proof that the representative is indeed the canonical one. We point out that the notion of canonical representative has no special mathematical meaning or structure.

We define the membership predicate of each P : 'poly_n as the membership predicate of its canonical representative. As expected, equality between two polyhedra of 'poly_n and extensional equality (denoted =i below) of their membership predicates are equivalent properties:

```
Lemma poly_eqP {P Q : 'poly_n} : (P = Q) <-> (P =i Q).
```

## 2.2   Operations over Polyhedra[3]

We first lift a number of basic primitives from the type 'hpoly_n to the quotient type 'poly_n, including the subset relation P `<=` Q and the intersection operation P `&` Q. The related properties are also lifted by using the fact that the membership predicate of any element of 'hpoly_n is extentionally equivalent to the membership predicate of its equivalence class in 'poly_n.

Even though we now work on the quotient type, we still need a way to build polyhedra from sets of inequalities. While H-polyhedra rely on inequality constraints under the matrix form, we choose now to be closer to the mathematical definition of polyhedra as the intersection of finitely many halfspaces. To this end, we introduce the type lrel[R]_n (or simply lrel_n when R is clear from the context), which is isomorphic to the cartesian product 'cV_n * R of vectors of size n and elements of R. This type is used to construct linear affine inequalities

---

[3] https://github.com/Coq-Polyhedra/Coq-Polyhedra/tree/IJCAR-2020/theories/polyhedron.v.

or equalities. In more detail, if e represents the pair $(\alpha, \beta) \in \mathbb{R}^n \times \mathbb{R}$, then the polyhedron [hs e] corresponds to the halfspace $\langle \alpha, x \rangle \geq \beta$:

Lemma **in_hs** (e : lrel_n) x : x \in [hs e] <-> ('[e.1,x] >= e.2).

Similarly, the element e is used to build a hyperplane denoted [hp e]:

Lemma **in_hp** (e : lrel_n) x : x \in [hp e] <-> ('[e.1,x] = e.2).

(In the last two statements, the terms e.1 and e.2 respectively stand for the first and second component of the pair formed by e, while '[.,.] stands for the scalar product between two vectors.)

We can now construct polyhedra defined by sets of inequalities. To this aim, we use the type {fset lrel_n} of finite sets of elements of type lrel_n. Then, given base : {fset lrel_n}, the polyhedron denoted by 'P(base) is defined as the intersection of the halfspaces [hs e] for e \in base. In particular, we introduce the empty polyhedron [poly0] and the full polyhedron [polyT], which are defined by the inequality $1 \leq 0$ and by no inequality respectively. As we shall see in Sect. 3, the formalization of faces requires us to manipulate polyhedra defined by systems mixing inequalities and equalities. We denote such a polyhedron by 'P^=(base; I), where both base and I are of type {fset lrel_n}. It represents the intersection of the polyhedron 'P(base) with the hyperplanes [hp e] for e \in I.

The cornerstone of more advanced constructions is the primitive proj0, which, given P : 'poly_(n.+1), builds its projection on the last n components. This is carried out by implementing Fourier–Motzkin elimination algorithm (see e.g. [24, Chapter 12]). In short, this algorithm starts from a system of linear inequalities, and constructs pairwise combinations of them in order to eliminate the first variable. The result is that the new system is a valid representation of the projected polyhedron. This is written as follows:

Theorem **proj0P** (P : 'poly_(n.+1)) :
  reflect (exists2 y : 'cV_(n.+1), x = row' 0 y & y \in P) (x \in proj0 P).

where row' 0 y : 'cV_n is the projection of y on the last n components, and reflect stands for a logical equivalence between the two properties. This projection primitive then allows us to construct many more polyhedra. For example, we can build the image of a polyhedron $\mathcal{P}$ by the linear map represented by a matrix $A \in \mathbb{R}^{k \times n}$. The latter is obtained by embedding $\mathcal{P}$ in a polyhedron over the variables $(x, y) \in \mathbb{R}^{n+k}$, intersecting it with the equality constraints $y = Ax$, and finally projecting it on the last $k$ components. The construction of the convex hull of finitely many points immediately follows. Indeed, the convex hull of a finite set $V = \{v^1, \ldots, v^p\} \subset \mathbb{R}^n$ can be defined as the image of the simplex $\Delta_p := \{\mu \in (\mathbb{R}_{\geq 0})^p : \sum_{i=1}^p \mu_i = 1\}$ by the linear map $\mu \mapsto \sum_{i=1}^p \mu_i v^i$. We denote the convex hull by conv V where V : {fset 'cV_n} represents a finite set of points, and we obtain (cf. Lemma **in_convP**) that x \in conv V if and only if x is a barycentric combination of the points of V. The convex hull constructor yields some other elementary yet very useful constructions, such as polyhedra

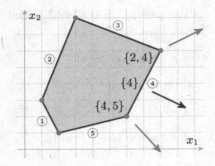

$$2x_1 + x_2 \geq 5 \qquad \text{①}$$
$$5x_1 - 2x_2 \geq -1 \qquad \text{②}$$
$$-2x_1 - 5x_2 \geq -46 \qquad \text{③}$$
$$-2x_1 + x_2 \geq -10 \qquad \text{④}$$
$$-x_1 + 4x_2 \geq 2 \qquad \text{⑤}$$

**Fig. 1.** A polyhedron, defined by the inequalities on the right, and its faces. The vertices (0-dim. faces) are represented by blue dots, while the edges (1-dim. faces) are depicted in black. Arrows correspond to linear functions associated with some of the faces, in the sense of Definition 1. We also indicate beside them the set $I$ of the defining inequalities turned into equalities, as in Theorem 1. (Color figure online)

reduced to a single point (denoted [pt x] where x : 'cV_n) or segments between two points (denoted [segm x; y] where x y : 'cV_n).

Finally, we recover some important results of the theory of polyhedra which were proved in [3]. In more detail, we lift a version of Farkas Lemma expressed on the type 'hpoly_n, and then obtain the Strong Duality Theorem, the complementary slackness conditions (which are conditions characterizing the optimality of solutions of linear programs), and some separation results. We refer to Section **Separation** and Section **Duality** for further details on these statements.

## 3    Representing Faces of Polyhedra[4]

### 3.1    Equivalent Definitions of Faces

Faces are commonly defined as sets of optimal solutions of linear programs, i.e. problems consisting in minimizing a linear function over a polyhedron.

**Definition 1.** *A set $\mathcal{F}$ is a* face *of the polyhedron $\mathcal{P} \subset \mathbb{R}^n$ if $\mathcal{F} = \emptyset$ or there exists $c \in \mathbb{R}^n$ such that $\mathcal{F}$ is the set of points of $\mathcal{P}$ minimizing the linear function $x \mapsto \langle c, x \rangle$ over $\mathcal{P}$.*

We note that $\mathcal{P}$ is a face of itself (take $c = 0$). Figure 1 provides an illustration of this definition.

In formal proving, the choice of the definition plays a major role on the ability to prove complex properties of the considered objects. A drawback of the previous definition is that it does not directly exhibit some of the most basic properties of faces: for instance, the fact that a face is itself a polyhedron, the fact that the intersection of two faces is a face, or the fact that a polyhedron

---

[4] https://github.com/Coq-Polyhedra/Coq-Polyhedra/tree/IJCAR-2020/theories/poly_base.v.

has finitely many faces. In contrast, these properties are straightforward consequences of the following characterization of faces:

**Theorem 1.** *Let* $\mathcal{P} = \{x \in \mathbb{R}^n : Ax \geq b\}$, *where* $A \in \mathbb{R}^{m \times n}$ *and* $b \in \mathbb{R}^m$. *A set* $\mathcal{F}$ *is face of* $\mathcal{P}$ *if and only if* $\mathcal{F} = \emptyset$ *or there exists* $I \subset \{1, \ldots, m\}$ *such that*

$$\mathcal{F} = \mathcal{P} \cap \{x \in \mathbb{R}^n : A_i x = b_i \text{ for all } i \in I\}. \tag{1}$$

Nevertheless, Theorem 1 is expressed in terms of a certain H-representation of the polyhedron $\mathcal{P}$, while the property of being a face is intrinsic to the set $\mathcal{P}$. This raises the problem of exploiting the most convenient representation of $\mathcal{P}$ to apply the characterization of Theorem 1. We illustrate this on the proof of the following property, which is used systematically (or even implicitly) in almost every proof of statements on faces:

**Proposition 1.** *If* $\mathcal{F}$ *is a face of* $\mathcal{P}$, *then any face of* $\mathcal{F}$ *is a face of* $\mathcal{P}$.

Assume $\mathcal{P}$ is represented by the inequality system $Ax \geq b$, and take $I$ as in (1). Let $\mathcal{F}'$ be a nonempty face of $\mathcal{F}$. We apply Theorem 1 with $\mathcal{F}$ as $\mathcal{P}$, by using the following H-representation of $\mathcal{F} : Ax \geq b$ and $-A_i x \geq -b_i$ for $i \in I$. We get that $\mathcal{F}' = \mathcal{F} \cap \{x \in \mathbb{R}^n : A_i x = b_i \text{ for all } i \in I'\}$ for a certain set $I' \subset \{1, \ldots, m\}$. We deduce that $\mathcal{F}' = \mathcal{P} \cap \{x \in \mathbb{R}^n : A_i x = b_i \text{ for all } i \in I \cup I'\}$, and conclude that $\mathcal{F}'$ is a face of $\mathcal{P}$ by applying Theorem 1. While the choice of the H-representation of $\mathcal{P}$ is irrelevant, we point out that the proof would not have been so immediate if we had initially chosen an arbitrary H-representation of $\mathcal{F}$.

## 3.2   Working Within a Fixed Ambient H-Representation

Theorem 1 leads us to the following strategy: when dealing with the faces of a polyhedron, and possibly with the faces of these faces, etc., we first set an H-representation of the top polyhedron, and then manipulate the subsequent H-representations of faces in which some inequalities are strengthened into equalities, like in (1).

The top H-representation will be referred to as the *ambient representation*, and is formalized as a term `base` of type `{fset lrel_n}` representing a finite set of inequalities. Then, we introduce the type `{poly base}`, which corresponds to the subtype of `'poly_n` whose inhabitants are the polyhedra Q satisfying the following property:

```
Definition has_base base Q :=
  (Q != [poly0]) -> exists I : {fsubset base}, Q = 'P^=(base; I).
```

where `{fsubset base}` is the type of subsets of `base`. We recall that `'P^=(base; I)` denotes the polyhedron defined by the inequalities in `base`, with the additional constraint that the inequalities in the subset `I` are satisfied with equality. This means that `{poly base}` corresponds to the polyhedra defined by equalities or inequalities in `base`. The choice of the name `base` is reminiscent of the terminology used in fiber bundles. Indeed, as we shall see in the next sections, several proofs will adopt the scheme of fixing a base locally, and then working on polyhedra of type {poly base}. Following this analogy, the latter may be thought of as a fiber.

We now present a first formalization of the set of faces relying on the subtype {poly base}:

```
Definition pb_face_set (P : {poly base}) :=
  [set Q : {poly base} | Q `<=` P].
```

It defines the set of faces of P : {poly base} as the set of elements of {poly base} contained in P. With this definition, some properties of faces come for free. For instance, the finiteness of the set of faces follows from the fact that there are only finitely many inhabitants of the type {fsubset base}, and subsequently of {poly base}. Another example is that Proposition 1 straightforwardly derives from the transitivity of the inclusion relation `<=`.

Some other properties come at the price of proving that a polyhedron inhabits the type {poly base}. As an example, if P : {poly base} and c : 'cV_n, the polyhedron argmin P c : 'poly_n is defined as the set of points of P minimizing the function fun x => '[c,x]. Showing that argmin P c is a face of P essentially amounts to proving the following property:

```
Lemma argmin_baseP (P : {poly base}) c : has_base base (argmin P c).
```

Indeed, the inclusion argmin P c `<=` P is immediate from the definition of the polyhedron argmin P c. However, even once Lemma **argmin_baseP** is proved, we cannot yet write a statement of the form argmin P c \in pb_face_set P due to the fact that argmin P c has type 'poly_n. In order to turn it into an element of the subtype {poly base}, we need to explain in more detail how this type is defined. The type {poly base} is a short-hand notation for the following inductive type:

```
Inductive poly_base base :=
  PolyBase { pval :> 'poly_n ; _ : has_base base pval }.
```

In other words, an element of type {poly base} is a record formed by an element pval : 'poly_n and a proof that the property has_base base pval holds. While we could construct the element PolyBase (argmin_baseP P c), we introduce a more general scheme to cast elements of type 'poly_n to {poly base} whenever possible. This scheme relies on CoQ canonical structures, which provide an automatic way to recover a term of record type from the head symbol. The association is declared as follows:

```
Canonical argmin_base (P : {poly base}) c := PolyBase (argmin_baseP P c).
```

One restriction of CoQ is that canonical structures are resolved only when unifying types, and not arbitrary terms. This is why our primitive poly_base_of, which casts a Q : 'poly_n to a {poly base}, encapsulates the value Q in a *phantom type*, i.e. a type isomorphic to the unit type, but with a dependency to Q.

```
Definition poly_base_of (Q : {poly base}) (_ : phantom 'poly_n Q) := Q.
Notation "Q %:poly_base" := (poly_base_of (Phantom _ Q)).
```

In consequence, writing (argmin P c)%:poly_base triggers the unification algorithm between the term argmin P c and a value of type {poly base}, which is resolved using the Canonical declared above. We finally end up with the following statement

```
Lemma argmin_pb_face_set base (P : {poly base}) c :
  (argmin P c)%:poly_base \in pb_face_set P.
```

whose proof is trivial: it just amounts to proving the inclusion argmin P c `<=` P.

We declare other canonical structures over elementary constructions for which the property has_base base _ can be shown to be satisfied. This includes the intersection P `&` Q of two elements P Q : {poly base}, the empty set [poly0 ], or polyhedra of the form 'P(base) or 'P^=(base; .). This allows us to cast complex terms to the type {poly base}, or, said differently, to prove automatically that they satisfy the property has_base base _. As an example, the term

```
('P^=(base; I) `&` argmin 'P(base) c)%:poly_base
```

typechecks thanks to multiple resolutions of canonical structures on the aforementioned declarations, without requiring extra proof from the user. We refer to [21] for the use of canonical structures in formal mathematics.

We point out that Lemma argmin_pb_face_set is a proof of one side of the equivalence between the definition of faces brought by pb_face_set and Definition 1 (i.e. the equivalence in Theorem 1). The other side can be written as follows:

```
Theorem pb_faceP base (P Q : {poly base}) :
  Q \in pb_face_set P -> Q != [poly0] ->
    exists c, Q = (argmin P c)%:poly_base.
```

When Q is nonempty, we use a set I such that Q = 'P^=(base, I), and we build c as the sum of the vectors -e.1 : 'cV_n for e \in I. The equality Q = argmin P c follows from a routine verification of the complementary slackness conditions.

## 3.3   Getting Free from Ambient Representations

So far, we have worked with a fixed ambient representation base, and restricted the formalization of faces to polyhedra that can be expressed as terms of type {poly base}. We now describe how to formalize the set of faces of any polyhedron of type 'poly_n as a finite set of polyhedra of the same type, without sacrificing the benefits brought by {poly base}.

First, we exploit the observation that for each polyhedron P : 'poly_n, there exists base : {fset lrel_n} and P' : {poly base} such that P = pval P' (recall that pval also stands for the coercion from the type {poly base} to 'poly_n). This can be proved by exploiting the definition of the quotient type 'poly_n.

Indeed, P admits a representative hrepr P : 'hpoly_n corresponding to a certain
H-representation, from which we can build a term base : {fset lrel_n} such
that P = pval 'P(base)%:poly_base.

Second, we introduce another quotient structure over the type 'poly_n, in
order to deal with the fact that a polyhedron may correspond to several elements of type {poly base} for different values of base. Our construction amounts
to showing that 'poly_n is isomorphic to the quotient of the dependent sum type
$\sum_{base}$ {poly base} by the equivalence relation in which Q1 : {poly base1} and
Q2 : {poly base2} are equivalent if pval Q1 = pval Q2. Given a polyhedron P of
type 'poly_n, this construction provides us a canonical ambient representation
denoted \repr_base P : {fset lrel_n}, and an associated canonical representative \repr P of type {poly (\repr_base P)} satisfying P = pval (\repr P).

We are now ready to define the set of faces of P in full generality:

```
Definition face_set (P : 'poly_n) :=
  [fset (pval F) | F in pb_face_set (\repr P)]%fset.
```

which corresponds to the image by the coercion pval of the face set of \repr P
(here, pval has type {poly (\repr_base P)} -> 'poly_n). Of course, we need to
check that this definition is independent of the choice of the representative of P
in this new quotient structure. This is written as follows:

```
Lemma face_set_morph (base : {fset lrel_n}) (P : {poly base}) :
  face_set P = [fset pval F | F in pb_face_set P]%fset.
```

The proof relies on the geometric properties of the elements of pb_face_set established in Sect. 3.2. Indeed, they imply that, regardless of the choice of the ambi-
ent representation, the set [fset pval F | F in pb_face_set P] always consists
of the empty set [poly0] and the polyhedra of the form argmin P c.

Now that this architecture is settled, we can prove some of the basic properties of faces. Most of the proof make use of the following elimination principle:

```
Lemma polybW (Pt : 'poly_n -> Prop) :
  (forall (base : {fset lrel_n}) (Q : {poly base}), Pt Q) ->
    (forall P : 'poly_n, Pt P).
```

which means that, given a property to be proved on any polyhedron P : 'poly_n,
it is sufficient to prove it over the type {poly base} for any choice of base. In
practice, Lemma polybW is used to introduce an ambient representation. Let us
illustrate it on the proof that the intersection of two faces of P is a face of P:

```
Lemma face_set_polyI (P Q1 Q2 : 'poly_n) :
  Q1 \in face_set P -> Q2 \in face_set P -> Q1 `&` Q2 \in face_set P.
Proof.
elim/polybW: P => base P.
case/face_setP => {}Q1 Q1_sub_P.
case/face_setP => {}Q2 Q2_sub_P.
by rewrite face_setE ?(poly_subset_trans poly_subsetIl) ?pvalP.
Qed.
```

The first line destructs P, and introduces the ambient representation base and an element still named P, now of type {poly base}. The second and third lines successively consume the assumptions that Q1 and Q2 are faces, then introduce two elements of type {poly base} having the same name and respectively satisfying Q1 `<=` P and Q2 `<=` P. Finally, the tactics rewrite face_setE replaces the goal Q1 `&` Q2 \in face_set P by the following two subgoals: Q1 `&` Q2 `<=` P and has_base base (Q1 `&` Q2). Since (Q1 `&` Q2) `<=` Q1 and Q1 `<=` P, the former is proved by transitivity of the subset relation. The latter is automatically provided by the canonical structure mechanism described in Sect. 3.2, triggered by the generic statement

```
Lemma pvalP base (P : {poly base}) : has_base base P.
```

## 3.4  From Faces to the Affine Hull and Dimension

We argue that the approach that we have introduced to represent faces of polyhedra also perfectly fits the formalization of the affine hull and dimension of polyhedra. Recall that the *affine hull* of a polyhedron refers to the smallest (inclusionwise) affine subspace of $\mathbb{R}^n$ containing it, and the *dimension* of the polyhedron is defined as the dimension of this subspace (i.e., the dimension of the underlying vector subspace).

To this end, given an ambient representation base and a polyhedron P of type {poly base}, we introduce the set of *active inequalities* of P, i.e. the set of e \in base such that the corresponding inequality is satisfied as equality over P. This is written as the inclusion P `<=` [hp e] (recall that [hp e] is the hyperplane '[e.1, x] = e.2). The active inequalities form a subset of base denoted {eq P}. Equivalently, when P is non-empty, {eq P} corresponds to the largest (inclusionwise) subset I such that P = 'P^=(base; I).

It is a classic property of polyhedra that the affine hull of a non-empty polyhedron is the affine subspace defined by the equalities in {eq P}. We take this property as a definition:

```
Definition pb_hull base (P : {poly base}) :=
  if P != [poly0] then affine << {eq P} >> else [poly0].
Definition hull (P : 'poly_n) := pb_hull (\repr P).
```

The second definition lifts the affine hull from {poly base} to 'poly_n. Of course, we show that the resulting affine subspace hull P is independent of the choice of base (cf. Lemma hullE). We establish that this definition is correct w.r.t. the usual mathematical definition discussed above, i.e.:

```
Lemma hullP P U : (P `<=` affine U) <-> (hull P `<=` affine U).
```

Here, U corresponds to a vector subspace of lrel_n, and the term affine U stands for the affine subspace given by the intersection of the affine equalities represented by the elements of U. (The term << {eq P} >> above corresponds to the vector subspace spanned by the elements of {eq P}.)

We follow the same scheme to formalize the dimension `dim P` of a polyhedron `P : 'poly_n`, which we define as one plus the co-dimension of the vector span of `{eq P}`. The shift by one originates from the fact that `dim P` ranges over the type `nat` of natural numbers. Therefore, we have to set the dimension of the empty set `[poly0]` to 0, while it is common to set it to −1 in the literature. As expected, we obtain the following statement:

```
Lemma dim_hull (P : 'poly_n) : dim P = dim (hull P).
```

Like in mathematics textbooks, Lemma `dim_hull` is the natural way to establish the basic statements concerning the dimension, i.e. by reducing to elementary proofs over vector spaces. For instance, we establish that the dimension is monotone (Lemma `dimS`), and compute the dimension for important classes of polyhedra. This includes the fact that segments of two distinct points have dimension 2 (remember the shift by one of our formalization):

```
Lemma dim_segm (x y : 'cV_n) : dim [segm x; y] = (x != y).+1.
```

and that, conversely, any compact polyhedron of dimension 2 is a segment of two points:

```
Lemma dim2P (P : 'poly_n) : compact P -> dim P = 2 ->
  exists x, exists2 y, P = [segm x; y] & x != y.
```

(We point out that `compact P` is simply defined as the fact that `P` is a bounded set, as polyhedra are topologically closed.) Similarly, we prove that polyhedra reduced to a single point are precisely the ones having dimension 1, that proper hyperplanes have codimension 1, etc. We refer to Section **Dimension** for a detailed account of our results.

# 4   The Face Lattice[5]

In this section, we illustrate how the framework that we have introduced in Sect. 3 serves as a foundation for formalizing the structural properties of faces. We refer to Fig. 2 for an example of the properties presented below.

At the core of the formalization lies the theory of ordered structures such as partial orders, semilattices and lattices. Some of these structures have been very recently introduced in the MathComp library – for instance, the non-distributive lattice structure has been introduced in early 2020. However, as we shall see in this section, the formalization of the face lattice requires to implement additional objects, such as graded lattices, sublattices, and lattice homomorphisms. This development is gathered in the module `xorder.v` of the Coq-Polyhedra project.

The first property that we can immediately formalize following the results of Sect. 3 is the finite lattice structure over the set `face_set P` for `P : 'poly_n`. The

---

[5] https://github.com/Coq-Polyhedra/Coq-Polyhedra/tree/IJCAR-2020/theories/poly_base.v.

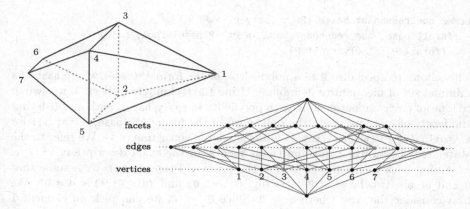

**Fig. 2.** A three-dimensional polytope (left) and the Hasse diagram of its face lattice (right). A interval of height 2 is depicted in blue. (Color figure online)

partial order is given by the polyhedron inclusion `<=`, the meet operator is the intersection `&` (as a consequence of Lemma **face_set_polyI**), while the bottom and top elements are respectively `[poly0]` and `P`. As a finite lattice, the join operator Q `|` Q' can be built as the meet of all the faces of P containing both Q and Q'.

## 4.1 Facets and Gradedness

Recall that a lattice $(L, \prec)$ is said to be *graded* if there exists a rank function $\Phi: L \to \mathbb{N}$ such that: (i) $\Phi(u) < \Phi(v)$ whenever $u \prec v$, (ii) $u \preceq v$ and $\Phi(u) + 1 < \Phi(v)$ implies that there exists $w \in L$ satisfying $u \prec w \prec v$. Equivalently, this is a lattice in which all maximal chains have the same length.

In the case of the face lattice, the rank function can be defined as the dimension of the face. Property (i) is proved as follows. If Q and Q' are both faces of P and Q `<` Q', then dim Q <= dim Q', as the dimension is monotone. Moreover, hull Q `<=` hull Q'. If we assume dim Q = dim Q', then we can prove that hull Q is equal to hull Q' (as affine subspaces of the same dimension). We conclude that Q = Q' by the fact that F = P `&` hull F for any face F of P.

The proof of Property (ii) (see Lemma **graded**) relies on the formalization of facets of polyhedra, and their combinatorial characterization in terms of active inequalities. We recall that a *facet* of a non-empty polyhedron $\mathcal{P}$ is a face of $\mathcal{P}$ of dimension dim $\mathcal{P} - 1$. A classical result states that when $\mathcal{P}$ is given by a *non-redundant* system of inequalities $Ax \geq b$ (i.e. the H-representation is minimal inclusionwise), the facets are precisely the polyhedra of the form $\mathcal{P} \cap \{x \in \mathbb{R}^n : A_i x = b_i\}$ for any $i$ such that $\mathcal{P} \not\subset \{x \in \mathbb{R}^n : A_i x = b_i\}$. The formalization of this statement first goes through the construction of non-redundant bases for any polyhedron, and the proof of the following elimination principle:

```
Lemma non_redundant_baseW (Pt : 'poly_n -> Prop) :
  (forall base, non_redundant base -> Pt 'P(base)%:poly_base) ->
    (forall P : 'poly_n, Pt P).
```

This allows to specialize P to a polyhedron of the form 'P(base) where base is a minimal set of inequalities defining P. Using the techniques of Sect. 3, we switch to a proof environment dealing with polyhedra in {poly base}, and establish that the facets of P are precisely the polyhedra of the form 'P^=(base; [fset i]) for i \notin {eq P} (where [fset i] is the singleton consisting of i). We refer to the statements Lemma dim_facet and Lemma facetP for the exact description.

Going back to the description of the proof of Property (ii), we assume that Q and Q' are two faces of P satisfying Q `<=` Q' and (dim Q).+1 < dim Q'. We first consider the case where Q' = P. Since Q `<` P, we can pick an element i in {eq Q} but not in {eq P}, and verify that the facet F := 'P^=(base; [fset i]) satisfies Q `<` F `<` P. The general case where Q' is a proper face of P is handled by using the fact that Q \in face_set P and Q `<=` Q' ensures that Q is a face of Q' (see Lemma face_set_of_face).

## 4.2  Vertices, Atomicity and Coatomicity

The *atoms* of a lattice $L$ are the elements $u \in L \setminus \{\bot\}$ such that there is no $v \in L$ satisfying $\bot \prec v \prec u$, where $\bot$ denotes the bottom element of $L$. In the face lattice of a polyhedron P, they correspond to the faces F of P such that dim F = 1, i.e. to the vertices of P (remember the shift by one of our formalization). This motivates the introduction of the vertex set of P, which satisfies the following two characteristic properties:

```
Lemma in_vertex_setP (P : 'poly_n) x :
  (x \in vertex_set P) <-> ([pt x] \in face_set P).
Lemma face_dim1 (P Q : 'poly_n) : Q \in face_set P -> dim Q = 1 ->
  exists2 x, Q = [pt x] & x \in vertex_set P.
```

A central property is that if P is compact, then it coincides with the convex hull of its vertices:

```
Theorem conv_vertex_set (P : 'poly_n) :
  compact P -> P = conv (vertex_set P).
```

Remark that this shows that any compact polyhedron is a polytope. Together with the converse statement (Lemma compact_conv in polyheron.v), this brings a proof of Minkowski Theorem.

The latter result allows us to prove that, when P is compact, its face lattice is atomistic, meaning that any face of P is the join of a finite set of atoms:

```
Lemma atomisticP (Q : face_set P) :
  reflect (exists2 S, (forall x, x \in S -> atom x) & Q = \join_(x in S) x)
          (atomistic Q).
Lemma face_atomistic (Q : face_set P) : atomistic Q.
```

To prove this statement for Q, we set S to the set of vertices of Q. The latter are vertices of P as well, and thus correspond to atoms of the face lattice of P. The inclusion Q `<=` \join_(x in vertex_set Q) x is established by substituting Q by conv (vertex_set Q) thanks to Lemma **conv_vertex_set**, which makes the statement obvious by construction of the convex hull and the join operator. The opposite inclusion Q `>=` \join_(x in vertex_set Q) x is trivial by property of the join operator, and this concludes the proof.

The *coatoms* of $L$ are defined dually: these are the elements $u \in L \setminus \{\top\}$ such that there is no $v \in L$ satisfying $u \prec v \prec \top$, where $\top$ denotes the top element of $L$. The coatomicity of face_set P means that any face of P is the intersection of facets of P. Our proof exploits the characterization of facets presented in Sect. 4.1. We refer to Lemma **face_coatomistic** for more details.

## 4.3  Closedness Under Taking Sublattices

The *closedness under sublattices* of the face lattices of polytopes states that if Q and Q' are two faces of a polytope P such that Q `<=` Q', then the interval '[< Q; Q' >], i.e. the sublattice formed by the faces F : face_set P satisfying Q `<=` F `<=` Q', is isomorphic to the face lattice of a polytope of dimension dim Q' - dim Q.

The interest of this property is that it allows involved induction schemes on the height of the face lattice. As an example, we can establish the so-called *diamond property*, namely that every sublattice of height 2 of the face lattice consists of precisely four faces ordered as ◇. Equivalently, this means that for any two faces $\mathcal{F}$ and $\mathcal{F}'$ of a polytope $\mathcal{P}$ such that $\dim \mathcal{F}' = \dim \mathcal{F}+2$ and $\mathcal{F} \subset \mathcal{F}'$, there are precisely two faces between them (see Lemma **diamond** for the statement, and Fig. 2 for an illustration). The proof exploits the closedness by sublattices, and the subsequent isomorphism of any interval of height 2 with the face lattice of a polytope P' verifying dim P' = 2. Lemma **dim2P** reduces it to the face lattice of a segment [segm x; y], which is given by the following characterization:

Lemma **face_set_segm** (x y : 'cV_n) :
  face_set [segm x; y] = [fset [poly0]; [pt x]; [pt y]; [segm x; y]].

The proof of the closedness by sublattices is done as follows. First, we reduce to the case where Q' = P, since the face lattice of Q' is isomorphic to the sublattice of the faces of P contained in Q'. We are left with the following statement:

Lemma **closed_by_interval_r** (Q : face_set P) :
exists (P' : 'compact_poly_n) (f : {omorphism '[< Q; P >] -> face_set P'}),
  bijective f.

The proof is done by induction on the dimension of Q. We restrict the exposition to the base case dim Q = 1, i.e. when Q corresponds to a vertex x of P, since the general case is just handled by iterating the process. When dim Q = 1, the construction of the polyhedron P' is achieved by the *vertex figure* method. It consists in slicing the polytope P with a hyperplane [hp e] separating the vertex

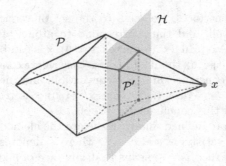

**Fig. 3.** The vertex figure construction, illustrated on the vertex $x$ of the polytope $\mathcal{P}$. The hyperplane $\mathcal{H}$ (light blue) separates $x$ from the other vertices of the polytope. In the sliced polytope $\mathcal{P}'$, the vertices (green) and edges (blue) are respectively in one-to-one correspondence with the edges and facets of $\mathcal{P}$ containing $x$. (Color figure online)

$x$ from the other vertices (see Fig. 3 for an illustration). We define `P'` as the sliced polytope. It has dimension (`dim P`)`-1`, and its face lattice can be shown to be isomorphic to the sublattice `'[< [pt x]; P >]`. Once again, the isomorphism is proved by exposing the polyhedron `P` to the subtype `{poly base}` for some ambient representation `base`, and reducing to basic manipulations of sets `{eq _}` of active inequalities of faces. Interestingly, two distinct ambient representations are used in the proof: `base` for the original polytope `P`, and its union `e +|` base` with the singleton `{e}` for the sliced polytope `P'`. Our use of canonical structures still applies to this setting, and provides the proof that any face of `P` sliced with the hyperplane `[hp e]` writes down over the base `e +|` base` of the sliced polytope `P'`.

## 5   Related Work

Many software developments related with convex polyhedra have been motivated by applications to formal verification. Several libraries have been developed for this purpose, e.g. [4,20], and, despite being informal, it is worth noting that they are also used by mathematicians to perform computation over polyhedra and polytopes, for instance in [16,27]. Initiatives on the development of formally verified polyhedral algorithms are more recent. The works of [26] and [8] in Isabelle/HOL aim at providing a formally proven yet practical and efficient algorithm to decide linear rational arithmetic for SMT-solving. The Micromega tactics [5] relies on polyhedra to prove automatically arithmetic goals over ordered rings in CoQ. The *Verified Polyhedral Library* [9,15] targets abstract interpretation, and brings the ability to verify polyhedral computations a posteriori in CoQ.

There are far fewer developments focusing on formal mathematics. Euler formula, which relates the number of vertices, edges and facets of three-dimensional polytopes, has been proved in [14] in COQ and in [1] in Mizar. The generalization to polytopes in arbitrary dimension, namely Euler–Poincaré formula, has been formally proved in HOL-Light [19], together with several intermediate properties of polyhedra and faces. In the intuitionistic setting, we are not aware of any work concerning faces and their properties. We point out that Fourier–Motzkin elimination has been formalized in COQ by [22].

## 6    Conclusion and Future Work

In this work, we have formalized a substantial part of the theory of polyhedra and their faces, which has allowed us to obtain some of the essential properties of face lattices. Beyond the mathematical results formally proven, a special attention has been paid to the usability of the library. This goes through a mechanism to bring the right representation of faces according to the context, and the automatic proof that these representations are valid thanks to the use of canonical structures.

This foundational work opens several perspectives. First, it has raised that an important development over ordered structures is still needed, in particular for the manipulation of ordered substructures such as sublattices, and the interplay between them through morphisms. The formalization of finite groups and subgroups in [17] may provide a possible source of inspiration to solve this problem. Second, there are many other interesting properties in relation with polyhedra and their faces to be formalized, such as getting upper bounds on the diameter of polytopes, or more generally, on the number of faces (the so-called f-vector theory). However, beyond the interest of formalizing already known mathematical results, we are even more interested in using proof assistants to help getting new ones. We think of mathematical results relying on computations that are not accessible by hand. To this extent, we aim at providing a way to compute with the objects introduced in this work, directly within the proof assistant, and to introduce all the needed mechanisms for the design and development of large scale reflection tactics. A basic goal is to compute the face lattice (or part of it) of a polyhedron defined by a set of inequalities. This requires us to formalize some algorithms based on faces, and to find a way to execute them on efficient data structures, in the spirit of the approach of [11].

**Acknowledgments.** We are grateful to Assia Mahboubi for helpful discussions on the subject. We thank the anonymous reviewers for their detailed comments and their suggestions to improve the presentation of the paper.

## References

1. Alama, J.: Euler's polyhedron formula in mizar. In: Fukuda, K., Hoeven, J., Joswig, M., Takayama, N. (eds.) ICMS 2010. LNCS, vol. 6327, pp. 144–147. Springer, Heidelberg (2010). https://doi.org/10.1007/978-3-642-15582-6_26

2. Allamigeon, X., Benchimol, P., Gaubert, S., Joswig, M.: Log-barrier interior point methods are not strongly polynomial. SIAM J. Appl. Algebra Geom. **2**(1), 140–178 (2018). https://doi.org/10.1137/17M1142132

3. Allamigeon, X., Katz, R.D.: A formalization of convex polyhedra based on the simplex method. J. Autom. Reason. **63**(2), 323–345 (2019). https://doi.org/10.1007/s10817-018-9477-1

4. Bagnara, R., Hill, P.M., Zaffanella, E.: The Parma Polyhedra Library: toward a complete set of numerical abstractions for the analysis and verification of hardware and software systems. Sci. Comput. Program. **72**(1–2), 3–21 (2008)

5. Besson, F.: Fast reflexive arithmetic tactics the linear case and beyond. In: Altenkirch, T., McBride, C. (eds.) TYPES 2006. LNCS, vol. 4502, pp. 48–62. Springer, Heidelberg (2007). https://doi.org/10.1007/978-3-540-74464-1_4

6. Bonifas, N., Di Summa, M., Eisenbrand, F., Hähnle, N., Niemeier, M.: On subdeterminants and the diameter of polyhedra. Discret. Comput. Geom. **52**(1), 102–115 (2014)

7. Borgwardt, S., De Loera, J.A., Finhold, E.: The diameters of network-flow polytopes satisfy the Hirsch conjecture. Math. Program. **171**(1), 283–309 (2018)

8. Bottesch, R., Haslbeck, M.W., Thiemann, R.: Verifying an incremental theory solver for linear arithmetic in Isabelle/HOL. In: Herzig, A., Popescu, A. (eds.) FroCoS 2019. LNCS (LNAI), vol. 11715, pp. 223–239. Springer, Cham (2019). https://doi.org/10.1007/978-3-030-29007-8_13

9. Boulmé, S., Maréchal, A., Monniaux, D., Périn, M., Yu, H.: The verified polyhedron library: an overview. In: 2018 20th International Symposium on Symbolic and Numeric Algorithms for Scientific Computing (SYNASC), September 2018, pp. 9–17 (2018). https://doi.org/10.1109/SYNASC.2018.00014

10. Cohen, C.: Pragmatic quotient types in COQ. In: Blazy, S., Paulin-Mohring, C., Pichardie, D. (eds.) ITP 2013. LNCS, vol. 7998, pp. 213–228. Springer, Heidelberg (2013). https://doi.org/10.1007/978-3-642-39634-2_17

11. Cohen, C., Dénès, M., Mörtberg, A.: Refinements for free!. In: Gonthier, G., Norrish, M. (eds.) CPP 2013. LNCS, vol. 8307, pp. 147–162. Springer, Cham (2013). https://doi.org/10.1007/978-3-319-03545-1_10

12. De Loera, J.: Algebraic and topological tools in linear optimization. Not. Am. Math. Soc. **66**, 1 (2019). https://doi.org/10.1090/noti1907

13. Deza, A., Terlaky, T., Zinchenko, Y.: Central path curvature and iteration-complexity for redundant Klee-Minty cubes. In: Gao, D., Sherali, H. (eds.) Advances in Applied Mathematics and Global Optimization. Advances in Mechanics and Mathematics, vol. 17, pp. 223–256. Springer, Boston (2009). https://doi.org/10.1007/978-0-387-75714-8_7

14. Dufourd, J.F.: Polyhedra genus theorem and Euler formula: a hypermap-formalized intuitionistic proof. Theor. Comput. Sci. **403**(2–3), 133–159 (2008)

15. Fouilhe, A., Boulmé, S.: A certifying frontend for (sub)polyhedral abstract domains. In: Giannakopoulou, D., Kroening, D. (eds.) VSTTE 2014. LNCS, vol. 8471, pp. 200–215. Springer, Cham (2014). https://doi.org/10.1007/978-3-319-12154-3_13

16. Gawrilow, E., Joswig, M.: polymake: a framework for analyzing convex polytopes. In: Kalai, G., Ziegler, G.M. (eds.) Polytopes—Combinatorics and Computation. DMV Seminar, vol. 29, pp. 43–73. Birkhäuser, Basel (2000). https://doi.org/10.1007/978-3-0348-8438-9_2

17. Gonthier, G., et al.: A machine-checked proof of the odd order theorem. In: Blazy, S., Paulin-Mohring, C., Pichardie, D. (eds.) ITP 2013. LNCS, vol. 7998, pp. 163–179. Springer, Heidelberg (2013). https://doi.org/10.1007/978-3-642-39634-2_14

18. Gonthier, G., Mahboubi, A., Tassi, E.: A small scale reflection extension for the Coq system. Research report RR-6455, Inria Saclay Ile de France (2016)
19. Harrison, J.: The HOL Light theory of Euclidean space. J. Autom. Reason. **50**, 173–190 (2013). https://doi.org/10.1007/s10817-012-9250-9
20. Jeannet, B., Miné, A.: APRON: a library of numerical abstract domains for static analysis. In: Bouajjani, A., Maler, O. (eds.) CAV 2009. LNCS, vol. 5643, pp. 661–667. Springer, Heidelberg (2009). https://doi.org/10.1007/978-3-642-02658-4_52
21. Mahboubi, A., Tassi, E.: Canonical structures for the working Coq user. In: Blazy, S., Paulin-Mohring, C., Pichardie, D. (eds.) ITP 2013. LNCS, vol. 7998, pp. 19–34. Springer, Heidelberg (2013). https://doi.org/10.1007/978-3-642-39634-2_5
22. Sakaguchi, K.: Vass (2016). https://github.com/pi8027/vass
23. Santos, F.: A counterexample to the Hirsch conjecture. Ann. Math. **176**(1), 383–412 (2012)
24. Schrijver, A.: Theory of Linear and Integer Programming. Wiley, New York (1986)
25. Smale, S.: Mathematical problems for the next century. Math. Intell. **20**(2), 7–15 (1998). https://doi.org/10.1007/BF03025291
26. Spasić, M., Marić, F.: Formalization of incremental simplex algorithm by stepwise refinement. In: Giannakopoulou, D., Méry, D. (eds.) FM 2012. LNCS, vol. 7436, pp. 434–449. Springer, Heidelberg (2012). https://doi.org/10.1007/978-3-642-32759-9_35
27. Stein, W., et al.: Sage Mathematics Software (Version 9.0). The Sage Development Team (2020). http://www.sagemath.org
28. Ziegler, G.M.: Lectures on Polytopes. Springer, New York (1995). https://doi.org/10.1007/978-1-4613-8431-1

# Algebraically Closed Fields
# in Isabelle/HOL

Paulo Emílio de Vilhena[1]([✉]) and Lawrence C. Paulson[2][iD]

[1] Inria, Paris, France
`paulo-emilio.de-vilhena@inria.fr`
[2] Computer Laboratory, University of Cambridge, Cambridge, UK
`lp15@cam.ac.uk`

**Abstract.** A fundamental theorem states that every field admits an algebraically closed extension. Despite its central importance, this theorem has never before been formalised in a proof assistant. We fill this gap by documenting its formalisation in Isabelle/HOL, describing the difficulties that impeded this development and their solutions.

## 1 Introduction

The *fundamental theorem of algebra* states that the field of complex numbers is algebraically closed: every nonconstant polynomial with complex coefficients has at least one complex root. By extending the field of real numbers with a single root of the polynomial $X^2 + 1$, we obtain a field (the complex numbers) where not only has $X^2 + 1$ a root but also every other polynomial.

At the beginning of the 20th century, this theorem about the reals raised the question of which other fields could similarly be extended to be algebraically closed. Curiously, the same mathematician to introduce the concept of a field, Ernst Steinitz, was the one to answer the question: *every* field admits an algebraically closed extension.

This result has important consequences. It guarantees that every polynomial has a splitting field; it links algebra and geometry. Kevin Buzzard comments:

> The theorems of local and global class field theory are one of the highlights of early 20th century mathematics. ... The Langlands Philosophy, one of the central questions in modern number theory, is a vast conjectural generalisation of these theorems, and one cannot even state the fundamental conjectures in this theory without mentioning algebraic closures. Wiles and Taylor proved an extremely small fragment of these conjectures in 1994 and deduced Fermat's Last Theorem [3].

Despite its importance, the existence of algebraic closures has never been formalised in a proof assistant. The gap suggests that the formal proof is challenging and that there might be ill-understood technical difficulties. Here we propose to settle this:

P. E. de Vilhena—The author was affiliated to École Polytechnique during the realisation of this work.

© Springer Nature Switzerland AG 2020
N. Peltier and V. Sofronie-Stokkermans (Eds.): IJCAR 2020, LNAI 12167, pp. 204–220, 2020.
https://doi.org/10.1007/978-3-030-51054-1_12

- We formally prove that every field has an algebraic closure: an equivalent way to state that every field admits an algebraically closed extension (Sect. 5).
- We discuss, from a mathematical perspective, how our proof relates to existing ones and how we planned the formalisation effort (Sect. 4).
- We describe a limitation of Isabelle's type system along with a general solution, where *identity* is replaced by *isomorphism* (Sect. 3).

Attempting such proof in Isabelle/HOL is aligned with our intent to investigate the hurdles of formalising algebraic reasoning in a simply-typed setting.

## 2  Background

Here we recall some key elements of algebra. We suppose familiarity with the definitions of rings, fields, homomorphisms, ideals and the quotient of a ring by an ideal. We start with the definition of *canonical surjections*, which show that every element of a quotient ring is accessible in a structure-preserving way.

**Definition 1 (Canonical surjection).** *Let $R$ be a ring and $I$ an ideal of $R$. The canonical surjection $\pi_{[R,I]} : R \to R/I$ is a surjective homomorphism from the ring $R$ to the quotient $R/I$ that associates an element $r$ of $R$ to its equivalence class in $R/I$, that is, $r \mapsto \{r + i \mid i \in I\}$.*

The most common application of the canonical surjection in this paper is when considering the quotient between the ring of polynomials with coefficients in some field $K$, denoted by $K[X]$, and the *ideal generated by* some polynomial $P \in K[X]$, defined as

$$(P) \triangleq \{ PQ \mid Q \in K[X] \}.$$

This gives rise to the canonical surjection $\pi_{[K[X],(P)]}$, which can sometimes be seen as a homomorphism from $K$ to $K[X]/(P)$ and not from $K[X]$ to $K[X]/(P)$. Justification comes from the usual abuse of notation of identifying elements of $K$ with constant polynomials in $K[X]$.

The next proposition elucidates why it is interesting to see $\pi_{[K[X],(P)]}$ as such: if $P$ only has trivial factors, that is, if $P$ is an *irreducible polynomial*, then the restriction of $\pi_{[K[X],(P)]}$ to $K$ is in fact a homomorphism of *fields*.

**Definition 2 (Irreducible polynomial).** *Let $K$ be a field and $P$ a nonconstant polynomial with coefficients in $K$. The polynomial $P$ is irreducible if for every $Q \in K[X]$, if $Q$ divides $P$ then $Q = k$ or $Q = kP$ for some $k \in K$.*

**Proposition 1.** *If $K$ is a field and $P$ is a polynomial with coefficients in $K$, then $P$ is irreducible iff the quotient of $K[X]$ by the ideal $(P)$ is a field.*

Given a homomorphism $\phi$ between two rings $A$ and $B$, and a polynomial $Q \in A[X]$, we can build the polynomial $Q^\phi \in B[X]$ by applying $\phi$ to each of the coefficients of $Q$:

$$Q^\phi = \left( \sum_i a_i X^i \right)^\phi \triangleq \sum_i \phi(a_i) X^i \text{ where } a_i \in A$$

It follows from this definition that $\phi$ maps the evaluation of $Q$ at an element $a \in A$ to the evaluation of $Q^\phi$ at $\phi(a)$:

$$\phi(Q(a)) = Q^\phi(\phi(a)). \tag{1}$$

Now, consider a field $K$ and a polynomial $P \in K[X]$. From our previous discussion, the coefficients of $P$ can be seen as constant polynomials in $K[X]$ so that $P$ can be seen as lying in $(K[X])[X]$. Therefore, the evaluation of $P$ at elements in $K[X]$ is meaningful. For instance, what happens if we evaluate $P$ at the monomial $X \in K[X]$? We recover $P$ itself: $P = P(X)$. Together with equation (1), we obtain the following result:

$$\pi(P) = \pi(P(X)) = P^\pi(\pi(X))$$

where $\pi$ stands for the canonical surjection $\pi_{[K[X],(P)]}$. Since the equivalence class of $P$ in $K[X]/(P)$ is the same as the zero polynomial, we have proved that $\pi(X)$ is a root of $P^\pi$. The next proposition rephrases this result using the terminology of a *field extension* when $P$ is an irreducible polynomial.

**Definition 3 (Field extension).** *Let $K$ and $L$ be fields and $\phi : K \to L$ a homomorphism. Since $K$ is a field, $\phi$ is either the trivial map $k \mapsto 0$ or an injective map. When it is injective, $L$ is called a* field extension *of $K$ under $\phi$.*

**Proposition 2.** *Let $K$ be a field and $P$ an irreducible polynomial in $K[X]$. The quotient of $K[X]$ by the ideal $(P)$ is a field extension of $K$ under the homomorphism $\pi_{[K[X],(P)]}$. Moreover, the polynomial $P^\pi$ admits $\pi(X)$ as a root in the field $K[X]/(P)$.*

Now let's see how to build a field isomorphic to the field of complex numbers. Consider the real polynomial $X^2 + 1$. It's irreducible, therefore by Proposition 2 the ring $\mathbb{R}[X]/(X^2 + 1)$ is actually a field extension of $\mathbb{R}$ under the homomorphism $\pi$. Furthermore, $\pi(X)$ is a root of the polynomial $(X^2+1)^\pi$. Thus, we have built a field where a square root of $-1$ exists. This fact can be formally stated by considering the map $\pi(a + bX) \mapsto a + bi$, which establishes an isomorphism from $\mathbb{R}[X]/(X^2 + 1)$ to the field $\mathbb{C}$ of complex numbers.

Related to the notion of a field extension is the notion of a subfield:

**Definition 4 (Subfield).** *Let $L$ be a field. A subset $K \subseteq L$ is called a* subfield *of $L$ if it verifies the axioms of a field when equipped with the same laws as $L$.*

This relation is actually bidirectional:

1. If $L$ is a field extension of $K$ under the homomorphism $\phi$, then the image $\phi(K)$ is a subfield of $L$.
2. If $K$ is a subfield of $L$, then $L$ is a field extension of $K$ under the identity homomorphism $k \mapsto k$.

It is important to notice the nuances of the definition of an irreducible polynomial in the context of subfields. If $K$ is a subfield of $L$, then a polynomial $P$ in $K[X]$ has its coefficients in both $K$ and $L$. When we say that $P$ is an irreducible polynomial we have to specify in which field: while $P$ may have no nontrivial factor with coefficients in $K$, it may have one with coefficients in $L$. To make the distinction unambiguous, we say that $P$ is irreducible *in* $K$ expressing that it only has trivial factors in the ring $K[X]$.

The next natural notion is that of an algebraic element:

**Definition 5 (Algebraic).** *Let $L$ be a field and $K$ a subfield of $L$. Then an element $l \in L$ is* algebraic over $K$ *if there exists a polynomial $P$ with coefficients in $K$ such that $P(l) = 0$, that is, $l$ is a root of $P$.*

The subset of elements of $L$ which are algebraic over $K$ is a subfield of $L$. Moreover, since the polynomial $X - k$ belongs to the ring $K[X]$ if $k \in K$, every element of the field $K$ is algebraic over $K$. In addition, the field $K$ is a subfield of the subset of algebraics of $L$, which gives the following inclusions:

$$K \subseteq \{l \in L \mid l \text{ is algebraic over } K\} \subseteq L$$

Finally, we introduce the formal definition of an algebraically closed field and the closely related notion of the algebraic closure.

**Definition 6 (Algebraically closed).** *A field $K$ is* algebraically closed *if every nonconstant polynomial with coefficients in $K$ has at least one root in $K$.*

**Definition 7 (Algebraic closure).** *Let $K$ and $L$ be two fields. Then $L$ is an* algebraic closure *of $K$ if there exists a homomorphism $\phi : K \to L$ such that*

1. *Every polynomial $P$ of degree $n$ with coefficients in the subfield $\phi(K)$ has $n$ roots in $L$: that is, $P$ splits in $L$.*
2. *Every element of $L$ is algebraic over the subfield $\phi(K)$.*

It is usual to refer to $L$ as *the* algebraic closure: they are unique up to isomorphism. An additional remark is that if $P$ splits in $L$, then $P$ only has trivial irreducible factors. That is, for every irreducible polynomial $Q \in L[X]$, if $Q$ divides $P$, then $Q$ must have degree 1. The converse also holds.

Our last claim in this section formally connects these notions:

**Proposition 3.** *Let $K$ be a field. The following three statements are equivalent:*

1. *There exists an algebraically closed field extension of $K$.*
2. *There exists an algebraic closure of $K$.*
3. *There exists a field extension $L$ of $K$ under a homomorphism $\phi$ such that every polynomial with coefficients in the subfield $\phi(K)$ splits in $L$.*

The way we establish that every field admits an algebraically closed extension in our formal development is by proving that (3) implies (2), that (2) implies (1) and that assertion (3) holds. The first two proofs are straightforward and our focus will be in describing how we prove (3).

# 3  Formalisation

The formalisation of algebra in the absence of dependent types can sometimes be challenging. Below we describe how we addressed these situations. We begin by presenting a technique for changing the underlying type of an algebraic structure. The application of this procedure is named the *induction of a structure*. In addition, we discuss our formalisation of multivariate polynomials. Both of these two features play an important role in the formal proof of the existence of the algebraic closure, and they surely have other applications.

## 3.1  Induced Structures

The type of monoids is formalised in the HOL-Algebra library as a record with three fields: the carrier, which has the polymorphic type $'a\,set$, the composition operator and finally the unit.

$$\texttt{record}\,'a\,monoid\ =\ carrier :: 'a\,set \quad mult :: 'a \Rightarrow 'a \Rightarrow 'a \quad one :: 'a$$

The binary or *direct product* of two monoids is defined as follows:

$$\texttt{definition}\ DirProd\,(\texttt{infix}\ \times\times)\ \texttt{where}$$
$$G \times\times H\ = (\!|\ carrier = carrier\ G \times carrier\ H;\ one = (one\ G, one\ H);$$
$$mult = \lambda(g_1, h_1)\,(g_2, h_2).\,(mult\ G\ g_1\ g_2, mult\ H\ h_1\ h_2)\ |\!)$$

This takes two monoids as arguments, $G$ and $H$. It returns a monoid whose carrier is the Cartesian product of the carriers of $G$ and $H$, with composition defined element-wise. Isabelle assigns *DirProd* the type

$$'a\,monoid \Rightarrow 'b\,monoid \Rightarrow ('a * 'b)\,monoid.$$

Now, let's say we want to define the direct product of a list of monoids $Gs$, that is, the $n$-ary product of monoids where $n$ is the length of $Gs$. Consider the following attempt to define this by recursion on the list:

$$\texttt{fun}\ DirProd\_list\ \texttt{where}$$
$$DirProd\_list\,(G\,\#\,Gs) = G \times\times DirProd\_list\ Gs$$
$$|\ DirProd\_list\,[] = (\!|\ carrier = \{[]\};\ mult = \lambda as\ bs.\,[];\ one = []\ |\!)$$

Such an attempt to define an $n$-ary product of monoids from a list of monoids must fail, as the result type would depend on the length of the list.

To solve this problem, we introduce the concept of *induced structures*. The idea is that, given the pair of an algebraic structure such as a monoid $G$ of type $'a\,monoid$ and an injective function $f$ of type $'a \Rightarrow 'b$, we can *induce* a monoid $H$ of type $'b\,monoid$ such that $f$ is an isomorphism between the monoids $G$ and $H$. Thus, $H$ has the same algebraic properties as $G$ but with the type we want.

**Definition 8 (Induced monoid).** *Let $G$ be a monoid, $H$ a set and $f : G \to H$ an injection from $G$ to $H$. Now the image $f(G)$ is a subset of $H$, which can be equipped with a monoid structure as defined below:*

$$\mathbf{1}_{f(G)} \triangleq f(\mathbf{1}_G), \qquad f(g_1) \otimes_{f(G)} f(g_2) \triangleq f(g_1 \otimes_G g_2).$$

*We obtain a monoid $f(G)$, called the monoid induced by $G$ and $f$. Furthermore, $f$ is an isomorphism between the monoids $G$ and $f(G)$.*

Here is the formal Isabelle definition (where ' denotes the image operator):

> **definition** *image_monoid* **where**
> $\quad$ *image_monoid $f$ $G$* = $(\!|$ *carrier* = $f$ ' *carrier $G$* ; *one* = $f$ (*one $G$*) ;
> $\quad\quad$ *mult* = $\lambda h_1 h_2.$ $f$ (*mult $G$* (*inv_into* (*carrier $G$*) $f$ $h_1$)
> $\quad\quad\quad\quad\quad\quad\quad\quad\quad\quad$ (*inv_into* (*carrier $G$*) $f$ $h_2$)) $|\!)$

Note that composition is defined as $f(f^{-1}h_1 \otimes_G f^{-1}h_2)$, using the inverse of the function $f$. We use the *inv_into* function (from Isabelle's standard library), which denotes the inverse of $f \upharpoonright A$ (the function $f$ restricted to domain $A$).

Let's return to the definition of the function *DirProd_list*. With this new tool, we are able to handle the inductive case:

$$DirProd\_list(G \# Gs) = image\_monoid\ (\lambda(a, as).\ a \# as)\ (G \times\times DirProd\_listGs)$$

Now, the definition is accepted by Isabelle, yielding the isomorphism

$$DirProd\_list(G \# Gs) \cong G \times\times DirProd\_list\,Gs.$$

Algebraically speaking, an isomorphism is enough: we do not need true equality.

More generally, given a monoid $G$ of type $'a$ *monoid* and an injective function $f$ of type $'a \Rightarrow {'b}$, we prove that the function $f$ is an isomorphism between the monoids $G$ and *image_monoid $f$ $G$*.

$$f \in iso\ G\ (image\_monoid\ f\ G)$$

Other algebraic structures such as groups, rings and fields can also benefit from this construction. The difference between groups and monoids in HOL-Algebra is only logical: they satisfy different axioms but both have the same type, so *image_monoid* can be applied to groups. The same idea works for other abstract mathematical structures, wherever isomorphism (as opposed to equality) is good enough.

The ability to choose the type of an algebraic structure while preserving its abstract properties can also be useful in the formalisation of certain proofs of existence. We illustrate this use case through the following theorem.

**Theorem 1.** *Let $K$ be a field and $P$ a polynomial with coefficients in $K$. Then there exists a field extension $L$ under a homomorphism $\phi : K \to L$ such that the polynomial $P^\phi$ splits in $L$.*

*Proof.* By induction on the degree of $P$. If $\deg P = 0$, then $P$ already splits in $K$. Thus, $K$ itself (with the identity homomorphism) is the required field extension.

If $\deg P = n+1$ for some $n$ then there exists an irreducible polynomial $Q$ with coefficients in $K$ such that $Q$ divides $P$. Since $Q$ is irreducible, by Proposition 2 we obtain the field extension $K[X]/(Q)$ where $Q^\pi$ has $\pi(X)$ as a root. Since $R \mapsto R^\pi$ is a homomorphism, $Q^\pi$ divides $P^\pi$. Consequently, $\pi(X)$ is also a root of $P^\pi$ and hence $X - \pi(X)$ divides $P^\pi$.

Let $R$ denote the division of $P^\pi$ by the polynomial $X - \pi(X)$, that is, $R$ is a polynomial with coefficients in $K[X]/(Q)$ such that $P^\pi = (X - \pi(X)) R$. Then $\deg(R) = n$ and by induction hypothesis we obtain a field $L$ and a homomorphism $\phi : K[X]/(Q) \to L$ such that $R^\phi$ splits in $L$. Clearly $L$ under the homomorphism $\phi \circ \pi$ is the required field extension.    □

There are a number of obstacles to the formalisation of this proof in Isabelle, starting with the statement itself. The theorem asserts the existence of a field, which in Isabelle has type $'a\ ring$, where $'a$ is the type of the elements of its carrier. We need to find a type for the field whose existence is being claimed.

In a dependent type setting, the problem could be avoided. It would suffice to quantify the type existentially: to announce the existence of both the type and the field. Then, we would be able to build the precise type during the proof.

This is not possible in Isabelle, but let's say that we have a type $'b$ that could satisfy the requirements of Theorem 1. Then, another problem appears: the application of the inductive hypothesis to the polynomial $R$ fails with a type unification issue, because its coefficients belong to $K[X]/(Q)$ while those of $P$ belong to $K$. In Isabelle, these fields have different types: if $K$ has type $'a\ ring$ and $'a\ poly$ is the type of polynomials with coefficients of type $'a$, then the quotient field $K[X]/(Q)$ would have the type $(('a\ poly)\ set)\ ring$. So even if we generalised the induction hypothesis with respect to the field, we would not be able to instantiate it with $K[X]/(Q)$. And again, dependent types are a solution: one could generalise the induction hypothesis with respect to the *type* of the field.

The solution we propose goes in another direction. The idea is to prove an intermediate result where we fix a well chosen type $'b$ for the elements of both the fields $K$ and $L$. Then, after we build the field $K[X]/(Q)$ of type $(('b\ poly)\ set)\ ring$ during the proof, we use our type-switching mechanism to induce a field of type $'b\ ring$ with the same properties of $K[X]/(Q)$. At this point, it will be possible to use the induction hypothesis and to conclude the proof.

The tricky part is choosing the type $'b$. In the definition of the function *DirProd_list*, we faced a similar problem. We had to come up with a type $'b$ such that an injective function of type $'a * 'b \Rightarrow 'b$ existed. Clearly, in that case, the type $'a\ list$ was sufficient. Now, the situation is less clear: we need to come up with a type $'b$ such that an injective function of type $('b\ poly)\ set \Rightarrow 'b$ exists.

Observe that the function only needs to be injective on the carrier of the structure we are planning to use as a model for the induction of a new one. Therefore, the definition of an injective function of type $('b\ poly)\ set \Rightarrow 'b$ does not constitute a violation of Cantor's theorem, since the injectivity only needs to hold in the subset of the elements of type $('b \, poly) \, set$ composed by those which belong to the carrier of the field $K[X]/(Q)$.

The type we conceived to satisfy these conditions is the type of *multivariate polynomials*, or, polynomials with indexed variables. Below we discuss how these are formalised and how they solve the problem for our development.

## 3.2  Multivariate Polynomials

Polynomials with coefficients in a field $K$ are usually treated as linear combinations of successive powers of a formal letter $X$. Multivariate polynomials follow the same idea, but, instead of dealing with the powers of a fixed formal letter $X^n$, we manipulate the linear combination of arbitrary expressions of the form $\prod_j \mathcal{X}_j^{n_j}$, where $j$ runs over a finite subset of a fixed indexing set $J$.

The formalisation of multivariate polynomials that we are about to present relies on the notion of finite maps. A finite map from the set $A$ to the set $B$ is a partial function from elements of $A$ to elements of $B$ whose support is finite. The set of finite maps from $A$ to $B$ is written $A \xrightarrow{\text{fin}} B$.

First, we formally define monomials over indexed variables: a monomial is uniquely identified by a finite map from elements of the set $J$ to positive integers $\mathbb{Z}_{>0}$. Imagine it as the choice of exponent for each indexed letter. If $Mon_J$ denotes the set of monomials over variables indexed by $J$, then we have

$$Mon_J \triangleq J \xrightarrow{\text{fin}} \mathbb{Z}_{>0}.$$

We define multivariate polynomials similarly, as finite maps from monomials $Mon_J$ to $K^*$, the set of nonzero elements of $K$. The elements in the support of the finite map are the monomials involved in the linear combination. The value in $K^*$ associated to each monomial is the choice of coefficient. Accordingly, if $K[J]$ denotes the set of polynomials over variables indexed by $J$ and coefficients in $K$, then

$$K[J] \triangleq Mon_J \xrightarrow{\text{fin}} K^*.$$

Isabelle's predefined type of *multisets* comprises the finite maps into $\mathbb{Z}_{>0}$. A multiset is a function denoting the number of occurrences of each element, and multisets are finite. In our development, monomials have the same type as multisets. So, if the indexing set $J$ has type $'c \, set$, then a monomial would have the type $'c \, multiset$.

For polynomials, there was no shortcut. They are modelled as functions from monomials to $K$ and we require that the image of this function is zero save for a finite set of monomials. Therefore, a multivariate polynomial has the type $'c \, multiset \Rightarrow 'a$, where, as usual, $'a$ is the type of the elements in the field $K$.

Now, we substantiate our claim that if the type $'b$ is instantiated with the type of multivariate polynomials, then there exists an injective function of type $('b \, poly) \, set \Rightarrow 'b$.

First, remember that the *set* constructor was introduced because it is the type of the elements of the quotient field $K[X]/(Q)$: equivalence classes are encoded as cosets of the form $\pi(P)$ for some $P \in K[X]$. However, these equivalence classes could also be uniquely identified by polynomials:

**Definition 9.** *Let $M$ be a field and $Q$ a polynomial with coefficients in $M$. We let $mod_{[M,Q]}$ denote the function which assigns the remainder of the euclidean division of $P$ by $Q$ to each equivalence class $\pi(P)$ for $P \in M[X]$. It is a well-defined injective function from $M[X]/(Q)$ to $M[X]$.*

We use the letter $M$ for an arbitrary field instead of the usual letter $K$, because we instantiate this definition with a field whose elements have type $'b$, that is, a field included in $K[J]$. Thus, back in Isabelle, we are able to define an injective function of type $('b\,poly)\,set \Rightarrow 'b\,poly$. The problem becomes simpler, since now it suffices to define an injective function of type $'b\,poly \Rightarrow 'b$.

In other words, given a polynomial $P$ with coefficients in $K[J]$, we need to define a unique way to recover an element of $K[J]$. We do this by replacing the formal letter $X$ from the polynomial $P$ with an indexed one, $\mathcal{X}_l$. However, in order to get injectivity, we constrain the coefficients of $P$ to not *use* the indexed variable $\mathcal{X}_l$. For this purpose, we introduce the notion of a *free* index:

**Definition 10 (Free index).** *Let $M$ be a subset of $K[J]$ and $l \in J$ be an index. Then $l$ is* free *in $M$ if $M$ is a subset of $K[J \setminus \{l\}]$.*

Intuitively, $l$ is free in $M$ if $\mathcal{X}_l$ does not appear in the writing of any term in $M$. The idea of replacing the formal letter $X$ with $\mathcal{X}_l$ is captured as follows:

**Definition 11 (Eval).** *Let $K$ be a field, $J$ be an indexing set and $l$ an index in $J$. We define $Eval_l$, an injective function from polynomials with coefficients in $K[J \setminus \{l\}]$ to elements in $K[J]$:*

$$Eval_l\left(\sum_k\left(\sum_i a_{ik}\prod_{j\in J_{ik}}\mathcal{X}_j^{n_{ijk}}\right)X^k\right) = \sum_{i,k} a_{ik}\prod_{j\in J_{ik}\cup\{l\}}\mathcal{X}_j^{n'_{ijk}}$$

$$where\ n'_{ijk} = \begin{cases} k & if\,j=l \\ n_{ijk} & otherwise \end{cases}$$

Finally, we are able to define an injective function of type $('b\,poly)\,set \Rightarrow 'b$ as the composition of *Eval* and *mod*. The following technical lemma exploits this function to induce a field isomorphic to $M[X]/(Q)$.

**Lemma 1.** *Let $K$ be a field, $J$ be an indexing set, $M$ be a field whose elements belong to $K[J]$ and $Q$ an irreducible polynomial in $M[X]$. If $l \in J$ is a free index in $M$, then the composition $Eval_l \circ mod_{[M,Q]}$ is an injective map from $M[X]/(Q)$ to $K[J]$.*

$$M[X]/(Q) \xrightarrow{mod_{[M,Q]}} M[X] \subseteq (K[J\setminus\{l\}])[X] \xrightarrow{Eval_l} K[J]$$

*Moreover, let $L$ denote the structure induced by $Eval_l \circ mod_{[M,Q]}$. It is a field whose elements belong to $K[J]$. Furthermore, $M$ is a subfield of $L$ and the indexed variable $\mathcal{X}_l$ is a root of $Q$ in $L$.*

*Proof.* Injectivity of the map $Eval_l \circ mod_{[M,Q]}$ comes from the composition of injective maps. Now, let $L$ denote the field induced by $Eval_l \circ mod_{[M,Q]}$. It is a field isomorphic to $M[X]/(Q)$ inheriting the same algebraic properties as such. For instance, there exists a root of $Q$ in $L$: it is the element that realises $\pi(X)$, that is, $(Eval_n \circ mod_{[M,Q]})(\pi(X))$. However, remember that $\pi(X)$ was a root of the polynomial $Q^\pi$ in $M[X]/(Q)$. So, the element

$$(Eval_n \circ mod_{[M,Q]})(\pi(X))$$

is actually a root of the polynomial $Q^{Eval_n \circ mod_{[M,Q]} \circ \pi}$. Fortunately,

$$(Eval_n \circ mod_{[M,Q]})(\pi(X)) = Eval_n(mod_{[M,Q]}(\pi(X)) = Eval_n(X) = \mathcal{X}_n$$

and $(Eval_n \circ mod_{[M,Q]} \circ \pi)(a) = a$ for $a \in M$.

Therefore, the field $M$ is a subfield of $L$, the polynomial $Q$ has its coefficients in the field $L$ and $\mathcal{X}_n$ is a root of $Q$ in $L$.  □

We can finally proceed to the proof of the intermediate result that enjoys a direct analogue in Isabelle as suggested in the previous subsection.

**Lemma 2.** *Let $M \subseteq K[\mathbb{N}]$ be a field and let $P$ be a polynomial with coefficients in $M$. Suppose that for every $j \in \{0, \dots, \deg P - 1\}$, the index $j$ is free in $M$. Under these hypotheses, there exists a field $L \subseteq K[\mathbb{N}]$, such that $M$ is a subfield of $L$ and $P$ splits in $L$.*

*Proof.* By induction on the degree of $P$. If $\deg P = 0$, then $P$ splits in $M$ and we are done. If $\deg P > 0$, then there exists an irreducible polynomial $Q \in M[X]$ such that $Q$ divides $P$. We can suppose that $\deg Q > 1$, otherwise $P$ have only trivial irreducible factors and would split in $M$.

With the polynomial $Q$ and the index $n$, we are in position to apply Lemma 1; let $S$ be the field induced by the injective map $Eval_n \circ mod_{[M,Q]}$ such that $M$ is a subfield of $S$ and $\mathcal{X}_n$ is a root of $Q$ in $S$.

Since the only free index in $M$ that becomes nonfree in $S$ is the index $n$, we are allowed to instantiate the induction hypothesis with the field $S$ and the polynomial $P/(X - \mathcal{X}_n)$, whose degree is equal to $n$. We obtain a field $L \subseteq K[\mathbb{N}]$, such that $S$ is a subfield of $L$ and the polynomial $P/(X - \mathcal{X}_n)$ splits in $L$. It is easy to see that $P$ splits in $L$ as well.  □

To recover the proper statement of Theorem 1, we only need to find a way to embed $K$ into the set $K[\mathbb{N}]$. If we have a homomorphism $\phi : K \to M$ such that $M$ is a subset of $K[\mathbb{N}]$, then the field we get from Lemma 2 together with $\phi$ will be the field extension satisfying the conditions assured by Theorem 1. The required embedding simply maps each $a \in K$ to the constant polynomial $a$. Using our representation of multivariate polynomials as finite maps, we can make this intuition precise:

**Definition 12.** *Let $K$ be a field and $J$ an indexing set. We define $const^{\#}$ a function from $K$ to the set $K[J]$:*

$$const^{\#}(a) = \begin{cases} \{\} & if\, a = 0 \\ \{\} \mapsto a & otherwise \end{cases}$$

*where $\_ \mapsto \_$ denotes the singleton map and $\{\}$, the empty one.*

*It is an injective map, therefore we can promote the function $const^{\#}$ to an isomorphism between $K$ and the induced field $const^{\#}(K)$, written $K^{\#}$ for brevity.*

**Corollary 1 (Same statement as Theorem 1).**

*Proof.* Consider the field $K^{\#} \subseteq K[\mathbb{N}]$. Observe that every index $n \in \mathbb{N}$ is free in $K^{\#}$. Indeed, an element of $K^{\#}$ is the image $const^{\#}(a)$ for some $a$ in $K$ and no indexed variable is involved in such terms. Lemma 2 applies. We obtain a field $L \subseteq K[\mathbb{N}]$ such that $K^{\#}$ is a subfield of $L$ and $P^{const^{\#}}$ splits in $L$. Rephrase it with $\phi = const^{\#}$: the field $L$ is a field extension of $K$ under the homomorphism $\phi$ such that the polynomial $P^{\phi}$ splits in $L$. $\qquad\square$

The proof of Lemma 2 follows the same structure as the one from Theorem 1: obtain an irreducible factor, build the quotient field and then apply induction. The major difference is that now we have instrumented the proof with technical arguments concerning indexes. These might seem artificial at first, but they do have their mathematical value. After all, we've built a splitting field of a polynomial in the same set from which we started. In other words, we had greater control over the field that was being built during the proof. Notice that all field extensions explicitly mentioned in the proof are those under the identity homomorphism.

A final remark: we do not define how multivariate polynomials compose with each other; we just show how they can be represented. We do not define the ring operations as we do not need their algebraic properties. We use only their set properties, which ease the definition of an injective function as described and allow the construction of an induced structure. Moreover, this procedure of inducing structures engenders the composition laws for free.

# 4    Conception

In this section we outline the main ideas behind our formal proof of the existence of an algebraic closure. We start with a short survey of existing proofs. Technical details are left for the next section.

We mentioned in the introduction that the existence of an algebraic closure was proved by Steinitz in 1910. However, because set theory was not yet well understood, constructions relying on infinite collections were unnecessarily involved and his proof was 20 pages long. The proof preferred by modern authors [2,9], due to Emil Artin, is much shorter and simpler:

*Proof.* Let $F$ be a field. Consider the ring $F[F[X]]$ of multivariate polynomials over variables indexed by polynomials in $F[X]$ and coefficients in $F$. Then the ring $F[F[X]]$ admits an ideal $I$ with two key properties: (1) the quotient $F[F[X]]/I$ is a field; (2) every nonconstant polynomial in $F[X]$ has a root in $F[F[X]]/I$. This yields a field extension of $F$ where every nonconstant polynomial with coefficients in $F$ has at least one root. Furthermore, the construction is parametric in $F$. Now, consider the chain of field extensions

$$K = E_0 \to E_1 \to E_2 \to \cdots \to E_n \to \cdots$$

where $K$ is the field for which we intend to find an algebraically closed extension and $E_{n+1}$ is the field obtained by instantiating $F$ with the field $E_n$. The sequence admits a limit. It is a field $E$ left invariant by the construction described above, that is, a fixed point: every nonconstant polynomial in $E[X]$ has a root in $E$. Thus, $E$ is an algebraically closed field.                                   □

There are two main components in Artin's proof: the construction of a field and its iteration. We might wonder which field we obtain if we change the first component. Would we still have an algebraically closed field? What would be the impact on the iteration part? Hernandez and Laszlo [6] present a variant of the proof where the construction is modified to build a field with stronger properties. They consider the larger ring of multivariate polynomials $K[K[X] \times \mathbb{N}]$ with the product $K[X] \times \mathbb{N}$ as the choice of indexing set and prove that a slightly different construction engenders an algebraically closed field extension of $K$ directly, without iteration.

Our proof goes in the opposite direction. We weaken the construction and strengthen the iteration method. If $F$ is the current field, we find an irreducible polynomial $Q$ with coefficients in $F$ and define the usual quotient $F[X]/(Q)$ of the ring $F[X]$ by the ideal $(Q)$ to be the next field in the sequence.

The increment now is small. Proposition 2 states that we add only one root at each step, while the previous construction found one root for every polynomial with coefficients in the current field. We compensate for this deficit through the use of Zorn's lemma, which guarantees the existence of a maximal element of a partially ordered set. Thus we abstract away from the process of iteration and leap immediately to the required limit.

In our proof, we need this ability. With the construction from Artin's proof, we were content to iterate "vertically" over the degree of each polynomial at the same time. Now, we have to iterate over each polynomial individually. A step of iteration considers one polynomial at a time. So, we iterate both over each polynomial and its degree, both "horizontally" and "vertically".

More precisely, given both a field $E_n$ in the sequence of Artin's proof and a polynomial $P$ with coefficients in the field $E_n$, we can find a $k$ such that $P$ splits in the field $E_{n+k}$. It suffices to take $k$ larger than the degree of $P$. With our new construction however, the distance between the field where a polynomial has its coefficients and the field in which the same polynomial splits might be larger than every natural number. Since our method iterates over polynomials, the subset of steps that interleave the two fields might have the same cardinality as the set of polynomials, which could be uncountable, depending on the the the cardinality of the field of coefficients.

This simplification eases the formalisation in a number of ways. The field $F[X]/(Q)$ depends on the ring of polynomials with coefficients in $F$ and the ideal generated by $Q$, two straightforward structures. The field $F[F[X]]/I$, on the other hand, depends on the *ring* of multivariate polynomials and the ideal $I$. While we don't see any issues with the definition of the composition laws for the set $F[F[X]]$, the proof that they satisfy the axioms of a commutative ring seems laborious. Moreover, the proof that $I$ enjoys the two key properties discussed in the proof sketch relies on the intermediate result that every field admits an extension that splits a finite set of polynomials (a corollary of Theorem 1). We would have to prove a lemma that would immediately be subsumed by our intended theorem: the existence of an algebraically closed extension.

The idea of coupling a simpler construction with Zorn's lemma is also seen in Jelonek [8]. But while he relies on set-theoretical arguments to obtain a set sufficiently large to host an algebraic closure, we exhibit this set explicitly.

## 5   The Proof

In this section, we explain our proof that every field admits an algebraic closure. First, let's recall order theory and Zorn's lemma:

**Definition 13 (Partial orders, etc.).** *Let $\leq$ be a binary relation on a set $S$. Then $(S, \leq)$ is a* partial order *if $\leq$ is transitive, reflexive and anti-symmetric. A* chain *of $S$ is a subset of $S$ for which every two of its elements are $\leq$-comparable. A* maximal *element is some $a \in S$ such that for all $x \in S$, if $a \leq x$ then $a = x$, while an* upper bound *of $S$ is some $a \in S$ for which $a \geq x$ for all $x \in S$.*

**Lemma 3 (Zorn).** *Let $(S, \leq)$ be a partial order where $S$ is nonempty. Suppose every chain $C \subseteq S$ has an upper bound. Then $(S, \leq)$ has a maximal element.*

Let $K$ be a field. We will search for an algebraic closure of $K$ in the set $K[J]$ of multivariate polynomials indexed by a well-chosen set $J$. Recall that an algebraic closure is a field extension, hence we have to search both for a field $L$ and for a homomorphism $\phi : K \to L$. We make some decisions to reduce the search space. We expect $L$ to be an extension of the induced field $K^{\#}$. Thus, we anticipate $\phi$ to be the natural homomorphism

$$const^{\#} : K \to K^{\#}$$

from $K$ to the induced field $K^{\#}$. Now, given a field $L$ embedded in $K[J]$ such that $K^{\#}$ is a subfield of $L$, we can simply put $L$ to be the field extension of $K$ under $const^{\#}$.

Let's simplify further. Let's search for a field $L \subseteq K[J]$ such that $K^{\#}$ is a subfield of $L$ and every polynomial with coefficients in $K^{\#}$ splits in $L$. Such a field is not necessarily an algebraic closure of $K^{\#}$, with each of its elements algebraic over $K^\#$. Still, exhibiting a field with these simpler characteristics is sufficient to prove the existence of an algebraic closure, as it proves assertion (3) from Proposition 3.

With these considerations in mind, let's define the concrete set of fields.

**Definition 14.** *For a field $K$, let $\mathcal{A}_K$ denote the set of fields $L \subseteq K[K[X] \times \mathbb{N}]$ satisfying the following properties:*

1. *$K^\#$ is a subfield of $L$.*
2. *For every $P \in K[X]$ and natural number $n$, if the pair $(P, n)$ is a nonfree index of $L$, then the indexed variable $\mathcal{X}_{(P,n)}$ is a root of $P^{const^\#}$ in $L$.*

The intuition for property (2) and for the choice of indexing set, $K[X] \times \mathbb{N}$, follows from the insights developed in the last section. Recall the main idea: to build a chain of fields where at each step we add a new root to the preceding field. What property (2) intuitively does is to ensure that the role of the new root is taken by a formal letter $\mathcal{X}_j$, where $j$ is some previously free index $j$. Additionally, it also ensures that the choice of index was judicious: if we choose to include the variable $\mathcal{X}_{(P,n)}$ in the next round of the iteration, then it must be in a way such that it is a root of the polynomial $P^{const^\#}$. Look both at the statement of Lemma 2 and its proof for an example of similar reasoning.

Next, we equip $\mathcal{A}_K$ with a partial ordering, aiming to use Zorn's lemma.

**Definition 15 (Subfield relation).** *Let $L_1$ and $L_2$ be two fields. We write $L_1 \lesssim L_2$ to denote that $L_1$ is a subfield of $L_2$.*

**Lemma 4.** *If $K$ is a field, then $\mathcal{A}_K$ has a maximal element with respect to the partial order $(\mathcal{A}_K, \lesssim)$.*

*Proof.* Clearly $(\mathcal{A}_K, \lesssim)$ is a partial order. Consider a chain $C \subseteq \mathcal{A}_K$ and let $E = \bigcup_{F \in C} F$. For any two elements $a$ and $b$ in $E$, there exists a field $L \in C$ such that $a, b \in L$. We define their composition in $E$ (addition and multiplication) to be the same as in $L$. Since every two fields in $C$ are comparable, if we obtain another field $L' \in C$ such that the elements $a$ and $b$ belong to $L'$, then the field $L'$ is either a subfield of $L$ or an extension of $L$. In both cases, the composition of $a$ and $b$ gives the same result either in $L$ or in $L'$. Hence, their composition is independent on the choice of the field $L$ and therefore well-defined. The set $E$ equipped with these laws satisfies the axioms of a field. It is also straightforward to check that $E$ is an upper bound of $C$ in $\mathcal{A}_K$.

Since every chain has an upper bound, the result holds by Zorn's lemma. □

Although the usual sequence of fields is hidden by the application of Zorn's lemma, we can think of the maximal element of $\mathcal{A}_K$ as the limit of that sequence. It must be a field that splits every polynomial in $K^\#[X]$, since otherwise there would still be space left to add another root, contradicting its maximality:

**Theorem 2.** *Let $K$ be a field and let $M$ be the maximal element of $\mathcal{A}_K$ for the subfield relation. Every polynomial with coefficients in $K^\#$ splits in $M$.*

*Proof.* For contradiction, suppose that there exists a polynomial with coefficients in $K^{\#}$ that does not split in $M$. Since the map $const^{\#}$ is an isomorphism between the fields $K$ and $K^{\#}$ we can suppose that the polynomial which fails to split in $M$ is of the form $P^{const^{\#}}$ where $P$ is some polynomial with coefficients in $K$.

Let $Q \in M[X]$ be an irreducible polynomial with degree greater than 1 such that $Q$ divides $P^{const^{\#}}$ and let $n$ be a natural number such that the pair $(P, n)$ is a free index in $M$. They both exist, otherwise the polynomial $P^{const^{\#}}$ would split in $M$. We are able to apply Lemma 1 instantiated with the polynomial $Q$ and the index $(P, n)$ to obtain a field $L$ such that $M$ is a subfield of $L$ and the indexed variable $\mathcal{X}_{(P,n)}$ is a root of $Q$ in $L$.

Clearly $L$ belongs to $\mathcal{A}_K$, and the indexed variable $\mathcal{X}_{(P,n)}$ is an element of $L$ that does not belong to $M$. So $L$ is an element of $\mathcal{A}_k$ that is strictly greater than $M$ for the subfield relation, contradicting the maximality of $M$.    □

**Corollary 2.** *Every field $K$ admits an algebraic closure.*

*Proof.* The maximal element of $\mathcal{A}_K$ is a field extension of $K$ under the homomorphism $const^{\#}$ satisfying statement 3 from Proposition 3. By the same theorem, this is sufficient to prove the existence of an algebraic closure of $K$.    □

# 6    Related Work

We believe that our formalisation of the existence of an algebraic closure of any field is novel. There is, however, a closely related theorem formalised by Gonthier as part of the Mathematical Components library: every *countable* field admits an algebraic closure [4]. Since the proof is carried out in Coq, it is especially interesting for its computational content. The restriction to countable fields, on the other hand, precludes the application of the theorem to uncountable fields such as the reals or the $p$-adic numbers. This is a problem for the $p$-adic numbers, which (unlike the reals) has no well-known algebraic closure construction.

Also in Coq, Mathematical Components supports multivariate polynomials [7] over finite indexing sets of the form $\{1, \ldots, n\}$.

Schwarzweller proved in Mizar that the real numbers and finite fields are not algebraically closed [11]. The Lean community maintains an online "Algebraic closure roadmap" [1].

Work has also been done in Isabelle. An AFP entry [12] uses lists to represent monomials and polynomials. This choice has a drawback—permutations of a list might denote the same monomial—but it allows the operations to be executable. Haftmann et al. [5] discuss different options for multivariate polynomials in Isabelle. They propose two representations, an abstract and a concrete one, and establish a formal correspondence between them. The abstract representation resembles ours; it is based on finite maps. However, the authors fix the indexing set to the integers. We cannot use either of these Isabelle libraries because both rely on type classes to model the algebraic properties satisfied by polynomial coefficients. While type classes are fine for sharing algebraic theorems between types [10], they are too restrictive for abstract algebra, forcing algebraic objects to be types.

# 7    Conclusion

We formalised a proof that every field has an algebraic closure, a fundamental theorem. We have given a precise description of the known proofs and demonstrated how Zorn's lemma led to a straightforward formalisation.

The obstacles that we encountered during the realisation of this work led to the investigation of some problems related to Isabelle's type-discipline: that all the objects in a collection must have the same type. As a solution, we introduced the notion of an *induced structure*, which yields an isomorphic algebraic structure having a specified type.

To complete this project, we had to extend the HOL-Algebra library with general-purpose topics such as multivariate polynomials, arithmetic on arbitrary rings, arithmetic on the ring of polynomials, subfields, finite extensions and much of the content in Sect. 2. The fundamental nature of this material strengthens the library as the basis for further developments of algebra in Isabelle/HOL. It comprises nearly 15,000 nonempty lines of code in 21 new theories. The work was completed within 5 months.

*Availability.* Our development is part of HOL-Algebra. It is included in the distribution of Isabelle, directory `src/HOL/Algebra` and can also be browsed online.[1]

**Acknowledgments.** This research was supported by the European Research Council: ERC Advanced Grant ALEXANDRIA (Project GA 742178). We thank Dr. Anthony Bordg for his invaluable guidance, and we also thank Martin Baillon who was directly involved in the formalisation effort and who was always available to having fruitful discussions, as a colleague and as a friend. Finally, we thank the anonymous reviewers for their extremely helpful feedback.

# References

1. Algebraic closure roadmap. https://github.com/leanprover-community/mathlib/wiki/Algebraic-closure-roadmap
2. Buzzard, K.: Existence of algebraic closure of a field (2016). https://wwwf.imperial.ac.uk/~buzzard/maths/teaching/15Aut/M3P11/algclosure.pdf
3. Buzzard, K.: Motivation for algebraically closed fields? (2019). Email dated 05 October 2019
4. Gonthier, G.: Closed fields. https://math-comp.github.io/htmldoc/mathcomp.field.closed_field.html
5. Haftmann, F., Lochbihler, A., Schreiner, W.: Towards abstract and executable multivariate polynomials in Isabelle. In: Nipkow, T., Paulson, L., Wenzel, M. (eds.) Isabelle Workshop (2014). https://www3.risc.jku.at/publications/download/risc_5012/IWS14.pdf
6. Hernandez, D., Laszlo, Y.: Introduction à la Théorie de Galois. Polytechnique (2014)

---

[1] https://isabelle.in.tum.de/dist/library/HOL/HOL-Algebra/.

7. Hivert, F., Thery, L.: Multinomials. https://github.com/math-comp/multinomials
8. Jelonek, Z.: A simple proof of the existence of the algebraic closure of a field. Univ. Iagell. Acta Math. **30**, 131–132 (1993). http://www2.im.uj.edu.pl/actamath/PDF/30-131-132.pdf
9. Lang, S.: Algebra. Springer, New York (2002). https://doi.org/10.1007/978-1-4613-0041-0
10. Paulson, L.C.: Organizing numerical theories using axiomatic type classes. J. Autom. Reason. **33**(1), 29–49 (2004)
11. Schwarzweller, C.: On roots of polynomials and algebraically closed fields. Formaliz. Math. **25**, 185–195 (2017)
12. Sternagel, C., Thiemann, R.: Executable multivariate polynomials. Archive of Formal Proofs (2010). https://www.isa-afp.org/entries/Polynomials.html

# Formalization of Forcing in Isabelle/ZF

Emmanuel Gunther[1], Miguel Pagano[1], and Pedro Sánchez Terraf[1,2(✉)]

[1] Facultad de Matemática, Astronomía, Física y Computación,
Universidad Nacional de Córdoba, Córdoba, Argentina
{gunther,pagano,sterraf}@famaf.unc.edu.ar
[2] Centro de Investigación y Estudios de Matemática (CIEM-FaMAF), Conicet,
Córdoba, Argentina

**Abstract.** We formalize the theory of forcing in the set theory framework of Isabelle/ZF. Under the assumption of the existence of a countable transitive model of $ZFC$, we construct a proper generic extension and show that the latter also satisfies $ZFC$. In doing so, we remodularized Paulson's `ZF-Constructibility` library.

**Keywords:** Forcing · Isabelle/ZF · Countable transitive models · Absoluteness · Generic extension · Constructibility

## 1 Introduction

The present work reports on the third stage of our project of formalizing the theory of forcing and its applications as presented in one of the more important references on the subject, Kunen's Set Theory [9] (a rewrite of the classical book [8]).

We work using the implementation of Zermelo-Fraenkel ($ZF$) set theory *Isabelle/ZF* by Paulson and Grabczewski [17]. In an early paper [3], we set up the first elements of the countable transitive model (ctm) approach, defining forcing notions, names, generic extensions, and showing the existence of generic filters via the Rasiowa-Sikorski lemma (RSL). In our second (unpublished) technical report [4] we advanced by presenting the first accurate *formal abstract* of the Fundamental Theorems of Forcing, and using them to show that the $ZF$ axioms apart from Replacement and Infinity hold in all generic extensions.

This paper contains the proof of Fundamental Theorems and complete proofs of the Axioms of Infinity, Replacement, and Choice in all generic extensions. In particular, we were able to fulfill the promised formal abstract for the Forcing Theorems almost to the letter. A requirement for Infinity and the absoluteness of forcing for atomic formulas, we finished the interface between our development and Paulson's constructibility library [15] which enables us to do well-founded recursion inside transitive models of an appropriate finite fragment of $ZF$. As a by-product, we finally met two long-standing goals: the fact that the generic filter

Supported by Secyt-UNC project 33620180100465CB and Conicet.

N. Peltier and V. Sofronie-Stokkermans (Eds.): IJCAR 2020, LNAI 12167, pp. 221–235, 2020.
https://doi.org/10.1007/978-3-030-51054-1_13

$G$ belongs to the extension $M[G]$ and $M \subseteq M[G]$. In order to take full advantage of the constructibility library we enhanced it by weakening the assumption of many results and also extended it with stronger results. Finally, our development is now independent of $AC$: We modularized RSL in such a way that a version for countable posets does not require choice.

In the course of our work we found it useful to develop Isar methods to automate repetitive tasks. Part of the interface with Paulson's library consisted in constructing formulas for each relativized concept; and actually Isabelle's Simplifier can synthesize terms for unbound schematic variables in theorems. The synthesized term, however, is not available outside the theorem; we introduced a method that creates a definition from a schematic goal. The second method is concerned with renaming of formulas: we improved our small library of bounded renamings with a method that given the source and target environments figures out the renaming function and produces the relevant lemmas about it.

The source code of our formalization, written for the 2019 version of Isabelle, can be browsed and downloaded at https://cs.famaf.unc.edu.ar/~pedro/forcing/.

We assume some familiarity with Isabelle and some terminology of set theory. The current paper is organized as follows. In Sect. 2 we comment briefly on the meta-theoretical implications of using Isabelle/ZF. In Sect. 3 we explain the use of relativized concepts and its importance for the ctm approach. The next sections cover the core of this report: In Sect. 4 we introduce the definition of the formula transformer `forces` and reasoning principles about it; in Sect. 5 we present the proofs of the fundamental theorems of forcing. We show in Sect. 6 a concrete poset that leads to a proper extension of the ground model. In Sect. 7 we complete the proof that every axiom and axiom scheme of ZFC is valid in any generic extension. Section 8 briefly discusses related works and we close the paper by noting the next steps in our project and drawing conclusions from this formalization.

## 2   Isabelle and (Meta)theories

Isabelle [14,18] is a general proof assistant based on fragment of higher-order logic called *Pure*. The results presented in this work are theorems of a version of *ZF* set theory (without the Axiom of Choice, *AC*) called *Isabelle/ZF*, which is one of the "object logics" that can be defined on top of Pure (which is then used as a language to define rules). Isabelle/ZF defines types i and o for sets and Booleans, resp., and the *ZF* axioms are written down as terms of type o.

More specifically, our results work under the hypothesis of the existence of a ctm of *ZFC*.[1] This hypothesis follows, for instance, from the existence of an inaccessible cardinal. As such, our framework is weaker than those found

---

[1] By Gödel's Second incompleteness theorem, one must assume at least the existence of some model of *ZF*. The countability is only used to prove the existence of generic filters and can be thus replaced in favor of this hypothesis.

usually in type theories with universes, but allows us to work "Platonistically"—assuming we are in a universe of sets (namely, i) and performing constructions there.

On the downside, our approach is not able to provide us with finitary consistency proofs. It is well known that, for example, the implication $\text{Con}(ZF) \implies \text{Con}(ZFC + \neg CH)$ can be proved in *primitive recursive arithmetic (PRA)*. To achieve this, however, it would have implied to work focusing on the proof mechanisms and distracting us from our main goal, that is, formalize the ctm approach currently used by many mathematicians.

It should be noted that Pure is a very weak framework and has no induction/recursion capabilities. So the only way to define functions by recursion is inside the object logic. (This works the same for Isabelle/HOL.) For this reason, to define the relation of forcing, we needed to resort to *internalized* first-order formulas: they form a recursively defined set `formula`. For example, the predicate of satisfaction `sats::i⇒i⇒i⇒o` (written $M, ms \models \varphi$ for a set $M$, $ms \in \text{list}(M)$ and $\varphi \in$ `formula`) had already been defined by recursion in Paulson [15].

# 3    Relativization, Absoluteness, and the Axioms

The concepts of relativization and absoluteness (due to Gödel, in his proof of the relative consistency of $AC$ [2]) are both prerequisites and powerful tools in working with transitive models. A *class* is simply a predicate $C(x)$ with at least one free variable $x$. The *relativization* $\varphi^C(\bar{x})$ of a set-theoretic definition $\varphi$ (of a relation such as "$x$ is a subset of $y$" or of a function like $y = \mathcal{P}(x)$) to a class $C$ is obtained by restricting all of its quantifiers to $C$.

$$x \subseteq^C y \equiv \forall z.\ C(z) \longrightarrow (z \in x \longrightarrow z \in y)$$

The new formula $\varphi^C(\bar{x})$ corresponds to what is obtained by defining the concept "inside" $C$. In fact, for a class corresponding to a set $c$ (i.e. $C(x) := x \in c$), the relativization $\varphi^C$ of a sentence $\varphi$ is equivalent to the satisfaction of $\varphi$ in the first-order model $\langle c, \in \rangle$.

It turns out that many concepts mean the same after relativization to a nonempty transitive class $C$; formally

$$\forall \bar{x}.\ C(\bar{x}) \longrightarrow (\varphi^C(\bar{x}) \longleftrightarrow \varphi(\bar{x}))$$

When this is the case, we say that the relation defined by $\varphi$ is *absolute for transitive models*.[2] As examples, the relation of inclusion $\subseteq$—and actually, any relation defined by a formula (equivalent to one) using only bounded quantifiers ($\forall x \in y$) and ($\exists x \in y$)—is absolute' for transitive models. On the contrary, this is not the case with the powerset operation.

A benefit of working with transitive models is that many concepts (pairs, unions, and fundamentally ordinals) are uniform across the universe i, a ctm

---

[2] Absoluteness of functions also requires the relativized definition to behave functionally over $C$.

(of an adequate fragment of *ZF*) *M* and any of its extensions *M*[*G*]. For that reason, then one can reason "externally" about absolute concepts, instead of "inside" the model; thus, one has at hand any already proved lemma about the real concept.

Paulson's formalization [15] of the relative consistency of AC by Gödel [2] already contains absoluteness results which were crucial to our project; we realized however that they could be refactored to be more useful. The main objective is to maximize applicability of the relativization machinery by adjusting the hypothesis of a greater part of its earlier development. Paulson's architecture had only in mind the consistency of *ZFC*, but, for instance, in order to apply it in the development of forcing, too much is assumed at the beginning; more seriously, some assumptions cannot be regarded as "first-order" (v.g. the Replacement Scheme) because of the occurrence of predicate variables. The version we present[3] of the constructibility library, `ZF-Constructible-Trans`, weakens the assumptions of many absoluteness theorems to that of a nonempty transitive class; we also include some stronger results such as the relativization of powersets.

Apart from the axiom schemes, the *ZFC* axioms are initially stated as predicates on classes (that is, of type $(i \Rightarrow o) \Rightarrow o$); this formulation allows a better interaction with `ZF-Constructible`. The axioms of Pairing, Union, Foundation, Extensionality, and Infinity are relativizations of the respective traditional first-order sentences to the class argument. For the Axiom of Choice we selected a version best suited for the work with transitive models: the relativization of a sentence stating that for every *x* there is surjection from an ordinal onto *x*. Finally, Separation and Replacement were treated separately to effectively obtain first-order versions afterwards. It is to be noted that predicates in Isabelle/ZF are akin to second order variables and thus do not correspond to first-order formulas. For that reason, Separation and Replacement are defined in terms of *the satisfaction* of an internalized formula $\varphi$. We improved their specification, with respect to our previous report [4], by lifting the arity restriction for the parameter $\varphi$; consequently we get rid of tupling and thus various proofs are now slicker.

A benefit of having class versions of the axioms is that we can apply our synthesis method to obtain their internal, first-order counterparts. For the case of the Pairing Axiom, the statement for classes is the following

*upair_ax(C)*==$\forall$ *x*[*C*]. $\forall$ *y*[*C*]. $\exists$ *z*[*C*]. *upair(C,x,y,z)*

where `upair` says that z is the unordered pair of x and y, relative to C.

The following schematic lemma synthesizes its internal version,[4]

**schematic_goal** *ZF_pairing_auto:*

---

[3] While preparing the final version of the present paper, our contributions were accepted as part of the official Isabelle 2020 release.

[4] The use of such schematic goals and the original definition of the collection of lemmas `sep_rules` are due to Paulson.

```
  "upair_ax(##A) ⟷ (A, [] ⊨ ?zfpair)"
unfolding upair_ax_def
  by ((rule sep_rules | simp)+)
```

and our **synthesize** method introduces a new term `ZF_pairing_fm` for it:

**synthesize** *"ZF_pairing_fm"* **from schematic** *"ZF_pairing_auto"*

the actual formula obtained is `Forall(Forall(Exists(upair_fm(2,1,0))))`.

## 4   The Definition of `forces`

The core of the development is showing the definability of the relation of forcing. As we explained in our previous report [4, Sect. 8], this comprises the definition of a function `forces` that maps the set of internal formulas into itself. It is the very reason of applicability of forcing that the satisfaction of a first-order formula $\varphi$ in all of the generic extensions of a ctm $M$ can be "controlled" in a definable way from $M$ (viz., by satisfaction of the formula `forces`($\varphi$)).

In fact, given a forcing notion $\mathbb{P}$ (i.e. a preorder with a top element) in a ctm $M$, Kunen defines the *forcing relation* model-theoretically by considering all extensions $M[G]$ with $G$ generic for $\mathbb{P}$. Then two fundamental results are proved, the Truth Lemma and the Definability Lemma; but the proof of the first one is based on the formula that witnesses Definability. To make sense of this in our formalization, we started with the internalized relation and then proved that it is equivalent to the semantic version ("`definition_of_forcing`," in the next section). For that reason, the usual notation of the forcing relation $p \Vdash \varphi$ *env* (for *env* a list of "names"), abbreviates in our code the satisfaction by $M$ of `forces`($\varphi$):

$$"p \Vdash \varphi \ env \quad \equiv \quad M, \ ([p,P,leq,one] \ @ \ env) \models forces(\varphi)"$$

The definition of `forces` proceeds by recursion over the set **formula** and its base case, that is, for atomic formulas, is (in)famously the most complicated one. Actually, newcomers can be puzzled by the fact that forcing for atomic formulas is also defined by (mutual) recursion: to know if $\tau_1 \in \tau_2$ is forced by $p$ (notation: `forces_mem`$(p, \tau_1, \tau_2)$), one must check if $\tau_1 = \sigma$ is forced for $\sigma$ moving in the transitive closure of $\tau_2$. To disentangle this, one must realize that this last recursion must be described syntactically: the definition of `forces`($\varphi$) for atomic $\varphi$ is then an internal definition of the alleged recursion on names.

Our aim was to follow the definition proposed by Kunen in [9, p. 257], where the following mutual recursion is given:

$$\mathbf{forces\_eq}(p, t_1, t_2) := \forall s \in \mathrm{domain}(t_1) \cup \mathrm{domain}(t_2). \ \forall q \preceq p.$$

$$\mathbf{forces\_mem}(q, s, t_1) \longleftrightarrow \mathbf{forces\_mem}(q, s, t_2), \quad (1)$$

$$\mathbf{forces\_mem}(p, t_1, t_2) := \forall v \preceq p. \ \exists q \preceq v.$$

$$\exists s. \ \exists r \in \mathbb{P}. \ \langle s, r \rangle \in t_2 \wedge q \preceq r \wedge \mathbf{forces\_eq}(q, t_1, s) \quad (2)$$

Note that the definition of forces_mem is equivalent to require the set

$$\{q \preceq p : \exists \langle s,r \rangle \in t_2.\ q \preceq r \wedge \texttt{forces\_eq}(q,t_1,s)\}$$

to be dense below $p$.

It was not straightforward to use the recursion machinery of Isabelle/ZF to define forces_eq and forces_mem. For this, we defined a relation frecR on 4-tuples of elements of $M$, proved that it is well-founded and, more important, we also proved an induction principle for this relation:[5]

**lemma** *forces_induction:*
  **assumes**
    "$\bigwedge \tau\ \vartheta.\ [\![ \bigwedge \sigma,\ \sigma \in domain(\vartheta) \implies Q(\tau,\sigma) ]\!] \implies R(\tau,\vartheta)$"
    "$\bigwedge \tau\ \vartheta.\ [\![ \bigwedge \sigma.\ \sigma \in domain(\tau)\ \cup\ domain(\vartheta) \implies R(\sigma,\tau)\ \wedge\ R(\sigma,\vartheta) ]\!]$
      $\implies Q(\tau,\vartheta)$"
  **shows**
    "$Q(\tau,\vartheta)\ \wedge\ R(\tau,\vartheta)$"

and obtained both functions as cases of a another one, forces_at, using a single recursion on frecR. Then we obtained (1) and (2) as our corollaries def_forces_eq and def_forces_mem.

Other approaches, like the one in Neeman [11] (and Kunen's older book [8]), prefer to have a single, more complicated, definition by simple recursion for forces_eq and then define forces_mem explicitly. On hindsight, this might have been a little simpler to do, but we preferred to be as faithful to the text as possible concerning this point.

Once forces_at and its relativized version is_forces_at were defined, we proceeded to show absoluteness and provided internal definitions for the recursion on names using results in ZF-Constructible. This finished the atomic case of the formula-transformer forces. The characterization of forces for negated and universal quantified formulas is given by the following lemmas, respectively:

**lemma** *sats_forces_Neg:*
  **assumes**
    "$p \in P$" "$env\ \in\ list(M)$" "$\varphi \in formula$"
  **shows**
    "$M,\ [p,P,leq,one]\ @\ env\ \models\ forces(Neg(\varphi))\ \longleftrightarrow$
      $\neg(\exists q \in M.\ q \in P\ \wedge\ is\_leq(\#\#M,leq,q,p)\ \wedge$
        $M,\ [q,P,leq,one]@env\ \models\ forces(\varphi))$"

**lemma** *sats_forces_Forall:*
  **assumes**
    "$p \in P$" "$env\ \in\ list(M)$" "$\varphi \in formula$"
  **shows**
    "$M,[p,P,leq,one]\ @\ env\ \models\ forces(Forall(\varphi))\ \longleftrightarrow$
      $(\forall x \in M.\ \ M,\ [p,P,leq,one,x]\ @\ env\ \models\ forces(\varphi))$"

---

[5] The logical primitives of *Pure* are $\implies$, &&&, and $\bigwedge$ (implication, conjunction, and universal quantification, resp.), which operate on the meta-Booleans prop.

Let us note in passing another improvement over our previous report: we made a couple of new technical results concerning recursive definitions. Paulson proved absoluteness of functions defined by well-founded recursion over a transitive relation. Some of our definitions by recursion (*check* and *forces*) do not fit in that scheme. One can replace the relation $R$ for its transitive closure $R^+$ in the recursive definition because one can prove, in general, that $F \restriction (R^{-1}(x))(y) = F \restriction ((R^+)^{-1}(x))(y)$ whenever $(x, y) \in R$.

## 5   The Forcing Theorems

After the definition of `forces` is complete, the proof of the Fundamental Theorems of Forcing is comparatively straightforward, and we were able to follow Kunen very closely. The more involved points of this part of the development were those where we needed to prove that various (dense) subsets of $\mathbb{P}$ were in $M$; for this, we have resorted to several ad-hoc absoluteness lemmas.

The first results concern characterizations of the forcing relation. Two of them are `Forces_Member`:

$$\text{(p } \Vdash \text{ Member(n,m) env) } \longleftrightarrow \text{ forces\_mem(p,t1,t2)},$$

where `t1` and `t1` are the nth resp. mth elements of `env`, and `Forces_Forall`:

$$\text{(p } \Vdash \text{ Forall}(\varphi) \text{ env) } \longleftrightarrow (\forall x \in M. \text{ (p } \Vdash \varphi \text{ ([x] @ env)))}.$$

Equivalent statements, along with the ones corresponding to `Forces_Equal` and `Forces_Nand`, appear in Kunen as the inductive definition of the forcing relation [9, Def. IV.2.42].

As with the previous section, the proofs of the forcing theorems have two different flavors: The ones for the atomic formulas proceed by using the principle of `forces_induction`, and then an induction on `formula` wraps the former with the remaining cases (`Nand` and `Forall`).

As an example of the first class, we can take a look at our formalization of [9, Lem. IV.2.40(a)]. Note that the context (a "locale," in Isabelle terminology, namely `forcing_data`) of the lemma includes the assumption of P being a forcing notion, and the predicate of being $M$-generic is defined in terms of P:

**lemma** *IV240a*:
  **assumes**
    `"M_generic(G)"`
  **shows**
    `"(`$\tau \in M \longrightarrow \vartheta \in M \longrightarrow$`(`$\forall p \in G.$`forces_eq(`$p,\tau,\vartheta$`) `$\longrightarrow$`val(`$G,\tau$`)=val(`$G,\vartheta$`)))`
    $\wedge$
    `(`$\tau \in M \longrightarrow \vartheta \in M \longrightarrow$`(`$\forall p \in G.$`forces_mem(`$p,\tau,\vartheta$`) `$\longrightarrow$`val(`$G,\tau$`)`$\in$`val(`$G,\vartheta$`)))"`

Its proof starts by an introduction of `forces_induction`; the inductive cases for each atomic type were handled before as separate lemmas (`IV240a_mem` and `IV240a_eq`). We illustrate with the statement of the latter.

**lemma** *IV240a_eq:*
  **assumes**
    *"M_generic(G)" "p∈G" "forces_eq(p,τ,ϑ)"*
  **and**
    *IH:"⋀q σ. q∈P ⟹ q∈G ⟹ σ∈domain(τ) ∪ domain(ϑ) ⟹*
      *(forces_mem(q,σ,τ) ⟶ val(G,σ) ∈ val(G,τ)) ∧*
      *(forces_mem(q,σ,ϑ) ⟶ val(G,σ) ∈ val(G,ϑ))"*
  **shows**
    *"val(G,τ) = val(G,ϑ)"*

Examples of proofs using the second kind of induction include the basic
`strengthening_lemma` and the main results in this section, the lemmas of Density (actually, its nontrivial direction `dense_below_imp_forces`) and Truth, which we state next.

**lemma** *density_lemma:*
  **assumes**
    *"p∈P" "φ∈formula" "env∈list(M)" "arity(φ)≤length(env)"*
  **shows**
    *"(p ⊩ φ env) ⟷ dense_below({q∈P. (q ⊩ φ env)},p)"*

**lemma** *truth_lemma:*
  **assumes**
    *"φ∈formula" "M_generic(G)"*
  **shows**
    *"⋀env. env∈list(M) ⟹ arity(φ)≤length(env) ⟹*
    *(∃p∈G. (p ⊩ φ env)) ⟷ M[G], map(val(G),env) ⊨ φ"*

From these results, the semantical characterization of the forcing relation (the "definition of ⊩" in [9, IV.2.22]) follows easily:

**lemma** *definition_of_forcing:*
  **assumes**
    *"p∈P" "φ∈formula" "env∈list(M)" "arity(φ)≤length(env)"*
  **shows**
    *"(p ⊩ φ env) ⟷*
    *(∀G. M_generic(G)∧ p∈G ⟶ M[G], map(val(G),env) ⊨ φ)"*

The present statement of the Fundamental Theorems is almost exactly the same of those in our previous report [4], with the only modification being the bound on arities and a missing typing constraint. This implied only minor adjustments in the proofs of the satisfaction of axioms.

# 6   Example of Proper Extension

Even when the axioms of *ZFC* are proved in the generic extension, one cannot claim that the magic of forcing has taken place unless one is able to provide some *proper* extension with the *same ordinals*. After all, one is assuming from the start a model $M$ of $ZFC$, and in some trivial cases $M[G]$ might end up to be exactly $M$; this is where *proper* enters the stage. But, for instance, in the presence of large cardinals, a model $M' \supsetneq M$ might be an end-extension of $M$—this is where we ask the two models to have the same ordinals, the same *height*.

Three theory files contain the relevant results. `Ordinals_In_MG.thy` shows, using the closure of $M$ under ranks, that $M$ and $M[G]$ share the same ordinals (actually, ranks of elements of $M[G]$ are bounded by the ranks of their names in $M$):

```
lemma rank_val: "rank(val(G,x)) ≤ rank(x)"
lemma Ord_MG_iff:
  assumes "Ord(α)"
  shows "α ∈ M ⟷ α ∈ M[G]"
```

To prove these results, we found it useful to formalize induction over the relation $\mathbf{ed}(x,y) := x \in \mathrm{domain}(y)$, which is key to arguments involving names.

`Succession_Poset.thy` contains our first example of a poset that interprets the locale `forcing_notion`, essentially the notion for adding one Cohen real. It is the set $2^{<\omega}$ of all finite binary sequences partially ordered by reverse inclusion. The sufficient condition for a proper extension is that the forcing poset is *separative*: every element has two incompatible ($\perp$s) extensions. Here, `seq_upd(f,x)` adds x to the end of the sequence f.

```
lemma seqspace_separative:
  assumes "f∈2^<ω"
  shows "seq_upd(f,0) ⊥s seq_upd(f,1)"
```

We prove in the theory file `Proper_Extension.thy` that, in general, every separative forcing notion gives rise to a proper extension.

# 7   The Axioms of Replacement and Choice

In our report [4] we proved that any generic extension preserves the satisfaction of almost all the axioms, including the separation scheme (we adapted those proofs to the current statement of the axiom schemes). Our proofs that Replacement and choice hold in every generic extension depend on further relativized concepts and closure properties.

## 7.1   Replacement

The proof of the Replacement Axiom scheme in $M[G]$ in Kunen uses the Reflection Principle relativized to $M$. We took an alternative pathway, following Neeman [11]. In his course notes, he uses the relativization of the cumulative hierarchy of sets.

The family of all sets of rank less than $\alpha$ is called $\mathtt{Vset}(\alpha)$ in Isabelle/ZF. We showed, in the theory file `Relative_Univ.thy` the following relativization

and closure results concerning this function, for a class $M$ satisfying the locale
M_eclose plus the Powerset Axiom and four instances of replacement.

**lemma** `Vset_abs:` `"⟦ M(i); M(V); Ord(i) ⟧ ⟹`
                `is_Vset(M,i,V) ⟷ V = {x∈Vset(i). M(x)}"`

**lemma** `Vset_closed:` `"⟦ M(i); Ord(i) ⟧ ⟹ M({x∈Vset(i). M(x)})"`

We also have the basic result

**lemma** `M_into_Vset:`
  **assumes** `"M(a)"`
  **shows** `"∃i[M]. ∃V[M]. ordinal(M,i) ∧ is_Vfrom(M,0,i,V) ∧ a∈V"`

stating that $M$ is included in $\bigcup\{\text{Vset}^M(\alpha) : \alpha \in M\}$ (actually they are equal).

For the proof of the Replacement Axiom, we assume that $\varphi$ is functional in
its first two variables when interpreted in $M[G]$ and the first ranges over the
domain $c \in M[G]$. Then we show that the collection of all values of the second
variable, when the first ranges over c, belongs to $M[G]$:

**lemma** `Replace_sats_in_MG:`
  **assumes**
    `"c∈M[G]"` `"env ∈ list(M[G])"`
    `"φ ∈ formula"` `"arity(φ) ≤ 2 #+ length(env)"`
    `"univalent(##M[G], c, λx v. (M[G], [x,v]@env ⊨ φ))"`
  **shows**
    `"{v. x∈c, v∈M[G] ∧ (M[G], [x,v]@env ⊨ φ)} ∈ M[G]"`

From this, the satisfaction of the Replacement Axiom in $M[G]$ follows very easily.

The proof of the previous lemma, following Neeman, proceeds as usual by
turning an argument concerning elements of $M[G]$ to one involving names lying
in $M$, and connecting both worlds by using the forcing theorems. In the case at
hand, by functionality of $\varphi$ we know that for every $x \in c \cap M[G]$ there exists
exactly one $v \in M[G]$ such that $M[G], [x,v]$ @ $env \models \varphi$. Now, given a name
$\pi' \in M$ for c, every name of an element of c belongs to $\pi := \text{domain}(\pi') \times \mathbb{P}$,
which is easily seen to be in $M$. We will use $\pi$ to be the domain in an application
of the Replacement Axiom in $M$. But now, obviously, we have lost functionality
since there are many names $\dot{v} \in M$ for a fixed $v$ in $M[G]$. To solve this issue, for
each $\rho p := \langle \rho, p \rangle \in \pi$ we calculate the minimum rank of some $\tau \in M$ such that
$p \Vdash \varphi(\rho, \tau, \dots)$ if there is one, or 0 otherwise. By Replacement in $M$, we can show
that the supremum ?sup of these ordinals belongs to $M$ and we can construct a
?bigname := {x∈Vset(?sup). x ∈ M} × {one} whose interpretation by (any
generic) $G$ will include all possible elements as $v$ above.

The previous calculation required some absoluteness and closure results
regarding the minimum ordinal binder, Least$(Q)$, also denoted $\mu x.Q(x)$, that
can be found in the theory file Least.thy.

## 7.2  Choice

A first important observation is that the proof of $AC$ in $M[G]$ only requires the assumption that $M$ satisfies (a finite fragment of) $ZFC$. There is no need to invoke Choice in the metatheory.

Although our previous version of the development used $AC$, that was only needed to show the Rasiowa-Sikorski Lemma (RSL) for arbitrary posets. We have modularized the proof of the latter and now the version for countable posets that we use to show the existence of generic filters does not require Choice (as it is well known). We also bundled the full RSL along with our implementation of the principle of dependent choices in an independent branch of the dependency graph, which is the only place where the theory ZF.AC is invoked.

Our statement of the Axiom of Choice is the one preferred for arguments involving transitive classes satisfying $ZF$:

$$\forall \texttt{x[M]}.\ \exists \texttt{a[M]}.\ \exists \texttt{f[M]}.\ \texttt{ordinal(M,a)} \wedge \texttt{surjection(M,a,x,f)}$$

The Simplifier is able to show automatically that this statement is equivalent to the next one, in which the real notions of ordinal and surjection appear:

$$\forall \texttt{x[M]}.\ \exists \texttt{a[M]}.\ \exists \texttt{f[M]}.\ \texttt{Ord(a)} \wedge \texttt{f} \in \texttt{surj(a,x)}$$

As with the forcing axioms, the proof of $AC$ in $M[G]$ follows the pattern of Kunen [9, IV.2.27] and is rather straightforward; the only complicated technical point being to show that the relevant name belongs to $M$. We assume that $\texttt{a} \neq \varnothing$ belongs to $M[G]$ and has a name $\tau \in M$. By $AC$ in $M$, there is a surjection $\texttt{s}$ from an ordinal $\alpha \in M$ ($\subseteq M[G]$) onto domain($\tau$). Now

$$\{\texttt{opair\_name(check}(\beta)\texttt{,s`}\beta)\texttt{.}\ \beta{\in}\alpha\} \ \times \ \{\texttt{one}\}$$

is a name for a function $\texttt{f}$ with domain $\alpha$ such that $\texttt{a}$ is included in its range, and where $\texttt{opair\_name}(\sigma, \rho)$ is a name for the ordered pair $\langle val(G, \sigma), val(G, \rho)\rangle$. From this, $AC$ in $M[G]$ follows easily.

## 7.3  The Main Theorem

With all these elements in place, we are able to transcript the main theorem of our formalization:

theorem $\textit{extensions\_of\_ctms}$:
  assumes
    "M $\approx$ nat" "Transset(M)" "M $\models$ ZF"
  shows
    "$\exists N$.
      $M \subseteq N \wedge N \approx$ nat $\wedge$ Transset($N$) $\wedge N \models$ ZF $\wedge M{\neq}N \wedge$
      ($\forall \alpha.$ Ord($\alpha$) $\longrightarrow$ ($\alpha \in M \longleftrightarrow \alpha \in N$)) $\wedge$
      (M, [] $\models$ AC $\longrightarrow$ N $\models$ ZFC)"

Here, $\approx$ stands for equipotency, nat is the set of natural numbers, and the predicate Transset indicates transitivity; and as usual, AC denotes the Axiom of Choice, and ZF and ZFC the corresponding subsets of formula.

# 8    Related Work

There are various formalizations of Zermelo-Fraenkel set theory in proof assistants (v.g. Mizar, Metamath, and recently Lean [10]) that proceed to different levels of sophistication. Isabelle/ZF can be regarded as a notational variant of NGB set theory [9, Sect. II.10], because the schemes of Replacement and Separation feature higher order (free) variables playing the role of formula variables. It cannot be proved that the axioms thus written correspond to first order sentences. For this reason, our relativized versions only apply to set models, where we restrict those variables to predicates that actually come from first order formulas. In that sense, the axioms of the locale M_ZF correspond more faithfully to the *ZF* axioms.

Traditional expositions of the method of forcing [7,9] are preceded by a study of relativization and absoluteness. For this reason, it was a natural choice at the beginning of this project to build on top of Paulson's formalization of constructibility on Isabelle/ZF, and that was one of the main early reasons to work on that logic instead of, e.g., HOL—below we discuss other reasons. In any case, our development of forcing does not depend on constructibility itself (in contrast to Cohen's original presentation, in which ground models are initial segments of the constructible hierarchy).

A natural question is whether Isabelle/HOL (with a far more solid framework to work with given its infrastructure and automation) would have been a better choice than Isabelle/ZF. In fact, there are two developments of Zermelo-Fraenkel set theory available on it: HOLZF by Obua [12] and ZFC_in_HOL by Paulson [16]. But these (logically equivalent) frameworks are higher in consistency strength than Isabelle/ZF. To elaborate on this, both ZF and HOL are axiomatized on top of Isabelle's metalogic *Pure*, which is a version of "intuitionistic higher order logic." In [13] Paulson proves that *Pure* is sound for intuitionistic first order logic, thus it does not add any strength to it. On top of this, the axiomatization of Isabelle/ZF results in a system equiconsistent with *ZFC*. On the other hand, showing the consistency of HOLZF (and thus ZFC_in_HOL) requires assuming the consistency of *ZFC* plus the existence of an inaccessible cardinal [12, Sect. 3]. We note, in contrast, that our extra running assumption of the existence of a countable transitive model is considerably weaker (directly and consistency-wise) than the existence of an inaccessible cardinal.

Concerning the formalization of the method of forcing, to the best of our knowledge there is only one other that deals with forcing for set theory: the recent *Flypitch* project by Han and van Doorn [5,6], which includes a formalization of the independence of CH using the Lean proof assistant. The Flypitch formalization is largely orthogonal to ours (it is based on Boolean-valued models, which are interpreted into type theory through a variant of the Aczel encoding of set theory), and this precludes a direct comparison of code. But we can highlight some conceptual differences between our development and the corresponding fraction of Flypitch.

A first observation concerns consistency strength. The consistency of Lean requires infinitely many inaccessibles. More precisely, let $Lean_n$ be the theory of CiC foundations of Lean restricted to $n$ type universes. Carneiro proved in his MSc thesis [1] the consistency of Lean$_n$ from $ZFC$ plus the existence of $n$ inaccessible cardinals. It is also reported in Carneiro's thesis that Werner's results in [19] can be adapted to show that Lean$_{n+2}$ proves the consistency of the latter theory. In that sense, although Flypitch includes proofs of unprovability results in first order logic, the meta-theoretic machinery used to obtain them is far heavier than the one we use to operate model-theoretically.

In second place, a formalization of forcing with general partial orders, generic filters and ctms has—in our opinion—the added value that this approach is used in an important (perhaps the greatest) fraction of the literature, both in exposition and in research articles and monographs. In verifying a piece of mature mathematics as the present one, representing the actual practice seems paramount to us.

Finally, as a matter of taste, one of the main benefits of using transitive models is that many fundamental notions are absolute and thus most of the concepts and statements can be interpreted transparently, as we have noted before. It also provides a very concrete way to understand generic objects: as sets that (in the non trivial case) are provably not in the original model; this dispels any mystical feel around this concept (contrary to the case when the ground model is the universe of all sets). In addition, two-valued semantics is closer to our intuition.

## 9    Conclusion and Future Work

We consider that the formalization of the definition of `forces` and its recursive characterization of forcing for atomic formulas is a turning point in our project; the reason for this is that all further developments will not involve such a daunting metamathematical component. Even the proofs of the Fundamental Theorems of Forcing turned out to follow rather smoothly after this initial setup was ready, the only complicated affair being to show that various dense sets belong to $M$. Actually, this is a point to be taken care of: For every new concept that is introduced, some lemmas concerning relativization and closure must be proved to be able to synthesize its internal definition. Further automation must be developed for this purpose.

In the course of obtaining internal formulas for the atomic case of forcing, a fruitful discussion concerning complementary perspectives on the role of proof assistants took place. An earlier approach relied more heavily in formula synthesis, thus making the Simplifier an indispensable main character. Following this line was quicker from the coding point of view since few new primitives were introduced and thus fewer lemmas concerning absoluteness and arities. On the downside, processing was a bit slower, the formulas synthesized were gigantic and the process on a whole was more error-prone. In fact, this approach was unsuccessful and we opted for a more detailed engineering, defining all intermediate steps. So the load on the assistant, in this part of the development, balanced from code-production to code-verification.

The next task in our path is pretty clear: To develop the forcing notions to obtain the independence of *CH* along with the prerequisite combinatorial results, v.g. the $\Delta$-system lemma. A development of cofinality is under way in a joint work with E. Pacheco Rodríguez, which is needed for a general statement of the latter. Once these developments are finished, we will be able to give a more thorough comparison between our project and the *Flypitch* approach using Boolean-valued models.

In a second release of `ZF-Constructible-Trans`, we intend to conform it to the lines of *Basic Set Theory (BST)* proposed by Kunen [9, I.3.1] in which elementary results have proofs using alternatively Powerset or Replacement. The interest in this arises because many natural set models (rank-initial segments of the universe or the family $H(\kappa)$ of sets of cardinality less than $\kappa$ hereditarily) satisfy one of those axioms and not the other. There are also still some older or less significant proofs written in tactical (**apply**) format; we hope we will find the time to translate them to Isar. Finally, the automation of formula synthesis is on an early stage of development; finishing that module will make writing our proofs of closure under various operations faster, and also turn the set theory libraries more usable to other researchers.

# References

1. Carneiro, M.: The type theory of Lean. Master's thesis, Carnegie Mellon University, April 2019. https://github.com/digama0/lean-type-theory/releases/tag/v1.0
2. Gödel, K.: The Consistency of the Continuum Hypothesis. Annals of Mathematics Studies, no. 3. Princeton University Press, Princeton (1940)
3. Gunther, E., Pagano, M., Sánchez Terraf, P.: First steps towards a formalization of forcing. In: Accattoli, B., Olarte, C. (eds.) Proceedings of the 13th Workshop on Logical and Semantic Frameworks with Applications, LSFA 2018, Fortaleza, Brazil, 26–28 September 2018. Electronic Notes in Theoretical Computer Science, vol. 344, pp. 119–136. Elsevier (2018). https://doi.org/10.1016/j.entcs.2019.07.008
4. Gunther, E., Pagano, M., Sánchez Terraf, P.: Mechanization of separation in generic extensions. arXiv e-prints arXiv:1901.03313, January 2019
5. Han, J.M., van Doorn, F.: A formalization of forcing and the unprovability of the continuum hypothesis. In: Harrison, J., O'Leary, J., Tolmach, A. (eds.) 10th International Conference on Interactive Theorem Proving (ITP 2019). Leibniz International Proceedings in Informatics (LIPIcs), vol. 141, pp. 19:1–19:19. Schloss Dagstuhl-Leibniz-Zentrum fuer Informatik, Dagstuhl (2019). https://doi.org/10.4230/LIPIcs.ITP.2019.19
6. Han, J.M., van Doorn, F.: A formal proof of the independence of the continuum hypothesis. In: Blanchette, J., Hritcu, C. (eds.) Proceedings of the 9th ACM SIG-PLAN International Conference on Certified Programs and Proofs, CPP 2020, New Orleans, LA, USA, 20–21 January 2020. ACM (2020). https://doi.org/10.1145/3372885
7. Jech, T.: Set Theory. The Millennium Edition. Springer Monographs in Mathematics, 3rd edn. Springer, Heidelberg (2002). Corrected fourth printing (2006)
8. Kunen, K.: Set Theory: An Introduction to Independence Proofs. Studies in Logic and the Foundations of Mathematics. Elsevier Science, Amsterdam (1980)

9. Kunen, K.: Set Theory. Studies in Logic, 2nd edn. College Publications, London (2011). Revised edition (2013)

10. de Moura, L., Kong, S., Avigad, J., van Doorn, F., von Raumer, J.: The Lean theorem prover (system description). In: Felty, A.P., Middeldorp, A. (eds.) CADE 2015. LNCS (LNAI), vol. 9195, pp. 378–388. Springer, Cham (2015). https://doi.org/10.1007/978-3-319-21401-6_26

11. Neeman, I.: Topics in set theory, forcing (2011). Course Lecture Notes https://bit.ly/2ErcbmI

12. Obua, S.: Partizan games in Isabelle/HOLZF. In: Barkaoui, K., Cavalcanti, A., Cerone, A. (eds.) ICTAC 2006. LNCS, vol. 4281, pp. 272–286. Springer, Heidelberg (2006). https://doi.org/10.1007/11921240_19

13. Paulson, L.C.: The foundation of a generic theorem prover. J. Autom. Reason. **5**(3), 363–397 (1989). https://doi.org/10.1007/BF00248324

14. Paulson, L.C. (ed.): Isabelle. LNCS, vol. 828. Springer, Heidelberg (1994). https://doi.org/10.1007/BFb0030541

15. Paulson, L.C.: The relative consistency of the axiom of choice mechanized using Isabelle/ZF. LMS J. Comput. Math. **6**, 198–248 (2003). https://doi.org/10.1112/S1461157000000449

16. Paulson, L.C.: Zermelo Fraenkel set theory in higher-order logic. Archive of Formal Proofs, October 2019. http://isa-afp.org/entries/ZFC_in_HOL.html. Formal proof development

17. Paulson, L.C., Grabczewski, K.: Mechanizing set theory. J. Autom. Reason. **17**(3), 291–323 (1996). https://doi.org/10.1007/BF00283132

18. Wenzel, M., Paulson, L.C., Nipkow, T.: The Isabelle framework. In: Mohamed, O.A., Muñoz, C., Tahar, S. (eds.) TPHOLs 2008. LNCS, vol. 5170, pp. 33–38. Springer, Heidelberg (2008). https://doi.org/10.1007/978-3-540-71067-7_7

19. Werner, B.: Sets in types, types in sets. In: Abadi, M., Ito, T. (eds.) TACS 1997. LNCS, vol. 1281, pp. 530–546. Springer, Heidelberg (1997). https://doi.org/10.1007/BFb0014566

# Reasoning About Algebraic Structures with Implicit Carriers in Isabelle/HOL

Walter Guttmann[✉]

Department of Computer Science and Software Engineering,
University of Canterbury, Christchurch, New Zealand
`walter.guttmann@canterbury.ac.nz`

**Abstract.** We prove Chen and Grätzer's construction theorem for Stone algebras in Isabelle/HOL. The development requires extensive reasoning about algebraic structures in addition to reasoning in algebraic structures. We present an approach for this using classes and locales with implicit carriers. This involves using function liftings to implement some aspects of dependent types and using embeddings of algebras to inherit theorems. We also formalise a theory of filters based on partial orders.

## 1 Introduction

There is an ongoing effort of formalising results from mathematics, computer science and other disciplines using various proof assistants such as ACL2, Agda, Coq, HOL, Isabelle, Lean, Mizar, Nuprl and PVS. These systems differ in the supported logics, type systems, libraries, automation, code generation capabilities and other ways. In this paper we focus on Isabelle/HOL, which is the higher-order logic instance of the generic proof assistant Isabelle [29]. It has very good proof automation facilities, in particular through the Sledgehammer integration of automated theorem provers and SMT solvers [4,30], but a somewhat limited type system, notably lacking dependent types.

Isabelle/HOL has a wide range of libraries that come with the system or the associated Archive of Formal Proofs at https://www.isa-afp.org/. They contain extensive theories of algebraic structures including groups, rings, lattices, Boolean algebras, Kleene algebras and many others. Algebras are frequently implemented in Isabelle/HOL using classes and locales, which offer means to package operations and axioms, arrange them in hierarchies, dynamically inherit results, and exhibit multiple instances [16,23]. Unlike classes, locales support multiple type parameters making them useful for applications such as describing homomorphisms between different structures.

Algebraic hierarchies in Isabelle/HOL come in two flavours: one which makes the carrier sets of the algebras explicit and one which leaves them implicit. The latter assumes a universe of discourse, which all operations, axioms, definitions, theorems and proofs implicitly refer to [16]. The former adds explicit constants for carrier sets; closure properties of operations and membership in the carrier sets must be stated explicitly [2,22].

© Springer Nature Switzerland AG 2020
N. Peltier and V. Sofronie-Stokkermans (Eds.): IJCAR 2020, LNAI 12167, pp. 236–253, 2020.
https://doi.org/10.1007/978-3-030-51054-1_14

Explicit carriers make statements more complex: theories are harder to read and to understand, and additional membership properties may create an overhead for automation [10]. It is therefore not surprising that large hierarchies of algebras have been developed with implicit carriers, which allow for convenient reasoning in algebraic structures. As development progressed over the years issues with this approach when reasoning about algebraic structures have become more apparent. For example, it is challenging to define subalgebras and to impose additional algebraic structure on a subset of an algebra. Two questions arise: to what extent can these issues be overcome while still using implicit carriers? How can hierarchies with implicit carriers be connected to hierarchies with explicit carriers while minimising redevelopment and maintenance efforts?

In the present paper we study Chen and Grätzer's construction theorem for Stone algebras [8] with the aim of providing some answers to the first of these questions. Briefly stated, every Stone algebra $S$ is isomorphic to a triple $(B, D, \varphi)$ comprising a Boolean algebra $B$, a distributive lattice $D$ with a greatest element and a bounded lattice homomorphism $\varphi$ from $B$ to filters of $D$. Stone algebras have applications ranging from topology [21] over rough sets for representing uncertainty [31] to modelling weighted graphs for algorithm verification [14,15]. The above theorem has not been formally proved before, is complex enough to push against some of Isabelle/HOL's limitations, is based on algebraic structures that have been developed using implicit carriers, but requires a significant amount of reasoning about algebras that would benefit from explicit carriers.

We present a proof development of the construction theorem based on a class hierarchy of pseudo-complemented algebras with implicit carriers. The aim is to demonstrate challenges when reasoning about algebras with implicit carriers and ways to deal with them. At the same time we prepare the ground for a proof development connecting algebras with implicit carriers and explicit carriers, which we will use to study the second of the above questions in future work.

The contributions of this paper are:

- A formal proof of Chen and Grätzer's construction theorem for Stone algebras. This result has not been formally proved so far.
- An Isabelle/HOL theory of filters based on partial orders. Existing theories of filters only apply to rings with a carrier and to sets of sets, respectively.
- Examples formalised in Isabelle/HOL of inheriting universal formulas using an embedding of algebras. This technique is well known in universal algebra but has not been formalised before.
- A function lifting technique based on universal algebra to circumvent the need for dependent types. This is a new way to get some aspects of dependent types in Isabelle/HOL.

The findings of this paper can be summarised as follows. Reasoning about algebraic structures can be carried out in Isabelle/HOL to a certain extent using implicit carriers, but would benefit from dependent types beyond. Function liftings can mitigate this problem to a certain extent, but they become complex and unnatural for more involved constructions.

In Sect. 2 we give basic definitions and state the construction theorem for Stone algebras. Section 3 discusses our theory of filters based on partial orders. The proof of the construction theorem for Stone algebras is described in Sect. 4. Sections 4.1 and 4.2 construct a triple from a Stone algebra and a Stone algebra from a triple. Sections 4.3 and 4.4 show that these constructions are mutually inverse up to isomorphism. The function lifting and embedding techniques are described in Sect. 4.2. Section 5 puts this work into context. The Isabelle/HOL theory files containing the results of this paper are available in the Archive of Formal Proofs [13].

## 2   Construction Theorem for Stone Algebras

This section states Chen and Grätzer's construction theorem for Stone algebras [8] and presents the necessary algebraic structures. In particular we discuss orders, lattices, pseudocomplemented algebras, homomorphisms, filters and triples. Further details about lattices and pseudocomplemented algebras can be found, for example, in Blyth's textbook [5].

A *partial order* $\sqsubseteq$ on a set $S$ is a reflexive, transitive and antisymmetric relation on $S$:

$$x \sqsubseteq x \qquad x \sqsubseteq y \wedge y \sqsubseteq z \Rightarrow x \sqsubseteq z \qquad x \sqsubseteq y \wedge y \sqsubseteq x \Rightarrow x = y$$

A *lattice* is a partially ordered set $(S, \sqsubseteq)$ where any two elements $x, y \in S$ have a least upper bound or join $x \sqcup y$ and a greatest lower bound or meet $x \sqcap y$:

$$x \sqsubseteq x \sqcup y \qquad y \sqsubseteq x \sqcup y \qquad x \sqsubseteq z \wedge y \sqsubseteq z \Rightarrow x \sqcup y \sqsubseteq z$$
$$x \sqcap y \sqsubseteq x \qquad x \sqcap y \sqsubseteq y \qquad z \sqsubseteq x \wedge z \sqsubseteq y \Rightarrow z \sqsubseteq x \sqcap y$$

The operations $\sqcup$ and $\sqcap$ are associative, commutative, idempotent and $\sqsubseteq$-isotone. The absorption laws $x \sqcup (x \sqcap y) = x = x \sqcap (x \sqcup y)$ hold. The order $\sqsubseteq$ is connected to join and meet by $x \sqcup y = y \Leftrightarrow x \sqsubseteq y \Leftrightarrow x \sqcap y = x$.

Equivalently, a lattice can be constructed from two operations $\sqcup$ and $\sqcap$ that are associative, commutative and satisfy the absorption laws. The relation $\sqsubseteq$ defined by either of the connection laws $x \sqsubseteq y \Leftrightarrow x \sqcup y = y$ or $x \sqsubseteq y \Leftrightarrow x \sqcap y = x$ and the operations $\sqcup$ and $\sqcap$ satisfy all properties of a lattice stated above.

A lattice is *bounded* if it has a least element $\bot$ and a greatest element $\top$:

$$\bot \sqsubseteq x \qquad\qquad\qquad x \sqsubseteq \top$$

These axioms are equivalent to $\bot \sqcup x = x = \top \sqcap x$.

A lattice is *distributive* if the following axioms hold:

$$x \sqcup (y \sqcap z) = (x \sqcup y) \sqcap (x \sqcup z) \qquad x \sqcap (y \sqcup z) = (x \sqcap y) \sqcup (x \sqcap z)$$

Either of these axioms implies the other in a lattice.

A *(distributive) p-algebra* [5] is a bounded (distributive) lattice with a unary pseudocomplement operation $^-$:

$$x \sqcap y = \bot \Leftrightarrow x \sqsubseteq \overline{y}$$

The pseudocomplement $\overline{y}$ of an element $y$ is the $\sqsubseteq$-greatest element whose meet with $y$ is $\bot$. An element $x$ in a p-algebra is *regular* if $\overline{\overline{x}} = x$ and *dense* if $\overline{x} = \bot$. Equivalently, a p-algebra is a bounded lattice with an operation $^-$ satisfying $\overline{\bot} = \top$ and $\overline{\top} = \bot$ and $x \sqcap \overline{x \sqcap y} = x \sqcap \overline{y}$.

A *Stone algebra* is a distributive p-algebra satisfying the following equation:

$$\overline{x} \sqcup \overline{\overline{x}} = \top$$

A *Boolean algebra* is a bounded distributive lattice with a complement $^-$:

$$x \sqcup \overline{x} = \top \qquad\qquad x \sqcap \overline{x} = \bot$$

Equivalently, a Boolean algebra is a Stone algebra whose elements are all regular. An example of a Stone algebra which is not a Boolean algebra is the three-element chain $\{\bot, a, \top\}$ where $\bot \sqsubseteq a \sqsubseteq \top$ and $\overline{\bot} = \top$ and $\overline{a} = \overline{\top} = \bot$.

A *bounded-lattice homomorphism* is a function $f : A \to B$ from a bounded lattice $A$ to a bounded lattice $B$ preserving join, meet, the least element and the greatest element:

$$f(x \sqcup_A y) = f(x) \sqcup_B f(y) \qquad\qquad f(\bot_A) = \bot_B$$
$$f(x \sqcap_A y) = f(x) \sqcap_B f(y) \qquad\qquad f(\top_A) = \top_B$$

A subset $X \subseteq S$ of a partially ordered set $(S, \sqsubseteq)$ is *up-closed* if all elements of $S$ above any element of $X$ are in $X$:

$$\forall x \in X : \forall y \in S : x \sqsubseteq y \Rightarrow y \in X$$

The set $X \subseteq S$ is *downward directed* if any two elements of $X$ have a lower bound in $X$:

$$\forall x, y \in X : \exists z \in X : z \sqsubseteq x \wedge z \sqsubseteq y$$

A *filter* of $S$ is a non-empty, downward directed, up-closed subset of $S$.

We give a general result about filters, which is necessary for the subsequent definition to make sense. Let $D$ be a distributive lattice with a greatest element $\top_D$. Then the filters of $D$ form a bounded distributive lattice $F(D)$ where the join of two filters $X$ and $Y$ is

$$X \sqcup Y = \{z \in D \mid \exists x \in X : \exists y \in Y : x \sqcap_D y \sqsubseteq_D z\},$$

meet is intersection, the greatest element is $D$, the least element is $\{\top_D\}$ and the lattice order is the subset order.

Following Chen and Grätzer [8], we use *triple* as a technical term rather than just for a collection of three components. A *triple* $(B, D, \varphi)$ comprises a Boolean algebra $B$, a distributive lattice $D$ with a greatest element, and a bounded-lattice homomorphism $\varphi : B \to F(D)$. Triples $(B_1, D_1, \varphi_1)$ and $(B_2, D_2, \varphi_2)$ are *isomorphic* if there is an isomorphism $b : B_1 \to B_2$ of Boolean algebras $B_1$ and $B_2$, and an isomorphism $d : D_1 \to D_2$ of lattices $D_1$ and $D_2$ with greatest elements such that $\varphi_2(b(x)) = d'(\varphi_1(x))$, where $d'(X) = \{d(x) \mid x \in X\}$ applies $d$ to all elements of $X$.

· In this paper we formally prove the construction theorem for Stone algebras [8,25], by which we understand the following collection of results:

1. Let $S$ be a Stone algebra. Consider the set $B = \{x \in S \mid \overline{\overline{x}} = x\}$ of regular elements of $S$, the set $D = \{x \in S \mid \overline{x} = \bot\}$ of dense elements of $S$, and the function $\varphi : B \to 2^D$ defined by $\varphi(x) = \{y \in D \mid \overline{x} \sqsubseteq_S y\}$. Then $(B, D, \varphi)$ is a triple, called the triple associated with $S$, where $B$ forms a subalgebra of $S$ and $D$ forms a subalgebra of the reduct of $S$ to $\sqcup, \sqcap$ and $\top$.
2. Let $(B, D, \varphi)$ be a triple. Consider the set

$$S = \{(x, Y) \in B \times F(D) \mid \exists z \in D : Y = \varphi(\overline{x}) \sqcup_{F(D)} {\uparrow}z\}$$

where ${\uparrow}z = \{y \in D \mid z \sqsubseteq_D y\}$ is the up-closure of $z$ in $D$. Then $S$ is a Stone algebra where

$$(x, Y) \sqsubseteq_S (z, W) \Leftrightarrow x \sqsubseteq_B z \wedge W \sqsubseteq_{F(D)} Y$$
$$\overline{(x, Y)} = (\overline{x}, \varphi(x))$$

It is called the Stone algebra associated with $(B, D, \varphi)$.
3. The Stone algebra associated with the triple $(B, D, \varphi)$ associated with a Stone algebra $S$ is isomorphic to $S$.
4. The triple associated with the Stone algebra $S$ associated with a triple $(B, D, \varphi)$ is isomorphic to $(B, D, \varphi)$.

There are a number of construction theorems using triples for similar algebraic structures such as Heyting semilattices and pseudocomplemented distributive lattices [24, 26, 28].

## 3   An Isabelle/HOL Theory of Filters Based on Orders

The construction theorem for Stone algebras is based on lattices, pseudocomplemented algebras and filters. In this section, we discuss the extent to which these foundations are supported in Isabelle/HOL.

Prior to this work, Isabelle/HOL had libraries for lattices and filters, but not for pseudocomplemented algebras. We reused the available theories for lattices. Only small extensions were necessary, in particular, for directed semilattices, bounded distributive lattices and lattice homomorphisms.

The available theories for filters could not be reused as they are too specific. The theory HOL/Algebra/Ideal.thy defines ring-theoretic ideals in locales with a carrier set. In the theory HOL/Filter.thy a filter is defined as a set of sets.

Filters based on orders and lattices abstract from the inner set structure; this approach is used in many texts [1,3,5,9,11]. Moreover, it is required for the construction theorem of Stone algebras, whence we implemented filters this way.

While theories were available for Boolean algebras, they did not cover pseudo-complemented algebras, which have the same signature but weaker axioms. We therefore developed a theory covering p-algebras, distributive p-algebras, Stone algebras, Heyting semilattices, Heyting algebras, Brouwer algebras and additional results for Boolean algebras. This theory has been used independently for modelling weighted graphs and verifying minimum spanning tree algorithms and has been described in this context [14,15].

In the remainder of this section, we describe our new theory of filters in more detail. It is structured by the assumptions on the underlying order. We consider filters based on partial orders, semilattices, lattices and distributive lattices. The following is a selection of results proved in this theory:

1. We generalise the ultrafilter lemma [9, Theorem 10.17] to orders with a least element. A *proper filter* of $S$ is a filter of $S$ that is different from $S$. An *ultrafilter* of $S$ is a $\subseteq$-maximal proper filter of $S$. The ultrafilter lemma states that every proper filter of $S$ is a subset of an ultrafilter of $S$.
   Actually, our proof does not need that $\sqsubseteq$ is a partial order, but also works if $\sqsubseteq$ is an arbitrary relation satisfying $\perp \sqsubseteq x$ for some element $\perp$ and all elements $x$ of the algebra (defining filters as in Sect. 2 but for arbitrary $\sqsubseteq$). The proof uses Isabelle/HOL's Zorn_Lemma, which requires closure under union of arbitrary (possibly empty) chains of sets.
2. We study the lattice structure of filters. A *meet-semilattice* is a partially ordered set $(S, \sqsubseteq)$ where all $x, y \in S$ have a greatest lower bound $x \sqcap y$. A meet-semilattice is *directed* if any two elements have an upper bound. A meet-semilattice is *bounded* if it has a greatest element. The results state:
   (a) The set of filters where the underlying order is a directed meet-semilattice forms a lattice with a greatest element.
   (b) The set of filters over a bounded meet-semilattice forms a bounded lattice.
   (c) The set of filters over a distributive lattice with a greatest element forms a bounded distributive lattice.
3. We connect ultrafilters and prime filters [9, Theorem 10.11]. A *prime filter* of $S$ is a proper filter $X$ of $S$ where $x \sqcup y \in X$ implies $x \in X$ or $y \in X$ for all $x, y \in S$. The result shows that in a distributive lattice every ultrafilter is a prime filter. The lattice does not need to be bounded [9, p. 234].
4. We prove a result about principal filters [12, Lemma II]. A *principal filter* of $S$ is a filter $X$ of $S$ such that $X = \uparrow x$ for some $x \in S$ where $\uparrow x = \{y \in S \mid x \sqsubseteq y\}$. The result shows that in a distributive lattice, if both join and meet of two filters are principal filters, both filters are principal filters.

## 4    The Construction Theorem for Stone Algebras

In this section, we describe the proof of the construction theorem for Stone algebras in Isabelle/HOL. Section 4.1 constructs the triple associated with a Stone algebra. Section 4.2 constructs the Stone algebra associated with a triple. It describes the function lifting technique for dependent types and the embedding technique, both based on universal algebra. Section 4.3 shows that the first construction followed by the second construction gives back the original Stone algebra up to isomorphism. Section 4.4 shows that the second construction followed by the first construction gives back the original triple, again up to isomorphism.

## 4.1  Constructing a Triple from a Stone Algebra

The first set of results concerns the construction of a triple from a Stone algebra $S$. Specifically, we show:

1. The regular elements of $S$ form a Boolean algebra $B$ that is a subalgebra of $S$.
2. The dense elements of $S$ form a distributive lattice $D$ with a greatest element, which is a subalgebra of the reduct of $S$ to $\sqcup$, $\sqcap$ and $\top$.

As shown in Sect. 3, it follows that the set of filters $F(D)$ of the dense elements of $S$ forms a bounded distributive lattice. Considering the function $\varphi : B \to 2^D$ defined by $\varphi(x) = \{y \in D \mid \overline{x} \sqsubseteq_S y\}$, we show:

3. $\varphi$ maps every regular element to a filter of $D$.
4. $\varphi$ is a bounded-lattice homomorphism from $B$ to $F(D)$.

Hence $(B, D, \varphi)$ is a triple.

We have implemented Stone algebras using classes in Isabelle/HOL, situating them between lattices and Boolean algebras in the existing class hierarchy provided by `HOL/Lattices.thy`. Every class has a single type parameter, which represents the carrier set of an algebra and is left implicit. For every operation of an algebra there is an additional class parameter. Axioms for these operations are provided as statements assumed to hold in the context of the class but not outside. Classes can be instantiated by providing a particular type, appropriate operations and proofs of these assumptions.

Using classes to implement algebraic structures makes it easy to extend the hierarchy. For example, p-algebras are defined as a subclass of the existing class for bounded lattices, extending the latter by a unary pseudocomplement operation satisfying the appropriate axiom:

**class** p_algebra = bounded_lattice + uminus +
   **assumes** pseudo_complement: "$x \sqcap y = \bot \longleftrightarrow x \sqsubseteq -y$"

Similarly, Stone algebras are introduced as a subclass of distributive p-algebras, which are introduced as a subclass of p-algebras. On top of that, the existing class for Boolean algebras forms a subclass of the new class for Stone algebras, which is proved as follows:

**context** boolean_algebra **begin**
   **subclass** stone_algebra
     – proof of axioms pseudo_complement and $\overline{x} \sqcup \overline{\overline{x}} = \top$ (omitted)

Existing subclass relationships are taken into account which avoids the need to repeat proofs. For example, the axioms of bounded distributive lattices, which are necessary for Stone algebras, follow automatically since the class for Boolean algebras is a subclass of the classes for bounded lattices and distributive lattices.

Reasoning in algebraic structures and proving subclass relationships of entire algebras is well supported this way. For the Stone construction theorem, however, we need to prove that the *subset* of regular elements forms a Boolean algebra. This cannot be done using the subclass mechanism in a class as it implicitly refers to the entire carrier sets of the related algebras. We therefore introduce a new type corresponding to the set of regular elements. New types have to be introduced outside a class at a global level, since otherwise they could refer to class parameters making them dependent types, which are not supported by HOL:

**typedef** 'a regular = "$\{x::('a::stone\_algebra) . x = --x\}$" **by** auto

Here 'a is a type parameter, which is constrained to being a subclass of the class for Stone algebras, and every element $x$ of the set underlying the new type is restricted to have type 'a. New type definitions require a proof that the set they are derived from is not empty, which is discharged by the auto proof method since any Stone algebra contains $\bot$ and $\top$. Such a type definition automatically introduces representation and abstraction functions between the new type and the set it is derived from, and automatically derives basic theorems about these functions including their inverse relationship. We then show that this type instantiates the class for Boolean algebras:

**instantiation** regular :: (stone_algebra) boolean_algebra **begin**
  – definitions of Boolean algebra operations and proofs of axioms (omitted)

The parentheses indicate the subclass constraint required for the type parameter, that is, the kind of algebra from which the regular elements are taken. A stronger constraint than required by the type can be provided in such an instantiation; for example, this is used to characterise the structure formed by the set of filters over various kinds of semilattice as mentioned in Sect. 3. The operations on the new type are derived from the operations on Stone algebras using the representation and abstraction functions of the type. Working with new types introduced like this often requires handling the representation and abstraction functions, which clutter definitions, statements and proofs. Some of this can be hidden using mechanisms from Isabelle/HOL's Lifting package [18].

Similarly, we introduce a new type for the dense elements of a Stone algebra and show that it forms a distributive lattice with a greatest element. We also introduce a new type for the filters of dense elements and show that it forms a bounded distributive lattice. The proofs of these instances are simple as the operations are derived from the underlying algebras without changes. The main work is proving that filters over a distributive lattice with a greatest element form a bounded distributive lattice, which was done generally in Sect. 3.

We next construct the function $\varphi$ mapping regular elements to sets of dense elements, where $\Rightarrow$ is the function type constructor:

**definition** stone_phi :: "'a::stone_algebra regular ⇒ 'a dense filter"
  **where** "stone_phi $x$ = Abs_filter $\{y \,.\, -\text{Rep\_regular } x \sqsubseteq \text{Rep\_dense } y\}$"

The representation functions convert elements $x$ and $y$ from the regular and dense types to the underlying Stone algebra, where they can be compared. The set of dense elements satisfying the given property is converted to a filter using the abstraction function of this type.

A triple consists of a Boolean algebra, a distributive lattice with a greatest element, and a structure map. The Boolean algebra and the distributive lattice are represented as HOL types with appropriate subclass constraints. Because both occur in the type of the structure map, the triple is determined simply by the structure map and its HOL type. The structure map needs to be a bounded lattice homomorphism. This information is collected in the following locale:

**locale** triple =
  **fixes** phi :: "'a::boolean_algebra ⇒ 'b::distrib_lattice_top filter"
  **assumes** hom: "bounded_lattice_homomorphism phi"

Unlike classes, locales support multiple type parameters such as 'a and 'b used here. Another difference is that a locale can have multiple instances for the same type. On the other hand, different instances of a class can share the same notation for an operation, which is closer to mathematical usage. It remains to show that $\varphi$ is indeed a bounded lattice homomorphism:

**interpretation** stone_phi: triple "stone_phi"
  – proof of homomorphism properties for ⊔, ⊓, ⊥, ⊤ (omitted)

The proof is cluttered with occurrences of the representation and abstraction functions for the types of regular and dense elements. The preservation of ⊔ has to be broken down into quite small steps, whereas Sledgehammer [30] is more helpful with automating the preservation of the other three operations. This is partly because the join of filters has the most complex definition, but the representation and abstraction functions cause further overhead.

Referees of this paper suggested to use the Lifting package to hide representation and abstraction functions in the definition of stone_phi and in other definitions in the following sections. The author has tried this but soon ran into problems which could not be solved during revision of the paper. One of the issues appeared to be the lack of suitable transfer rules for the composite type "'a dense filter" even though transfer rules are automatically generated for the dense and filter type constructors.

## 4.2   Constructing a Stone Algebra from a Triple

Next, from a triple $(B, D, \varphi)$ such that $B$ is a Boolean algebra, $D$ is a distributive lattice with a greatest element and $\varphi : B \to F(D)$ is a bounded lattice homomorphism, we construct a Stone algebra $S$. The elements of $S$ are pairs taken from $B \times F(D)$ following the construction of Katriňák [25]. This set and the operations making it a Stone algebra can be defined in the locale for triples:

**context** triple **begin**
  **definition** pairs :: "('a × 'b filter) set"
    **where** "pairs = $\{(x,y)\ .\ \exists z\ .\ y = \mathrm{phi}(-x) \sqcup \mathrm{Abs\_filter}(\uparrow z)\}$"
  **fun** pairs_uminus :: "('a × 'b filter) $\Rightarrow$ ('a × 'b filter)"
    **where** "pairs_uminus $(x,y) = (-x, \mathrm{phi}(x))$"
  **fun** pairs_sup :: "('a × 'b filter) $\Rightarrow$ ('a × 'b filter) $\Rightarrow$ ('a × 'b filter)"
    **where** "pairs_sup $(x,y)\ (z,w) = (x \sqcup z, y \sqcap w)$"
  – definitions of further operations on pairs (omitted)

We need to represent the set of pairs as a type to be able to instantiate the Stone algebra class. Because the definition of this set depends on $\varphi$, which is a parameter of the triple locale, this would require dependent types. Since Isabelle/HOL does not have dependent types, we use a function lifting instead (which is unrelated to the Lifting package). Similarly to the 'lambda lifting' technique in functional programming [20], our function lifting makes a conceptually local entity global by capturing its free variables as parameters. However, in our case the result is a global type not a global function.

We initially describe the process for the application at hand. Because it applies more generally we summarise the ideas at the end of this section.

We first define a type to capture the parameter $\varphi$ of the triple locale. This parameter is the structure map that occurs in the definition of the set of pairs. The set of all structure maps is the set of all bounded lattice homomorphisms (of appropriate type):

**typedef** ('a,'b) phi = "$\{f::('a::\mathrm{non\_trivial\_boolean\_algebra} \Rightarrow$
  'b::distrib_lattice_top filter) . bounded_lattice_homomorphism $f\}$"
  – proof that the type is not empty (omitted)

In order to make the set a HOL type, we need to show that at least one such structure map exists. To this end we use the ultrafilter lemma shown in Sect. 3: the required bounded lattice homomorphism is essentially the characteristic map of an ultrafilter, but the latter must exist. In particular, the underlying Boolean algebra must contain at least two elements, which we guarantee by introducing a suitable subclass of the class for Boolean algebras.

We then implement the type that represents the set of pairs depending on structure maps. It uses functions from structure maps to pairs with the requirement that, for each structure map, the corresponding pair is contained in the set of pairs constructed for a triple with that structure map:

**typedef** ('a,'b) lifted_pair = "$\{p::('a::\mathrm{non\_trivial\_boolean\_algebra},$
  'b::distrib_lattice_top) phi $\Rightarrow$ ('a × 'b filter) . $\forall f\ .\ p(f) \in \mathrm{triple.pairs}\,(\mathrm{Rep\_phi}\,f)\}$"

If this type could be defined in the locale triple and instantiated to Stone algebras in this locale, there would be no need for the lifting and we could work with triples directly. Since the type needs to be defined outside the triple locale at global level, we supply the type parameter (Rep_phi $f$) when referring to the set of pairs defined in the locale. The function lifting allows us to express the dependence on the locale parameter at the type level.

The lifted pairs form a Stone algebra, where the operations are lifted point-wise from pairs to functions:

**instantiation** lifted_pair :: (non_trivial_boolean_algebra,distrib_lattice_top)
  stone_algebra **begin**
  – definitions of Stone algebra operations and proof of axioms (omitted)

The proofs of the Stone algebra axioms are again quite low-level as a consequence of having to deal with the function lifting in addition to various representation functions. Apart from these technical issues, at this stage the development deviates from the original statements. We are here constructing a Stone algebra of functions from structure maps to pairs, whereas the original construction yields a Stone algebra of pairs for any given structure map. However, any Stone algebra of pairs obtained for a given structure map is isomorphic to a subalgebra of the Stone algebra of functions. While this relationship cannot be expressed directly as it would again require dependent types, we can prove a special case of it.

To this end, we specialise the construction to start with the triple associated with a Stone algebra, that is, the triple obtained in Sect. 4.1. For that particular structure map stone_phi (as for any other particular structure map) the resulting type of pairs is no longer a dependent type. It is just the set of pairs obtained for the given structure map:

**typedef** 'a stone_phi_pair =
  "triple.pairs (stone_phi::('a::stone_algebra regular $\Rightarrow$ 'a dense filter))"

It could be proved directly that this type is a Stone algebra. To demonstrate how a technique of universal algebra can be realised in Isabelle/HOL, we choose a different approach: we embed the type of pairs into the lifted type. The embedding injects a pair $x$ into a function as the value at the given structure map; this makes the embedding injective. The value of the function at any other structure map is carefully chosen to make the resulting function a Stone algebra homomorphism. We use $\overline{\overline{x}}$, which is essentially a projection to the regular element component of $x$, whence the range of $\lambda x.\overline{\overline{x}}$ has the structure of a Boolean algebra:

**fun** stone_phi_embed :: "'a::non_trivial_stone_algebra stone_phi_pair $\Rightarrow$
  ('a regular,'a dense) lifted_pair"
  **where** "stone_phi_embed $x$ = Abs_lifted_pair ($\lambda f$ .
    **if** Rep_phi $f$ = stone_phi **then** Rep_stone_phi_pair $x$
      **else** triple.pairs_uminus (Rep_phi $f$) (triple.pairs_uminus (Rep_phi $f$)
      (Rep_stone_phi_pair $x$)))"

Again, since we reason outside the triple locale at a global level, we supply the locale parameter, in this case to both occurrences of the pseudocomplement operation triple.pairs_uminus.

We then show that stone_phi_embed is an embedding, that is, it preserves $\sqcup$, $\sqcap$, $\bot$, $\top$, $^-$ and it is injective. Hence all Stone algebra axioms can be inherited using the embedding. This is because the axioms are universal formulas, that is, first-order formulas in prenex form where all quantifiers are universal [6]. We also show that stone_phi_embed is an order-isomorphism, which allows us to inherit inequalities without transforming them to equations. It follows that the pairs form a Stone algebra:

**instantiation** stone_phi_pair :: (non_trivial_stone_algebra) stone_algebra **begin**
– definitions of Stone algebra operations and proof of axioms (omitted)

When proving the Stone algebra axioms, Sledgehammer automatically finds proofs using the embedding property of stone_phi_embed and the corresponding axioms of the underlying Stone algebra.

**Generalising the Function-Lifting Construction.** We discuss the ideas of this section as more general recipes. Mathematically speaking we wish to show that a set $S$ depending on a parameter $p$ forms an algebra. For instance, assume the algebra has an operation $F : S \to S \to S$ and a relation $R : S \to S \to$ bool, whose definitions may also depend on $p$.

In Isabelle/HOL, we have a locale $L$ with parameter $p$ of type $A$. In this locale, we define the set $S$ of elements of type $B$ and show $S$ is not empty. We also define $F$ and $R$ and show they satisfy the axioms of the algebra. However, we cannot instantiate the class that implements the algebra. This requires a type instead of the set $S$ and the type cannot be defined in $L$ where it might depend on $p$.

We therefore simulate a dependent type outside $L$. Observe that for any $p$ satisfying the assumptions of $L$, the locale yields a set $S_p$; in Isabelle/HOL it is obtained by $L.S\ p$. We construct the (infinite) direct product of these sets $\prod_p S_p$. We represent each value in this product by a function indexed with $p$.

Technically, we first create a type $T'$ from the set of all possible values of the locale parameter $p$. This set contains the elements of type $A$ subject to assumptions about $p$ in $L$; it is not empty (or else the locale could not be instantiated). We then create the type $T$ of functions $f : T' \to B$ such that $f(p) \in S_p$ for all $p \in T'$. This is the direct product; it exists because no $S_p$ is empty.

We show that $T$ instantiates the algebra where $F$ and $R$ are lifted pointwise from $S$ to $T$. Specifically, $F_T\ f\ g = \lambda p \ . \ F\ (f\ p)\ (g\ p)$ is the lifting of $F$ and $R_T\ f\ g = \forall p \ . \ R\ (f\ p)\ (g\ p)$ is the lifting of $R$. Because the constituent sets $S_p$ satisfy the axioms of the algebra so does their direct product $T$ in many cases. In particular, direct products preserve universal Horn formulas [17]. Hence it suffices if all axioms are universally quantified conditional equations. This works not just for Stone algebras but for many others such as groups, rings, fields, lattices, Boolean algebras, dioids, Kleene algebras and action algebras.

We next specialise the product to a given value $q$ for the locale parameter. That is, we create a type $T_q$ for the set $S_q$ and show that it instantiates the algebra. To this end, we embed $T_q$ in $T$ using a function $H : T_q \to T$. Specifically, $H\ x = \lambda p \ .$ if $p = q$ then $x$ else $h\ p\ x$, where $h\ p$ is any homomorphism from $S_q$ to $S_p$. The latter condition for $h$ is sufficient for $H$ to be an embedding. While we do not give a general scheme for how to obtain $h$, we note that all algebras $S_p$ are defined using the same pattern, which can help. If a suitable $h$ is available, the axioms of the algebra can be derived for $T_q$ via the embedding $H$. This works for arbitrary universal formulas, which covers even more algebras than listed above.

We explain how to derive universal formulas via embeddings using universal algebra [6]. Let $X$ and $Y$ be algebras with the same signature and let $e$ be an embedding of $X$ into $Y$. Let $U$ be the universe of $X$. Induction over the structure of terms yields $e(t^X(x_1,\ldots,x_n)) = t^Y(e(x_1),\ldots,e(x_n))$ for all terms $t$ over the signature (with interpretations $t^X, t^Y$ in $X, Y$) and all $x_i \in U$. Since $e$ is injective, $t^X(x_1,\ldots,x_n) = s^X(x'_1,\ldots,x'_m) \Leftrightarrow t^Y(e(x_1),\ldots,e(x_n)) = s^Y(e(x'_1),\ldots,e(x'_m))$ for all terms $s,t$ and all $x_i, x'_j \in U$. Consider a formula $P$ with $k$ free variables, which is constructed by combining such equalities between terms with propositional connectives. Induction over the structure of $P$ yields $P^X(x_1,\ldots,x_k) \Leftrightarrow P^Y(e(x_1),\ldots,e(x_k))$ for all $x_i \in U$. Hence, if $\forall y_1,\ldots,y_k : P^Y(y_1,\ldots,y_k)$ holds, so does $\forall x_1,\ldots,x_k : P^X(x_1,\ldots,x_k)$. This argument generalises from algebras to first-order structures (with relations in the signature). It does not extend to existential quantifiers as the asserted element may lie outside the range of $e$.

### 4.3    The Stone Algebra of a Triple of a Stone Algebra

Next, we show that the Stone algebra constructed in Sect. 4.2 from the triple constructed in Sect. 4.1 from a Stone algebra $S$ is isomorphic to $S$. We give explicit mappings in both directions:

**abbreviation** sa_iso :: "'a::non_trivial_stone_algebra $\Rightarrow$ 'a stone_phi_pair"
    **where** "sa_iso = $\lambda x$ . Abs_stone_phi_pair (Abs_regular $(--x)$,
      stone_phi (Abs_regular $(-x)$) $\sqcup$ Abs_filter ($\uparrow$ Abs_dense $(x \sqcup -x)$)))"

**abbreviation** sa_iso_inv :: "'a::non_trivial_stone_algebra stone_phi_pair $\Rightarrow$ 'a"
    **where** "sa_iso_inv = $\lambda p$ . Rep_regular (fst (Rep_stone_phi_pair $p$)) $\sqcap$
      Rep_dense (triple.rho_pair stone_phi (Rep_stone_phi_pair $p$))"

Without the necessary representation and abstraction functions, the first mapping is $\lambda x.(\overline{\overline{x}}, \varphi(\overline{x}) \sqcup \uparrow(x \sqcup \overline{x}))$. The second of the above mappings extracts from a pair a dense element using the following function defined in the locale triple:

**fun** rho_pair :: "'a $\times$ 'b filter $\Rightarrow$ 'b"
    **where** "rho_pair $(x,y)$ = (SOME $z$ . Abs_filter $(\uparrow z)$ = phi$(x) \sqcap y$)"

The Hilbert choice construct SOME $z$ . $P(z)$ yields some element $z$ that satisfies $P(z)$. This works because the intersection of $\varphi(x)$ with a principal filter is a principal filter, which we prove using a result shown in Sect. 3 [12, Lemma II].

We then show that sa_iso_inv and sa_iso are mutually inverse and that sa_iso is a homomorphism of Stone algebras. The proofs of these results are cluttered with representation and abstraction functions as the above definitions indicate.

### 4.4    The Triple of a Stone Algebra of a Triple

Finally, we show that the triple constructed in Sect. 4.1 from the Stone algebra constructed in Sect. 4.2 from a triple $(B, D, \varphi)$ is isomorphic to $(B, D, \varphi)$. This requires an isomorphism of Boolean algebras, an isomorphism of distributive lattices with a greatest element, and a commuting diagram involving the structure maps. We give explicit mappings of the Boolean algebra isomorphism and the distributive lattice isomorphism in both directions.

We first define and prove the isomorphism of Boolean algebras. Because the Stone algebra of a triple is implemented as a lifted pair, we also lift the Boolean algebra using a parameter of the same type as that used by the lifted pairs:

**typedef** ('a,'b) lifted_boolean_algebra =
"{$f$::(('a::non_trivial_boolean_algebra,'b::distrib_lattice_top) phi $\Rightarrow$ 'a).True}"

The resulting function type forms a Boolean algebra with operations lifted pointwise:

**instantiation** lifted_boolean_algebra ::
(non_trivial_boolean_algebra,distrib_lattice_top) boolean_algebra

We can now define the mappings between the two lifted structures:

**abbreviation** ba_iso :: "('a::non_trivial_boolean_algebra,'b::distrib_lattice_top) lifted_pair regular $\Rightarrow$ ('a,'b) lifted_boolean_algebra"
**where** "ba_iso = $\lambda p$ . Abs_lifted_boolean_algebra ($\lambda f$ .
fst (Rep_lifted_pair (Rep_regular $p$) $f$))"

**abbreviation** ba_iso_inv :: "('a::non_trivial_boolean_algebra,
'b::distrib_lattice_top) lifted_boolean_algebra $\Rightarrow$ ('a,'b) lifted_pair regular"
**where** "ba_iso_inv = $\lambda x$ . Abs_regular (Abs_lifted_pair ($\lambda f$ .
(Rep_lifted_boolean_algebra $x$ $f$,
Rep_phi $f$ ($-$Rep_lifted_boolean_algebra $x$ $f$))))"

We then show that ba_iso_inv and ba_iso are mutually inverse and that ba_iso is a homomorphism of Boolean algebras.

We carry out a similar development for the isomorphism of distributive lattices with greatest elements. Again, the original distributive lattice with a greatest element needs to be lifted to match the lifted pairs. The resulting function type forms a distributive lattice with a greatest element with operations lifted pointwise. The mappings between the two lifted structures are:

**abbreviation** dl_iso :: "('a::non_trivial_boolean_algebra,'b::distrib_lattice_top) lifted_pair dense $\Rightarrow$ ('a,'b) lifted_distrib_lattice_top"
**where** "dl_iso = $\lambda p$ . Abs_lifted_distrib_lattice_top (get_dense $p$)"

**abbreviation** dl_iso_inv :: "('a::non_trivial_boolean_algebra,
'b::distrib_lattice_top) lifted_distrib_lattice_top $\Rightarrow$ ('a,'b) lifted_pair dense"
**where** "dl_iso_inv = $\lambda x$ . Abs_dense (Abs_lifted_pair ($\lambda f$ .
($\top$,Abs_filter($\uparrow$ Rep_lifted_distrib_lattice_top $x$ $f$))))"

The first mapping uses the following function to extract the least element of the filter of a dense pair, which turns out to be a principal filter:

**fun** get_dense :: "('a::non_trivial_boolean_algebra,'b::distrib_lattice_top) lifted_pair dense $\Rightarrow$ ('a,'b) phi $\Rightarrow$ 'b"
**where** "get_dense $p$ $f$ = (SOME $z$ . Rep_lifted_pair (Rep_dense $p$) $f$ = ($\top$,Abs_filter($\uparrow z$)))"

We then show that dl_iso_inv and dl_iso are mutually inverse and that dl_iso preserves ⊔, ⊓ and ⊤.

We finally show that the isomorphisms are compatible with the structure maps. This involves lifting the distributive lattice isomorphism to filters of distributive lattices as these are the targets of the structure maps. To this end, we show that the lifted isomorphism preserves filters. The compatibility of isomorphisms states that the same result is obtained in two ways by starting with a regular lifted pair $p$:

- apply the Boolean algebra isomorphism to the pair; then apply a structure map $f$ to obtain a filter of dense elements; or,
- apply the structure map stone_phi to the pair; then apply the distributive lattice isomorphism lifted to the resulting filter.

This commutativity property is formally stated as follows:

**lemma** phi_iso: "Rep_phi $f$ (Rep_lifted_boolean_algebra (ba_iso $p$) $f$) =
    Abs_filter (($\lambda q$ . Rep_lifted_distrib_lattice_top (dl_iso $q$) $f$) '
    Rep_filter (stone_phi $p$))"

Here $g$ ' $X$ is the union of the sets $g(x)$ taken over all $x \in X$.

Apart from the many representation and abstraction functions occurring in these definitions and proofs, the development again deviates from the original statements. We have to artificially lift the constituent algebras of triples – a Boolean algebra and a distributive lattice with a greatest element – to functions. This is necessary in order to establish the isomorphisms to lifted pairs, which are parameterised in $\varphi$. The same parameter $\varphi$ is introduced for the Boolean algebra and the distributive lattice, even though it is not needed for these, simply to get a matching cardinality for the isomorphism. This is a flow-on effect from lifting to functions in Sect. 4.2 due to the lack of dependent types there.

## 5    Discussion

We put this work into context by discussing a number of questions.

Are the encountered issues self-inflicted and could they have been prevented by choosing other proof assistants? Possibly; for example, some proof assistants such as Agda, Coq and Lean support dependent types. Aspects of universal algebra have been formalised in Coq [7,32]. The present paper does not aim to find the 'best' system for reasoning about algebraic structures. Proof assistants differ in many dimensions; a particular choice will typically involve trade-offs. There may be various reasons (such as external requirements, existing libraries, automation support) for choosing Isabelle/HOL despite the fact it does not support dependent types. By studying a sufficiently complex example, this paper provides a data point to inform such compromises.

Are the encountered issues well known in the Isabelle community? Difficult to say. Some issues have been briefly noted [10]. We are not aware of a sufficiently complex case study exploring the limitations of reasoning about algebras with implicit carriers. Without such case studies newcomers to a community might take a very long time to appreciate the issues. This paper also provides new means to overcome some of the limitations.

Do we allow Isabelle and not the mathematics to dictate the approach? Yes and no. To formally verify any result a system has to be chosen; staying 'close' to the mathematics is one aspect of the trade-off mentioned above and discussed in this paper. By choosing Isabelle/HOL, its capabilities necessarily constrain the approach. The trade-off between implicit and explicit carrier sets has been discussed in the introduction. Within the setting of implicit carriers in Isabelle/HOL we tried to follow the mathematics. Of course, Isabelle's capabilities might be extended to bring the approach closer to mathematics. Here too, this paper provides a data point to inform about what might be useful.

What concrete lessons could an Isabelle user learn? Users who wish to reason about algebraic structures can learn about the consequences of working with implicit carriers to inform their choice between them and explicit carriers. Users could find some of the techniques such as function lifting for dependent types and inheriting theorems via embeddings helpful for their work. Since the underlying universal algebra can be applied in many settings, the techniques are potentially of wider interest.

Are the experiences reported in this paper largely negative? Not in the author's opinion. There are negative aspects and positive aspects. This paper attempts to present a balanced view to inform choices.

# 6    Conclusion

The proof of Chen and Grätzer's construction theorem shows that reasoning about algebraic structures can be carried out in Isabelle/HOL using algebraic structures with implicit carrier sets. At the same time, the lack of dependent types leads to the introduction of function liftings. To remain compatible with this lifting, some other types also need to be lifted to functions even though they do not have an actual dependence. Overall this makes the constructions more complex and less related to the original proof.

An alternative way to reason about algebras is to explicitly represent their carrier sets in the corresponding classes and locales. Algebraic structures with explicit carriers are defined, for example, in the theories HOL/Algebra/*.thy. Previous work [10] briefly compares algebras with implicit and explicit carriers. It is desirable to automatically connect hierarchies of algebras with implicit and explicit carriers to ensure consistency and avoid duplication in maintenance and evolution. Isabelle/HOL's types-to-sets framework [19,27] offers a promising approach using local type definitions, which we will explore in future work.

The function liftings in this paper are used to work around the dependence on locale parameters. There is no claim that this gives a full-fledged implementation of dependent types. Kammüller [22] describes an approach to represent modular structures by dependent types constructed as sets in Isabelle/HOL.

**Acknowledgement.** I thank the anonymous referees for helpful comments.

# References

1. Balbes, R., Dwinger, P.: Distributive Lattices. University of Missouri Press, Columbia (1974)
2. Ballarin, C.: Locales: a module system for mathematical theories. J. Autom. Reason. **52**(2), 123–153 (2014)
3. Birkhoff, G.: Lattice Theory, Colloquium Publications, vol. XXV, 3rd edn. American Mathematical Society, Providence (1967)
4. Blanchette, J.C., Böhme, S., Paulson, L.C.: Extending Sledgehammer with SMT solvers. In: Bjørner, N., Sofronie-Stokkermans, V. (eds.) CADE 2011. LNCS (LNAI), vol. 6803, pp. 116–130. Springer, Heidelberg (2011). https://doi.org/10.1007/978-3-642-22438-6_11
5. Blyth, T.S.: Lattices and Ordered Algebraic Structures. Springer, London (2005). https://doi.org/10.1007/b139095
6. Burris, S., Sankappanavar, H.P.: A Course in Universal Algebra. Springer, New York (1981)
7. Capretta, V.: Universal algebra in type theory. In: Bertot, Y., Dowek, G., Hirschowitz, A., Paulin, C., Théry, L. (eds.) TPHOLs 1999. LNCS, vol. 1690, pp. 131–148. Springer, Heidelberg (1999). https://doi.org/10.1007/3-540-48256-3_10
8. Chen, C.C., Grätzer, G.: Stone lattices. I: Construction theorems. Can. J. Math. **21**, 884–894 (1969)
9. Davey, B.A., Priestley, H.A.: Introduction to Lattices and Order, 2nd edn. Cambridge University Press, Cambridge (2002)
10. Foster, S., Struth, G., Weber, T.: Automated engineering of relational and algebraic methods in Isabelle/HOL. In: de Swart, H. (ed.) RAMiCS 2011. LNCS, vol. 6663, pp. 52–67. Springer, Heidelberg (2011). https://doi.org/10.1007/978-3-642-21070-9_5
11. Grätzer, G.: Lattice Theory: First Concepts and Distributive Lattices. W. H. Freeman and Co., New York (1971)
12. Grätzer, G., Schmidt, E.T.: On ideal theory for lattices. Acta Sci. Math. **19**(1–2), 82–92 (1958)
13. Guttmann, W.: Stone algebras. Archive of Formal Proofs (2016)
14. Guttmann, W.: An algebraic framework for minimum spanning tree problems. Theor. Comput. Sci. **744**, 37–55 (2018)
15. Guttmann, W.: Verifying minimum spanning tree algorithms with Stone relation algebras. J. Log. Algebraic Methods Program. **101**, 132–150 (2018)
16. Haftmann, F., Wenzel, M.: Constructive type classes in Isabelle. In: Altenkirch, T., McBride, C. (eds.) TYPES 2006. LNCS, vol. 4502, pp. 160–174. Springer, Heidelberg (2007). https://doi.org/10.1007/978-3-540-74464-1_11
17. Horn, A.: On sentences which are true of direct unions of algebras. J. Symbol. Log. **16**(1), 14–21 (1951)
18. Huffman, B., Kunčar, O.: Lifting and transfer: a modular design for quotients in Isabelle/HOL. In: Gonthier, G., Norrish, M. (eds.) CPP 2013. LNCS, vol. 8307, pp. 131–146. Springer, Cham (2013). https://doi.org/10.1007/978-3-319-03545-1_9
19. Immler, F., Zhan, B.: Smooth manifolds and types to sets for linear algebra in Isabelle/HOL. In: Mahboubi, A., Myreen, M.O. (eds.) CPP 2019, pp. 65–77. ACM (2019)
20. Johnsson, T.: Lambda lifting: transforming programs to recursive equations. In: Jouannaud, J.-P. (ed.) FPCA 1985. LNCS, vol. 201, pp. 190–203. Springer, Heidelberg (1985). https://doi.org/10.1007/3-540-15975-4_37

21. Johnstone, P.T.: Stone Spaces. Cambridge University Press, Cambridge (1982)
22. Kammüller, F.: Modular structures as dependent types in Isabelle. In: Altenkirch, T., Naraschewski, W., Reus, B. (eds.) TYPES 1998. LNCS, vol. 1657, pp. 121–133. Springer, Heidelberg (1999). https://doi.org/10.1007/3-540-48167-2_9
23. Kammüller, F., Wenzel, M., Paulson, L.C.: Locales: a sectioning concept for Isabelle. In: Bertot, Y., Dowek, G., Hirschowitz, A., Paulin, C., Théry, L. (eds.) TPHOLs 1999. LNCS, vol. 1690, pp. 149–165. Springer, Heidelberg (1999). https://doi.org/10.1007/3-540-48256-3_11
24. Katriňák, T.: Die Kennzeichnung der distributiven pseudokomplementären Halbverbände. Journal für die reine und angewandte Mathematik **241**, 160–179 (1970)
25. Katriňák, T.: A new proof of the construction theorem for Stone algebras. Proc. Am. Math. Soc. **40**(1), 75–78 (1973)
26. Katriňák, T., Mederly, P.: Constructions of p-algebras. Algebra Univers. **17**(1), 288–316 (1983)
27. Kunčar, O., Popescu, A.: From types to sets by local type definition in higher-order logic. J. Autom. Reason. **62**(2), 237–260 (2018)
28. Nemitz, W.C.: Implicative semi-lattices. Trans. Am. Math. Soc. **117**, 128–142 (1965)
29. Nipkow, T., Paulson, L.C., Wenzel, M.: Isabelle/HOL: A Proof Assistant for Higher-Order Logic. LNCS, vol. 2283. Springer, Heidelberg (2002). https://doi.org/10.1007/3-540-45949-9
30. Paulson, L.C., Blanchette, J.C.: Three years of experience with Sledgehammer, a practical link between automatic and interactive theorem provers. In: Sutcliffe, G., Ternovska, E., Schulz, S. (eds.) Proceedings of the 8th International Workshop on the Implementation of Logics, pp. 3–13 (2010)
31. Polkowski, L.: Rough Sets: Mathematical Foundations. Springer, Heidelberg (2002). https://doi.org/10.1007/978-3-7908-1776-8
32. Spitters, B., van der Weegen, E.: Type classes for mathematics in type theory. Math. Struct. Comput. Sci. **21**(4), 795–825 (2011)

# Formal Proof of the Group Law
# for Edwards Elliptic Curves

Thomas Hales[1(✉)] and Rodrigo Raya[2]

[1] University of Pittsburgh, Pittsburgh, USA
hales@pitt.edu
[2] Technical University of Munich, Munich, Germany

**Abstract.** This article gives an elementary computational proof of the group law for Edwards elliptic curves. The associative law is expressed as a polynomial identity over the integers that is directly checked by polynomial division. Unlike other proofs, no preliminaries such as intersection numbers, Bézout's theorem, projective geometry, divisors, or Riemann Roch are required. The proof of the group law has been formalized in the Isabelle/HOL proof assistant.

## 1 Introduction

Elliptic curve cryptography is a cornerstone of mathematical cryptography. Many cryptographic algorithms (such as the Diffie-Hellman key exchange algorithm which inaugurated public key cryptography) were first developed in the context of the arithmetic of finite fields. The preponderance of finite-field cryptographic algorithms have now been translated to an elliptic curve counterpart. Elliptic curve algorithms encompass many of the fundamental cryptographic primitives: pseudo-random number generation, digital signatures, integer factorization algorithms, and public key exchange.

One advantage of elliptic curve cryptography over finite-field cryptography is that elliptic curve algorithms typically obtain the same level of security with smaller keys than finite-field algorithms. This often means more efficient algorithms.

Elliptic curve cryptography is the subject of major international cryptographic standards (such as NIST). Elliptic curve cryptography has been implemented in widely distributed software such as NaCl [BLS12]. Elliptic curve algorithms appear in nearly ubiquitous software applications such as web browsers and digital currencies.

The same elliptic curve can be presented in different ways by polynomial equations. The different presentations are known variously as the Weierstrass curve ($y^2$ = cubic in $x$), Jacobi curve ($y^2$ = quartic in $x$), and Edwards curve (discussed below).

The set of points on an elliptic curve forms an abelian group. Explicit formulas for addition are given in detail below. The Weierstrass curve is the most familiar presentation of an elliptic curve, but it suffers from the shortcoming

© Springer Nature Switzerland AG 2020
N. Peltier and V. Sofronie-Stokkermans (Eds.): IJCAR 2020, LNAI 12167, pp. 254–269, 2020.
https://doi.org/10.1007/978-3-030-51054-1_15

that the group law is not given by a uniform formula on all inputs. For example, special treatment must be given to the point at infinity and to point doubling: $P \mapsto 2P$. Exceptional cases are bad; they are the source of hazards such as side-channel attacks (timing attacks) by adversaries and implementation bugs [BJ02].

Edwards curves have been widely promoted for cryptographic algorithms because their addition law avoids exceptional cases and their hazards. Every elliptic curve (in characteristic different from 2) is isomorphic to an elliptic curve in Edwards form (possibly after passing to a quadratic extension). Thus, there is little loss of generality in considering elliptic curves in Edwards form. For most cryptographic applications, Edwards curves suffice.

The original contributions of this article are both mathematical and formal. Our proof that elliptic curve addition satisfies the axioms of an abelian group is new (but see the literature survey below for prior work). Our proofs were designed with formalization specifically in mind. To our knowledge, our proof of associativity in Sect. 3.3 is the most elementary proof that exists anywhere in the published literature (in a large mathematical literature on elliptic curves extending back to Euler's work on elliptic integrals). Our proof avoids the usual machinery found in proofs of associativity (such as intersection numbers, Bézout's theorem, projective geometry, divisors, or Riemann Roch). Our algebraic manipulations require little more than multivariate polynomial division with remainders, even avoiding Gröbner bases in most places. Based on this elementary proof, we give a formal proof in the Isabelle/HOL proof assistant that every Edwards elliptic curve (in characteristic other than 2) satisfies the axioms of an abelian group.[1]

It is natural to ask whether the proof of the associative law also avoids exceptional cases (encountered in Weierstrass curves) when expressed in terms of Edwards curves. Indeed, this article gives a two-line proof of the associative law for so-called *complete* Edwards curves that avoids case splits and all the usual machinery.

By bringing significant simplification to the fundamental proofs in cryptography, our paper opens the way for the formalization of elliptic curve cryptography in many proof assistants. Because of its extreme simplicity, we hope that our approach might be widely replicated and translated into many different proof assistants.

## 2    Published Literature

A number of our calculations are reworkings of calculations found in Edwards, Bernstein, Lange et al. [Edw07], [BBJ+08], [BL07]. A geometric interpretation of addition for Edwards elliptic curves appears in [ALNR11].

---

[1] Mathematica calculations are available at https://github.com/thalesant/publications-of-thomas-hales/tree/master/cryptography/group_law_edwards.
The Isabelle/HOL formalization is available at https://github.com/rjraya/Isabelle/blob/master/curves/Hales.thy.

Working with the Weierstrass form of the curve, Friedl was the first to give a proof of the associative law of elliptic curves in a computer algebra system (in Cocoa using Gröbner bases) [Fri98], [Fri17]. He writes, "The verification of some identities took several hours on a modern computer; this proof could not have been carried out before the 1980s." These identities were later formalized in Coq with runtime one minute and 20 s [The07]. A non-computational Coq formalization based on the Picard group appears in [BS14]. By shifting to Edwards curves, we have eliminated case splits and significantly improved the speed of the computational proof.

An earlier unpublished note contains more detailed motivation, geometric interpretation, pedagogical notes, and expanded proofs [Hal16]. The earlier version does not include formalization in Isabelle/HOL. Our formalization uncovered and corrected some errors in the ideal membership problems in [Hal16] (reaffirming the pervasive conclusion that formalization catches errors that mathematicians miss).

Other formalizations of elliptic curve cryptography are found in Coq and ACL2 by different methods [Rus17]. After we posted our work to the arXiv, another formalization was given in Coq along our same idea [Erb17,EPG+17]. It goes further by including formalization of implementation of code, but it falls short of our work by not including the far more challenging and interesting case of projective curves.

We do not attempt to survey the various formalizations of cryptographic algorithms built on top of elliptic curves. Because of the critical importance of cryptography to the security industry, the formalization of cryptographic algorithms is rightfully a priority within the formalization community.

## 3   Group Axioms

This section gives an elementary proof of the group axioms for addition on Edwards curves (Theorem 1). We include proofs, because our approach is not previously published.

Our definition of Edwards curve is more inclusive than definitions stated elsewhere. Most writers prefer to restrict to curves of genus one and generally call a curve with $c \neq 1$ a twisted Edwards curve. We have interchanged the $x$ and $y$ coordinates on the Edwards curve to make it consistent with the group law on the circle.

### 3.1   Rings and Homomorphisms

In this section, we work algebraically over an arbitrary field $k$. We assume a basic background in abstract algebra at the level of a first course (rings, fields, homomorphisms, and kernels). We set things up in a way that all of the main identities to be proved are identities of polynomials with integer coefficients.

All rings are assumed to be commutatative with identity $1 \neq 0$. If $R$ is an integral domain and if $\delta \in R$, then we write $R[\frac{1}{\delta}]$ for the localization of $R$ with respect to the multiplicative set $S = \{1, \delta, \delta^2, \ldots\}$; that is, the set of fractions with numerators in $R$ and denominators in $S$. We will need the well-known fact that if $\phi : R \to A$ is a ring homomorphism sending $\delta$ to a unit in $A$, then $\phi$ extends uniquely to a map $R[\frac{1}{\delta}] \to A$ that maps a fraction $r/\delta^i$ to $\phi(r)\phi(\delta^i)^{-1}$.

**Lemma 1 (kernel property).** *Suppose that an identity $r = r_1 e_1 + r_2 e_2 + \cdots + r_k e_k$ holds in a commutative ring $R$. If $\phi : R \to A$ is a ring homomorphism such that $\phi(e_i) = 0$ for all $i$, then $\phi(r) = 0$.*

*Proof.* $\phi(r) = \sum_{i=1}^{k} \phi(r_i)\phi(e_i) = 0$.  $\square$

We use the following rings: $R_0 := \mathbb{Z}[c, d]$ and $R_n := R_0[x_1, y_1, \ldots, x_n, y_n]$. We introduce the polynomial for the Edwards curve. Let

$$e(x, y) = x^2 + cy^2 - 1 - dx^2 y^2 \in R_0[x, y]. \tag{1}$$

We write $e_i = e(x_i, y_i)$ for the image of the polynomial in $R_j$, for $i \le j$, under $x \mapsto x_i$ and $y \mapsto y_i$. Set $\delta_x = \delta^-$ and $\delta_y = \delta^+$, where

$$\delta^{\pm}(x_1, y_1, x_2, y_2) = 1 \pm dx_1 y_1 x_2 y_2 \quad \text{and}$$

$$\delta(x_1, y_1, x_2, y_2) = \delta_x \delta_y \in R_2.$$

We write $\delta_{ij}$ for its image of $\delta$ under $(x_1, y_1, x_2, y_2) \mapsto (x_i, y_i, x_j, y_j)$. So, $\delta = \delta_{12}$.

## 3.2 Inverse and Closure

We write $z_i = (x_i, y_i)$. We define a pair of rational functions that we denote using the symbol $\oplus_0$:

$$z_1 \oplus_0 z_2 = \left( \frac{x_1 x_2 - c y_1 y_2}{1 - dx_1 x_2 y_1 y_2}, \frac{x_1 y_2 + y_1 x_2}{1 + dx_1 x_2 y_1 y_2} \right) \in R_2[\frac{1}{\delta}] \times R_2[\frac{1}{\delta}]. \tag{2}$$

When specialized to $c = 1$ and $d = 0$, the polynomial $e(x, y) = x^2 + y^2 - 1$ reduces to a circle, and (2) reduces to the standard group law on a circle. Commutativity is a consequence of the subscript symmetry $1 \leftrightarrow 2$ evident in the pair of rational functions:

$$z_1 \oplus_0 z_2 = z_2 \oplus_0 z_1.$$

If $\phi : R_2[\frac{1}{\delta}] \to A$ is a ring homomorphism, we also write $P_1 \oplus_0 P_2 \in A^2$ for the image of $z_1 \oplus_0 z_2$. We write $e(P_i) \in A$ for the image of $e_i = e(z_i)$ under $\phi$. We often mark the image $\bar{r} = \phi(r)$ of an element with a bar accent.

Let $\iota(z_i) = \iota(x_i, y_i) = (x_i, -y_i)$. The involution $z_i \to \iota(z_i)$ gives us an inverse with properties developed below.

There is an obvious identity element $(1, 0)$, expressed as follows. Under a homomorphism $\phi : R_2[\frac{1}{\delta}] \to A$, mapping $z_1 \mapsto P$ and $z_2 \mapsto (1, 0)$, we have

$$P \oplus_0 (1, 0) = P. \tag{3}$$

**Lemma 2 (inverse).** *Let* $\phi : R_2[\frac{1}{\delta}] \to A$, *with* $z_1 \mapsto P$, $z_2 \mapsto \iota(P)$. *If* $e(P) = 0$, *then* $P \oplus_0 \iota(P) = (1, 0)$.

*Proof.* Plug $P = (a, b)$ and $\iota P = (a, -b)$ into (2) and use $e(P) = 0$.     □

**Lemma 3 (closure under addition).** *Let* $\phi : R_2[\frac{1}{\delta}] \to A$ *with* $z_i \mapsto P_i$. *If* $e(P_1) = e(P_2) = 0$, *then*

$$e(P_1 \oplus_0 P_2) = 0.$$

*Proof.* This proof serves as a model for several proofs that are based on multi-variate polynomial division. We write

$$e(z_1 \oplus_0 z_2) = \frac{r}{\delta^2},$$

for some polynomial $r \in R_2$. It is enough to show that $\phi(r) = 0$. Polynomial division gives

$$r = r_1 e_1 + r_2 e_2, \tag{4}$$

for some polynomials $r_i \in R_2$. Concretely, the polynomials $r_i$ are obtained as the output of the one-line Mathematica command

$$\text{PolynomialReduce}[r, \{e_1, e_2\}, \{x_1, x_2, y_1, y_2\}].$$

The result now follows from the kernel property and (4); $e(P_1) = e(P_2) = 0$ implies $\phi(r) = 0$, giving $e(P_1 \oplus_0 P_2) = 0$.     □

Mathematica's `PolynomialReduce` is an implementation of a naive multi-variate division algorithm [CLO92]. In particular, our approach does not require the use of Gröbner bases until Sect. 5.3. We write

$$r \equiv r' \mod S,$$

where $r - r'$ is a rational function and $S$ is a set of polynomials, to indicate that the numerator of $r - r'$ has zero remainder when reduced by polynomial division with respect to $S$ using `PolynomialReduce`. We also require the denominator of $r - r'$ to be invertible in the localized polynomial ring. The zero remainder will give $\phi(r) = \phi(r')$ in each application. We extend the notation to $n$-tuples

$$(r_1, \ldots, r_n) \equiv (r'_1, \ldots, r'_n) \mod S,$$

to mean $r_i \equiv r'_i \mod S$ for each $i$. Using this approach, most of the proofs in this article almost write themselves.

### 3.3   Associativity

This next step (associativity) is generally considered the hardest part of the verification of the group law on curves. Our proof is two lines and requires little more than polynomial division. The polynomials $\delta_x, \delta_y$ appear as denominators in the addition rule. The polynomial denominators $\Delta_x, \Delta_y$ that appear when we add twice are more involved. Specifically, let $(x_3', y_3') = (x_1, y_1) \oplus_0 (x_2, y_2)$, let $(x_1', y_1') = (x_2, y_2) \oplus_0 (x_3, y_3)$, and set

$$\Delta_x = \delta_x(x_3', y_3', x_3, y_3)\delta_x(x_1, y_1, x_1', y_1')\delta_{12}\delta_{23} \in R_3.$$

Define $\Delta_y$ analogously.

**Lemma 4 (generic associativity).** *Let $\phi : R_3[\frac{1}{\Delta_x \Delta_y}] \to A$ be a homomorphism with $z_i \mapsto P_i$. If $e(P_1) = e(P_2) = e(P_3) = 0$, then*

$$(P_1 \oplus_0 P_2) \oplus_0 P_3 = P_1 \oplus_0 (P_2 \oplus_0 P_3).$$

*Proof.* By polynomial division in the ring $R_3[\frac{1}{\Delta_x \Delta_y}]$

$$((x_1, y_1) \oplus_0 (x_2, y_2)) \oplus_0 (x_3, y_3) \equiv (x_1, y_1) \oplus_0 ((x_2, y_2) \oplus_0 (x_3, y_3)) \mod \{e_1, e_2, e_3\}.$$

□

## 3.4 Group Law for Affine Curves

**Lemma 5 (affine closure).** *Let $\phi : R_2 \to k$ be a homomorphism into a field $k$. If $\phi(\delta) = e(P_1) = e(P_2) = 0$, then either $\bar{d}$ or $\bar{c}\bar{d}$ is a nonzero square in $k$.*

The lemma is sometimes called completeness, in conflict with the usual definition of *complete* varieties in algebraic geometry. To avoid possible confusion, we avoid this terminology. We use the lemma in contrapositive form to give conditions on $\bar{d}$ and $\bar{c}\bar{d}$ that imply $\phi(\delta) \neq 0$.

*Proof.* Let $r = (1 - cdy_1^2 y_2^2)(1 - dy_1^2 x_2^2)$. We have

$$r = d^2 y_1^2 y_2^2 x_2^2 e_1 + (1 - dy_1^2)\delta - dy_1^2 e_2. \tag{5}$$

This forces $\phi(r) = 0$, which by the form of $r$ implies that $\bar{c}\bar{d}$ or $\bar{d}$ is a nonzero square. □

We are ready to state and prove one of the main results of this article. This group law is expressed generally enough to include the group law on the circle and ellipse as a special case $\bar{d} = 0$.

**Theorem 1 (group law).** *Let $k$ be a field, let $\bar{c} \in k$ be a square, and let $\bar{d} \notin k^{\times 2}$. Then*

$$C = \{P \in k^2 \mid e(P) = 0\}$$

*is an abelian group with binary operation $\oplus_0$.*

*Proof.* This follows directly from the earlier results. For example, to check associativity of $P_1 \oplus_0 P_2 \oplus_0 P_3$, where $P_i \in C$, we define a homomorphism $\phi : R_3 \to k$ sending $z_i \mapsto P_i$ and $(c, d) \mapsto (\bar{c}, \bar{d})$. By a repeated use of the affine closure lemma, $\phi(\Delta_y \Delta_x)$ is nonzero and invertible in the field $k$. The universal property of localization extends $\phi$ to a homomorphism $\phi : R_3[\frac{1}{\Delta_y \Delta_x}] \to k$. By the associativity lemma applied to $\phi$, we obtain the associativity for these three (arbitrary) elements of $C$. The other group axioms follow similarly from the lemmas on closure, inverse, and affine closure. □

The Mathematica calculations in this section are fast. For example, the associativity certificate takes about 0.12 s to compute on a 2.13 GHz processor.

# 4  Formalization in Isabelle/HOL

In this section, we describe the proof implementation in Isabelle/HOL. We have formalized the two main theorems (Theorem 1 and Theorem 2). Formalization uses two different locales: one for the affine and one for the projective case. (The projective case will be discussed in Sect. 5.)

Let $k$ be the underlying curve field. $k$ is introduced as the type class *field* with the assumption that $2 \neq 0$ (characteristic different from 2). This is not included in the simplification set, but used when needed during the proof. The formalized theorem is slightly less general than then informal statement, because of this restriction.

## 4.1  Affine Edwards Curves

The formal proof fixes the curve parameters $c, d \in k$ (dropping the bar accents from notation). The group addition $\oplus_0$ (of Eq. 2) can be written as in Fig. 1. In Isabelle's division ring theory, the result of division by zero is defined as zero. This has no impact on validity of final results, but gives cleaner simplifications in some proofs.

```
add :: 'a × 'a ⇒ 'a × 'a ⇒ 'a × 'a
add (x₁,y₁) (x₂,y₂) = ((x₁*x₂ - c*y₁*y₂) div (1-d*x₁*y₁*x₂*y₂),
                       (x₁*y₂+y₁*x₂) div (1+d*x₁*y₁*x₂*y₂))
```

**Fig. 1.** Definition of $\oplus_0$ in Isabelle/HOL

Most of the proofs in this section are straight-forward. The only difficulty was to combine the Mathematica certificates of computation, into a single process in Isabelle.

In Fig. 2, we show an excerpt of the proof of associativity. We use the following abbreviations:

$$e_i = x_i^2 + c * y_i^2 - 1 - d * x_i^2 * y_i^2$$

where $e_i = 0$, since the involved points lie on the curve and

$$\text{gxpoly} = ((p_1 \oplus_0 p_2) \oplus_0 p_3 - p_1 \oplus_0 (p_2 \oplus p_3))_1 * \Delta_x$$

which stands for a normalized version of the associativity law after clearing denominators. We say that points are *summable*, if the rational functions defining their sum have nonzero denominators. Since the points $p_i$ are assumed to be summable, $\Delta_x \neq 0$. As a consequence, the property stated in Fig. 2 immediately implies that associativity holds in the first component of the addition.

Briefly, the proof unfolds the relevant definitions and then normalizes to clear denominators. The remaining terms of $\Delta_x$ are then distributed over addends. The unfolding and normalization of addends is repeated in the lemmas *simp1gx* and

```
have "∃ r1 r2 r3. gxpoly = r1 * e1 + r2 * e2 + r3 * e3"
   unfolding gxpoly_def gₓ_def Deltaₓ_def
   apply(simp add: assms(1,2))
   apply(rewrite in "_ / ⊓" delta_minus_def[symmetric])+
   apply(simp add: divide_simps assms(9,11))
   apply(rewrite left_diff_distrib)
   apply(simp add: simp1gx simp2gx)
   unfolding delta_plus_def delta_minus_def
             e1_def e2_def e3_def e_def
   by algebra
```

**Fig. 2.** An excerpt of the proof of associativity

*simp2gx*. Finally, the resulting polynomial identity is proved using the *algebra* method. Note that no computation was required from an external tool.

The *rewrite* tactic, which can modify a goal with various rewrite rules in various locations (specified with a pattern), is used to normalized terms [NT14]. Rewriting in the denominators is sufficient for our needs.

For proving the resulting polynomial expression, the *algebra* proof method is used [CW07,Cha08,Wen19]. Given $e_i(x)$, $p_{ij}(x)$, $a_i(x) \in R[x_1, \ldots, x_n]$, where $R$ is a commutative ring and $x = (x_1, \ldots, x_n)$, the method verifies formulas

$$\forall x. \bigwedge_{i=1}^{L} e_i(x) = 0 \rightarrow \exists y. \bigwedge_{i=1}^{M} \left( a_i(x) = \sum_{j=1}^{N} p_{ij}(x) y_j \right)$$

The method is complete for such formulas that hold over all commutative rings with unit [Har07].

## 5   Group Law for Projective Edwards Curves

By proving the group laws for a large class of elliptic curves, Theorem 1 is sufficiently general for many applications to cryptography. Nevertheless, to achieve full generality, we push forward.

This section shows how to remove the restriction $\bar{d} \notin k^{\times 2}$ that appears in the group law in the previous section. By removing this restriction, we obtain a new proof of the group law for all elliptic curves in characteristics different from 2. Unfortunately, in this section, some case-by-case arguments are needed, but no hard cases are hidden from the reader. The level of exposition here is less elementary than in the previous section. Again, we include proofs, because our approach is designed with formalization in mind and has not been previously published.

The basic idea of our construction is that the projective curve $E$ is obtained by gluing two affine curves $E_{aff}$ together. The associative property for $E$ is a consequence of the associative property on affine pieces $E_{aff}$, which can be expressed as polynomial identities.

## 5.1    Definitions

In this section, we assume that $c \neq 0$ and that $c$ and $d$ are both squares. Let $t^2 = d/c$. By a change of variable $y \mapsto y/\sqrt{c}$, the Edwards curve takes the form

$$e(x, y) = x^2 + y^2 - 1 - t^2 x^2 y^2. \tag{6}$$

We assume $t^2 \neq 1$. Note if $t^2 = 1$, then the curve degenerates to a product of intersecting lines, which cannot be a group. We also assume that $t \neq 0$, which only excludes the circle, which has already been fully treated. Shifting notation for this new setting, let

$$R_0 = \mathbb{Z}[t, \frac{1}{t^2 - 1}, \frac{1}{t}], \quad R_n = R_0[x_1, y_1, \ldots, x_n, y_n].$$

As before, we write $e_i = e(z_i)$, $z_i = (x_i, y_i)$, and $e(P_i) = \phi(e_i)$ when a homomorphism $\phi$ is given.

Define rotation by $\rho(x, y) = (-y, x)$ and inversion $\tau$ by

$$\tau(x, y) = (1/(tx), 1/(ty)).$$

Let $G$ be the abelian group of order eight generated by $\rho$ and $\tau$.

## 5.2    Extended Addition

We extend the binary operation $\oplus_0$ using the automorphism $\tau$. We also write $\delta_0$ for $\delta$, $\nu_0$ for $\nu$ and so forth.

Set

$$z_1 \oplus_1 z_2 := \tau((\tau z_1) \oplus_0 z_2) = \left( \frac{x_1 y_1 - x_2 y_2}{x_2 y_1 - x_1 y_2}, \frac{x_1 y_1 + x_2 y_2}{x_1 x_2 + y_1 y_2} \right) = (\frac{\nu_{1x}}{\delta_{1x}}, \frac{\nu_{1y}}{\delta_{1y}}) \tag{7}$$

in $R_2[\frac{1}{\delta_1}]^2$ where $\delta_1 = \delta_{1x} \delta_{1y}$.

We have the following easy identities of rational functions that are proved by simplification of rational functions:

$$\textit{inversion invariance:} \qquad \tau(z_1) \oplus_i z_2 = z_1 \oplus_i \tau z_2; \tag{8}$$

$$\textit{rotation invariance:} \qquad \begin{aligned} \rho(z_1) \oplus_i z_2 &= \rho(z_1 \oplus_i z_2); \\ \delta_i(z_1, \rho z_2) &= \pm \delta_i(z_1, z_2); \end{aligned} \tag{9}$$

$$\textit{inverses for } \sigma = \tau, \rho: \qquad \begin{aligned} \iota\sigma(z_1) &= \sigma^{-1}\iota(z_1); \\ \iota(z_1 \oplus_i z_2) &= (\iota z_1) \oplus_i (\iota z_2). \end{aligned} \tag{10}$$

$$\textit{coherence:} \qquad \begin{aligned} z_1 \oplus_0 z_2 &\equiv z_1 \oplus_1 z_2 \quad \mod \{e_1, e_2\}; \\ e(z_1 \oplus_1 z_2) &\equiv 0 \quad \mod \{e_1, e_2\}. \end{aligned} \tag{11}$$

The first identity of (11) inverts $\delta_0 \delta_1$, and the second inverts $\delta_1$. Proofs of (11) use polynomial division.

## 5.3   Projective Curve and Dichotomy

Let $k$ be a field of characteristic different from two. We let $E_{aff}$ be the set of zeros of Eq. (6) in $k^2$. Let $E^\circ \subset E_{aff}$ be the subset of $E_{aff}$ with nonzero coordinates $x, y \neq 0$.

We construct the projective Edwards curve $E$ by taking two copies of $E_{aff}$, glued along $E^\circ$ by isomorphism $\tau$. We write $[P, i] \in E$, with $i \in \mathbb{Z}/2\mathbb{Z} = \mathbb{F}_2$, for the image of $P \in E_{aff}$ in $E$ using the $i$th copy of $E_{aff}$. The gluing condition gives for $P \in E^\circ$:

$$[P, i] = [\tau P, i + 1]. \tag{12}$$

The group $G$ acts on the set $E$, specified on generators $\rho, \tau$ by $\rho[P, i] = [\rho(P), i]$ and $\tau[P, i] = [P, i + 1]$.

We define addition on $E$ by

$$[P, i] \oplus [Q, j] = [P \oplus_\ell Q, i + j], \quad \text{if } \delta_\ell(P, Q) \neq 0, \quad \ell \in \mathbb{F}_2 \tag{13}$$

We will show that the addition is well-defined, is defined for all pairs of points in $E$, and that it gives a group law with identity element $[(1, 0), 0]$. The inverse is $[P, i] \mapsto [\iota P, i]$, which is well-defined by the inverse rules (10).

**Lemma 6.** *$G$ acts without fixed point on $E^\circ$. That is, $gP = P$ implies that $g = 1_G \in G$.*

*Proof.* Write $P = (x, y)$. If $g = \rho^k \neq 1_G$, then $gP = P$ implies that $2x = 2y = 0$ and $x = y = 0$ (if the characteristic is not two), which is not a point on the curve. If $g = \tau\rho^k$, then the fixed-point condition $gP = P$ leads to $2txy = 0$ or $tx^2 = ty^2 = \pm 1$. Then $e(x, y) = 2(\pm 1 - t)/t \neq 0$, and again $P$ is not a point on the curve. $\qquad\square$

The domain of $\oplus_i$ is

$$E_{aff,i} := \{(P, Q) \in E_{aff}^2 \mid \delta_i(P, Q) \neq 0\}.$$

Whenever we write $P \oplus_i Q$, it is always accompanied by the implicit assertion of summability; that is, $(P, Q) \in E_{aff,i}$.

There is a group isomorphism $\langle \rho \rangle \to E_{aff} \setminus E^\circ$ given by

$$g \mapsto g(1, 0) \in \{\pm(1, 0), \pm(0, 1)\} = E_{aff} \setminus E^\circ.$$

**Lemma 7 (dichotomy).** *Let $P, Q \in E_{aff}$. Then either $P \in E^\circ$ and $Q = g\iota P$ for some $g \in \tau\langle\rho\rangle$, or $(P, Q) \in E_{aff,i}$ for some $i$. Moreover, assume that $P \oplus_i Q = (1, 0)$ for some $i$, then $Q = \iota P$.*

*Proof.* We start with the first claim. We analyze the denominators in the formulas for $\oplus_i$. We have $(P, Q) \in E_{aff,0}$ for all $P$ or $Q \in E_{aff} \setminus E^\circ$. That case completed, we may assume that $P, Q \in E^\circ$. Assuming

$$\delta_0(P, Q) = \delta_{0x}(P, Q)\delta_{0y}(P, Q) = 0, \quad \text{and} \quad \delta_1(P, Q) = \delta_{1x}(P, Q)\delta_{1y}(P, Q) = 0,$$

we show that $Q = g\iota P$ for some $g \in \tau\langle\rho\rangle$. Replacing $Q$ by $\rho Q$ if needed, which exchanges $\delta_{0x} \leftrightarrow \delta_{0y}$, we may assume that $\delta_{0x}(P,Q) = 0$. Set $\tau Q = Q_0 = (a_0, b_0)$ and $P = (a_1, b_1)$.

We claim that

$$(a_0, b_0) \in \{\pm(b_1, a_1)\} \subset \langle\rho\rangle\iota P. \tag{14}$$

We describe the main polynomial identity that must be verified. Write $\delta', \delta_+, \delta_-$ for $x_0 y_0 \delta_{0x}$, $t x_0 y_0 \delta_{1x}$, and $t x_0 y_0 \delta_{1y}$ respectively, each evaluated at $(P, \tau(Q_0)) = (x_1, y_1, 1/(tx_0), 1/(ty_0))$. The nonzero factors $x_0 y_0$ and $t x_0 y_0$ have been included to clear denominators, leaving us with polynomials.

We have two cases $\pm$, according to $\delta_\pm = 0$. In each case, let

$$S_\pm = \text{Gröbner basis of } \{e_1, e_2, \delta', \delta_\pm\}.$$

We have

$$(x_0^2 - y_1^2, y_0^2 - x_1^2, x_0 y_0 - x_1 y_1) \equiv (0, 0, 0) \mod S_+$$
$$(2x_0 y_0(x_0^2 - y_1^2), 2(1 - t^2)x_0 y_0(y_0^2 - x_1^2), x_0 y_0 - x_1 y_1) \equiv (0, 0, 0) \mod S_-.$$
$$\tag{15}$$

In fact, $\delta' = x_0 y_0 - x_1 y_1$, so that the ideal membership for this polynomial is immediate. The factors $2, 1 - t^2$, and $x_0 y_0$ are nonzero and can be removed from the left-hand side. These equations then immediately yield $(a_0, b_0) = \pm(b_1, a_1)$. This gives the needed identity: $\tau Q = Q_0 = (a_0, b_0) = g\iota P$, for some $g \in \langle\rho\rangle$. Then $Q = \tau g \iota P$.

The second statement of the lemma has a similar proof. Polynomial division gives for $i \in \mathbb{F}_2$:

$$(x_1 - x_2, y_1 + y_2) \equiv (0, 0) \mod \text{Gröbner}\{e_1, e_2, q_x \delta_{ix} - 1, q_y \delta_{iy} - 1, \nu_{iy}, \nu_{ix} - \delta_{ix}\}.$$

In fact, both $x_1 - x_2$ and $y_1 + y_2$ (which specify the condition $Q = \iota P$) are already members of the Gröbner basis. The fresh variables $q_x, q_y$ force the denominators $\delta_{ix}$ and $\delta_{iy}$ to be invertible. Here the equations $\nu_{iy} = \nu_{ix} - \delta_{ix} = 0$ specify the sum $(1, 0) = (\nu_{ix}/\delta_{ix}, \nu_{iy}/\delta_{iy})$ of $Q$ and $P$.  $\square$

**Lemma 8 (covering).** *The rule (13) defining $\oplus$ assigns at least one value for every pair of points in $E$.*

*Proof.* If $Q = \tau\rho^k\iota P$, then $\tau Q$ does not have the form $\tau\rho^k\iota P$ because the action of $G$ is fixed-point free. By dichotomy,

$$[P, i] \oplus [Q, j] = [P \oplus_\ell \tau Q, i + j + 1] \tag{16}$$

works for some $\ell$. Otherwise, by dichotomy $P \oplus_\ell Q$ is defined for some $\ell$.  $\square$

**Lemma 9 (well-defined).** *Addition $\oplus$ given by (13) on $E$ is well-defined.*

*Proof.* The right-hand side of (13) is well-defined by coherence (11), provided we show well-definedness across gluings (12). We use dichotomy. If $Q = \tau \rho^k \iota P$, then by an easy simplification of polynomials,

$$\delta_0(z, \tau \rho^k \iota z) = \delta_1(z, \tau \rho^k \iota z) = 0.$$

so that only one rule (16) for $\oplus$ applies (up to coherence (11) and inversion (8)), making it necessarily well-defined. Otherwise, coherence (11), inversion (8), and (7)) give when $[Q, j] = [\tau Q, j+1]$:

$$[P \oplus_k \tau Q, i+j+1] = [\tau(P \oplus_k \tau Q), i+j] = [P \oplus_{k+1} Q, i+j] = [P \oplus_\ell Q, i+j].$$

$\square$

## 5.4   Group Law

**Theorem 2.** *E is an abelian group.*

*Proof.* We have already shown the existence of an identity and inverse.

We prove associativity. Both sides of the associativity identity are clearly invariant under shifts $[P, i] \mapsto [P, i+j]$ of the indices. Thus, it is enough to show

$$[P, 0] \oplus ([Q, 0] \oplus [R, 0]) = ([P, 0] \oplus [Q, 0]) \oplus [R, 0].$$

By polynomial division, we have the following associativity identities

$$(z_1 \oplus_k z_2) \oplus_\ell z_3 \equiv z_1 \oplus_i (z_2 \oplus_j z_3) \quad \mod \{e_1, e_2, e_3\} \tag{17}$$

in the appropriate localizations, for $i, j, k, \ell \in \mathbb{F}_2$.

Note that $(g[P_1, i]) \oplus [P_2, j] = g([P_1, i] \oplus [P_2, j])$ for $g \in G$, as can easily be checked on generators $g = \tau, \rho$ of $G$, using dichotomy, (13), and (9). We use this to cancel group elements $g$ from both sides of equations without further comment.

We claim that

$$([P, 0] \oplus [Q, 0]) \oplus [\iota Q, 0] = [P, 0]. \tag{18}$$

The special case $Q = \tau \rho^k \iota(P)$ is easy. We reduce the claim to the case where $P \oplus_\ell Q \neq \tau \rho^k Q$, by applying $\tau$ to both sides of (18) and replacing $P$ with $\tau P$ if necessary. Then by dichotomy, the left-hand side simplifies by affine associativity 17 to give the claim.

Finally, we have general associativity by repeated use of dichotomy, which reduces in each case to (17) or (18). $\square$

## 5.5   Formalization in Isabelle/HOL of Projective Edwards Curves

Following the change of variables performed in Sect. 5.1, it is assumed that $c = 1$ and $d = t^2$ where $t \neq -1, 0, 1$. The resulting formalization is more challenging. In the following, some key insights are emphasized.

**Gröbner Basis.** The proof of Lemma 7 (dichotomy) requires solving particular instances of the ideal membership problem. Formalization caught and corrected some ideal membership errors in [Hal16], which resulted from an incorrect interpretation of computer algebra calculations. For instance, a goal

$$\exists r_1\, r_2\, r_3\, r_4.\; y_0^2 - x_1^2 = r_1 e(x_0, y_0) + r_2 e(x_1, y_1) + r_3 \delta' + r_4 \delta_-$$

(derived from [Hal16]) had to be corrected to

$$\exists r_1\, r_2\, r_3\, r_4.\; 2x_0 y_0 (y_0^2 - x_1^2) = r_1 e(x_0, y_0) + r_2 e(x_1, y_1) + r_3 \delta' + r_4 \delta_-$$

to prove (15). In another subcase, it was necessary to strengthen the hypothesis $\delta_+ = 0$ to $\delta_- \neq 0$. Eventually, after some reworking, *algebra* solved the required ideal membership problems.

**Definition of the Group Addition.** We defined the addition in three stages. This is convenient for some lemmas like covering (Lemma 8). First, we define the addition on projective points (Fig. 3). Then, we add two classes of points by applying the basic addition to any pair of points coming from each class. Finally, we apply the gluing relation and obtain as a result a set of classes with a unique element, which is then defined as the resulting class (Fig. 4).

The definitions use Isabelle's ability to encode partial functions. However, it is possible to obtain an equivalent definition more suitable for execution. In

```
type_synonym ('b) ppoint = ⟨((('b × 'b) × bit)⟩

p_add :: 'a ppoint ⇒ 'a ppoint ⇒ 'a ppoint where
  p_add ((x₁, y₁), 1) ((x₂, y₂), j) = (add (x₁, y₁) (x₂, y₂), 1+j)
  if delta x₁ y₁ x₂ y₂ ≠ 0 ∧ (x₁, y₁) ∈ e'_aff ∧ (x₂, y₂) ∈ e'_aff
| p_add ((x₁, y₁), 1) ((x₂, y₂), j) = (ext_add (x₁, y₁) (x₂, y₂), 1+j)
  if delta' x₁ y₁ x₂ y₂ ≠ 0 ∧ (x₁, y₁) ∈ e'_aff ∧ (x₂, y₂) ∈ e'_aff
```

**Fig. 3.** Definition of $\oplus$ on points

```
type_synonym ('b) pclass = ⟨('b) ppoint set⟩

proj_add_class :: ('a) pclass ⇒ ('a) pclass ⇒ ('a) pclass set
  proj_add_class c₁ c₂ =
        (p_add ' {(((x₁, y₁), i),((x₂, y₂), j)).
                 (((x₁, y₁), i),((x₂, y₂), j)) ∈ c₁ × c₂ ∧
                 ((x₁, y₁), (x₂, y₂)) ∈ e'_aff_0 ∪ e'_aff_1}) // gluing
  if c₁ ∈ e_proj and c₂ ∈ e_proj

proj_addition c₁ c₂ = the_elem (proj_add_class c₁ c₂)
```

**Fig. 4.** Definition of $\oplus$ on classes

particular, it is easy to compute the gluing relation (see lemmas e_proj_elim_1, e_proj_elim_2 and e_proj_aff in the formalization scripts).

Finally, since projective addition works with classes, we had to show that its definition does not depend on the representative used.

**Table 1.** List of $\delta$ relations

$$\delta \, \tau P_1 \, \tau P_2 \neq 0 \implies \delta \, P_1 \, P_2 \neq 0$$
$$\delta' \, \tau P_1 \, \tau P_2 \neq 0 \implies \delta' \, P_1 \, P_2 \neq 0$$
$$\delta \, P_1 \, P_2 \neq 0, \quad \delta \, P_1 \, \tau P_2 \neq 0 \implies \delta' \, P_1 \, P_2 \neq 0$$
$$\delta' \, P_1 \, P_2 \neq 0, \quad \delta' \, P_1 \, \tau P_2 \neq 0 \implies \delta \, P_1 \, P_2 \neq 0$$

$$\delta' \, (P_1 \oplus_1 P_2) \, \tau\iota P_2 \neq 0 \implies \delta \, (P_1 \oplus_1 P_2) \, \iota P_2 \neq 0$$
$$\delta \, P_1 \, P_2 \neq 0, \quad \delta \, (P_1 \oplus_0 P_2) \, \tau\iota P_2 \neq 0 \implies \delta'(P_1 \oplus_0 P_2) \, \iota P_2 \neq 0$$
$$\delta \, P_1 \, P_2 \neq 0, \quad \delta' \, (P_0 \oplus_0 P_1) \, \tau\iota P_2 \neq 0 \implies \delta \, (P_0 \oplus_0 P_1) \, \iota P_2 \neq 0$$
$$\delta' \, P_1 \, P_2 \neq 0, \quad \delta \, (P_0 \oplus_1 P_1) \, \tau\iota P_2 \neq 0 \implies \delta' \, (P_0 \oplus_1 P_1) \, \iota P_2 \neq 0$$

**Proof of Associativity.** During formalization, we found several relations between $\delta$ expressions (see Table 1). While they were proven in order to show associativity, the upper group can rather be used to establish the independence of class representative and the lower group is crucial to establish the associativity law.

In particular, the lower part of the table is fundamental to the formal proof of Eq. (18). In more detail, the formal proof development showed that it was necessary to perform a dichotomy (Lemma 7) three times. The first dichotomy is performed on $P$, $Q$. The non-summable case was easy. Therefore, we set $R = P \oplus Q$. On each of the resulting branches, a dichotomy on $R$, $\iota Q$ is performed. This time the summable cases were easy, but the non-summable case required a third dichotomy on $R, \tau\iota Q$. The non-summable case was solved using the no-fixed-point theorem but for the summable subcases the following expression is obtained:

$$([P, 0] \oplus [Q, 0]) \oplus [\tau\iota Q, 0] = [(P \oplus Q) \oplus \tau\iota Q, 0]$$

Here we cannot invoke associativity because $Q$, $\tau\iota Q$ are non-summable (lemma not_add_self). Instead, we use the equations from the lower part of the table and the hypothesis of the second dichotomy to get a contradiction.

# 6    Conclusion

We have shown that Isabelle can encompass the process of defining, computing and certifying intensive algebraic calculations. The encoding in a proof-assistant allows a better comprehension of the methods used and helps to clarify its structure.

# References

[ALNR11] Arene, C., Lange, T., Naehrig, M., Ritzenthaler, C.: Faster computation of the Tate pairing. J. Number Theory **131**(5), 842–857 (2011)

[BBJ+08] Bernstein, D.J., Birkner, P., Joye, M., Lange, T., Peters, C.: Twisted Edwards curves. In: Vaudenay, S. (ed.) AFRICACRYPT 2008. LNCS, vol. 5023, pp. 389–405. Springer, Heidelberg (2008). https://doi.org/10.1007/978-3-540-68164-9_26

[BJ02] Brier, É., Joye, M.: Weierstraß elliptic curves and side-channel attacks. In: Naccache, D., Paillier, P. (eds.) PKC 2002. LNCS, vol. 2274, pp. 335–345. Springer, Heidelberg (2002). https://doi.org/10.1007/3-540-45664-3_24

[BL07] Bernstein, D.J., Lange, T.: Faster addition and doubling on elliptic curves. In: Kurosawa, K. (ed.) ASIACRYPT 2007. LNCS, vol. 4833, pp. 29–50. Springer, Heidelberg (2007). https://doi.org/10.1007/978-3-540-76900-2_3

[BLS12] Bernstein, D.J., Lange, T., Schwabe, P.: The security impact of a new cryptographic library. In: Hevia, A., Neven, G. (eds.) LATINCRYPT 2012. LNCS, vol. 7533, pp. 159–176. Springer, Heidelberg (2012). https://doi.org/10.1007/978-3-642-33481-8_9

[BS14] Bartzia, E.-I., Strub, P.-Y.: A formal library for elliptic curves in the Coq proof assistant. In: Klein, G., Gamboa, R. (eds.) ITP 2014. LNCS, vol. 8558, pp. 77–92. Springer, Cham (2014). https://doi.org/10.1007/978-3-319-08970-6_6

[Cha08] Chaieb, A.: Automated methods for formal proofs in simple arithmetics and algebra. Ph.D. thesis, Technische Universität München (2008)

[CLO92] Cox, D., Little, J., O'Shea, D.: Ideals, Varieties, and Algorithms, vol. 3. Springer, New York (1992). https://doi.org/10.1007/978-1-4757-2181-2

[CW07] Chaieb, A., Wenzel, M.: Context aware calculation and deduction. In: Kauers, M., Kerber, M., Miner, R., Windsteiger, W. (eds.) Calculemus/MKM -2007. LNCS (LNAI), vol. 4573, pp. 27–39. Springer, Heidelberg (2007). https://doi.org/10.1007/978-3-540-73086-6_3

[Edw07] Edwards, H.: A normal form for elliptic curves. Bull. Am. Math. Soc. **44**(3), 393–422 (2007)

[EPG+17] Erbsen, A., Philipoom, J., Gross, J., Sloan, R., Chlipala, A.: Systematic generation of fast elliptic curve cryptography implementations. Technical report, MIT, Cambridge, MA, USA (2017)

[Erb17] Erbsen, A.: Crafting certified elliptic curve cryptography implementations in Coq. Ph.D. thesis, Massachusetts Institute of Technology (2017)

[Fri98] Friedl, S.: An elementary proof of the group law for elliptic curves. The Group Law on Elliptic Curves (1998)

[Fri17] Friedl, S.: An elementary proof of the group law for elliptic curves. Groups Complex. Cryptol. **9**(2), 117–123 (2017)

[Hal16] Hales, T.: The group law for Edwards curves. arXiv preprint arXiv:1610.05278 (2016)

[Har07] Harrison, J.: Automating elementary number-theoretic proofs using Gröbner bases. In: Pfenning, F. (ed.) CADE 2007. LNCS (LNAI), vol. 4603, pp. 51–66. Springer, Heidelberg (2007). https://doi.org/10.1007/978-3-540-73595-3_5

[NT14] Noschinski, L., Traut, C.: Pattern-based subterm selection in Isabelle. In: Proceedings of Isabelle Workshop (2014)

[Rus17]  Russinoff, D.M.: A computationally surveyable proof of the group properties of an elliptic curve. arXiv preprint arXiv:1705.01226 (2017)
[The07]  Théry, L., Hanrot, G.: Primality proving with elliptic curves. In: Schneider, K., Brandt, J. (eds.) TPHOLs 2007. LNCS, vol. 4732, pp. 319–333. Springer, Heidelberg (2007). https://doi.org/10.1007/978-3-540-74591-4_24
[Wen19]  Wenzel, M.: The Isabelle/Isar reference manual (2019)

# Verifying Faradžev-Read Type Isomorph-Free Exhaustive Generation

Filip Marić[✉][iD]

Faculty of Mathematics, University of Belgrade, Belgrade, Serbia
filip@matf.bg.ac.rs

**Abstract.** Many applications require generating catalogues of combinatorial objects, that do not contain isomorphs. Several efficient abstract schemes for this problem exist. One is described independently by I. A. Faradžev and R. C. Read and has since been applied to catalogue many different combinatorial structures. We present an Isabelle/HOL verification of this abstract scheme. To show its practicality, we instantiate it on two concrete problems: enumerating digraphs and enumerating union-closed families of sets. In the second example abstract algorithm specification is refined to an implementation that can quite efficiently enumerate all canonical union-closed families over a six element universe (there is more than 100 million such families).

**Keywords:** Isomorph-free exhaustive generation · Orderly · Software verification · Isabelle/HOL

## 1 Introduction

Cataloguing finite combinatorial structures (e.g., subsets, partitions, words, Latin squares, graphs, designs, codes) described by certain specified properties is required in many application domains. It is very desirable that such catalogues are exhaustive and isomorph-free i.e., to contain exactly one representative of each class of isomorphic structures. Often it is not enough to count objects (to enumerate them), but it is needed to generate them explicitly.

Efficient isomorph-free cataloguing algorithms are often divided into three types: (i) Faradžev–Read-type orderly algorithms based on canonical representatives [8,20], (ii) McKay-type algorithms based on canonical orderings [1], and (iii) algorithms based on the homomorphism principle for group actions [10]. These are applied to a wide variety of problems (according to Google Scholar, McKay's paper has more than 500 citations, most of which describe its concrete applications in mathematics, computer science, chemistry, biology etc.).

In this paper we present a formal verification of Faradžev-Read cataloguing scheme within Isabelle/HOL. Verified cataloguing of combinatorial structures is often used in formal proofs (e.g., enumeration of Tame Graphs given by Nipkow et al. [18] was an important part of the Flyspeck proof of Kepler conjecture). We

© Springer Nature Switzerland AG 2020
N. Peltier and V. Sofronie-Stokkermans (Eds.): IJCAR 2020, LNAI 12167, pp. 270–287, 2020.
https://doi.org/10.1007/978-3-030-51054-1_16

advocate that verifying general isomorph-free catalouging schemes might facilitate verifying enumeration algorithms needed for concrete applications. Author's personal motivation for these algorithms comes from his previous and current work in formalizing combinatorics and finite geometry [15,16].

To demonstrate its usefulness, we applied our general framework on two concrete problems: cataloguing all directed graphs on $n$ nodes (this was the first problem analyzed in the original Read's paper [20]) and cataloguing families of subsets of an $n$-element domain, closed under unions. A solution for the second problem was described by Brinkmann and Deklerck in 2018 [4] and it combines Faradžev-Read type orderly generation [8,20] and the homomorphism principle [3]. For $n = 6$, their C implementation found around 100 million such families in several seconds, while for $n = 7$, it found around $2 \cdot 10^{15}$ in around 10 to 12 CPU years (on a cluster computer). We refine an abstract algorithm specification to an efficient implementation (still purely functional) and show that it can solve the case $n = 6$ within Isabelle/HOL in a matter of minutes.

In the current paper we focus mainly on presenting definitions (in the most cases proofs are not discussed). It is assumed that the reader is familiar with functional programming and Isabelle/HOL [19]. Some definitions are slightly simplified, to make them more comprehensible. Proof documents are available in the Downloads section at http://argo.matf.bg.ac.rs/ and are going to be submitted to the Archive of Formal Proofs.

**Contributions.** Our central contribution is the verification of the abstract Faradžev-Read scheme that can be instantiated for many concrete applications. Other contributions are:

– a verified algorithm for cataloguing digraphs and other similar objects [20];
– a verified efficient algorithm for generating union-closed families [4];
– a small verified library for generating basic combinatorial objects (permutations and combinations);
– verified bitwise representation of sets, set operations and families of sets by unsigned integers and some common "bit-hacks".

**Related Work.** Literature on fast computer-based enumeration of various combinatorial objects is vast, but it seems that there are not many formally verified algorithms and tools. As a part of Flyspeck project, Nipkow et al. used Isabelle/HOL to verify an algorithm for enumerating tame graphs [17,18]. Bowles and Caminati used Isabelle/HOL to verify an algorithm for enumerating event structures and, as a byproduct, all preorders and partial orders [2]. Giorgetti et al. used Why3 and Coq to generate basic combinatorial objects, used in software testing [7,9]. We are not aware of any verified general methods for isomorph-free exhaustive combinatorial enumeration.

## 2   General Faradžev-Read Scheme

Algorithms for generating combinatorial objects are usually based on recursive schemes that build larger objects by augmenting smaller ones. We shall assume that the set of all objects $S$ is divided into its subsets $S_0, S_1, S_2, \ldots$ grouped by object "size" (e.g., the size can be the number of edges in a graph or the number of sets in a family). Objects in $S_{q+1}$ are produced by augmenting objects in $S_q$.

A classic, naive algorithm for isomorphism rejection maintains a list $L_q$ of objects of $S_q$ produced so far, and compares the current object with all objects in that list, adding it to the list $L_q$ only if $L_q$ does not contain its isomorph. That assumes that there is an efficient isomorphism test and is doomed to be inefficient when the list becomes long. Efficient schemes (including the Faradžev-Read's) avoid comparing the current object with the previous ones an can deduce whether it should be added to the list only by examining the object itself.

A central component of Faradzev-Read's scheme is a *linear order* (we shall denote it by $<$) in which objects are produced (the scheme is sometimes called *orderly generation*). $L_q$ shall always be sorted wrt. that order. Also, it is assumed that for each isomorphism class there is a single *canonical* object, that for each object we can test if it is canonical and that lists $L_q$ shall contain only canonical objects. We specify this in an Isabelle/HOL locale (use of locales for stepwise implementation is described by Nipkow [17]).

**locale** *FaradzevRead'* =
**fixes** $S$ :: *"nat $\Rightarrow$ ('s::linorder) set"*
**fixes** *equiv* :: *"'s $\Rightarrow$ 's $\Rightarrow$ bool"*
**fixes** *is_canon* :: *"'s $\Rightarrow$ bool"*
**fixes** *is_canon_test* :: *"'s $\Rightarrow$ bool"*
**fixes** *augment* :: *"'s $\Rightarrow$ 's list"*
**assumes** *"$\bigwedge$ q. equivp_on (S q) equiv"*
**assumes** *"$\bigwedge$ s s' q. $[\![$equiv s s'; s $\in$ S q$]\!]$ $\Longrightarrow$ s' $\in$ S q"*
**assumes** *"$\bigwedge$ s q. s $\in$ S q $\Longrightarrow$ $\exists$! s$_c$. equiv s s$_c$ $\wedge$ is_canon s$_c$"*
**assumes** *"$\bigwedge$ s s' q. $[\![$is_canon s; s' $\in$ set (augment s)$]\!]$ $\Longrightarrow$ s $\in$ S q $\longleftrightarrow$ s' $\in$ S (q + 1)"*
**assumes** *"$\bigwedge$ s s' q. $[\![$s $\in$ S q; is_canon s; s' $\in$ set (augment s)$]\!]$ $\Longrightarrow$*
$\qquad\qquad\qquad$ *is_canon_test s' $\longleftrightarrow$ is_canon s"'*

First two assumptions ensure that *equiv* is an equivalence (isomorphism) relation on every $S_q$ (note that Faradžev-Read scheme does not require to have an executable isomorphism test). The third one ensures that each isomorphism class contains exactly one canonical representative. The fourth one describes the augmentation procedure that builds a list containing possible extensions of a given canonical object (their dimension is always increased by one). Definition of a canonical representative should be as simple as possible, since it is used in proofs. It need not be executable and if it is executable it need not be efficient. We provide another function *is_canon_test* that is used to test for canonicity. It does not need to match the abstract *is_canon* definition in general, but they need to match on the objects obtained by augmenting canonical objects.

Faradžev-Read algorithm iterates through a sorted catalogue $L_q$ of canonical objects of $S_q$ and builds a sorted catalogue $L_{q+1}$ for $S_{q+1}$. For each object $p$ it iterates trough a list of objects $s$ that augment it. If an object $s$ is non-canonical it is eliminated. If it is canonical, then it is appended at the end of $L_{q+1}$ only if it does not violate the list order. This procedure is specified as follows.

*order_test s res = (if res = [] ∨ s > hd res then s # res else res)*
*step L = (let cs = filter is_canon_test (concat (map augment L))*
                  **in** *rev (fold order_test cs []))*

Three conditions are sufficient for the correctness of the previous procedure. First, it must be possible to obtain each canonical object $s$ in $S_{q+1}$ by augmenting at least one canonical object in $S_q$. If that holds, then all canonical objects in $S_{q+1}$ will be enumerated at least once and we need to guarantee that they will survive the order test exactly once. Since the ordering of $L_{q+1}$ is strict, a canonical object cannot survive the order test more than once. The first appearance of a canonical object $s$ will be eliminated by the order test iff the list $L_{q+1}$ constructed so far contains an object $s'$ such that $s' > s$. Element $s'$ could be produced either by the same $p$ that produced $s$ or by some element $p'$ of $L_q$ that precedes $p$. The former cannot happen if the augmentation procedure always gives elements $s$ in sorted order. Let $f(s) = p$ be the first element of $L_q$ which produces $s$, and let $f(s') = p'$ be the first element of $L_q$ which produces $s'$. To forbid that $s'$ is produced by some element $p'$ of $L_p$ that precedes $p$, we must forbid that both $s' > s$ and $f(s') < f(s)$ hold i.e., we must require that $s' > s$ implies $f(s') \geq f(s)$. To formalize this, we first define the function $f$ (we call it the *minimal parent* function).

*parent p s* ⟷ *is_canon p ∧ s ∈ set (augment p)*
*min_parent p s* ⟶ *parent p s ∧ (∀ p'. parent p' s ⟶ p ≤ p')*

Then we extend the locale by requiring the following three conditions.

**locale** *FaradzevRead = FaradzevRead' +*
**assumes** "⋀ s q. ⟦s ∈ S (q + 1); is_canon s⟧ ⟹ (∃ p ∈ S q. parent p s)"
**assumes** "⋀ s q. ⟦s ∈ S q; is_canon s⟧ ⟹ sorted (filter is_canon (augment s))"
**assumes** "⋀ s s' p p' q.
                ⟦s ∈ S (q + 1); is_canon s; min_parent p s;
                s' ∈ S (q + 1); is_canon s'; min_parent p' s'; s < s'⟧ ⟹ p ≤ p'"

Now we can formulate and prove algorithm correctness. We define *catalogue* for $S_q$ as a strictly sorted list of canonical elements of $S_q$ that contains an element for each isomorphism class.

*catalogue L q* ⟷ *sorted L ∧ distinct L ∧ set L ⊆ S q ∧ (∀ s ∈ set L. is_canon s) ∧ (∀ s ∈ S q. ∃ s_c ∈ set L. is_canon s_c ∧ equiv s s_c)*

It is easily shown that catalogue for every $q$ is unique and that it always exists if $Sq$ is finite (we denote it by *the_catalogue q*).

The central theorem states that the *step* function when applied to a catalogue for $S_q$ produces a catalogue for $S_{q+1}$.

**theorem** *"catalogue L q ⟹ catalogue (step L) (q + 1)"*

The original proof given by Read is as follows. "Condition 1 ensures that every canonical configuration $X$ in $S_{q+1}$ is produced at least once. Condition 2 ensures that when $X$ is produced for the first time from $f(X)$ there cannot be an entry $Y$ produced from $f(Y) \neq f(X)$ which follows $X$ in $L_{q+1}$ and whose presence will therefore block the addition of $X$ to the list. Condition 3 ensures the same thing when $f(X) = f(Y)$." This informal proof sketch had to be expanded to around 300 lines long Isar proof which employed nested reverse list induction.

**Strict Faradžev/Read Conditions.** In some concrete instances stronger conditions are met that make possible to skip the order test within *step* function. Although that does not make the implementation much more efficient (the order test is usually quite fast), this can significantly simplify the depth-first variant of the procedure.

**locale** *FaradzevReadStrict = FaradzevRead'* +
**assumes** "$\bigwedge s\ q.\ [\![s \in S\ (q+1);\ is\_canon\ s]\!] \implies (\exists\ p \in S\ q.\ parent\ p\ s)$"
**assumes** "$\bigwedge s\ q.\ [\![s \in S\ q;\ is\_canon\ s]\!] \implies sorted\ (augment\ s) \wedge distinct\ (augment\ s)$"
**assumes** "$\bigwedge p\ s\ p'\ s'\ q.\ [\![p \in S\ q;\ parent\ p\ s;\ p' \in S\ q;\ parent\ p'\ s';\ p < p']\!] \implies s < s'$"

We formally showed that the order test can be skipped.

**lemma**
**assumes** "*distinct L*" "*sorted L*" "*set $L \subseteq S\ q$*" "$\forall\ x \in set\ L.\ is\_canon\ x$"
**shows** "*step L = filter is_canon_test (concat (map augment L))*" "*distinct (step L)*"

We have also shown that strict conditions imply the original conditions (by showing that *FaradzevReadStrict* is a sublocale of *FaradzevRead*).

**Depth First Variant.** When *step* function is iterated, objects are generated in breadth first fashion. A serious concern about such procedure is its memory usage, since at each step it needs to store both the whole list $L_q$ and the elements of $L_{q+1}$ that are generated. We have defined a procedure that makes the catalogue in depth first fashion, and that usually consumes significantly less memory. Note that such procedure could have been defined in the non-strict Faradžev/Read locale, but it would be more complicated, since, to be able to perform the order tests, it would have to store the largest element in $L_q$ for all recursion levels $q$. In many concrete applications, including both our case studies, strict conditions hold, so we opted only for the simpler variant. We have defined a function *fold_dfs* that "folds" the elements of the Faradžev-Read tree, enumerated in the DFS order, by some given accumulating function. This tree can be formed by augmenting each node in each possible way (using the *augment* function) and retaining only the canonical descendants (filtered by the *is_canon_test* function), but it is not explicitly built in the memory. The *lvl* parameter guarantees termination by controlling the depth of the generated tree. The definition of *fold_dfs* is quite technical.

*fold_dfs lvl f i ss =*
  (**if** *lvl = 0* **then** *i*
    **else** *fold ($\lambda$ s' x. f s' (fold_dfs (lvl - 1) f x (filter is_canon_test (augment s')))) ss i*)

Elements of the tree are usually folded by the following functions. The function *catalogue_dfs* computes the catalogue by collecting all tree nodes in a list, while the function *count_dfs* only counts nodes, without keeping their list.

*catalogue_dfs lvl ss = fold_dfs lvl (λ s x. s # x) [] ss*

*count_dfs lvl ss = fold_dfs lvl (λ s x. x + 1) 0 ss*

If the procedure *catalogue_dfs* starts from a catalogue for $S_q$, then it traverses over elements of all catalogues from $S_q$ to $S_{q+lvl-1}$ (although in different order than the traversal based on the *step* function). This is formalized by the following theorem (where *mset* denotes the multiset of list elements).

**theorem**
    **assumes** *"catalogue L q"*
    **shows** *"mset (catalogue_dfs lvl L) = mset (concat (map the_catalogue [q..<q+lvl]))"*

## 3   Cataloguing Digraphs

The first case-study used to test our general scheme was cataloguing all loopless directed graphs (digraphs) with $n$ nodes. It was the first problem described by Read [20] and we directly follow his approach. As this was just a toy-example, we did not invest much effort into low-level implementation details (e.g., we have used lists which are the simplest data structures)—additional refinement step that would introduce more efficient data structures and some other algorithmic enhancements could make the enumeration much more efficient.

**Objects.** Following [20], digraphs are represented by their adjacency matrices. As only loopless digraphs with $n$ nodes are considered, the diagonal can be excluded from the matrix and by concating matrix rows an $n \times (n-1)$ vector (a list) representation can be obtained. Graphs will be augmented by adding branches, so we define sets $S_0, S_1, \ldots$ in the following way (the number of nodes $n$ is fixed when interpreting the locale).

$S\ n\ q = \{l.\ length\ l = n\ *\ (n\ -\ 1) \land set\ l \subseteq \{0,\ 1\} \land sum\_list\ l = q\}$

Example graph, its matrix and list representation are shown on Fig. 1.

$$\begin{pmatrix} 0\ 0\ 0 \\ 1\ 0\ 0 \\ 1\ 0\ 0 \end{pmatrix} \qquad [0, 0, 1, 0, 1, 0]$$

**Fig. 1.** A graph represented graphically, by a matrix and by a list

**Equivalence.** Two graphs are equivalent if there is a permutation of nodes that would map one graph onto another. Permutations are represented by lists of length $n$ (e.g., $[2, 0, 1]$ denotes a permutation that maps 0 to 2, 1 to 0 and 2 to 1). For example, if nodes in the Fig. 1 are ordered 0, 1, 2 instead of 1, 0, 2, then the graph would be represented by the list $[1, 0, 0, 0, 0, 1]$. A direct (but not the most efficient) way to define action of node permutation to a list representing a digraph is to convert it to a matrix, permute the rows and columns of the matrix and then convert the matrix back to a digraph list.

*permute_matrix p M = permute_list p (map (permute_list p) M)*
*permute_dig p n l $\longleftrightarrow$ mat2dig (permute_matrix p (dig2mat n l))*

Equivalence is often defined by using permutations, so we introduce it abstractly in a separate locale and prove its properties.

**locale** *Permute* =
**fixes** *invar* :: "*nat $\Rightarrow$ 'a $\Rightarrow$ bool*"
**fixes** *permute* :: "*nat $\Rightarrow$ nat list $\Rightarrow$ 'a $\Rightarrow$ 'a*"
**assumes** "$\bigwedge$ *a p n.* ⟦*invar n a; is_perm n p*⟧ $\Longrightarrow$ *invar n (permute n p a)*"
**assumes** "$\bigwedge$ *a n. invar n a $\Longrightarrow$ permute n (perm_id n) a = a*"
**assumes** "$\bigwedge$ *a $p_1$ $p_2$ n.* ⟦*invar n a; is_perm n $p_1$; is_perm n $p_2$*⟧ $\Longrightarrow$
    *permute n (perm_comp $p_1$ $p_2$) a = permute n $p_1$ (permute n $p_2$ a))*"
**assumes** "$\bigwedge$ *a p n.* ⟦*invar n a; is_perm n p*⟧ $\Longrightarrow$
    *permute n (perm_inv p) (permute n p a) = a*"

Predicate *is_perm* abbreviates the condition $p <\sim\sim> [0.. < n]$, where $<\sim\sim>$ is the permutation relation from the Isabelle/HOL library. Permutations are applied on objects of some abstract type *'a* (e.g., to lists that represent digraphs) that may satisfy some given invariant (e.g., that the list length is $n(n-1)$). The function *permute* is the action of permutations on the objects of type *'a*. If it respects the permutation group operations (identity *perm_id*, inverse *perm_inv*, and composition *perm_comp*), then we can use it to define equivalent objects and to prove that it is an equivalence relation.

*equiv n $F_1$ $F_2$ $\longleftrightarrow$ ($\exists$ p. is_perm n p $\land$ $F_2$ = permute p $F_1$)*

**Ordering.** The ordering is very simple – the lexicographic order on lists used to represent graphs is used, except that the order of list elements is reversed (1 is treated as less than 0).

**Canonical Objects.** Permutations are also used to define canonical objects. An object is canonical if it is minimal among all its possible permutations. For example, the list $[1, 0, 0, 0, 0, 1]$ is canonical for the graph shown in Fig. 1. This definition is also generic and can be specified within the previous locale (a linear order on the type *'a* is assumed). Then it can be proved that each equivalence class contains a single canonical representative (what is needed for Faradzev-Read enumeration).

"*is_canon n F $\longleftrightarrow$ ($\forall$ p. is_perm n p $\longrightarrow$ F $\leq$ permute p F)*"
**lemma** "*inv n a $\Longrightarrow$ $\exists$! c. equiv n a c $\land$ is_canon n c*"

An optimization can be made when checking canonicity of a digraph. By the definition of ordering, the list that starts with as most ones as possible will be always less than lists that have zeros at that initial positions. Therefore only permutations that put a maximal degree node at the beginning and nodes that it is connected to after it need to be considered. This is the essence of our *is_canon_test* definition (that we do not show here).

**Augmentation.** Graphs are augmented by adding an edge i.e., by changing one 0 in the list to 1. If the list contains some elements 1, then only zeros behind the last 1 can be changed (otherwise any zero can be changed). For example, the list $[1, 0, 0, 1, 0, 0]$ can be augmented to $[1, 0, 0, 1, 1, 0]$ and $[1, 0, 0, 1, 0, 1]$. This can be formalized as a relation between two lists[1].

*is_last_one i xs* $\longleftrightarrow$ *xs ! i = 1* $\wedge$ *($\forall$ i'. i < i'* $\wedge$ *i' < length xs* $\longrightarrow$ *xs ! i' = 0)*
*all_zeros xs* $\longleftrightarrow$ *($\forall$ i < length xs. xs ! i = 0)*
*increment_after_last_one xs ys* $\longleftrightarrow$ *($\exists$ j. j < length xs* $\wedge$ *ys = xs [j := 1]* $\wedge$
       *(all_zeros xs* $\vee$ *($\exists$ i. is_last_one xs i* $\wedge$ *i < j)))*

All required properties of the augmentation procedure are proved using this abstract definition, and only then its concrete implementation is given (it is quite technical, so we do not show it here). It must return digraphs in sorted order, which is ensured by sequentially incrementing every 0, after the last 1, one by one.

**Results.** The naive implementation we defined can catalogue all 1 540 944 digraphs with 6 nodes in 276 seconds (on an 2.4GHz, Intel Core i5, 8GB RAM laptop). Interestingly, the original paper reports only 1 540 744 digraphs [20]. Cataloguing more than 800 million digraphs with 7 nodes is possible, but would require significant improvements of the implementation.

# 4    Cataloguing Union-Closed Families

Families of sets closed under unions have gotten a lot of research attention due to the famous conjecture by Péter Frankl, claiming that in each such family there is an element occurring in at least half of the sets. Although quite elementary, the conjecture is still open [5,16]. Recently Brinkmann and Deklerck applied Faradžev-Read type algorithm to catalogue union-closed and intersection-closed families [4]. We have formalized their procedure in Isabelle/HOL.

## 4.1    Abstract Procedure Specification

**Objects.** The most natural way to model sets of natural numbers in Isabelle/HOL is to use the built-in *nat set* type. The type *nat set set* could

---

[1] By following Read [20], we formalized a slightly more general case where the lists can contain larger numbers than 1 (so at some future point multigraphs can also be considered).

be used for families of sets. However, in order to apply Faradžev-Read enumeration, we need to define a very specific total order of families (based on a specific ordering of sets). We cannot change the default ordering of sets on the type $'a$ *set* nor the ordering of families on the type $'a$ *set set*. Additionally, only finite sets can be ordered, so we must introduce the following two types.

**typedef** *Set* = "{ $s :: nat\ set.\ finite\ s$ }" **morphisms** *elems Set*
**typedef** *Family* = "{ $s :: Set\ set.\ finite\ s$ }" **morphisms** *sets Family*

The union-closed property is defined as follows.

*union s1 s2 = Set (elems s1 $\cup$ elems s2)*
*union_closed F $\longleftrightarrow$ ($\forall$ A $\in$ sets F. $\forall$ B $\in$ sets F. union A B $\in$ sets F)*

We want to enumerate all families closed for union whose largest set is $\{0, 1, \ldots, n-1\}$. Since the empty set does not affect union-closedness, when enumerating union-closed families it is usually excluded from all families. Enumeration starts from the family $\{\{0, 1, \ldots, n-1\}\}$, and extends it by adding sets with less elements. We define the dimension of a family, as the number of its sets without this largest set. Therefore, we define collections $S_0, S_1, S_2, \ldots$ by the following definition.

"*S n q = {F. ($\forall$ s $\in$ sets F. elems s $\subseteq$ {0..$<n$}) $\wedge$ card (sets F) = q+1*"
      *Set {0..$<n$} $\in$ sets s $\wedge$ Set {} $\notin$ sets F $\wedge$ union_closed F}*

**Equivalence.** The function *permute_family* permutes every set in a family by applying the *permute_set* function which permutes a set by applying the given permutation to each member. The function *permute_family* interprets the locale *Permute* (with the invariant that all elements if family sets are less than $n$) and the definition and properties of equivalence given in that locale are used.

**Ordering.** The ordering of families is based on an ordering of sets. Sets are ordered first by their cardinality (sets with more elements are declared to precede sets with less elements). Sets of the same cardinality are ordered by lexicographically comparing reverse-sorted lists of their elements. For example, the following is a strictly increasing chain of sets $\{0, 1, 2\} < \{0, 1\} < \{0, 2\} < \{1, 2\} < \{0\} < \{1\} < \{2\} < \{\}$.

*less_Set (Set $s_1$) (Set $s_2$) $\longleftrightarrow$*
   (**let** $n_1$ = *card* $s_1$; $n_2$ = *card* $s_2$
    **in** $n_1 > n_2 \vee (n_1 = n_2 \wedge rev$ (*sorted_list_of_set* $s_1$) $> rev$ (*sorted_list_of_set* $s_2$)))

Families are ordered by lexicographically comparing sorted lists of their sets (wrt. the previous ordering of sets).

*less_Family (Family F1) (Family F2) $\longleftrightarrow$ sorted_list_of_set F1 < sorted_list_of_set F2*

**Canonical Objects.** Canonical families are also defined by using permutations, by reusing definitions and statements from the *Permute* locale—two families are equivalent if there is a permutation that transforms one family to the other, and a family is canonical if it is the least one (wrt. the ordering of families) among all its permutations. We have formalized an efficient method for testing if a given family is canonical. The crucial insight is that if a family is obtained by augmenting a canonical family (and that is always the case in the Faradzev-Read scheme), then it is certainly less than all families obtained by permutations that change some of its sets with cardinality greater than minimal. Therefore, it is enough to check only the permutations that fix such sets. For example, when extending the family $\{\{0,1,2\},\{0,1\},\{0,2\},\{0\}\}$, the permutation $0 \mapsto 1, 1 \mapsto 2, 2 \mapsto 0$ needs not to be considered since it maps non-minimal cardinality sets $\{0,1,2\},\{0,1\},\{0,2\}$ to $\{0,1,2\},\{0,1\},\{1,2\}$, thus always yielding a greater family. Only permutations that map 0 to 0 need to be analyzed. Note that this is one of the crucial components of the algorithm, since it tremendously reduces the number of permutations that have to be applied to check if a given family is canonical (as the number of sets in families is increased, that number very quickly drops to just a couple of permutations).

$min\_card\ F = Min\ (set\_card\ `\ (sets\ F))$
$above\_card\_sets\ F\ c = \{s.\ s \in sets\ F \wedge set\_card\ s > c\}$
$perm\_fixes\ F \longleftrightarrow (\forall\ s \in F.\ permute\_set\ p\ s \in F)$
$filter\_perms\ ps\ F = (\textbf{let}\ F' = above\_card\_sets\ F\ (min\_card\ F)$
$\qquad\qquad \textbf{in}\ filter\ (\lambda\ p.\ perm\_fixes\ p\ F')\ ps)$
$is\_canon\_test\ n\ F \longleftrightarrow (\forall\ p \in set\ (filter\_perms\ (permute\ [0..<n])\ F).$
$\qquad\qquad F \leq permute\_family\ p\ F)$

**Augmentation.** A family is augmented by adding sets that are larger than its largest set (wrt. the ordering of sets).

$augment\_set\ n\ s = \{s'.\ elems\ s' \neq \{\} \wedge elems\ s' \subseteq \{0..<n\} \wedge s' > s\}$

$augment\ n\ F = (\textbf{let}\ Fs = \{add\_set\ F\ s \mid s.\ s \in augment\_set\ n\ (Max\ (sets\ F))\}$
$\qquad\qquad \textbf{in}\ sorted\_list\_of\_set\ \{F' \in Fs.\ union\_closed\ F'\})$

Testing if a family is union-closed requires analyzing all pairs of sets. However, since families are generated by adding sets to smaller union-closed families, we only need to find unions of the new set $s$ with the sets present in the family $F$ that is being augmented. Since $s$ is larger than all sets in $F$, the procedure can be optimized. It suffices to check only those sets of $F$ that do not contain subsets in $F$ (those sets form the *reduction* of $F$).

$reduction\ F = \{s \in sets\ F.\ \neg\ (\exists\ s' \in sets\ F.\ elems\ s' \subset elems\ s)\}$
**lemma**
   **assumes** "$union\_closed\ F$" "$\forall\ s' \in sets\ F.\ s > s'$"
   **shows** "$union\_closed\ (add\_set\ F\ s) \longleftrightarrow (\forall\ s' \in reduction\ F.\ union\ s'\ s \in sets\ F)$"

Note that many previous definitions are not efficient or even not executable (e.g., in the augmentation procedure it is not specified how to construct sets larger than the given one, and the required sorted order of the resulting list of families is ensured by explicitly sorting the list, which would be inefficient in a real implementation). However, abstract specification like this one are very convenient for proving algorithm correctness, while efficient executable implementation can be defined later.

## 4.2  Implementation

The abstract procedure specification already contains two very important optimizations: filtering permutations when checking canonicity and filtering sets when checking union-closedness. However, there are many additional optimizations that should be done in order to get an executable, efficient implementation of the procedure:

- sets and families must be represented using efficient data structures;
- objects should be generated in-order, and a-posteriori sorting must be avoided;
- computations that are redundantly repeated many times should be avoided by applying memorization and storing results in lookup tables.

Unlike abstract specification that is stateless, an efficient implementation must be stateful. There are many methods to handle state in functional programs, and we use the simplest one: it is explicitly passed trough function calls.

**Objects.** Using bitwise representation is the best choice for representing sets and families. A set can be represented by an unsigned integer that has the bit $i$ equal to 1 iff the set contains the element $i$. For example, if 8-bit words are used, the set $\{0, 2, 5\}$ can be represented by 00100101, i.e., by 37. Similarly, a family can be represented by unsigned integer that has the bit $i$ equal to 1 iff the family contains the set represented by $i$. Since there are $2^{2^n}$ families over $\{0, \ldots, n-1\}$, 64-bit words can be used to represent families over $\{0, \ldots, 5\}$.

However, since we wanted to make a very clear separation between the high-level algorithm correctness and low-level bit-twiddling hacks, we have introduced another layer of abstraction. We have introduced another locale, parametrised by the type $'s$ for representing sets and $'f$ for representing families, and by some primitive operations over these types.

For example, a type $'s$ that represents sets must support a constant for the empty set, must support reading the list of set elements, checking if the set contains an element, adding an element to a set, finding union of two sets, determining the cardinality of a set, finding the list of all possible subsets of $\{0, \ldots, n-1\}$, etc. It must be linearly ordered and that order must be compliant with the lexicographic order of reversed lists of set elements. Since only elements up to a certain size must be represented, all assumptions in our locale are guarded by the condition $n \leq n_{max}$, where $n_{max}$ is a locale parameter. Based on such primitives, algorithm-specific set operations are defined (e.g., ordering of sets is defined based on *card* and $<$, and permuting sets is defined by traversing the list of elements and inserting their permuted images into a resulting set).

A type *'f* must support a constant for the empty family, must support reading the list of family sets, adding and removing set from the given family etc. Again, value $n_{max}$ assures that all families can be properly represented.

**Caching Information About Families.** Each family $F$ must contain information about all sets that it contains (and this is represented by a value of type *'f*). However, in order to avoid repeating computations, we shall associate some additional data with each family. For augmentation of $F$ we need to know the maximal set and the reduction of a family (so that we can efficiently check union-closure). For testing canonicity we need to know a list of permutations that fix sets in $F$ with cardinality above minimal. We store all these in a record (permutations are represented by numbers from 0 to $n! - 1$).

**datatype** *('f, 's) FamilyRecord =*
  *FamilyRecord (all_sets : 'f) (max_set : 's) (reduction : 'f) (perms : "nat list")*

**Ordering.** Ordering families is a bit tricky in the general case. If bitwise representation is used, the order of family codes need not necessarily comply with our abstract ordering of families (which takes into account set cardinality). However, within the enumeration we only compare families with their permuted variants for permutations that fix all sets except those with the minimal cardinality. Therefore, it suffices just to extract sets with minimal cardinality and compare two families based only on those sets. When bitwise representation is used, since all other bits will be the same, it suffices just to compare family codes.

**Canonical Objects.** Due to a relatively low number of subsets of $\{0, \ldots, n - 1\}$ (for $n = 6$, there are only 64 such sets) and a relatively low number of permutations of $[0, \ldots, n - 1]$ (for $n = 6$, there are only 720 such permutations), the action of all permutations on all sets can be precomputed and stored into a lookup table (we use a RBT Mapping available from the Isabelle/HOL library). The function that initializes the lookup table can easily be defined and it need not be very efficient (it is called only once at the very beginning of the procedure).

**type_synonym** *'s SetPerms = "((nat × 's), 's) mapping"*

*init_set_perms n =*
  *(**let** ps = permute [0..<n]; ss = powerset n;*
    *keys = concat (map (λ p. map (λ s. (p, s)) ss) [0..<length ps])*
  **in** *Mapping.tabulate keys (λ (p, s). permute_set n (ps ! p) s))*

In the previous code, the function *permute* is defined within our small library for generating basic combinatorial objects and it generates all permutations of the given list. The function that permutes a given family then just looks up permuted sets from the *set_perms* mapping.

*permute_family F p =*
  *foldl (λ F' s. add F' (the (Mapping.lookup set_perms (p, s)))) empty_family (sets F)*

Now the canonicity check can easily be implemented. The list of relevant permutations is stored within the family record. The first permutation in that list is always the identity permutation and it does not need to be checked (so in many cases no family permutations at all need to be made).

*is_canon_test set_perms F =*
  *list_all (λ p. less_eq_family (all_sets F) (permute_family set_perms p (all_sets F)))*
    *(tl (perms F))*

**Augmentation.** Implementing augmentation has several important parts. First, we need to know how to enumerate all augmenting sets for a given set, then we need to check if adding an augmenting set to a family would leave it union-closed and finally, when the set is added we need to update the list of permutations that need to be tested when checking if the family is canonical, to update the reduction of the family and to update its maximal set.

The function that finds all possible augmentations for a given set might be implemented in the following way (again, it does not need to be much efficient, since it is also called only once).

*augment_set n s = filter (λ s'. s < s') (set_of (combine [0..<n] (card s))) @*
      *concat (map (λ c'. set_of (combine [0..<n] c')) (rev [1..<card s]))*

The function *combine* is also defined within our small library for generating basic combinatorial objects and *combine l k* computes all $k$-element sublists of the given list $l$. Augmenting sets of a set $s$ first contain sets with the same cardinality as $s$ that are larger than it, and then, all sets of each cardinality less than the cardinality of $s$, in decreasing order (this gives a sorted list of all augmenting sets wrt. our set order).

Since the same sets are augmented over and over again (as they occur in different families), results of *augment_set* for each $s$ in *powerset n* are stored in a lookup table and that lookup table becomes a parameter of the family augmentation procedure *augment*.

Each augmenting set is analyzed and it is checked if adding it to the family leaves it union-closed. This is done by examining only the sets from the reduced family (which are stored within the family record).

*is_union_closed F s = list_all (λ s'. contains_set F (union s' s)) (sets (reduction F))*

If adding the set $s$ to the family $F$ leaves it union closed, then a new family record is created. The set is added to the collection of all sets using the primitive operation and it is set as the maximal set of the extended family (since the augmenting sets are always larger than all sets in the family). The reduction of the extended family is obtained by analyzing all sets in the reduction of the original family $F$, removing those that contain $s$ (by means of the primitive operation), and by adding $s$ to the reduction (as the maximal set it has the minimal cardinality and the family cannot contain its subset).

```
update_reduction Rs s =
   (let Rs' = foldl (λ Rs s'. if is_subset s s' then remove Rs s' else Rs) Rs (sets Rs)
    in add Rs' s)

add_set F s = FamilyRecord (add (all_sets F) s)
                          (update_reduction (reduction F) s) s (perms F)
```

Finally, the augmented set is added to the family and if its cardinality is strictly less than cardinality of other sets in the family, the set of permutations is filtered (permutations that do not fix sets above minimal cardinality are removed).

```
perm_fixes set_perms F p ⟷
   list_all (λ s. contains_set F (the (Mapping.lookup set_perms (p, s)))) (sets F)

filter_perms n set_perms F c =
   (let perms' = filter (perm_fixes set_perms (sets_of_card F c)) (perms F)
    in FamilyRecord (all_sets F) (reduction F) (last_set F) perms')

extend_family n set_perms F s =
   (let F' = add_set F s; c = card (max_set F); c' = card s
    in if c ≠ c' then filter_perms n set_perms c F' else F')
```

With these functions available, we define the augmenting procedure.

```
augment n augmenting_sets powerset_by_card set_perms F =
   map (extend_family n powerset_by_card set_perms F)
      (filter (λ s. is_union_closed F s)
         (the (Mapping.lookup augmenting_sets (max_set F))))
```

**Correctness Proof.** The correctness proof reduces to showing that this stateful implementation corresponds to the abstract specification. Functions *abs_set* and *abs_family* that convert *'s* to *Set* and *'f* to *Family* are easily defined and it is easily shown (by using the locale assumptions) that primitive operations given in a locale are in accordance with operations on sets (the real burden of showing this is when interpreting the locale by bitwise representation). Then, a set of lemmas is proved that connects each implemented function with its abstract counterpart. For example, the lemma that establishes the connection between the abstract test for canonicity and its implementation is the following.

**lemma**
    **assumes** *"$n \leq n\_max$" "inv_f n (all_sets F)"*
                *"set_perms_OK set_perms n" "perms_filtered F n" "hd_perms F"*
    **shows** *"FamilyImpl.is_canon_test n set_perms F ⟷*
        *FamilyAbs.is_canon_test n (abs_fam (all_sets F))"*

The assumptions require that all sets in the family record satisfy all required representation invariants (for example, this guarantees that all sets in $F$ are subsets of $\{0, \ldots, n-1\}$), that the lookup table *set_perms* contains permutations of all sets, that the family record contains exactly those permutations that fix all sets of $F$ with more elements than the minimal set of $F$, and that the first element in the list of those permutations is the identity permutation. In many cases such lemmas are proved almost immediately (by using similar lemmas about functions called in the current function definition). However, in some cases there is more work that should be done (e.g., we need to show that our *augment_set* implementation builds a sorted and distinct list of sets that covers every set that is larger than the one being augmented).

It is also necessary to show that functions preserve invariants. All lookup tables are initialized before the enumeration starts and we prove that functions that initialize lookup tables do that correctly. For example, we show that *init_set_perms* builds a lookup table that for each set $s$ in *powerset* $n$ and each permutation index $p$ from 0 to $n! - 1$ returns the set obtained by permuting $s$ by the $p$-th permutation in the lexicographic ordering of permutations of $[0, \ldots, n-1]$. Other invariants characterize data in the family record. For example, one such invariant claims that *max_set* $F$ is always a set in $F$ that is the largest among all sets of $F$. Since the family record is updated only in the *augment* function, the major challenge is to show that it preservers all such invariants.

### 4.3  Bitwise Set Representation

Finally, we used the bitwise representation to represent sets and families, based on the Native word library [14]. Sets are represented by the type uint8, while families are represented by the type uint64. Primitive operations on sets are implemented using the bitwise operations. For example, adding element and removing element from a set, union and intersection of sets is defined by

*add x e = x OR (1 << k)*    *remove x e = x AND NOT (1 << k)*
*union x1 x2 = x1 OR x2*    *inter x1 x2 = x1 AND x2*

We have also implemented an efficient function for finding the cardinality of a set, by using the parallel bit-count algorithm.

*card s0 =* (**let** *s1 = (s0 AND 0x55) + ((s0 >> 1) AND 0x55);*
         *s2 = (s1 AND 0x33) + ((s1 >> 2) AND 0x33);*
         *s3 = (s2 AND 0x0F) + ((s2 >> 4) AND 0x0F)*
      **in** *nat_of_uint8 s3)*

However, since we calculate cardinality only for 8-bit numbers, it turns out that there is no much benefit to using a naive, sequential bit-testing algorithm.

Similarly, a list of sets in a family could be determined by a naive, sequential test of each of 64 bits. For families that do not contain many sets, it is more efficient to iterate only trough the bits that are set. Many hardware architectures offer *count trailing zeros* (ctz) instruction that is used to find the last set bit. Clearing last set bit can be achieved by calculating x & (x-1). Unfortunately, it seems that ctz instruction is not available from functional languages. It can be implemented by a binary search approach, yielding a six-step algorithm for 64-bit words, but our experiments reveal that using such implementation is less efficient than the naive algorithm.

## 4.4   Results

Our verified implementation exported to Haskell catalogues all 108 281 182 union-closed families in around 11 min (on an 2.4GHz, Intel Core i5, 8GB RAM laptop). Our fastest, unverified implementation in C++ that uses the same algorithm, but is based on arrays, does it in around 28 seconds. Profiling shows that the verified implementation spends more than 60% of the time in RBT lookup. Replacing $O(\log n)$ RBT with $O(1)$ lookup array reduces the time to less than 5 min (for this we can use the Isabelle Collections Framework [11] or Imperative/HOL [6]). Unfortunately, a range-check is performed with each verified bitwise operation, and there is no direct access from Isabelle/HOL to all hardware implemented bitwise operations (e.g., `__builtin_ctzl` in GCC), so its hard to expect that C++ runtimes could be reached with standard Isabelle code generator. When families are only counted using the depth-first variant of the algorithm, memory consumption is not an issue.

## 5   Conclusions and Further Work

We have formalized the general Faradžev-Read scheme for making exhaustive, isomorph-free catalogues of combinatorial objects within Isabelle/HOL and have shown its applicability by instantiating in on two different problems: cataloguing directed graphs and cataloguing union-closed families. In the second case study we have created an efficient implementation capable of generating more than one hundred million union-closed families over a six-element domain.

Our experience shows that even with the general scheme verified, there is still much work to do for each concrete application, especially if efficient implementation is required (our rough estimate is that verifying the general scheme is around 30–50% of the effort needed to verify a concrete efficient algorithm). Still, a verified general scheme does save a significant amount of work in each concrete instance, and, more importantly, guides us towards elements that should be defined in order to get an efficient algorithm.

Specification and the correctness proof of the abstract Faradžev-Read scheme contains 3 locales with around 10 assumptions, 10 definitions and 40 lemmas, consuming around 1200 lines of code (LOC). The case study of digraphs contains around 25 definitions and 100 lemmas, consuming around 4000 LOC. The case study of union-closed families contains around 105 definitions and 350 lemmas, consuming around 8000 LOC (5000 LOC are devoted to efficient implementation). Some definitions and lemmas are shared between both case studies.

We use refinement based on Isabelle/HOL locales to separate reasoning about abstract procedure properties and concrete implementation details. Using a framework (e.g., Isabelle refinement framework [12]) might give us better proof automation and easier introduction of imperative features [13], so we plan to use it in our future work.

There are other general cataloguing schemes, some more efficient than Faradžev-Read's. Most notable of them is McKay's canonical path generation [1]. In our further work we plan to formalize it, too. We hope that some parts of the developed theory could be reused (e.g., the definition of isomorphism based on permutations and their action and the definition of the catalogue). On the other hand, Faradzev/Read and McKay's approach are substantially different so we are not too optimistic that any parts of Faradzev/Read algorithm specification would be useful for computing canonical labellings. A prerequisite for McKay's algorithm trusted implementation is an efficient, trusted graph isomorphism testing algorithm which we plan to construct (either by its verification within a theorem prover, or by some kind of certificate checking).

# References

1. McKay, B.D.: Isomorph-free exhaustive generation. J. Algorithms **26**(2), 306–324 (1998)
2. Bowles, J., Caminati, M.B.: A verified algorithm enumerating event structures. In: Geuvers, H., England, M., Hasan, O., Rabe, F., Teschke, O. (eds.) CICM 2017. LNCS (LNAI), vol. 10383, pp. 239–254. Springer, Cham (2017). https://doi.org/10.1007/978-3-319-62075-6_17
3. Brinkmann, G.: Isomorphism rejection in structure generation programs. In: Discrete Mathematical Chemistry (1998)
4. Brinkmann, G., Deklerck, R.: Generation of union-closed sets and Moore families. J. Integer Sequences **21**(1), 9–18 (2018)
5. Bruhn, H., Schaudt, O.: The journey of the union-closed sets conjecture. Graphs Comb. **31**(6), 2043–2074 (2015)
6. Bulwahn, L., Krauss, A., Haftmann, F., Erkök, L., Matthews, J.: Imperative functional programming with Isabelle/HOL. In: Mohamed, O.A., Muñoz, C., Tahar, S. (eds.) TPHOLs 2008. LNCS, vol. 5170, pp. 134–149. Springer, Heidelberg (2008). https://doi.org/10.1007/978-3-540-71067-7_14
7. Erard, C., Giorgetti, A.: Bounded exhaustive testing with certified and optimized data enumeration programs. In: Gaston, C., Kosmatov, N., Le Gall, P. (eds.) ICTSS 2019. LNCS, vol. 11812, pp. 159–175. Springer, Cham (2019). https://doi.org/10.1007/978-3-030-31280-0_10
8. Faradzev, I.A.: Constructive enumeration of combinatorial objects. Colloques Int. CNRS **260**, 131–135 (1978)
9. Giorgetti, A., Dubois, C., Lazarini, R.: Combinatoire formelle avec why3 et coq. In: Journées Francophones des Langages Applicatifs (JFLA 2019), pp. 139–154, Les Rousses, France (2019)
10. Kerber, A., Laue, R.: Group actions, double cosets, and homomorphisms: unifying concepts for the constructive theory of discrete structures. Acta Applicandae Mathematicae **52**, 63–90 (1998). https://doi.org/10.1023/A:1005998722658
11. Lammich, P.: Collections framework. Archive of Formal Proofs. Formal proof development, November 2009. http://isa-afp.org/entries/Collections.html
12. Lammich, P.: Refinement for monadic programs. Archive of Formal Proofs. Formal proof development, January2012. http://isa-afp.org/entries/Refine_Monadic.html
13. Lammich, P.: Refinement to imperative HOL. J. Autom. Reason. **62**(4), 481–503 (2019). https://doi.org/10.1007/s10817-017-9437-1
14. Lochbihler, A.: Native word. Archive of Formal Proofs. Formal proof development, September 2013. http://isa-afp.org/entries/Native_Word.html
15. Maric, F.: Fast formal proof of the Erdős-Szekeres conjecture for convex polygons with at most 6 points. J. Autom. Reasoning **62**, 301–329 (2017)

16. Marić, F., Živković, M., Vučković, B.: Formalizing Frankl's conjecture: FC-families. In: Jeuring, J., et al. (eds.) CICM 2012. LNCS (LNAI), vol. 7362, pp. 248–263. Springer, Heidelberg (2012). https://doi.org/10.1007/978-3-642-31374-5_17

17. Nipkow, T.: Verified efficient enumeration of plane graphs modulo isomorphism. In: van Eekelen, M., Geuvers, H., Schmaltz, J., Wiedijk, F. (eds.) ITP 2011. LNCS, vol. 6898, pp. 281–296. Springer, Heidelberg (2011). https://doi.org/10.1007/978-3-642-22863-6_21

18. Nipkow, T., Bauer, G., Schultz, P.: Flyspeck I: tame graphs. In: Furbach, U., Shankar, N. (eds.) IJCAR 2006. LNCS (LNAI), vol. 4130, pp. 21–35. Springer, Heidelberg (2006). https://doi.org/10.1007/11814771_4

19. Nipkow, T., Paulson, L.C., Wenzel, M.: Isabelle/HOL—A Proof Assistant for Higher-Order Logic. LNCS, vol. 2283. Springer, Heidelberg (2002). https://doi.org/10.1007/3-540-45949-9

20. Read, R.C.: Every one a winner or how to avoid isomorphism search when cataloguing combinatorial configurations. In: Alspach, B., Hell, P., Miller, D. (eds.) Algorithmic Aspects of Combinatorics, Annals of Discrete Mathematics, vol. 2, pp. 107–120. Elsevier (1978)

# Verification

# Verified Approximation Algorithms

Robin Eßmann[1] , Tobias Nipkow[1] , and Simon Robillard[2(✉)]

[1] Technische Universität München, Munich, Germany
[2] IMT Atlantique, Nantes, France

**Abstract.** We present the first formal verification of approximation algorithms for NP-complete optimization problems: vertex cover, independent set, load balancing, and bin packing. We uncover incompletenesses in existing proofs and improve the approximation ratio in one case.

## 1 Introduction

Approximation algorithms for NP-complete problems [12] are a rich area of research untouched by automated verification. We present the first formal verifications of three classical and one lesser known approximation algorithm. Three of these algorithms had been verified on paper by program verification experts [3,4]. We found that their claimed invariants need additional conjuncts before they are strong enough to be real invariants. That is, their proofs are incomplete. The fourth algorithm only comes with a sketchy informal proof.

To put an end to this situation we formalized the correctness proofs of four approximation algorithms for fundamental NP-complete problems in the theorem prover Isabelle/HOL [9,10]. We verified (all proofs are online [6]) that

- the classic approximation algorithm for a minimal vertex cover is a $k$-approximation algorithm for rank $k$ hypergraphs;
- Wei's algorithm for a maximal independent set [13] is a $\Delta$-approximation algorithm for graphs with maximum degree $\Delta$;
- the greedy algorithm for the load balancing problem is a $\frac{3}{2}$-approximation algorithm if job loads are sorted and a 2-approximation algorithm if job loads are unsorted [8];
- the bin packing algorithm by Berghammer and Reuter [4] is a $\frac{3}{2}$-approximation algorithm.

Isabelle not only helped finding mistakes in pen-and-paper proofs but also encouraged proof refactoring that led to simpler proofs, and in one case, to a stronger result: The invariant given by Berghammer and Müller for Wei's algorithm [3] is sufficient to show that the algorithm has an approximation ratio of $\Delta + 1$. We managed to simplify their argument significantly which lead to an improved approximation ratio of $\Delta$.

All algorithms are expressed in a simple imperative *WHILE*-language. In each case we show that the approximation algorithm computes a valid solution

N. Peltier and V. Sofronie-Stokkermans (Eds.): IJCAR 2020, LNAI 12167, pp. 291–306, 2020.
https://doi.org/10.1007/978-3-030-51054-1_17

that is at most a constant factor worse than an optimum solution. The polynomial running time of the approximation algorithm is easy to see in each case.

## 2  Isabelle/HOL and Imperative Programs

Isabelle/HOL is largely based on standard mathematical notation but with some differences and extensions.

Type variables are denoted by $'a, 'b$, etc. The notation $t :: \tau$ means that term $t$ has type $\tau$. Except for function types $'a \Rightarrow 'b$, type constructors follow postfix syntax, e.g. $'a$ set is the type of sets of elements of type $'a$. Function some :: $'a\ set \Rightarrow 'a$ picks an arbitrary element from a set; the result is unspecified if the set is empty.

The types nat and real represent the sets $\mathbb{N}$ and $\mathbb{R}$. In this paper we drop the coercion function real :: nat $\Rightarrow$ real. The set $\{m..n\}$ is the closed interval $[m, n]$.

The Isabelle/HOL distribution comes with a simple implementation of Hoare logic where programs are annotated with pre- and post-conditions and invariants (all in HOL) as in this example, where all variables are of type nat:

$$\{m = 0 \wedge p = 0\}$$
$$WHILE\ m \neq a\ INV\ \{p = m * b\}\ DO\ p := p + b;\ m := m + 1\ OD$$
$$\{p = a * b\}$$

The box around the program means that it has been verified. All our proofs employ a VCG and essentially reduce to showing the preservation of the invariants.

## 3  Vertex Cover

We verify the proof in [3] that the classic greedy algorithm for vertex cover is a 2-approximation algorithm. In fact, we generalize the setup from graphs to hypergraphs. A hypergraph is simply a set of edges $E$, where an edge is a set of vertices of type $'a$. A vertex cover for $E$ is a set of vertices $C$ that intersects with every edge of $E$:

$$vc :: 'a\ set\ set \Rightarrow 'a\ set\ \Rightarrow\ bool$$
$$vc\ E\ C\ = (\forall\ e \in E.\ e \cap C \neq \emptyset)$$

A matching (matching :: $'a\ set\ set\ \Rightarrow\ bool$) is a set of pairwise disjoint sets. The following is a key property that relates vc and matching:

$$finite\ C\ \wedge\ matching\ M\ \wedge\ M \subseteq E\ \wedge\ vc\ E\ C\ \longrightarrow\ |M| \leq |C|$$

We fix a rank-$k$ hypergraph $E :: 'a\ set\ set$ assuming $\emptyset \notin E$, finite $E$ and $e \in E \longrightarrow finite\ e\ \wedge\ |e| \leq k$.

We have verified the well known greedy algorithm that computes a vertex cover $C$ for $E$. It keeps picking an arbitrary edge that is not covered by $C$ yet until all vertices are covered. The final $C$ has at most $k$ times as many vertices as any vertex cover of $E$ (which is essentially optimal [1]).

$$\{True\}$$
$$C := \emptyset;\ F := E;$$
$$WHILE\ F \neq \emptyset\ INV\ \{invar\ C\ F\}$$
$$DO\ C := C \cup some\ F;\ F := F - \{e' \in F \mid some\ F \cap e' \neq \emptyset\}\ OD$$
$$\{vc\ E\ C\ \wedge (\forall C'.\ finite\ C'\ \wedge vc\ E\ C' \longrightarrow |C| \leq k * |C'|)\}$$

where *invar* is the following invariant:

$$invar :: 'a\ set\ \Rightarrow\ 'a\ set\ set\ \Rightarrow\ bool$$
$$invar\ C\ F =$$
$$(F \subseteq E \wedge vc\ (E - F)\ C \wedge finite\ C \wedge (\exists M.\ inv\_matching\ C\ F\ M))$$

$$inv\_matching\ C\ F\ M =$$
$$(matching\ M \wedge M \subseteq E \wedge |C| \leq k * |M| \wedge (\forall e \in M.\ \forall f \in F.\ e \cap f = \emptyset))$$

The key step in the program proof is that the invariant is invariant:

**Lemma 1.** $F \neq \emptyset \wedge invar\ C\ F \longrightarrow$
$\quad invar\ (C \cup some\ F)\ (F - \{e' \in F \mid some\ F \cap e' \neq \emptyset\})$

Our invariant is stronger than the one in [3] which lacks $F \subseteq E$. But without $F \subseteq E$ the claimed invariant is not invariant (as acknowledged by Müller-Olm).

## 4   Independent Set

As in the previous section, a graph is a set of edges. An independent set of a graph $E$ is a subset of its vertices such that no two vertices are adjacent.

$$iv :: 'a\ set\ set\ \Rightarrow\ 'a\ set\ \Rightarrow\ bool$$
$$iv\ E\ S = (S \subseteq \bigcup E \wedge (\forall v_1\ v_2.\ v_1 \in S \wedge v_2 \in S \longrightarrow \{v_1,\ v_2\} \notin E))$$

We fix a finite graph $E :: 'a\ set\ set$ such that all edges of $E$ are sets of cardinality 2. The set of vertices $\bigcup E$ is denoted $V$, and the maximum number of neighbors for any vertex in $V$ is denoted $\Delta$. We show that the greedy algorithm proposed by Wei is a $\Delta$-approximation algorithm. The proof is inspired by one given in [3]. In particular, the proof relies on an auxiliary variable $P$, which is not needed for the execution of the algorithm, but is used for bookkeeping in the proof. In [3], $P$ is initially a program variable and is later removed from the program and turned into an existentially quantified variable in the invariant. We directly use the latter representation.

$\{\, True\, \}$
$S := \emptyset;\ X := \emptyset;$
$WHILE\ X \neq V\ INV\ \{\ \exists P.\ inv\_partition\ S\ X\ P\ \}$
$DO\ x := some\,(V - X);\ S := S \cup \{x\};\ X := X \cup neighbors\ x \cup \{x\}\ OD$
$\{\ iv\ E\ S\ \wedge\ (\forall S'.\ iv\ E\ S'\ \longrightarrow\ |S'| \leq |S| * \Delta)\ \}$

To keep the size of definitions manageable, we split the invariant in two. The first part is not concerned with $P$, but suffices to prove the functional correctness of the algorithm, i.e. that it outputs an independent set of the graph:

$inv\_iv :: {}'a\ set\ \Rightarrow\ {}'a\ set\ \Rightarrow\ bool$
$inv\_iv\ S\ X\ =$
$(iv\ E\ S \wedge\ X \subseteq V \wedge\ (\forall v_1 \in V\ -\ X.\ \forall v_2 \in S.\ \{v_1,\ v_2\}\ \notin\ E)\ \wedge\ S\ \subseteq\ X)$

This invariant is taken almost verbatim from [3], except that in [3] it says that $S$ is an independent set of the subgraph generated by $X$. This is later used to show that the $x$ picked at each iteration from $V - X$ is not already in $S$. Defining subgraphs adds unnecessary complexity to the invariant. We simply state $S \subseteq X$, together with the fact that $S$ is an independent set of the whole graph.

We now extend the invariant with properties of the auxiliary variable $P$.

$inv\_partition :: {}'a\ set\ \Rightarrow\ {}'a\ set\ \Rightarrow\ {}'a\ set\ set\ \Rightarrow\ bool$
$inv\_partition\ S\ X\ P\ =$
$(inv\_iv\ S\ X\ \wedge$
$\bigcup P = X \wedge\ (\forall\, p \in P.\ \exists s \in V.\ p = \{s\}\ \cup\ neighbors\ s)\ \wedge\ |P|\ =\ |S|\ \wedge\ finite\ P)$

We can view the set $P$ as an auxiliary program variable. In order to satisfy the invariant, $P$ would be initially empty and the loop body would include the assignement $P := P \cup \{neighbors\ x\ \cup\ \{x\}\}$. Intuitively, $P$ contains the sets of vertices that are added to $X$ at each iteration (or more precisely, an over-approximation, since some vertices in $neighbors\ x$ may have been added to $X$ in a previous iteration). Instead of adding an unnecessary variable to the program, we only use the existentially quantified invariant. The assignments described above correspond directly to instantiations of the quantifier that are needed to solve proof obligations. This is illustrated with the following lemma, which corresponds to the preservation of the invariant:

**Lemma 2.** $(\exists P.\ inv\_partition\ S\ X\ P) \wedge x \in V - X \longrightarrow$
$(\exists P'.\ inv\_partition\ (S\ \cup\ \{x\})\ (X\ \cup\ neighbors\ x\ \cup\ \{x\})\ P')$

The existential quantifier in the antecedent yields a witness $P$. After instantiating the quantifier in the succedent with $P \cup \{neighbors\ x\ \cup\ \{x\}\}$, the goal can be solved straightforwardly. Finally, the following lemma combines the invariant and the negated post-condition to prove the approximation ratio:

**Lemma 3.** $inv\_partition\ S\ V\ P\ \longrightarrow\ (\forall S'.\ iv\ E\ S'\ \longrightarrow\ |S'|\ \leq\ |S|\ *\ \Delta)$

To prove it, we observe that any set $p \in P$ consists of a vertex $x$ and its neighbors, therefore an independent set $S'$ can contain at most $\Delta$ of the vertices in $p$, thus $|S'| \leq |P| * \Delta$. Furthermore, as indicated by the invariant, $|P| = |S|$.

Compared to the proof in [3], our invariant describes the contents of the set $P$ more precisely, and thus yields a better approximation ratio. In [3], the invariant merely indicates that $X = \bigcup P$, together with two cardinality properties: $\forall p \in P$. $|p| \leq \Delta + 1$ and $|P| \leq |S|$. Taken with the negated post-condition, this invariant can be used to show that for any independent set $S'$, we have $|S'| \leq |S| * (\Delta + 1)$. The proof of this lemma makes use of the following (in)equalities: $|S'| \leq |V|, |V| = |\bigcup P|, |\bigcup P| \leq |P| * (\Delta + 1)$ and finally $|P| * (\Delta + 1) \leq |S| * (\Delta + 1)$. Note that this only relies on the trivial fact that an independent set cannot contain more vertices than the graph. By contrast, our own argument takes into account information regarding the edges of the graph.

Although this proof results in a weaker approximation ratio than our own, it yields a useful insight: an approximation ratio is given by the cardinality of the largest set $p \in P$ (i.e., the largest number of vertices added to $X$ during any given iteration). In the worst case, this is equal to $\Delta + 1$, but in practice the number may be smaller. This suggests a variant of the algorithm that stores that value in a variable $r$, as described in [3]. At every iteration, the variable $r$ is assigned the value $max\ r\ |\{x\} \cup neighbors\ x - X|$. Ultimately, the algorithm returns both the independent set $S$ and the value $r$, with the guarantee that $|S'| \leq |S| * r$ for any independent set $S'$.

We also formalized this variant and proved the aforementioned property. The proof follows the idea outlined above, but does away with the variable $P$ entirely: instead, the invariant simply maintains that $inv\_iv\ S\ X \wedge |X| \leq |S| * r$, and the proof of preservation is adapted accordingly. Indeed, this demonstrates that the argument used in [3] does not require an auxiliary variable nor an existentially quantified invariant. For the proof of the approximation ratio $\Delta$, a similar simplification is not as easy to obtain, because the argument relies on a global property of the graph (a constraint that edges impose on independent sets) that is not easy to summarize in an inductive invariant.

So far, we have only considered an algorithm where the vertex $x$ is picked non-deterministically. An obvious heuristic is to pick, at every iteration, the vertex with the smallest number of neighbors among $V - X$. Halldórsson and Radhakrishnan [7] prove that this heuristic achieves an approximation ratio of $(\Delta + 2) / 3$. However their proof is far more complex than the arguments presented here. It is also not given as an inductive invariant, instead relying on case analysis for different types of graphs. This is beyond the scope of our paper.

## 5   Load Balancing

Our starting point for the load balancing problem is [8, Chapter 11.1]. We need to distribute $n :: nat$ jobs on $m :: nat$ machines with $0 < m$. A job $j \in \{1..n\}$ has a load $t(j) :: nat$. Variables $m$, $n$, and $t$ are fixed throughout this section. A solution is described by a function $A$ that maps machines to sets of jobs: $k \in \{1..m\}$ has job $j$ assigned to it iff $j \in A(k)$. The sum of job loads on a machine is given by a function $T$ that is derived from $t$ and $A$: $(\sum j \in A\ k.\ t\ j) = T\ k$. Predicate $lb$ defines when $T$ and $A$ are a partial solution for $j \leq n$ jobs:

$lb :: (nat \Rightarrow nat) \Rightarrow (nat \Rightarrow nat\ set) \Rightarrow nat \Rightarrow bool$
$lb\ T\ A\ j =$
$((\forall x \in \{1..m\}.\ \forall y \in \{1..m\}.\ x \neq y \longrightarrow A\ x\ \cap\ A\ y\ =\ \emptyset)\ \wedge$
$(\bigcup_{x \in \{1..m\}}\ A\ x)\ =\ \{1..j\}\ \wedge\ (\forall x \in \{1..m\}.\ (\sum y \in A\ x.\ t\ y)\ =\ T\ x))$

It consists of three conjuncts. The first ensures that the sets returned by $A$ are pairwise disjoint, thus, no job appears in more than one machine. The second conjunct ensures that every job $x \in \{1..j\}$ is contained in at least one machine. It also ensures that only jobs $\{1..j\}$ have been added. The final conjunct ensures that $T$ is correctly defined to be the total load on a machine. To ensure that jobs are distributed evenly, we need to consider the machine with maximum load. This load is referred to as the *makespan* of a solution (where $f\ `\ I$ is the image of $f$ over $I$):

$makespan :: (nat \Rightarrow nat) \Rightarrow nat$
$makespan\ T = Max\ (T\ `\{1..m\})$

The greedy approximation algorithm outlined in [8] relies on the ability to determine the machine $k \in \{1..m\}$ that has a minimum combined load. As the goal is to approximate the optimum in polynomial time, a linear scan through $T$ suffices to find the machine with minimum load. However, other methods may be considered to further improve time complexity. To determine the machine with minimum load, we will use the following function:

$min_k :: (nat \Rightarrow nat) \Rightarrow nat \Rightarrow nat$
$min_k\ T\ 0\ =\ 1$
$min_k\ T\ (x + 1)\ =$
$(let\ k\ =\ min_k\ T\ x\ in\ if\ T\ (x + 1)\ <\ T\ k\ then\ x + 1\ else\ k)$

We will focus on the approximation factor of $\frac{3}{2}$, which can be proved if the job loads are assumed to be sorted in descending order. The proof for the approximation factor of 2 if jobs are unsorted is very similar and we describe the differences at the end. We say that $j$ jobs are sorted in descending order if *sorted* holds:

$sorted :: nat \Rightarrow bool$
$sorted\ j\ =\ (\forall x \in \{1..j\}.\ \forall y \in \{1..x\}.\ t\ x \leq t\ y)$

Below we prove the following conditional Hoare triple that expresses the approximation factor and functional correctness of the algorithm given in [8]:

$$
\begin{array}{l}
sorted\ n \longrightarrow \\
\{True\} \\
T := (\lambda \_.\ 0);\ A := (\lambda \_.\ \emptyset);\ j := 0; \\
WHILE\ j < n\ INV\ \{inv_2\ T\ A\ j\} \\
DO\ i := min_k\ T\ m;\ j := j + 1; \\
\quad A := A(i := A(i) \cup \{j\});\ T := T(i := T(i) + t(j)) \\
OD \\
\{lb\ T\ A\ n\ \wedge \\
\quad (\forall T'\ A'.\ lb\ T'\ A'\ n \longrightarrow makespan\ T \le 3\ /\ 2 * makespan\ T')\}
\end{array}
$$

Property *sorted n* is not part of the precondition because it is not influenced by the algorithm and thus there is no need to prove that it remains unchanged. Therefore we made *sorted n* an assumption of the whole Hoare triple. The notation $f(a := b)$ denotes an updated version of function $f$ that maps $a$ to $b$ and behaves like $f$ otherwise. Thus an assignment $f := f(i := b)$ is nothing but the conventional imperative array update notation $f[i] := b$.

Functional correctness follows because each iteration extends a partial solution for $j$ jobs to one for $j + 1$ jobs:

**Lemma 4.** $lb\ T\ A\ j \wedge x \in \{1..m\} \longrightarrow$
$\quad lb\ (T(x := T\ x + t\ (j + 1)))\ (A(x := A\ x \cup \{j + 1\}))\ (j + 1)$

Moreover, it is easy to see that the initialization establishes $lb\ T\ A\ j$.

To prove the approximation factor in both the sorted and unsorted case, the following lower bound is important:

**Lemma 5.** $lb\ T\ A\ j \longrightarrow (\sum_{x=1}^{j} t\ x)\ /\ m \le makespan\ T$

This is a result of $\sum_{x=1}^{m} T(x) = \sum_{x=1}^{j} t(x)$ together with this general property of sums: $finite\ A \wedge A \ne \emptyset \longrightarrow (\sum a \in A.\ f\ a) \le |A| * Max\ (f\ `\ A)$.

A similar observation applies to individual jobs. Any job must be a lower bound on some machine, as it is assigned to one and, by extension, it must also be a lower bound of the makespan:

**Lemma 6.** $lb\ T\ A\ j \longrightarrow Max_0\ (t`\{1..j\}) \le makespan\ T$

As any job load is a lower bound on the makespan over the machines, the job with maximum load must also be such a lower bound. Note that $Max_0$ returns 0 for the empty set.

When jobs are sorted in descending order, a stricter lower bound for an individual job can be established. We observe that an added job is at most as large as the jobs preceding it. Therefore, if a machine contains at least two jobs, this added job is only at most *half* as large as the makespan. We can use this observation by assuming the machines to be filled with more than $m$ jobs, as this will ensure that some machine must contain at least two jobs.

**Lemma 7.** $lb\ T\ A\ j \wedge m < j \wedge sorted\ j \longrightarrow 2 * t\ j \le makespan\ T$

Note that this lower bound only holds if there are strictly more jobs than machines. One must, however, also consider how the algorithm behaves in the other case. One may intuitively see that the algorithm will be able to distribute the jobs such that every machine will only have at most one job assigned to it, making the algorithm trivially optimal. To prove this, we need to show the following behavior of $min_k$:

**Lemma 8.**

1. $x \in \{1..m\} \wedge T\ x = 0 \longrightarrow T\ (min_k\ T\ m)\ =\ 0$
2. $x \in \{1..m\} \wedge T\ x = 0 \longrightarrow min_k\ T\ m \leq x$

They can be shown by induction on the number of machines $m$.

As the proof in [8] is only informal, Kleinberg and Tardos do not provide any loop invariant. We propose the following invariant for sorted jobs:

$inv_2 :: (nat \Rightarrow nat) \Rightarrow (nat \Rightarrow nat\ set) \Rightarrow nat \Rightarrow bool$
$inv_2\ T\ A\ j =$
$(lb\ T\ A\ j \wedge j \leq n\ \wedge$
$(\forall T'\ A'.\ lb\ T'\ A'\ j\ \longrightarrow\ makespan\ T \leq 3\ /\ 2 * makespan\ T') \wedge$
$(\forall x > j.\ T\ x = 0) \wedge (j \leq m\ \longrightarrow\ makespan\ T\ =\ Max_0\ (t\ '\ \{1..j\})))$

The final two conjuncts relate to the trivially optimal behavior of the algorithm if $j \leq m$. The penultimate conjunct shows that only as many machines can be occupied as there are available jobs. The final conjunct ensures that every job is distributed on its own machine, making the makespan equivalent to the job with maximum load.

It should be noted that if the makespan is sufficiently large, an added job may not increase the makespan at all, as the machine with minimum load combined with the job may not exceed the previous makespan. As such, we will also consider the possibility that an added job can simply be ignored without affecting the overall makespan.

**Lemma 9.** $makespan\ (T(x := T\ x + y)) \neq T\ x + y \longrightarrow$
$makespan\ (T(x := T\ x + y)) = makespan\ T$

To make use of this observation, we need to be able to relate the makespan of a solution with the added job to the makespan of a solution without it. One can easily show the following by removing $j + 1$ from the solution:

**Lemma 10.** $lb\ T\ A\ (j + 1) \longrightarrow$
$(\exists T'\ A'.\ lb\ T'\ A'\ j \wedge makespan\ T' \leq makespan\ T)$

We can now prove the preservation of $inv_2$. Let $i\ =\ min_k\ T\ m$ be the machine with minimum load. We define:

$$T_g := T\ (i := T(i) + t(j + 1)) \qquad A_g := A\ (i := A(i)\ \cup\ \{j + 1\})$$

We begin with a case distinction. If $j + 1 \leq m$, we can make use of the additional conjuncts to prove the trivially optimal behavior. We first note *in-range*: $j + 1 \in \{1..m\}$. Moreover, from the penultimate conjunct, $T(j + 1) = 0$. Combining this with Lemma 8.1, we can see that $T(i) = 0$. Therefore $T_g(i) = t(j+1)$ and with the final conjunct of the assumed invariant, the makespan of $T_g$ remains equivalent to the job with maximum load. To prove that the penultimate conjunct is preserved, we again use *in-range*, $T(j + 1) = 0$, and Lemma 8.2 to prove that $i \leq j + 1$. Moreover, $T_g$ only differs from $T$ by the modification of machine $i$. Thus, the penultimate conjunct for $j + 1$ jobs is preserved as well. From Lemma 6 we can then see that, as the makespan of $T_g$ is equivalent to the job with maximum load, it must be trivially optimal. Functional correctness can be shown using Lemma 4, and proving the preservation of the remaining conjunct is trivial. We now come to the case $j + 1 > m$. We first show that the penultimate conjunct is preserved (the final conjunct can be ignored, as $\neg\ j + 1 \leq m$). This follows from the correctness of $min_k$, as the index returned by it has to be in $\{1..m\}$ as long as $m > 0$. Therefore, we can simply show this from the penultimate conjunct of the assumed invariant. We now come to the proof of the approximation factor:

$$\forall T'\ A'.\ lb\ T'\ A'\ (j+1) \longrightarrow makespan\ T_g \leq 3\ /\ 2 * makespan\ T'$$

To prove it, we fix $T_1$ and $A_1$ such that $lb\ T_1\ A_1\ (j + 1)$. Using Lemma 10, one can now obtain $T_0$ and $A_0$ such that $lb\ T_0\ A_0\ j$ and $MK: makespan\ T_0 \leq makespan\ T_1$. From the assumed loop invariant, we can now show:

$$makespan\ T \leq \frac{3}{2} makespan\ T_0 \qquad\qquad \text{by } inv_2\text{-def}$$

$$\leq \frac{3}{2} makespan\ T_1 \qquad\qquad \text{by } MK$$

To prove the makespan for $j + 1$ jobs, there are now two cases to consider: The added job $j + 1$ contributes to the makespan or it does not. The case in which it does not can be shown by combining the previous calculation with Lemma 9. For the first case, we may then assume that $makespan\ T_g = T(i) + t(j + 1)$. Like in Lemma 5, we note that $sum\text{-}eq$: $(\sum_{x=1}^{m} T\ x) = (\sum_{x=1}^{j} t\ x)$. Moreover, $min\text{-}avg$: $m * T\ (min_k\ T\ m) \leq (\sum_{i=1}^{m} T\ i)$. This allows us to calculate the following lower bound for $T(i)$:

$$m * T(i) \leq \sum_{i=1}^{m} T(i) = \sum_{i=1}^{j} t(i) \qquad\qquad \text{by } min\text{-}avg \text{ and } sum\text{-}eq$$

$$\Longleftrightarrow T(i) \leq \frac{\sum_{i=1}^{j} t(i)}{m} \qquad\qquad \text{because } m > 0$$

$$\leq makespan\ T_0 \leq makespan\ T_1 \qquad \text{by Lemma 5 and } MK$$

From Lemma 7 we can also show that $t(j + 1)$ is a lower bound for $\frac{1}{2}$ of the makespan of $T_1$. Therefore:

$$makespan\ T_g = T(i) + t(j + 1) \leq makespan\ T_1 + \frac{makespan\ T_1}{2}$$

$$= \frac{3}{2} makespan\ T_1$$

The proof of functional correctness and remaining conjuncts is again trivial.

Let us now consider the unsorted case where one can still show an approximation factor of 2. The algorithm is identical but the invariant is simpler:

$$inv_1\ T\ A\ j =$$
$$(lb\ T\ A\ j \wedge j \leq n \wedge (\forall\ T'\ A'.\ lb\ T'\ A'\ j \longrightarrow makespan\ T \leq 2 * makespan\ T'))$$

The proof for this invariant is a simpler version of the proof above: We do not need the initial case distinction (case $j + 1 \leq m$ need not be considered separately) and to prove the approximation factor we use Lemma 6 instead of Lemma 7 to obtain a bound for $t(j + 1)$.

# 6    Bin Packing

We finally consider the linear time $\frac{3}{2}$-approximation algorithm for the bin packing problem proposed by Berghammer and Reuter [4]. The bin packing problem is similar to the load balancing problem described in the previous section. The main distinction is that in the load balancing problem, the number of machines is fixed, while the load a single machine can hold is unbounded. With the bin packing problem, this is essentially reversed. The maximum capacity a single bin can hold is limited by some fixed $c$. However, we are free to use as many bins as necessary to achieve a solution. The goal is now to minimize this number of bins used instead of the maximum capacity of a bin.

For the bin packing problem we are given a finite, non-empty set of objects $U :: 'a\ set$, whose *weights* are given by a function $w :: 'a \Rightarrow real$. Note that in this paper *nat*s are implicitly converted to *real*s if needed. The weight of an object in $U$ is strictly greater than zero, but bounded by a maximum capacity $c :: nat$. The abbreviation $W(B) \equiv \sum_{u \in B} w(u)$ denotes the weight of a bin $B \subseteq U$. The set $U$ can also be separated into *small* and *large* objects. An object $u$ is considered small if $w(u) \leq \frac{c}{2}$. An object is large if the opposite is the case. We will begin by assuming that all small objects in $U$ can be found in a set $S$, and large objects in $U$ can be found in a set $L$, such that $S \cup L = U$ and $S \cap L = \emptyset$. Of course $L$ and $S$ can also be computed from $U$ in linear time. Variables $U$, $w$, $c$, $L$, and $S$ are fixed throughout this section.

A solution $P$ to the bin packing problem is then defined as follows:

$$bp :: 'a\ set\ set \Rightarrow bool$$
$$bp\ P = (partition\_on\ U\ P \wedge (\forall\ B \in P.\ W\ B \leq c))$$

$P$ contains all the bins necessary such that it is a correct partition of $U$. To check for this, we use a function $partition\_on :: 'a\ set \Rightarrow 'a\ set\ set \Rightarrow bool$ which can be found in the Isabelle HOL-Library. We add the final conjunct to ensure that no bin $B \in P$ exceeds the maximum capacity $c$.

The idea behind the algorithm proposed by Berghammer and Reuter is to split the solution $P$ into two partial solutions $P_1$ and $P_2$. At every step of the algorithm we consider two bins $B_1$ and $B_2$ which we try to fill with remaining objects from $V \subseteq U$ that have not been assigned yet. If adding the object to

$B_1$ or $B_2$ would cause it to exceed its maximum capacity, the bin is moved into the partial solution $P_1$ or $P_2$ respectively and cleared. Once there are no small objects left, the solution is the union of the partial solutions $P_1$ and $P_2$, the bins $B_1$ and $B_2$ (if they still contain objects), and the remaining large objects, which each receive their own bin, as no two large objects can fit into a single bin. To ensure that no empty bins are added to the solution, we define:

$$\llbracket \cdot \rrbracket :: {}'a\ set \Rightarrow {}'a\ set\ set$$
$$\llbracket B \rrbracket = (if\ B\ = \emptyset\ then\ \emptyset\ else\ \{B\})$$

The final union can now be written as $P_1\ \cup\ \llbracket B_1 \rrbracket\ \cup\ P_2\ \cup\ \llbracket B_2 \rrbracket\ \cup\ \{\{v\} \mid v \in V\}$ where $V$ contains the remaining large elements. The algorithm can be specified by the following Hoare triple:

---

$\{True\}$
$P_1 := \emptyset;\ P_2 := \emptyset;\ B_1 := \emptyset;\ B_2 := \emptyset;\ V := U;$
$WHILE\ V \cap S \neq \emptyset\ INV\ \{inv_3\ P_1\ P_2\ B_1\ B_2\ V\}\ DO$
$IF\ B_1 \neq \emptyset\ THEN\ u := some\ (V \cap\ S)$
$ELSE\ IF\ V \cap L \neq \emptyset\ THEN\ u := some\ (V\ \cap\ L)$
$\qquad ELSE\ u := some\ (V \cap S)\ FI\ FI;$
$V\ :=\ V - \{u\};$
$IF\ W(B_1) + w(u) \leq c\ THEN\ B_1 :=\ B_1 \cup \{u\}$
$ELSE\ IF\ W(B_2)\ +\ w(u) \leq c\ THEN\ B_2 := B_2 \cup \{u\}$
$\qquad ELSE\ P_2 := P_2 \cup \llbracket B_2 \rrbracket;\ B_2 := \{u\}\ FI;$
$\qquad P_1 := P_1 \cup \llbracket B_1 \rrbracket;\ B_1 := \emptyset\ FI$
$OD;$
$P := P_1 \cup \llbracket B_1 \rrbracket \cup P_2 \cup \llbracket B_2 \rrbracket \cup \{\{v\} \mid v \in V\}$
$\{bp\ P \wedge (\forall Q.\ bp\ Q \longrightarrow |P| \leq 3\ /\ 2\ *\ |Q|)\}$

---

Berghammer and Reuter prove functional correctness using a simplified version of this algorithm where an arbitrary element of $V$ is assigned to $u$. This allows for fewer case distinctions, as the first $IF-THEN-ELSE$ block can be ignored. One needs to find a loop invariant that implies functional correctness and prove that it is preserved in the following cases:

**Case 1.** The object fits into $B_1$:

$inv_1\ P_1\ P_2\ B_1\ B_2\ V \wedge u \in V \wedge W\ B_1 + w\ u \leq c\ \longrightarrow$
$inv_1\ P_1\ P_2\ (B_1 \cup \{u\})\ B_2\ (V - \{u\})$

**Case 2.** The object fits into $B_2$:

$inv_1\ P_1\ P_2\ B_1\ B_2\ V \wedge u \in V \wedge W\ B_2 + w\ u \leq c\ \longrightarrow$
$inv_1\ (P_1 \cup \llbracket B_1 \rrbracket)\ P_2\ \emptyset\ (B_2 \cup \{u\})\ (V - \{u\})$

**Case 3.** The object fits into neither bin:

$inv_1\ P_1\ P_2\ B_1\ B_2\ V \wedge u \in V \longrightarrow$
$inv_1\ (P_1 \cup \llbracket B_1 \rrbracket)\ (P_2 \cup \llbracket B_2 \rrbracket)\ \emptyset\ \{u\}\ (V - \{u\})$

Berghammer and Reuter [4] define the following predicate as their loop invariant:

$$inv_1 \; P_1 \; P_2 \; B_1 \; B_2 \; V \; = \; bp \; (P_1 \cup [\![B_1]\!] \cup P_2 \cup [\![B_2]\!] \cup \{\{v\} \mid v \in V\})$$

As it turns out, this invariant is too weak. Assume $inv_1 \; P_1 \; P_2 \; B_1 \; B_2 \; V$. Suppose $P_1$ (alternatively $P_2$) already contains the non-empty bin $B_1$. Note that this does not violate the invariant because $P_1 \cup [\![B_1]\!] = P_1$. Now, if the algorithm modifies $B_1$ by adding an element from $V$ such that $B_1$ becomes some $B_1'$ then $B_1 \cap B_1' \neq \emptyset$ and $B_1 \in P_1$, i.e., $B_1'$ is no longer disjoint from the elements of $P$. The same issue arises with the added object $u \in V$, if $\{u\}$ is already in $P_1$ or $P_2$. To account for such cases, we will require additional conjuncts:

$$inv_1 :: \, 'a \; set \; set \; \Rightarrow \, 'a \; set \; set \; \Rightarrow \, 'a \; set \; \Rightarrow \, 'a \; set \; \Rightarrow \, 'a \; set \; \Rightarrow \; bool$$
$$inv_1 \; P_1 \; P_2 \; B_1 \; B_2 \; V \; =$$
$$(bp \; (P_1 \cup [\![B_1]\!] \cup P_2 \cup [\![B_2]\!] \cup \{\{v\} \mid v \in V\}) \; \wedge$$
$$\bigcup \; (P_1 \cup [\![B_1]\!] \cup P_2 \cup [\![B_2]\!]) \; = \; U - V \; \wedge$$
$$B_1 \notin P_1 \cup P_2 \cup [\![B_2]\!] \; \wedge$$
$$B_2 \notin P_1 \cup [\![B_1]\!] \cup \; P_2 \; \wedge$$
$$(P_1 \cup [\![B_1]\!]) \cap (P_2 \cup [\![B_2]\!]) \; = \; \emptyset)$$

There are different ways to strengthen the original $inv_1$. We use the above additional conjuncts as they can be inserted in existing proofs with little modification, and their preservation in the invariant can be proved quite trivially. The first additional conjunct ensures that no element still in $V$ is already in a bin or partial solution. The second and third additional conjuncts ensure distinctness of the bins $B_1$ and $B_2$ with the remaining solution. The final conjunct ensures that the partial solutions with their added bins are disjoint from each other. It should be noted that the last conjunct is *not* necessary to prove functional correctness. It will, however, be needed in later proofs, and as its preservation in this invariant for the simplified algorithm can be used in the proof of the full algorithm, one can save redundant case distinctions by proving it now. Another advantage of proving it now is that later invariants can remain identical to the invariants proposed in the paper.

We now prove the preservation of $inv_1$ in all three cases. As we assume the invariant to hold before the execution of the loop body, we can see from the first additional conjunct $\bigcup \; (P_1 \cup [\![B_1]\!] \cup P_2 \cup [\![B_2]\!]) = U - V$ and the assumption $u \in V$ that *not-in*: $\forall B \in P_1 \cup [\![B_1]\!] \cup P_2 \cup [\![B_2]\!]. \; u \notin B$ holds. This will be needed for all three cases. Now, we can begin with Case 1. We first show

$$bp \; (P_1 \cup [\![B_1 \cup \{u\}]\!] \cup P_2 \cup [\![B_2]\!] \cup \{\{v\} \mid v \in V - \{u\}\})$$

One can see that this union does not contain the empty set. The object $u$ is now moved from a singleton set into $B_1$. Therefore, the union of all bins will again return $U$. To show that this union remains pairwise disjoint, we can use *not-in* and the second additional conjunct of $inv_1$ to show that $u$ is not yet contained in the partial solution and $B_1$ is *distinct* from any other bin. Therefore, combined with the assumption that the union was pairwise disjoint before the modification, the union remains pairwise disjoint. To prove the preservation of the second conjunct of $bp$, we need to show that the bin weights do not exceed their maximum capacity $c$. The only bin that was changed in this step is $B_1$, which has increased its weight by $w(u)$. As we are in Case 1, we can assume that $u$ fits into $B_1$, $W(B_1) + w(u) \leq c$. Therefore, this conjunct holds as well. Now, one only needs to show that the additional conjuncts are preserved. For the first additional conjunct, we can again use $not\text{-}in$ to show:

$$\bigcup \left(P_1 \cup \llbracket B_1 \cup \{u\}\rrbracket \cup P_2 \cup \llbracket B_2\rrbracket\right) = U - (V - \{u\})$$

$$\Longleftrightarrow \quad \bigcup \left(P_1 \cup \llbracket B_1\rrbracket \cup P_2 \cup \llbracket B_2\rrbracket\right) \cup \{u\} = U - (V - \{u\}) \qquad \text{by } not\text{-}in$$

$$\Longleftrightarrow \quad \bigcup \left(P_1 \cup \llbracket B_1\rrbracket \cup P_2 \cup \llbracket B_2\rrbracket\right) \cup \{u\} = U - V \cup \{u\} \qquad \text{by } u \in U$$

Using the first additional conjunct of the assumed invariant, one can see that this must hold. The remaining conjuncts

$B_1 \cup \{u\} \notin P_1 \cup P_2 \cup \llbracket B_2\rrbracket$
$B_2 \notin P_1 \cup \llbracket B_1 \cup \{u\}\rrbracket \cup P_2$
$(P_1 \cup \llbracket B_1 \cup \{u\}\rrbracket) \cap (P_2 \cup \llbracket B_2\rrbracket) = \emptyset$

can be automatically proved in Isabelle using $not\text{-}in$ and the assumption that the conjuncts of $inv_1\ P_1\ P_2\ B_1\ B_2\ V$ held before the modification. The proof for Case 2 is almost identical to that of Case 1. The main difference is that the focus now lies on $B_2$ and the fact that $B_1$ is now emptied and the previous contents added to the partial solution $P_1$. One therefore has to show that

$$bp\ (P_1 \cup \llbracket B_1\rrbracket \cup \llbracket\emptyset\rrbracket \cup P_2 \cup \llbracket B_2 \cup \{u\}\rrbracket \cup \{\{v\} \mid v \in V - \{u\}\})$$

holds. As $\llbracket\emptyset\rrbracket$ can be ignored, one can see that the act of emptying $B_1$ and adding it to the partial solution will not otherwise affect the proof. The proof of $bp$ in Case 3 is trivial, as the modifications made in this step can simply be undone by applying the following steps:

$P_1 \cup \llbracket B_1\rrbracket \cup \llbracket\emptyset\rrbracket \cup (P_2 \cup \llbracket B_2\rrbracket) \cup \llbracket\{u\}\rrbracket \cup \{\{v\} \mid v \in V - \{u\}\}$
$= P_1 \cup \llbracket B_1\rrbracket \cup P_2 \cup \llbracket B_2\rrbracket \cup \{\{u\}\} \cup \{\{v\} \mid v \in V - \{u\}\} \qquad \text{by } \llbracket \cdot \rrbracket - def$
$= P_1 \cup \llbracket B_1\rrbracket \cup P_2 \cup \llbracket B_2\rrbracket \cup \{\{v\} \mid v \in V\} \qquad \text{by } u \in V$

Now, one only needs to show that the remaining additional conjuncts hold. This can again be shown automatically using $not\text{-}in$ and the fact that $inv_1\ P_1\ P_2\ B_1\ B_2\ V$ held before the modifications. Therefore, $inv_1$ is preserved in all three cases.

To prove the approximation factor, we proceed as in [4] and establish suitable lower bounds. The first lower bound

**Lemma 11.** $bp\ P \longrightarrow |L| \leq |P|$

holds because a bin can only contain at most one large object, and every large object needs to be in the solution. To prove this in Isabelle, we first make the observation that for every large object there exists a bin in $P$ in which it is contained. Therefore, we may obtain a function $f$ that returns this bin for every $u \in L$. Using the fact that any bin can hold at most one large object, we can show that this function has to be injective, as every large object must map to a unique bin. Hence, the number of large objects is equal to the number of bins $f$ maps to. Moreover, the image of $f$ has to be a subset of $P$. Thus, the number of large objects has to be a lower bound on the number of bins in $P$.

As it turns out, the algorithm will ensure that there is always at least one large object in a bin for the first partial solution as long as large objects are available. This means we can assume that:

$$V \cap L \neq \emptyset \longrightarrow (\forall B \in P_1 \cup [\![B_1]\!]. \ B \cap L \neq \emptyset)$$

Therefore, we can use the previous lower bound to show the following:

**Lemma 12.** $bp \ P \wedge inv_1 \ P_1 \ P_2 \ B_1 \ B_2 \ V \wedge (\forall B \in P_1 \cup [\![B_1]\!]. \ B \cap L \neq \emptyset) \longrightarrow |P_1 \cup [\![B_1]\!] \cup \{\{v\} \mid v \in V \cap L\}| \leq |P|$

Another easy lower bound is this one:

**Lemma 13.** $bp \ P \longrightarrow (\sum_{u \in U} w \ u) \leq c * |P|$

The next lower bound arises from the fact that an object is only ever put into $B_2$, and therefore $P_2$, if it would have caused $B_1$ to overflow. As a result of this, we can define a bijective function $f$ that maps every bin of $P_1$ to the object in $P_2 \cup [\![B_2]\!]$ that would have caused the bin to overflow. We define:

$bij\_exists :: \ 'a \ set \ set \ \Rightarrow \ 'a \ set \ \Rightarrow \ bool$
$bij\_exists \ P \ V \ = \ (\exists f. \ bij\_betw \ f \ P \ V \wedge (\forall B \in P. \ c \ < \ W \ B \ + \ w \ (f \ B)))$

From this, we can make the observation that the number of bins in $P_1$ is a *strict* lower bound on the number of bins of any correct bin packing $P$:

**Lemma 14.** $bp \ P \wedge inv_1 \ P_1 \ P_2 \ B_1 \ B_2 \ V \wedge bij\_exists \ P_1 \ (\bigcup (P_2 \cup [\![B_2]\!])) \longrightarrow |P_1| \ + \ 1 \ \leq \ |P|$

Unlike the proof outlined in [4], we begin with a case distinction on $P_1$. The reasoning behind this is that if $P_1$ is empty, the strict nature of the lower bound cannot be shown from the calculation that Berghammer and Reuter make. Therefore, we consider the case where $P_1$ is empty separately. If $P_1$ is empty, our goal is to prove that 1 is a lower bound on the number of bins in $P$. This follows from the fact that $U$ is non-empty, and therefore any correct bin packing must contain at least one bin. For the other case, we may now assume that $P_1$ is non-empty. In the following proof, we will need the final conjunct of $inv_1$, $(P_1 \cup [\![B_1]\!]) \cap (P_2 \cup [\![B_2]\!]) = \emptyset$, which we can transform into *disjoint*: $P_1 \cap (P_2 \cup [\![B_2]\!]) = \emptyset$. We also obtain the bijective function $f$ and observe that, as the object obtained from $f$ for a bin $B \in P_1$ caused $B$ to exceed its capacity, *exceed*: $c \ < \ W(B) \ + \ w(f(B))$ must hold. We can now perform the following calculation:

$$c|P_1| = \sum_{B \in P_1} c$$

$$< \sum_{B \in P_1} W(B) + \sum_{B \in P_1} w(f(B)) \qquad \text{by } P_1 \neq \emptyset \text{ and } exceed$$

$$= \sum_{B \in P_1} W(B) + \sum_{B \in P_2 \cup [\![B_2]\!]} W(B) \qquad \text{by } f \text{ bijective}$$

$$= \sum_{B \in P_1 \cup P_2 \cup [\![B_2]\!]} W(B) \qquad \text{by } disjoint$$

$$\leq \sum_{u \in U} w(u) \leq c|P| \qquad \text{by } inv_1 \text{ and Lemma 13}$$

Therefore $|P_1| < |P|$ and, by extension, $|P_1| + 1 \leq |P|$.

We only sketch the rest of the proof because it is almost identical to that in [4]. First we need two extensions of $inv_1$ to show the approximation ratio:

$inv_2\ P_1\ P_2\ B_1\ B_2\ V =$
$(inv_1\ P_1\ P_2\ B_1\ B_2\ V\ \wedge$
$(V \cap L \neq \emptyset \longrightarrow (\forall B \in P_1 \cup [\![B_1]\!].\ B \cap L \neq \emptyset)) \wedge$
$bij\_exists\ P_1\ (\bigcup (P_2 \cup [\![B_2]\!])) \wedge 2 * |P_2| \leq |\bigcup P_2|)$
$inv_3\ P_1\ P_2\ B_1\ B_2\ V = (inv_2\ P_1\ P_2\ B_1\ B_2\ V\ \wedge\ B_2 \subseteq S)$

The motivation for the last conjunct in $inv_2$ is the following lower bound:
$inv_1\ P_1\ P_2\ B_1\ B_2\ V \wedge 2 * |P_2| \leq |\bigcup P_2| \wedge bij\_exists\ P_1\ (\bigcup (P_2 \cup [\![B_2]\!])) \longrightarrow$
$2 * |P_2 \cup [\![B_2]\!]| \leq |P_1| + 1$.

The main lower bound lemma (Theorem 4.1 in [4]) is the following:

**Lemma 15.** $V \cap S = \emptyset \wedge inv_2\ P_1\ P_2\ B_1\ B_2\ V \wedge bp\ P \longrightarrow$
$|P_1 \cup [\![B_1]\!] \cup P_2 \cup [\![B_2]\!] \cup \{\{v\} \mid v \in V\}| \leq 3/2 * |P|$

From this lower bound the postcondition of the algorithm follows easily under the assumption that $inv_2$ holds at the end of the loop. This in turn follows because $inv_3$ can be shown to be a loop invariant.

# 7 Conclusion

In the first application of theorem proving to approximation algorithms we have verified three classical and one less well-known approximation algorithm for fundamental NP-complete problems, have corrected purported invariants from the literature and could even strengthen the approximation ratio in one case. Although we have demonstrated the benefits of formal verification of approximation algorithms, we have only scratched the surface of this rich theory. The next step is to explore the subject more systematically. As a large fraction of the theory of approximation algorithms is based on linear programming, this is a promising and challenging direction to explore. Some linear programming theory has been formalized in Isabelle already [5,11]. Approximation algorithms can also be formulated as relational programs, and verified accordingly. This approach was explored in [2], with some support from theorem provers, but has yet to be fully formalized.

**Acknowledgement.** Tobias Nipkow is supported by DFG grant NI 491/16-1.

# References

1. Bansal, N., Khot, S.: Inapproximability of hypergraph vertex cover and applications to scheduling problems. In: Abramsky, S., Gavoille, C., Kirchner, C., Meyer auf der Heide, F., Spirakis, P.G. (eds.) ICALP 2010. LNCS, vol. 6198, pp. 250–261. Springer, Heidelberg (2010). https://doi.org/10.1007/978-3-642-14165-2_22
2. Berghammer, R., Höfner, P., Stucke, I.: Cardinality of relations and relational approximation algorithms. J. Log. Algebraic Methods Program. **85**(2), 269–286 (2016)
3. Berghammer, R., Müller-Olm, M.: Formal development and verification of approximation algorithms using auxiliary variables. In: Bruynooghe, M. (ed.) LOPSTR 2003. LNCS, vol. 3018, pp. 59–74. Springer, Heidelberg (2004). https://doi.org/10.1007/978-3-540-25938-1_6
4. Berghammer, R., Reuter, F.: A linear approximation algorithm for bin packing with absolute approximation factor 3/2. Sci. Comput. Program. **48**(1), 67–80 (2003). https://doi.org/10.1016/S0167-6423(03)00011-X
5. Bottesch, R., Haslbeck, M.W., Thiemann, R.: Verifying an incremental theory solver for linear arithmetic in Isabelle/HOL. In: Herzig, A., Popescu, A. (eds.) FroCoS 2019. LNCS (LNAI), vol. 11715, pp. 223–239. Springer, Cham (2019). https://doi.org/10.1007/978-3-030-29007-8_13
6. Eßmann, R., Nipkow, T., Robillard, S.: Verified approximation algorithms. Archive of Formal Proofs, Formal proof development, January 2020. http://isa-afp.org/entries/Approximation_Algorithms.html
7. Halldórsson, M.M., Radhakrishnan, J.: Greed is good: approximating independent sets in sparse and bounded-degree graphs. Algorithmica **18**(1), 145–163 (1997)
8. Kleinberg, J.M., Tardos, É.: Algorithm Design. Addison-Wesley (2006)
9. Nipkow, T., Klein, G.: Concrete Semantics with Isabelle/HOL. Springer, Heidelberg (2014). http://concrete-semantics.org
10. Nipkow, T., Wenzel, M., Paulson, L.C. (eds.): Isabelle/HOL. LNCS, vol. 2283. Springer, Heidelberg (2002). https://doi.org/10.1007/3-540-45949-9
11. Parsert, J., Kaliszyk, C.: Linear programming. Archive of Formal Proofs, Formal proof development, August 2019. http://isa-afp.org/entries/Linear_Programming.html
12. Vazirani, V.V.: Open problems. Approximation Algorithms, pp. 334–343. Springer, Heidelberg (2003). https://doi.org/10.1007/978-3-662-04565-7_30
13. Wei, V.: A lower bound for the stability number of a simple graph. Technical Memorandum 81–11217-9, Bell Laboratories (1981)

# Efficient Verified Implementation
# of Introsort and Pdqsort

Peter Lammich[✉]

The University of Manchester, Manchester, UK
peter.lammich@manchester.ac.uk

**Abstract.** Sorting algorithms are an important part of most standard
libraries, and both, their correctness and efficiency is crucial for many
applications.

As generic sorting algorithm, the GNU C++ Standard Library implements the introsort algorithm, a combination of quicksort, heapsort, and
insertion sort. The Boost C++ Libraries implement pdqsort, an extension of introsort that achieves linear runtime on inputs with certain patterns.

We verify introsort and pdqsort in the Isabelle LLVM verification
framework, closely following the state-of-the-art implementations from
GNU and Boost. On an extensive benchmark set, our verified implementations perform on par with the originals.

## 1 Introduction

Sorting algorithms are an important part of any standard library. The GNU
C++ Library (libstdc++) [15] implements Musser's introspective sorting algorithm (introsort) [28]. It is a combination of quicksort, heapsort, and insertion
sort, which has the fast average case runtime of quicksort and the optimal
$O(n \log(n))$ worst-case runtime of heapsort. The Boost C++ Libraries [6] provide
a state-of-the-art implementation of *pattern-defeating quicksort* (pdqsort) [29],
an extension of introsort to achieve better performance on inputs that contain
certain patterns like already sorted sequences. Verification of these algorithms
and their state-of-the-art implementations is far from trivial, but turns out to
be manageable when handled with adequate tools.

Sorting algorithms in standard libraries have not always been correct. The
timsort [30] algorithm in the Java standard library has a history of bugs[1], the
(hopefully) last of which was only found by a formal verification effort [10]. Also,
many real-world mergesort implementations suffered from an overflow bug [5].
Finally, LLVM's libc++ [26] implements a different quicksort based sorting algorithm. While it may be functionally correct, it definitely violates the C++ standard by having a quadratic worst-case run time[2].

---

[1] See https://bugs.java.com/bugdatabase/view_bug.do?bug_id=8011944.
[2] See https://bugs.llvm.org/show_bug.cgi?id=20837. This has not been fixed by April
2020.

© Springer Nature Switzerland AG 2020
N. Peltier and V. Sofronie-Stokkermans (Eds.): IJCAR 2020, LNAI 12167, pp. 307–323, 2020.
https://doi.org/10.1007/978-3-030-51054-1_18

In this paper, we present efficient implementations of introsort and pdqsort that are verified down to their LLVM intermediate representation [27]. The verification uses the Isabelle Refinement Framework [24], and its recent Isabelle-LLVM backend [23]. We also report on two extensions of Isabelle-LLVM, to handle nested container data structures and to automatically generate C-header files to interface the generated code. Thanks to the modularity of the Isabelle Refinement Framework, our verified algorithms can easily be reused in larger verification projects.

While sorting algorithms are a standard benchmark for theorem provers and program verification tools, verified real-world implementations seem to be rare: apart from our work, we are only aware of two verified sorting algorithms [3,10] from the Java standard library.

The complete Isabelle/HOL formalization and the benchmarks are available at http://www21.in.tum.de/~lammich/isabelle_llvm/.

## 2    The Introsort and Pdqsort Algorithms

The introsort algorithm by Musser [28] is a generic unstable sorting algorithm that combines the good average-case runtime of quicksort [18] with the optimal $O(n \log(n))$ worst-case complexity of heapsort [1]. The basic idea is to use quicksort as main sorting algorithm, insertion sort for small partitions, and heapsort when the recursion depth exceeds a given limit, usually $2\lfloor \log_2 n \rfloor$ for $n$ elements.

```
1: procedure INTROSORT(xs, l, h)
2:     if h − l > 1 then
3:         INTROSORT_AUX(xs, l, h, 2⌊log₂(h − l)⌋)
4:         FINAL_INSERT(xs, l, h)
5: procedure INTROSORT_AUX(xs, l, h, d)
6:     if h − l > threshold then
7:         if d = 0 then HEAPSORT(xs, l, h)
8:         else
9:             m ← PARTITION_PIVOT(xs, l, h)
10:            INTROSORT_AUX(xs, l, m, d − 1)
11:            INTROSORT_AUX(xs, m, h, d − 1)
```

**Algorithm 1:** Introsort

Algorithm 1 shows our implementation of introsort, which closely follows the implementation in libstdc++ [15]. The function INTROSORT sorts the slice from index $l$ (inclusive) up to index $h$ (exclusive) of the list[3] $xs$. If there is more than one element (line 2), it initializes a depth counter and calls the function INTROSORT_AUX (line 3), which partially sorts the list such that every element

---

[3] Our formalization initially uses lists to represent the sequence of elements to be sorted, and refines them to arrays later (cf. Sect. 4).

is no more than threshold positions away from its final position in the sorted list. The remaining sorting is then done by insertion sort (line 4). The function INTROSORT_AUX implements a recursive quicksort scheme: recursion stops if the slice becomes smaller than the threshold (line 6). If the maximum recursion depth is exhausted, heapsort is used to sort the slice (line 7). Otherwise, the slice is partitioned (line 9), and the procedure is recursively invoked for the two partitions (line 10–11). Here, PARTITION_PIVOT moves the pivot element to the first element of the left partition, and returns the start index of the right partition.

Note that we do not try to invent our own implementation, but closely follow the existing (and hopefully well-thought) libstdc++ implementation. This includes the slightly idiosyncratic partitioning scheme, which leaves the pivot-element as first element of the left partition. Moreover, the libstdc++ implementation contains a manual tail-call optimization, replacing the recursive call in line 11 by a loop. While we could easily add this optimization in an additional refinement step, it turned out to be unnecessary, as LLVM recognizes and eliminates this tail call automatically.

```
 1: procedure PDQSORT(xs, l, h)
 2:     if h − l > 1 then PDQSORT_AUX(true, xs, l, h, log(h − l))
 3: procedure PDQSORT_AUX(lm, xs, l, h, d)
 4:     if h − l < threshold then INSORT(lm, xs, l, h)
 5:     else
 6:         PIVOT_TO_FRONT(xs, l, h)
 7:         if ¬lm ∧ xs[l − 1] ≮ xs[l] then
 8:             m ← PARTITION_LEFT(xs, l, h)
 9:             assert m + 1 ≤ h
10:             PDQSORT_AUX(false, xs, m + 1, h, d)
11:         else
12:             (m, ap) ← PARTITION_RIGHT(xs, l, h)
13:             if m − l < ⌊(h − l)/8⌋ ∨ h − m − 1 < ⌊(h − l)/8⌋ then
14:                 if −−d = 0 then HEAPSORT(xs,l,h); return
15:                 SHUFFLE(xs,l,h,m)
16:             else if ap ∧ MAYBE_SORT(xs, l, m) ∧ MAYBE_SORT(xs, m + 1, h) then
17:                 return
18:             PDQSORT_AUX(lm, xs, l, m, d)
19:             PDQSORT_AUX(false, xs, m + 1, h, d)
```

**Algorithm 2:** Pdqsort

Algorithm 2 shows our implementation of pdqsort. As for introsort, the wrapper PDQSORT just initializes a depth counter, and then calls the function PDQSORT_AUX (line 2), which, in contrast to introsort, completely sorts the list, such that no final insertion sort is necessary. Again, the PDQSORT_AUX function implements a recursive quicksort scheme, however, with a few additional optimizations. Slices smaller than the threshold are sorted with insertion sort (line 4). If the current slice is not the leftmost one of the list, as indicated by the parameter $lm$, the element before the start of the slice is guaranteed to be smaller than any element of the slice itself. This can be exploited to omit a comparison in the inner loop of insertion sort (cf. Sect. 3.3). If the slice is not smaller than the threshold, a pivot element is selected and moved to the front of the slice (line 6). If the pivot is equal to the element before the current slice (line 7), this indicates a lot of equal elements. The PARTITION_LEFT function (line 8) will put them in the left partition, and then only the right partition needs to be sorted recursively (line 10). Otherwise, PARTITION_RIGHT (line 12) places elements equal to the pivot in the right partition. Additionally, it returns a flag $ap$ that indicates that the slice was already partitioned. Next, we check for a highly unbalanced partitioning (line 13), i.e., if one partition is less than 1/8th of the overall size. After encountering a certain number of highly unbalanced partitionings, pdqsort switches to heapsort (line 14). Otherwise, it will shuffle some elements in both partitions, trying to break up patterns in the input (line 15). If the input was already partitioned wrt. the selected pivot (indicated by the flag $ap$), pdqsort will optimistically try to sort both partitions with insertion sort (line 17). However, these insertion sorts abort if they cannot sort the list with a small number of swaps, limiting the penalty for being too optimistic. Finally, the two partitions are recursively sorted (lines 18–19).

Our implementation of pdqsort closely follows the implementation we found in Boost [6]. Again, we omitted a manual tail call optimization that LLVM does automatically. Moreover, for certain comparison functions, Boost's pdqsort uses a special branch-aware partitioning algorithm [11]. We leave its verification to future work, but note that it will easily integrate in our existing formalization.

While introsort and pdqsort are based on the same idea, this presentation focuses on the more complex pdqsort: apart from the more involved PDQ-SORT_AUX function, PIVOT_TO_FRONT uses Tukey's 'ninther' pivot selection [4], while introsort uses the simpler median-of-three scheme. It has two partitioning algorithms used in different situations, and the PARTITION_RIGHT algorithm also checks for already partitioned slices. Finally, with INSORT and MAYBE_SORT, it uses two different versions of insertion sort.

## 3    Verification

We use the Isabelle Refinement Framework [23,24] to formally verify our algorithms. It provides tools to develop algorithms by stepwise refinement, and generates code in the LLVM intermediate representation [27].

A program returns an element of the following datatype:

$$\alpha\ nres \equiv \texttt{fail} \mid \texttt{spec}\ (\alpha \Rightarrow bool)$$

Here **fail** represents possible non-termination or assertion violation, and **spec** $P$ a result nondeterministically chosen to satisfy predicate $P$. Note that we use $\equiv$ to indicate defining equations. We define a *refinement ordering* on *nres* by spec $P \leq$ spec $Q \equiv \forall x.\ P\,x \implies Q\,x$ $\qquad$ fail $\not\leq$ spec $Q$ $\qquad$ $m \leq$ fail

Intuitively, $m_1 \leq m_2$ means that $m_1$ returns fewer possible results than $m_2$, and may only fail if $m_2$ may fail. Note that $\leq$ is a complete lattice, with top element fail. The *monad combinators* are then defined as

return $x \equiv$ spec $y.\ y{=}x$
bind (spec $P$) $f \equiv \bigsqcup\{f\,x \mid P\,x\}$ $\qquad$ bind fail $f \equiv$ fail

Here, return $x$ deterministically returns $x$, and bind $m\,f$ chooses a result of $m$ and then applies $f$ to it. If $m$ may fail, then the bind may also fail. We write $x{\leftarrow}m;\ f\,x$ for bind $m\ (\lambda x.\ f\,x)$, and $m_1;\ m_2$ for bind $m_1\ (\lambda_-.\ m_2)$.

Arbitrary recursive programs can be defined via a fixed-point construction [20]. An *assertion* fails if its condition is not met, otherwise it returns the unit value:

assert $P \equiv$ if $P$ then return () else fail;

Assertions are used to express that a program $m$ satisfies the Hoare triple with precondition $P$ and postcondition $Q$:

$m \leq$ assert $P;$ spec $x.\ Q\,x$

If the precondition is false, the right hand side is fail, and the statement trivially holds. Otherwise, $m$ cannot fail, and every possible result $x$ of $m$ must satisfy $Q$.

While the Isabelle Refinement Framework provides some syntax to express programs, for better readability, we use the slightly more informal syntax that we have already used in Algorithms 1 and 2. In particular, we treat lists as if they were updated in place, while our actual formalization is purely functional, i.e., generates a new version of the list on each update, which is explicitly threaded through the program. Destructively updated arrays will only be introduced in a later refinement step (cf. Sect. 4).

## 3.1 Specification of Sorting Algorithms

The first step to verify a sorting algorithm is to specify the desired result. We specify a sorting algorithm as follows:

*sort_spec xs l h*
$\equiv$ assert $l{\leq}h \wedge h{\leq}|xs|;$ spec $xs'.\ xs =_{l,h} xs' \wedge sorted\ (xs'[l..{<}h])$

here $|xs|$ is the length of the list $xs$ and $xs[I]$ is the slice of the list $xs$ for indexes in the interval $I$[4]. The equivalence relation $xs =_{l,h} xs'$ relates lists $xs$ and $xs'$ iff they are equal outside the slice $l..{<}h$ and $xs$ is a permutation of $xs'$. To simplify the presentation, we assume a linear ordering on the elements. Note that both C++ and our actual formalization support arbitrary weak orderings [19].

---

[4] An interval from index $l$ (inclusive) to $h$ (exclusive) is denoted as $l..{<}h$. If both indexes are exclusive, we write $l{<}..{<}h$.

## 3.2  Quicksort Scheme

We split a call of PDQSORT_AUX into phases, described by the following predicates:

$pvt\ xs \equiv a_0 =_{l,h} xs \wedge (\exists i \in l<..<h.\ xs[i] \leq xs[l]) \wedge (\exists i \in l<..<h.\ xs[i] \geq xs[l])$

$part\ m\ xs \equiv a_0 =_{l,h} xs \wedge l \leq m \wedge m<h$
$\qquad\qquad \wedge\ (\forall i \in l..<m.\ xs[i] \leq xs[m]) \wedge (\forall i \in m<..<h.\ xs[m] \leq xs[i])$

$sortl\ m\ xs \equiv part\ m\ xs \wedge sorted\ (xs[l..<m])$

$sortr\ m\ xs \equiv sortl\ m\ xs \wedge sorted\ (xs[m<..<h])$

Let $a_0$ denote the original list. First, a pivot element is selected and moved to the beginning of the slice (phase $pvt$). The pivot is selected in a way such there is at least one smaller ($\leq$) and one greater ($\geq$) element, e.g., by a median-of-three selection. This knowledge can later be exploited to optimize the inner loops of the partitioning algorithm. After the partitioning (phase $part\ m$), $m$ points to the pivot element, and all elements before $m$ are smaller, and all elements after $m$ are greater. Then, first the left (phase $sortl\ m$), and then the right (phase $sortr\ m$) partition gets sorted, while the list remains partitioned around $m$.

This approach allows us to prove correct the algorithm, without assuming too many details of the underlying subroutines. The following is all we need to know about the subroutines:

(a) $lm \vee notleft\ xs\ l\ h \implies insert\ lm\ xs\ l\ h \leq sort\_spec\ xs\ l\ h$

(b) $l+4<h \implies pivot\_to\_front\ xs\ l\ h \leq \mathbf{spec}\ xs'.\ pvt\ xs'$

(c) $pvt\ xs \implies partition\_right\ xs\ l\ h \leq \mathbf{spec}\ (xs',m,\_).\ part\ m\ xs'$
$\qquad\qquad \wedge\ partition\_left\ xs\ l\ h\quad \leq \mathbf{spec}\ (xs',m,\_).\ part\ m\ xs'$

(d) $heapsort\ xs\ l\ h \leq sort\_spec\ xs\ l\ h$

(e) $part\ m\ xs \implies shuffle\ xs\ l\ h\ m \leq \mathbf{spec}\ xs'.\ part\ m\ xs'$

(f) $i \leq j \wedge j \leq |xs|$
$\qquad \implies maybe\_sort\ xs\ i\ j \leq \mathbf{spec}\ (b,xs').\ xs =_{i,j} xs' \wedge (\neg b \vee sorted\ xs'[i..<j])$

where $notleft\ xs\ l\ h \equiv 0<l \wedge \forall i \in l..<h.\ xs[l-1] \leq xs[i]$ states that the element $xs[l-1]$ before the slice is smaller than any element of the slice. Note that we explicitly mention the changed list $xs'$ in these specifications, while we left the list changes implicit in the algorithm description.

Intuitively, (a,d,f) state correctness of the sorting subroutines, (b) states that pivot selection goes to phase $pvt$, (c) states that partitioning transitions from phase $pvt$ to phase $part$, and (e) states that shuffling preserves phase $part$. From the above, we easily prove the following lemmas:

(g) $part\ m\ xs \wedge notleft\ xs\ l\ h \wedge xs[m] \leq xs[l-1] \implies sorted\ xs[l..<m]$

(h) $part\ m\ xs \wedge (xs =_{l,m} xs' \vee xs =_{m+1,h} xs') \implies part\ m\ xs'$

(i) $sortl\ m\ xs \wedge xs =_{m+1,h} xs' \implies sortl\ m\ xs'$

(j) $part\ m\ xs \implies sort\_spec\ xs\ l\ m \leq \mathbf{spec}\ xs'.\ sortl\ m\ xs'$

(k) $sortl\ m\ xs \implies sort\_spec\ xs\ (m+1)\ h \leq \mathbf{spec}\ xs'.\ sortr\ m\ xs'$

(l) $sortr\ m\ xs \implies sorted\ xs[l..<h]$

The correctness statement for PDQSORT_AUX is:

$$lm \lor notleft \; xs \; l \; h \implies pdqsort\_aux \; lm \; xs \; l \; h \; d \leq sort\_spec \; xs \; l \; h$$

The proof is done by using the Refinement Framework's verification condition generator, and then discharging the generated VCs using the above lemmas. The line numbers in the following brief sketch refer to Algorithm 2. As termination measure for the recursion, we use the size $h - l$ of the slice to be sorted. If we switch to insertion sort in line 4, (a) implies that the slice gets sorted, and we are done. Otherwise, we select a pivot in line 6, going to phase *pvt* (b). When the equals optimization is triggered in line 7, we transition to phase *part* (c), and the left partition is already sorted[5] (g), such that we can transition to phase *sortl* (j), and, via a recursive call in line 10 to phase *sortr* (k). This implies that the slice is sorted (l), and we are done. When the equals optimization is not used, (c) shows that we transition to phase *part* in line 12. If the partition is unbalanced, we either use heapsort (line 14) to directly sort the slice (d), or shuffle the elements (line 15) and stay in phase *part* (e). In line 17, the algorithm may attempt to sort the slice. If this succeeds, we are done (f). Otherwise, we stay in phase *part* (h), and the recursive calls in lines 18 and 19 will take us to phase *sortr* (j,k), which implies sortedness of the slice (l).

Using the above statement, and an analogous statement for introsort, we can prove the main correctness theorem:

**Theorem 1.**     $pdqsort \; xs \; l \; h \leq sort\_spec \; xs \; l \; h$
          **and**  $introsort \; xs \; l \; h \leq sort\_spec \; xs \; l \; h$

Note that we could prove the correctness of our algorithm with only minimal assumptions about the used subroutines. This decoupling of the algorithm from its subroutines simplifies the proof, as it is not obfuscated with unnecessary details. For example, correctness of the algorithm does not depend on the exact partitioning scheme being used, as long as it implements a transition from the *pvt* to the *part* phase. It also simplifies changing the subroutines later, e.g., adding further optimizations such as branch-aware partitioning [11].

Breaking down an algorithm into small and decoupled modules is often the key to its successful verification. Note that the original implementation in Boost is more coarse grained, inlining much of the functionality into the main algorithm. After having proved correct an algorithm, we can always do the inlining in a later refinement step, or rely on the LLVM optimizer to do the inlining for us. In our formalization, we use the inlining feature of Isabelle-LLVM's preprocessor.

---

[5] Actually all elements in the left partition are equal to the pivot.

```
1:  procedure INSORT(G, xs, l, h)        7:  procedure INSERT(G, xs, l, i)
2:     if l = h then return              8:     t ← xs[i]
3:     i ← l + 1                         9:     while (¬G ∨ l < i) ∧ t < xs[i−1] do
4:     while i < h do                   10:        xs[i] ← xs[i−1]
5:        INSERT(G, xs, l, i)           11:        −−i
6:        ++i                           12:     xs[i] ← t
```

**Algorithm 3:** Insertion Sort

## 3.3  Insertion Sort

Algorithm 3 shows our implementation of insertion sort. The INSORT procedure repeatedly calls INSERT to add elements to a sorted prefix of the list. The flag $G$ controls the *unguarded* optimization: if it is false, we assume that INSERT will hit a smaller element before underflowing the list index $i$, and thus omit the comparison $l < i$ (line 9) in the inner loop. We later specialize the INSORT algorithm for the two cases of $G$, and simplify the loop conditions accordingly.

Again, we split the insertion sort algorithm into two smaller parts, which are proved separately via the following specification for INSERT:

assert *sorted* $xs[l..<i] \wedge l{\leq}i \wedge i{<}|xs| \wedge (G \vee xs[l{-}1] \leq xs[i])$;
spec $xs'$. $xs{=}_{l,i+1}xs' \wedge$ *sorted* $xs[l..<i+1]$

This captures the intuition that INSERT goes from a slice that is sorted up to index $i$ to one that is sorted up to index $i + 1$.

## 3.4  The Remaining Subroutines

The proofs of the remaining subroutines follow a similar plot, and are not displayed here in full. Most of them were straightforward, and we could use existing Isabelle proofs as guideline [16,22,25]. For the SHUFFLE and PIVOT_TO_FRONT procedures, which contain a large number of indexing and update operations, we ran into a scalability problem: the many partially redundant in-bound statements for the indexes overwhelmed the linear arithmetic solver that is hard-wired into the simplifier. We worked around this problem by introducing auxiliary definitions, which hide the in-bound statements from the simplifier, and allow us to precisely control when it sees them.

Finally, we point out another interesting application of refinement: the SIFT_DOWN function of heapsort restores the heap property by *floating down* an element[6]. A straightforward implementation swaps the element with one of its children, until the heap property is restored (Algorithm 4 (left)). However, the element that is written to $xs[\text{right}(i)]$ or $xs[\text{left}(i)]$ by the swap will get overwritten in the next iteration. A common optimization to save half of the writes is to store the element to be moved down in a temporary variable, and only assign it to its final position after the loop (Algorithm 4 (right)). Note that the

---

[6] See, e.g., [9, Ch. 6] or [33, Ch. 2.4] for a description of heapsort.

---

**procedure** SIFT_DOWN($xs, i$)

    **while** has_right($i$) **do**
      **if** $xs[\text{left}(i)] < xs[\text{right}(i)]$ **then**
      **if** $xs[i] < xs[\text{right}(i)]$ **then**
        SWAP($xs[i], xs[\text{right}(i)]$)
        $i \leftarrow \text{right}(i)$
      **else return**
      **else if** $xs[i] < xs[\text{left}(i)]$ **then**
        SWAP($xs[i], xs[\text{left}(i)]$)
        $i \leftarrow \text{left}(i)$
      **else return**
    **if** has_left($i$) $\wedge$ $xs[i] < xs[\text{left}(i)]$
**then**
      SWAP($xs[i], xs[\text{left}(i)]$)

**procedure** SIFT_DOWN_OPT($xs', i$)
  $t \leftarrow xs'[i]$
    **while** has_right($i$) **do**
      **if** $xs'[\text{left}(i)] < xs'[\text{right}(i)]$ **then**
      **if** $t < xs'[\text{right}(i)]$ **then**
        $xs'[i] \leftarrow xs'[\text{right}(i)]$
        $i \leftarrow \text{right}(i)$
      **else return**
      **else if** $t < xs[\text{left}(i)]$ **then**
        $xs'[i] \leftarrow xs'[\text{left}(i)]$
        $i \leftarrow \text{left}(i)$
      **else return**
    **if** has_left($i$) $\wedge$ $t < xs'[\text{left}(i)]$ **then**
      $xs'[i] \leftarrow xs'[\text{left}(i)]$
      $i \leftarrow \text{left}(i)$
    $xs'[i] \leftarrow t$

**Algorithm 4:** The standard (left) and optimized (right) sift-down function.

INSERT procedure of insertion sort (cf. Algorithm 3) does a similar optimization. However, for INSERT, it was feasible to prove the optimization together with the actual algorithm. For the slightly more complicated sift-down procedure, we first prove correct the simpler algorithm with swaps, and then refine it to the optimized version. Inside the loop, the refinement relation between the abstract list $xs$ and the concrete list $xs'$ is $xs = xs'[i := t]$. Using the tool support of the Isabelle Refinement Framework, the proof that the optimized version refines the version with swaps requires only about 20 lines of straightforward Isabelle script.

# 4 Imperative Implementation

We have presented a refinement based approach to verify the introsort and pdqsort algorithms, including most optimizations we found in their libstdc++ and Boost implementations. However, the algorithms are still expressed as nondetermistic programs on functional lists and unbounded natural numbers. In this section, we use the Isabelle-LLVM framework [23] to (semi-)automatically refine them to LLVM programs on arrays and 64 bit integers.

## 4.1 The Sepref Tool

The Sepref tool [21,23] symbolically executes an abstract program in the *nres*-monad, keeping track of refinements for every abstract variable to a concrete representation, which may use pointers to dynamically allocated memory. During the symbolic execution, the tool synthesizes an imperative Isabelle-LLVM program, together with a refinement proof. The synthesis is automatic, but usually requires some program-specific setup and boilerplate. For a detailed discussion of Sepref and Isabelle-LLVM, we refer the reader to [21,23].

316     P. Lammich

Sepref comes with standard setup to refine lists to arrays. List updates are
refined to destructive array updates, as long as the old version of the list is
not used after the update. It also provides setup to refine unbounded natural
numbers to bounded integers. It tries to discharge the resulting in-bounds proof
obligations automatically. If this is not possible, it relies on hints from the user.

A common technique to provide such hints is to insert additional assertions
into the abstract program. Usually, these can be proved easily. For example,
in the PDQSORT_AUX algorithm (Algorithm 2, line 9), the assertion $m + 1 \leq h$
ensures that the addition $m + 1$ in the next line cannot overflow. This assertion adds a proof obligation to the correctness proof of PDQSORT_AUX, which is
easily discharged (we are in phase *part*, which guarantees $m < h$). When refining PDQSORT_AUX to an implementation with bounded integers, one can assume
$m + 1 \leq h$ to discharge the non-overflow proof obligation. Note that re-proving
$m + 1 \leq h$ when doing the refinement would require duplicating large parts of
the correctness proof. Thus, assertions provide a convenient tool to pass properties down the refinement chain. Our actual formalization contains multiple such
assertions, which we have omitted in this presentation for the sake of readability.

```
ug_insert_impl ≡ λa l i. doM {
    x ← array_nth a i;
    (a, i) ← llc_while (λ(a, i). doM {
        bi ← ll_sub i 1;
        t ← array_nth a bi;
        ll_icmp_ult x t
    }) (λ(a, i). doM {
        i' ← ll_sub i 1;
        t ← array_nth a i';
        a ← array_upd a i t;
        i ← ll_sub i 1;
        return (a, i)
    }) (a, i);
    array_upd a i x
}
```

**Fig. 1.** Implementation of the INSERT procedure for 64 bit unsigned integer elements
and $G = false$, which is generated by the Sepref tool. This definition lies within the
executable fragment of Isabelle-LLVM, i.e., the Isabelle LLVM code generator can
translate it to LLVM intermediate representation. Note that the function does not
depend on the lower bound parameter $l$ any more, as this was only required in the
guarded version. Inlining will remove this bogus parameter.

Using the Sepref tool, it is straightforward to refine the sorting algorithms
and their subroutines to an Isabelle-LLVM program. For example, Fig. 1 shows
the Isabelle-LLVM code that is generated for the INSERT procedure for unsigned
64 bit integer elements and $G = false$ (cf. Sect. 3.3). Moreover, the Sepref tool
proves that the generated program actually implements the abstract one:

$(ug\_insert\_impl, \; insert \; False) \; : \; arr^d \times nat_{64}^k \times nat_{64}^k \rightarrow arr$

This specifies the refinement relations for the parameters and the result, where
$arr$ relates arrays with lists, and $nat_{64}$ relates 64-bit integers with natural numbers. The $\cdot^d$ annotation means that the parameter will be *destroyed* by the
function call, while $\cdot^k$ means that the parameter is *kept*. Here, the insertion is
done in place, such that the original array is destroyed.

The final correctness statement for our implementations is:

**Theorem 2.** $(introsort\_impl, \; sort\_spec) \; : \; arr^d \times nat_{64}^k \times nat_{64}^k \rightarrow arr$
$\quad\quad$ *and* $\;(pdqsort\_impl, \; sort\_spec) \;\; : \; arr^d \times nat_{64}^k \times nat_{64}^k \rightarrow arr$

Here, *introsort_impl* and *pdqsort_impl* are the Isabelle-LLVM programs generated
by Sepref from INTROSORT and PDQSORT (Algorithms 1 and 2). The theorem is
easily proved by combining Theorem 1 with the theorems generated by Sepref.

## 4.2   Separation Logic and Ownership

Internally, the Sepref tool represents the symbolic state that contains all abstract
variables and their refinements to concrete variables as an assertion in separation
logic [8,31]. Thus, two variables can never reference the same memory. This is a
problem for nested container data structures like arrays of strings: when indexing
the array, both the array element and the result of the indexing operation would
point to the same string. In the original Sepref tool [21], which targeted Standard
ML, we worked around this problem by always using functional data types (e.g.
lists) to represent the inner type of a nested container. This workaround is no
longer applicable for the purely imperative LLVM, such that we could not use
Sepref for nested container data structures[7].

We now describe an approach towards solving this problem for Sepref.
Abstractly, we model an array by the type $\alpha$ *option list*[8], where *None* means that
the array does currently not own the respective element. The abstract indexing
operation then moves the element from the list to the result:

$move \; xs \; i \equiv$ **assert** $i < |xs| \; \wedge \; xs[i] \neq None;$ **return** $(the \; (xs[i]), \; xs[i:=None])$

As no memory is shared between the result and the array, we can show the
following refinement:

$(\lambda a \; i. \; (a[i], a), \; \lambda xs \; i. \; move \; xs \; i) \; : \; oarr^d \times nat_{64}^k \rightarrow A \times oarr$

where *oarr* is the relation between an array and an $\alpha$ *option list*, and $A$ is the
relation for the array elements. Note that this operation does not change the
concrete array $a$. The movement of ownership is a purely abstract concept, which
results in no implementation overhead.

---

[7] We could still reason about such structures on a lower level.

[8] Here, $\alpha$ *option = None | Some $\alpha$* is Isabelle's option datatype, and *the (Some x)* $\equiv x$
is the corresponding selector function.

The transition from $\alpha$ *list* to $\alpha$ *option list* can typically be done in an additional refinement step, and thus does not obfuscate the actual correctness proofs, which are still done on plain $\alpha$ *list*. Moreover, the $\alpha$ *option list* representation is only required for subroutines where extracted array elements are actually visible. For example, we define an operation to compare two array elements:

$$cmp\_idxs\ xs\ i\ j \equiv \mathtt{assert}\ i{<}|xs| \wedge j{<}|xs|;\ \mathtt{return}\ xs[i] < xs[j]$$

Inside this operation, we have to temporarily extract the elements $i$ and $j$ from the array, requiring an intermediate refinement step to $\alpha$ *option list*. However, at the start and end of this operation, the array owns all its elements. For the whole operation, we thus get a refinement on plain arrays:

$$(cmp\_impl,\ cmp\_idxs) : arr^k \times nat_{64}^k \times nat_{64}^k \to bool$$

In our case, we only have to explicitly refine INSERT and SIFT_DOWN. The other subroutines use *cmp_idxs* and *swap* operations on plain lists.

### 4.3    The Isabelle-LLVM Code Generator

The programs that are generated by Sepref (cf. Fig. 1) lie in the fragment for which Isabelle-LLVM [23] can generate LLVM text. For example, the pdqsort algorithm for strings yields an LLVM function with the signature:

$\{$ i64, $\{$ i64, i8* $\}$ $\}$* @str_pdqsort($\{$ i64, $\{$ i64, i8* $\}$ $\}$*, i64, i64)

Here, the type $\{$ i64, $\{$ i64, i8* $\}$ $\}$ represents dynamic arrays of characters[9], represented by length, capacity, and a data pointer.

The generated LLVM text is then compiled to machine code using the LLVM toolchain. To make the generated program usable, one has to link it to a C wrapper, which handles parsing of command line options and printing of results. However, the original Isabelle-LLVM framework provides no support for interfacing the generated code from C: one has to manually write a C header file, which hopefully matches the object file generated by the LLVM compiler. If it doesn't, the program has undefined behaviour[10].

To this end, we extended Isabelle-LLVM to also generate a header file for the exported functions. For example, the Isabelle command

**export_llvm** *str_sort_introsort_impl*
  **is** *llstring\* str_introsort(llstring\*, int64_t, int64_t)*
  **defines** ‹ *typedef struct* {
    *int64_t size; struct {int64_t capacity; char \*data;};*
  } *llstring;* ›

---

[9] For strings, we use the verified dynamic array implementation provided by Isabelle-LLVM. Note that C++ uses a similar representation, with an additional optimization for small strings.

[10] In practice, this means it will probably SEGFAULT. However, it also might return wrong results, or be prone to various kinds of exploits.

will check that the specified signature actually matches the Isabelle definition, and generate the following C declarations:

**typedef struct** { *int64_t size;* **struct** { *int64_t capacity;* **char** *∗data;* }; } *llstring;*
*llstring∗ str_introsort(llstring∗, int64_t, int64_t);*

## 5   Benchmarks

The Boost library comes with a sorting algorithm benchmark suite, which we extended with further benchmarks indicated in [4]: apart from sorting random lists of elements that are mostly different (random), we also sort lists of length $n$ that contain only $n/10$ different elements (random-dup-10), and lists of only two different elements (random-boolean), as well as lists where all elements are equal (equal). We also consider already sorted sequences (sorted, rev-sorted), as well as a sequence of $n/2$ elements in ascending order, followed by the same elements in descending order (organ-pipe). We also consider sorted sequences where we applied $pn/100$ random swap operations (almost-sorted-$p$). Finally, we consider sorted sequences with $pn/100$ random elements inserted at the end or in the middle ([rev-]sorted-end-$p$, [rev-]sorted-middle-$p$).

We sorted integer arrays with $n = 10^8$ elements, and string arrays with $n = 10^7$ elements. For strings, all implementations use the same data structure and compare function. For integers, we disable pdqsorts branch-aware partitioning, which we have not yet verified. For strings, it does not apply anyway.

We compile both, the verified and unverified algorithms with clang-6.0.0, and run them on a laptop with an Intel(R) Core(TM) i7-8665U CPU and 32 GiB of RAM, as well as on a server machine with 24 AMD Opteron 6176 cores and 128 GiB of RAM. Ideally, the same algorithm should take exactly the same time when repeatedly run on the same data and machine. However, in practice, we encountered some noise up to 17%. Thus, we have repeated each experiment at least ten times, and more often to confirm outliers where the verified and unverified algorithms' run times differ significantly. Assuming that the noise only slows down an algorithm, we take the fastest time measured over all repetitions. The results are displayed in Fig. 2.

They indicate that both our pdqsort and introsort implementations are competitive. There is one outlier for pdqsort for already sorted integer arrays on the laptop. We have not yet understood its exact reason. The remaining cases differ by less than 20%, and in many cases our verified algorithm is actually faster.

320    P. Lammich

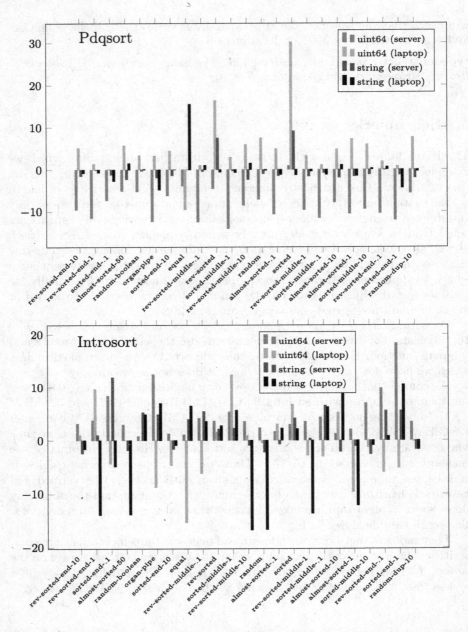

**Fig. 2.** Benchmarking our verified implementations against the unverified originals. For each element type, machine, and distribution, the value $(t_1/t_2 - 1) * s$ is shown, where $t_1$ is the slower time, $t_2$ is the faster time, and $s = 100$ if our implementation is slower, and $s = -100$ if the original implementation is slower. That is, a positive value $p$ indicates that our implementation is slower, requiring $100 + p$ percent of the run time of the original. Analogously, a negative value $-p$ means that the original implementation is slower, requiring $100 + p$ percent of our implementation's run time.

# 6  Conclusions

We have presented the first verification of the introsort and pdqsort algorithms. We verified state-of-the-art implementations, down to LLVM intermediate representation. On an extensive set of benchmarks, our verified implementations perform on par with their unverified counterparts from the GNU C++ and Boost C++ libraries. Apart from our work, the only other verified real-world implementations of sorting algorithms are Java implementations of timsort and dual-pivot quicksort that have been verified with KeY [3,10].

Compared to other program verification methods, the trusted code base of our approach is small: apart from the well-tested and widely used LLVM compiler, it only includes Isabelle's logical inference kernel, and the relatively straightforward Isabelle-LLVM semantics and code generation [23]. In contrast, deductive verification tools like KeY [2] depend on the correct axiomatization of the highly complex Java semantics, as well as on several automatic theorem provers, which, themselves, are highly complex and optimized C programs.

Our verified algorithms can readily be used in larger verification projects, and we have already replaced a naive quicksort implementation that caused a stack overflow in an ongoing SAT-solver verification project [13]. For fixed element types and containers based on arrays (e.g. std::vector), we can use our verified algorithms as a drop-in replacement for C++'s std::sort. A direct verification of the C++ code of std::sort, however, would require a formal semantics of C++, including templates and the relevant concepts from the standard template library (orderings, iterators, etc.). To the best of our knowledge, such a semantics has not yet been formalized, let alone been used to verify non-trivial algorithms.

The verification of introsort took us about 100 person hours. After we had set up most of the infrastructure for introsort, we could verify the more complex pdqsort in about 30 h. The development consists of roughly 8700 lines of Isabelle text, of which 2400 lines are for introsort and 2200 lines for pdqsort. The rest is boilerplate and libraries shared between both algorithms, among them 1500 lines for the verification of heapsort.

## 6.1  Related Work

Sorting algorithms are a standard benchmark for program verification tools, such that we cannot give an exhaustive overview here. Nevertheless, we discuss a few notable examples: the arguably first formal proof of quicksort was given by Foley and Hoare himself [14], though, due to the lack of powerful enough theorem provers at these times, it was only done on paper.

One of the first mechanical verifications of imperative sorting algorithms is by Filliâtre and Magaud [12], who prove correct imperative versions of quicksort, heapsort, and insertion sort in Coq. However, they use a simplistic partitioning scheme, do not report on code generation or benchmarking, nor do they combine their separate algorithms to get introsort.

The timsort algorithm, which was used in the Java standard library, has been verified with the KeY tool [10]. A bug was found and fixed during the verification.

Subsequently, KeY has been used to also verify the dual-pivot quicksort algorithm from the Java standard library [3]. This time, no bugs were found.

## 6.2  Future Work

An obvious next step is to verify a branch-aware partitioning algorithm [11]. Thanks to our modular approach, this will easily integrate with our existing formalization. We also plan to extend our work to stable sorting algorithms. Recently, we have extended the Refinement Framework to support reasoning about algorithmic complexity [17]. Once this work has been integrated with Isabelle-LLVM, we can also prove that our implementations have a worst-case complexity of $O(n \log(n))$, as required by the C++ standard. Finally, we proposed an explicit ownership model for nested lists. We plan to extend this to more advanced concepts like read-only shared ownership, inspired by Rust's [32] ownership system. Formally, this could be realized with fractional permission separation logic [7].

**Acknowledgements.** We received funding from DFG grant LA 3292/1 "Verifizierte Model Checker" and VeTSS grant "Formal Verification of Information Flow Security for Relational Databases".

## References

1. Williams, J.W.J.: Algorithm 232: heapsort. Commun. ACM **7**(6), 347–349 (1964)
2. Beckert, B., Hähnle, R., Schmitt, P.H.: Verification of Object-oriented Software: The KeY Approach. Springer, Heidelberg (2007). https://doi.org/10.1007/978-3-540-69061-0
3. Beckert, B., Schiffl, J., Schmitt, P.H., Ulbrich, M.: Proving JDK's dual pivot quicksort correct. In: Paskevich, A., Wies, T. (eds.) VSTTE 2017. LNCS, vol. 10712, pp. 35–48. Springer, Cham (2017). https://doi.org/10.1007/978-3-319-72308-2_3
4. Bentley, J.L., McIlroy, M.D.: Engineering a sort function. Softw. Pract. Exp. **23**(11), 1249–1265 (1993)
5. Bloch, J.: Extra, extra - read all about it: nearly all binary searches and mergesorts are broken
6. Boost C++ libraries (2011)
7. Bornat, R., Calcagno, C., O'Hearn, P., Parkinson, M.: Permission accounting in separation logic. In: Proceedings of the 32nd ACM SIGPLAN-SIGACT Symposium on Principles of Programming Languages, POPL 2005, pp. 259–270. ACM, New York (2005)
8. Calcagno, C., O'Hearn, P., Yang, H.: Local action and abstract separation logic. In: LICS 2007, pp. 366–378, July 2007
9. Cormen, T.H., Leiserson, C.E., Rivest, R.L., Stein, C.: Introduction to Algorithms, 3rd edn. The MIT Press, Cambridge (2009)
10. de Gouw, S., Rot, J., de Boer, F.S., Bubel, R., Hähnle, R.: OpenJDK's Java.utils.Collection.sort() is broken: the good, the bad and the worst case. In: Kroening, D., Păsăreanu, C.S. (eds.) CAV 2015. LNCS, vol. 9206, pp. 273–289. Springer, Cham (2015). https://doi.org/10.1007/978-3-319-21690-4_16

11. Edelkamp, S., Weiß, A.: Blockquicksort: how branch mispredictions don't affect quicksort. CoRR, abs/1604.06697 (2016)
12. Filliâtre, J.-C., Magaud, N.: Certification of sorting algorithms in the Coq system (1999)
13. Fleury, M., Blanchette, J.C., Lammich, P.: A verified SAT solver with watched literals using Imperative HOL. In: Proceedings of CPP, pp. 158–171 (2018)
14. Foley, M., Hoare, C.A.R.: Proof of a recursive program: quicksort. Comput. J. **14**(4), 391–395 (1971)
15. The GNU C++ library. Version 7.4.0
16. Griebel, S.: Binary heaps for imp2. Archive of Formal Proofs, June 2019. http://isa-afp.org/entries/IMP2_Binary_Heap.html. Formal proof development
17. Haslbeck, M., Lammich, P.: Refinement with time - refining the run-time of algorithms in Isabelle/HOL. In: ITP2019: Interactive Theorem Proving, June 2019
18. Hoare, C.A.R.: Algorithm 64: quicksort. Commun. ACM **4**(7), 321 (1961)
19. Josuttis, N.M.: The C++ Standard Library: A Tutorial and Reference, 2nd edn. Addison-Wesley Professional, Boston (2012)
20. Krauss, A.: Recursive definitions of monadic functions. In: Proceedings of the PAR, vol. 42, pp. 1–13 (2010)
21. Lammich, P.: Refinement to Imperative/HOL. In: Urban, C., Zhang, X. (eds.) ITP 2015. LNCS, vol. 9236, pp. 253–269. Springer, Cham (2015). https://doi.org/10.1007/978-3-319-22102-1_17
22. Lammich, P.: Refinement based verification of imperative data structures. In: Avigad, J., Chlipala, A. (eds.) CPP 2016, pp. 27–36. ACM (2016)
23. Lammich, P.: Generating verified LLVM from Isabelle/HOL. In: Harrison, J., O'Leary, J., Tolmach, A. (eds.) 10th International Conference on Interactive Theorem Proving (ITP 2019), Leibniz International Proceedings in Informatics (LIPIcs), Dagstuhl, Germany, vol. 141, pp. 22:1–22:19. Schloss Dagstuhl-Leibniz-Zentrum fuer Informatik (2019)
24. Lammich, P., Tuerk, T.: Applying data refinement for monadic programs to Hopcroft's algorithm. In: Beringer, L., Felty, A. (eds.) ITP 2012. LNCS, vol. 7406, pp. 166–182. Springer, Heidelberg (2012). https://doi.org/10.1007/978-3-642-32347-8_12
25. Lammich, P., Wimmer, S.: Imp2 – simple program verification in Isabelle/HOL. Archive of Formal Proofs, January 2019. http://isa-afp.org/entries/IMP2.html. Formal proof development
26. "libc++" C++ standard library
27. LLVM language reference manual
28. Musser, D.R.: Introspective sorting and selection algorithms. Softw. Pract. Exp. **27**(8), 983–99 (1997)
29. Pattern-defeating quicksort
30. Peters, T.: Original description of timsort. Accessed 21 Oct 2019
31. Reynolds, J.C.: Separation logic: a logic for shared mutable data structures. In Proceedings of Logic in Computer Science (LICS), pp. 55–74. IEEE (2002)
32. The rust programming language
33. Sedgewick, R., Wayne, K.: Algorithms, 4th edn. Addison-Wesley Professional, Boston (2011)

# A Fast Verified Liveness Analysis in SSA Form

Jean-Christophe Léchenet$^{(\boxtimes)}$ (ID), Sandrine Blazy(ID), and David Pichardie

Univ Rennes, Inria, CNRS, IRISA, Rennes, France
{jean-christophe.lechenet,sandrine.blazy,david.pichardie}@irisa.fr

**Abstract.** Liveness analysis is a standard compiler analysis, enabling several optimizations such as deadcode elimination. The SSA form is a popular compiler intermediate language allowing for simple and fast optimizations. Boissinot et al. [7] designed a fast liveness analysis by combining the specific properties of SSA with graph-theoretic ideas such as depth-first search and dominance. We formalize their approach in the Coq proof assistant, inside the CompCertSSA verified C compiler. We also compare experimentally this approach on CompCert's benchmarks with respect to the classic data-flow-based liveness analysis, and observe performance gains.

**Keywords:** Liveness analysis · SSA form · Dominance · Verified compilation

## 1 Introduction

In order to be precise, several important compiler analyses need to know the lifetime of variables. This is of course the case with deadcode elimination and register allocation, but also for instance with software pipelining and trace scheduling. Computing this information efficiently is thus of utmost importance. This is the purpose of *liveness analysis*.

Given a program and a variable, liveness analysis consists in determining the points of the program where this variable is needed, i.e. the points from which an execution can reach an instruction where this variable is used. At such points, this variable is said to be *live*. Like many other semantic properties, this property is undecidable and is classically over-approximated by its syntactic counterpart which considers, instead of real executions, paths in the control flow graph (CFG) of the program.

Traditionally, liveness information is computed by a backward data-flow analysis that computes monolithically the liveness status of all program variables at all program points [2]. In 2008, Boissinot et al. [7] described another method to compute this information, with two particularities. Firstly, their technique is applicable only to programs in SSA form, an intermediate language adopted by most of the modern compilers, e.g. LLVM [11]. Indeed, their approach relies on one of the key properties of SSA, that they combine with graph-theoretic

N. Peltier and V. Sofronie-Stokkermans (Eds.): IJCAR 2020, LNAI 12167, pp. 324–340, 2020.
https://doi.org/10.1007/978-3-030-51054-1_19

notions. Secondly, it is not designed to compute the whole liveness information of the program, but instead to answer so-called liveness queries, of the form "is variable a live at point $q$?". They call this approach, considering only one variable and one program point at a time, "liveness checking". Since this approach computes only limited information compared to the data-flow based one, they claim that it outperforms it as long as the number of asked queries is low, which their experiments confirm.

In this paper, we focus on liveness checking, as presented in [7], from the point of view of formally verified compilation. In this context, an implementation of liveness checking should not only be efficient, as usual in compilation, but also needs to be formally proved correct.

We tackle this problem in the context of CompCert [12,13], a verified C compiler written in the Coq proof assistant, and its fork with an SSA middle-end, CompCertSSA [3]. CompCert and CompCertSSA already contain several liveness analyses (e.g. in module Liveness), but all of them, like in the other verified compilers (e.g. CakeML [15]), are data-flow based. Our goal is to implement liveness checking on the SSA form of CompCertSSA, taking into account the particularities of Coq and CompCertSSA, and carefully enough so that we observe the expected performance improvement w.r.t. the data-flow based approach.

After describing liveness checking, as well as the required background, in detail in Sect. 2, we present the following contributions:

- an implementation of liveness checking in CompCertSSA (Sect. 3) adapting the ideas of [7] to CompCertSSA, including some advanced optimizations;
- a proof of correctness of this algorithm (Sect. 4) showing the validity of Boissinot et al.'s subtle graph-theoretic arguments;
- experiments on CompCert's benchmarks (Sect. 5) showing that two variants of the liveness checking algorithm compare favorably w.r.t. the data-flow based approach.

The formalization and the experiments are available online [1].

## 2  Background

We first recall some notions from graph theory and compilers in Sect. 2.1, then we give the idea of liveness checking in Sect. 2.2, before describing it in detail in Sects. 2.3 and 2.4.

### 2.1  Basic Concepts in Graphs and Compilers

**Depth-First Search.** DFS classifies the edges of a graph into four categories (cf. Fig. 1): the *tree edges* that form a spanning tree, the *forward edges* connecting a node to one of its descendants in the spanning tree, the *cross edges* connecting a node to an unrelated node in the spanning tree, and the *back edges* connecting a node to one of its ancestors in the spanning tree.

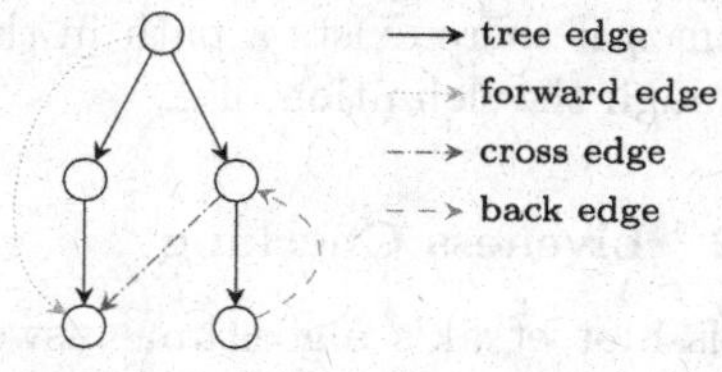

Fig. 1. Example of edge classification

**Encoding Reachability in a Tree.** It is possible to label each node of a tree with a pair of integers, allowing to determine whether a node is an ancestor of another node just by comparing their labels. One possible labeling is based on a DFS preorder numbering, the first integer of a node being its preorder number and the second one being the maximum preorder number in the subtree rooted at that node. An example of such a labeling is provided in Fig. 2b.

**Dominance.** The dominance relation is traditionally defined on a flow graph, i.e. a graph with a distinguished node *entry* such that every vertex is reachable from that node. We say that a node $u$ dominates a node $v$ if every path from *entry* to $v$ goes through $u$; $u$ strictly dominates $v$ if $u$ dominates $v$ and $u$ and $v$ are distinct. Dominance is an order relation, i.e. it is reflexive, transitive and antisymmetric. Moreover, each node $u$ distinct from *entry* has a unique strict dominator dominated by all the strict dominators of $u$, showing that dominance can be encoded as a tree, called the *dominance tree*.

**SSA Form.** The SSA form, standing for *Static Single Assignment*, is a program representation where each variable is textually defined at most once. To turn a non-SSA representation into SSA, variables that are assigned to multiple times are renamed so that each renamed version is associated to one definition point only. When two flows of the program, carrying two different versions of the same initial variable, merge at a so-called *join point*, we need a way to express which version is selected. SSA introduces special nodes for this, called $\phi$-nodes. The $\phi$-function inside the $\phi$-node takes as many arguments as the number of predecessors of the node. When the flow comes from the $i^{th}$ predecessor, the $\phi$-function returns the $i^{th}$ argument, thus selecting the version of the variable corresponding to that predecessor. $\phi$-nodes must be handled with care in terms of where they use and define variables. In this paper, each argument of a $\phi$-function is considered used at the corresponding predecessor of the $\phi$-node. The variables defined by the $\phi$-node are treated normally. An example SSA program is shown in Fig. 2, along with its dominance tree.

A program in *strict* SSA form is a program where each use of a variable is preceded by its definition (unique per definition of SSA). A program in strict SSA form obeys the *dominance property* [7], stating that each use of a variable is dominated by its definition.

**Liveness.** In this paper, by "live", we mean "live-in", which in the context of a program in strict SSA form can be defined as follows. A variable a is live-in at point $q$ if there exists a path in the CFG from $q$ to a use of a that does not go through the definition of a.

## 2.2   Liveness Checking

Boissinot et al.'s algorithm answers liveness queries efficiently based on some precomputed information. The algorithm is thus composed of two parts: a pre-

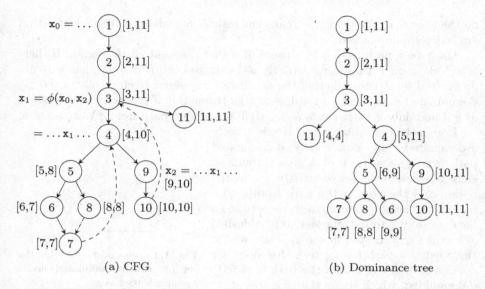

**Fig. 2.** The CFG and dominance tree of an SSA program, both labeled with reachability intervals based on preorder numberings

computation part that captures information about the CFG structure and an online part that answers the liveness queries based on this information.

This architecture has two main advantages compared to the classic one. Firstly, the precomputation step is faster than the full liveness analysis. Thus, if the number of queries is rather small, this algorithm is faster than the classic one. Secondly, since the precomputation step depends on the CFG structure and not on liveness information, its result remains correct if the program is modified by some transformations that preserve its structure. In this sense, precomputed information is more robust than the liveness one.

Actually, the classic liveness analysis approach can also be seen as being made of a precomputation part (the analysis), followed by an online part (reading in the liveness table). From this point of view, Boissinot et al.'s algorithm just chooses a different trade-off than the classic approach: a faster precomputation at the cost of slower queries. As mentioned above, this compromise is interesting if the number of queries is low.

### 2.3  Precomputation

Let us consider the following liveness query: "is variable a live at point $q$?". This query amounts to checking whether a path exists between $q$ and a use of a that does not go through the definition $d$ of a. Note that, by the dominance property, we know that all uses of a are dominated by $d$. It is possible that a is used at $d$, but since in this paper by "live" we mean "live-in", such a case has no impact on the answer to the query. We can thus restrict ourselves to the uses of a that are strictly $d$-dominated.

Let $\pi$ be a path from $q$ to a use $u$ of a that does not go through $d$. If there is a node $x$ on $\pi$ that is not strictly $d$-dominated, we can show that $u$ is not dominated by $d$, contradicting the dominance property. Reciprocally, a strictly $d$-dominated path from $q$ to $u$ does not go through $d$. This shows that a is live at $q$ if and only if there exists a strictly $d$-dominated path from $q$ to a use of a.

Boissinot et al. show more. If $q$ is strictly $d$-dominated, any non-strictly $d$-dominated path from $q$ to a use $u$ of a goes through $d$, since it reenters the set of strictly dominated nodes, and the part of the path from $q$ to $d$ contains a back edge (intuitively, we need to go back up from $q$ to $d$), represented as a dashed arrow in Fig. 3. Stated in the opposite way, if there exists a path from $q$ to $u$ that does not contain a back edge, then the path is strictly $d$-dominated, which shows that a is live at $q$.

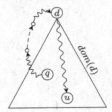

**Fig. 3.** Leaving and reentering the set of strictly dominated nodes requires a back edge.

Based on this observation, Boissinot et al.'s main idea is that back edges must be dealt with separately from the other edges. They suggest to decompose the reachability in the original graph into two relations, called $R$ and $T$. Relation $R$ captures the reachability in the reduced graph $\widetilde{G}$, the acyclic graph obtained by removing the back edges from the original graph. Relation $T$ associates to each program point both itself and a set of interesting back edge targets.

Formally, $T$ is the reflexive and transitive closure of $T^\uparrow$, where $T_t^\uparrow$ (cf. Definition 1)[1] is the set of back edge targets not reduced reachable (i.e. reachable in the reduced graph) from node $t$ but whose source is reduced reachable from $t$. For instance, in Fig. 2a, $T_5^\uparrow = \{4\}$, $T_4^\uparrow = \{3\}$, and thus $T_5 = \{3, 4, 5\}$.

**Definition 1** ($T^\uparrow$ and $T$).

$$T_t^\uparrow = \{t' \in V \setminus R_t \mid \exists s' \in R_t \wedge (s', t') \in E^\uparrow\} \text{ and } T = (T^\uparrow)^*$$

where $E^\uparrow$ is the set of back edges, $*$ is the reflexive and transitive closure.

## 2.4   Online Part

The online part leverages precomputed and dominance information to answer liveness queries efficiently. Boissinot et al.'s algorithm ([7, Algorithm 1]) is reproduced as Algorithm 1. Given a variable a and a program point $q$, the algorithm filters the content of $T_q$ to keep only the set $T_{(q,\mathbf{a})}$ of points that are strictly dominated by the definition point of a (line 2). Then it tests whether one of these points can reach a use of a in the reduced graph (lines 3–4). If one test succeeds, then it returns *true* (line 4), the variable a is live at $q$, otherwise it returns *false* (line 5), the variable a is not live at $q$. In Fig. 2a, $T_{(5,\mathbf{x}_1)} = \{4, 5\}$, $uses(\mathbf{x}_1) = \{4, 9\}$, $4 \in R_4$, thus $\mathbf{x}_1$ is live at 5. $T_{(5,\mathbf{x}_2)} = \emptyset$, thus $\mathbf{x}_2$ is not live at 5. $T_{10} = \{3, 10\}$, $T_{(10,\mathbf{x}_1)} = \{10\}$, $R_{10} \cap uses(\mathbf{x}_1) = \emptyset$, thus $\mathbf{x}_1$ is not live at 10.

---

[1] Given a relation $R$, $R_x$ denotes the set of elements related to $x$ in $R$.

```
1  Function IsLiveIn(variable a, node q):
2  │   T(q,a) ← Tq ∩ sdom(def(a))
3  │   for t ∈ T(q,a) do
4  │   │   if Rt ∩ uses(a) ≠ ∅ then return true
5  │   return false
```
**Algorithm 1:** Online part of Boissinot et al.'s algorithm

## 3   Formalization

Our Coq implementation follows approximately the same structure as the algorithm described in Sect. 2. In particular, it is divided into two parts: the precomputation and the online parts.

### 3.1   Precomputation

As highlighted in Sect. 2.2, the precomputation step depends only on the CFG structure. Thus, we can abstract the specific features of the SSA form and only work at the graph-theoretic level. We model the CFG as a map of type `graph = map (list node)`[2] associating to each node the list of its successors, and a node `entry` representing the entry point of the CFG. Moreover, to implement the second optimization described in Sect. 3.2, we need to model the preorder numbering on the dominance tree. We assume that we are given a function `dom_pre : node → Z` associating to each node the corresponding number.

As proposed in [7], the precomputation step itself is split into two parts. In [7], the first one computes $R$, while the second one computes $T$ based on $R$. We slightly adapted both parts. In our implementation, the first part computes $R$ and $T^\uparrow$, and the second part computes $T$ in a different way than in [7].

**Precomputation of $R$ and $T^\uparrow$ .** Boissinot et al. [7] suggest encoding the set of reduced reachable nodes from node $t$, $R_t$, as a set (using bitsets or sorted arrays). But they assume, as is the case for most compilers, that the nodes in the CFG represent blocks of instructions, which means that the CFG is not really large. CompCertSSA's peculiarity is that, like CompCert, each node in the CFG represents only one instruction, and thus the CFG is noticeably bigger. To avoid manipulating large sets, we decided to encode $R$ differently, drawing our inspiration from Boissinot et al.'s idea to treat back edges specially. We choose to treat cross edges specially, and to break down reachability in the reduced graph into reachability in the spanning tree from sets of cross edge targets. This decomposition seems to forget forward edges, but as far as only reachability is concerned, they can be safely ignored, as they are just shortcuts of tree edges.

We introduce the relations $\tilde{R}$ that denotes the reachability in the spanning tree, and $C$ that associates to each program point both itself and a set of cross

---

[2] `node` is an alias for `positive`, a binary encoding of strictly positive integers; `map` is implemented using `PTree.t`, an associative map whose keys are `positive` and which is used pervasively in CompCertSSA.

```
1    Record state := {
2      gr: graph;                       (* current graph, without already-visited nodes *)
3      wrk: list (node * positive * list node * (set * list (node * Z)));
4        (* worklist: node, label, children to be treated, results from treated children *)
5      next: positive;                  (* number to use for next numbering *)
6      r : map itv;                     (* reachability relation using intervals *)
7      c : map set;                     (* cross nodes to test for reduced reachability *)
8      t_up : map (list (node * Z));    (* sorted list of back nodes to test for reachability *)
9      back : list (node * node)        (* back edges *)
10   }.
11   (* result is the expected type of the returned tuple (r, c, t_up, back) *)
12   Definition result := map itv * map set * map (list (node * Z)) * list(node * node).
13
14   Definition transition (dom_pre : node → Z) (s: state) : result + state :=
15     match s.(wrk) with
16     | [] ⇒ inl (s.(r), s.(c), s.(t_up), s.(back)) (* end of the DFS *)
17     | (u, n, [], (s_c, s_t)) :: wrk' ⇒ (* end of processing of node u *)
18        let r' := update u (n, Pos.pred s.(next)) s.(r) in
19        let s_c' := filter (fun v _ ⇒ negb (is_directly_included r' u v)) s_c in
20        let s_c'' := add u s_c' in
21        let c' := update u s_c'' s.(c) in
22        let s_t' := List.filter (fun '(v, _) ⇒ negb (is_cross_included r' c' u v)) s_t in
23        let t_up' := update u s_t' s.(t_up) in
24        inr {| s with wrk := wrk'; r := r'; c := c'; t_up := t_up' |}
25     | (u, n, v :: succs_u, (s_c, s_t)) :: wrk' ⇒ (* processing of child v of node u *)
26        match s.(gr) ! v with (* "!" is the lookup operator in maps *)
27        | None ⇒ (* v has already been discovered *)
28           match s.(r) ! v with
29           | None ⇒ (* back edge *)
30              let s_t' := merge [(v, dom_pre v)] s_t in (* merge is order-preserving *)
31              let back' := (u, v) :: s.(back) in
32              inr {| s with wrk := (u, n, succs_u, (s_c, s_t')) :: wrk'; back := back' |}
33           | Some _ ⇒ (* processed tree edge, forward edge or cross edge *)
34              let s_c' := match s.(c) ! v with | None ⇒ s_c | Some s ⇒ union s s_c end in
35              let s_t' := match s.(t_up) ! v with | None ⇒ s_t | Some s ⇒ merge s s_t end in
36              inr {| s with wrk := (u, n, succs_u, (s_c', s_t')) :: wrk' |}
37           end
38        | Some succs_v ⇒ (* new tree edge *)
39           inr {| s with gr := remove v s.(gr);
40                        (* v is left in the worklist so that we can propagate the result *)
41                        wrk := (v, s.(next), succs_v, (empty, [])) :: s.(wrk);
42                        next := Pos.succ s.(next) |}
43        end
44     end.
45
46   Definition precompute_r_t_up (g: graph) (root: node) (dom_pre : node → Z) : result :=
47     WfIter.iterate (transition dom_pre)
48                    lt_state lt_state_wf (transition_decreases dom_pre)
49                    (init_state g root).
```

**Fig. 4.** Function `precompute_r_t_up` implements the first part of the precomputation.

edge targets that are interesting for checking reduced reachability at this point. Like $T$, $C$ is defined as the reflexive and transitive closure of $C^{\uparrow}$, where $C_t^{\uparrow}$ (cf. Definition 2) associates to node $t$ the set of cross edge targets not tree reachable (i.e. reachable in the spanning tree) from $t$ but whose source is tree reachable from $t$. In Fig. 2a, only $C_8^{\uparrow} = \{7\}$ is non-empty. We have thus $C_8 = \{7, 8\}$.

**Definition 2** ($C^{\uparrow}$ and $C$).

$$C_t^{\uparrow} = \{t' \in V \setminus \widetilde{R}_t \mid \exists s' \in \widetilde{R}_t \wedge (s', t') \in \widetilde{E}^{\uparrow}\} \text{ and } C = (C^{\uparrow})^*$$

where $\widetilde{E}^{\uparrow}$ designates the set of cross edges.

Moreover, since the spanning tree is a tree, we can use the technique mentioned in Sect. 2.1, i.e. encode $\widetilde{R}$ as a labeling of each node in the spanning tree with a pair of integers representing an interval. We can then answer reachability queries in the spanning tree efficiently by testing inclusion of those intervals.

In the Coq development, function `precompute_r_t_up`, shown in Fig. 4, implements this first step of precomputation. For the sake of clarity, the Coq code was a little prettified. In particular, the notation {| .. with .. := .. |}, allowing to update only some fields of a record, is not a valid Coq expression. `precompute_r_t_up` returns a quadruple (`r`, `c`, `t_up`, `back`), where:

- `r` : `map itv` encodes $\widetilde{R}$ by associating to each node an interval of `positive`;
- `c` : `map set`[3] implements $C$;
- `t_up` : `map (list (node * Z))` encodes $T^\uparrow$ (`t_up` associates to each node a list of pairs (`u`, `n`) where `u` is a node and `n` is just `dom_pre u`; this list is sorted on the second component (see Sect. 3.2); that second component is not really needed, it is a slight optimization that allows to reduce the number of calls to `dom_pre u` by storing its result next to `u` the first time it is called);
- `back` : `list (node * node)` is the list of identified back edges.

Function `precompute_r_t_up` performs a DFS traversal of the CFG. In the style of module `Postorder` of CompCert, it calls iteratively a transition function (l. 47) that updates a state (initialized l. 49) with the guarantee that the iterations eventually terminate (l. 48). The state aggregates seven fields (l. 1). Four fields (`r`, `c`, `t_up` and `back`) correspond to the final results. The three other fields are used to implement the DFS: `gr` remembers whether a node has already been seen during the traversal; `next` is the current value of the counter used to number the encountered nodes; and `wrk` is a worklist of nodes to be treated. Each element of `wrk` is a quadruple (`u`, `n`, `succs`, (`s_c`, `s_t`)), where `u` is a node labeled with number `n`, `succs` is the list of successors of `u` yet to be treated, and `s_c` and `s_t` (detailed below) are pieces of information, retrieved from the successors of `u` that have already been treated, and used to compute the value attached to `u` in `c` and `t_up` respectively.

Function `transition` begins with checking whether the worklist is empty. If so (l. 16), it is the last iteration and the appropriate fields of the state are returned. If not, it analyzes the status of the first node `u` of the worklist. If it has still children to be treated (l. 25), it checks the status of the first child `v`. If `v` is new to the DFS (l. 38), it is given number `s.(next)`, and is explored recursively by extending the worklist (l. 41). If `v` has already been seen before during the DFS (l. 27), we retrieve from it the pieces of information that need to be propagated to `u`, and we update `s_c` and `s_t` accordingly depending on the type of edge connecting `u` and `v` (ll. 28–37). Note that, in the first case (l. 38), `v` is intentionally left as a child of `u` in the worklist (l. 41), so that it can be seen again in the second case (l. 33), and results can be propagated from `v` to `u`. If all the children of `u` (l. 17) have been treated, we use the data available in the state and the worklist to update maps `r`, `c` and `t_up` at key `u`.

---

[3] `set` = `map unit` is a map where only keys are meaningful.

To update $r$, we attach to $u$ (l. 18) an interval based on the number $n$ associated to $u$ when it was discovered (l. 41) and the current value of the counter $\texttt{next}$. The update of $c$ relies on the following equation: $C_u = \{u\} \cup \left[\bigcup_{(u,v) \in \widetilde{E}} C_v\right] \setminus \widetilde{R}_u$. $C_u$ is computed from the sets $C_v$ of its children in the reduced graph (i.e. children $v$ where $(u,v)$ is not a back edge). The union of these sets (l. 34) is filtered (l. 19), so that only nodes that are not already tree reachable from $u$ are kept. Finally, node $u$ is added to the set (l. 20). The update of $\texttt{t\_up}$ relies on a similar equation: $T_u^{\uparrow} = \left[\bigcup_{(u,v) \in E^{\uparrow}} \{v\} \cup \bigcup_{(u,v) \in \widetilde{E}} T_v^{\uparrow}\right] \setminus R_u$. $T_u^{\uparrow}$ is computed from its children in the graph. If $(u,v)$ is a back edge, then the contribution of $v$ is $\{v\}$ (l. 30). If $(u,v)$ is not a back edge, then the contribution of $v$ is $T_v^{\uparrow}$ (l. 35). These sets are merged in an order-preserving way, and then filtered so that only nodes that are not already reduced reachable from $u$ are kept (l. 22).

The edges in $\texttt{back}$ are classically identified during the DFS (l. 31) as the edges from the current node to nodes already discovered but not fully processed.

In terms of structure, our code is really close to the code of module $\texttt{Postorder}$. There are two key differences, though. The first one is that we need to remember some information between the time a node is discovered and the time it is fully processed (the preorder number $n$). The second one is that we need to propagate some information during the traversal (the sets $\texttt{s\_c}$ and $\texttt{s\_t}$). This implied the two following changes. Firstly, the tuples in our worklist are more complex, since they contain the additional data. In $\texttt{Postorder}$, the worklist has the simpler type $\texttt{list (node * list node)}$. Secondly, as mentioned above, a node that is discovered is left in the worklist as a child of its parent, so that some information can be propagated to its parent the second time it is seen.

**Precomputation of $T$.** The second part of the precomputation consists in computing $T$ from $T^{\uparrow}$, i.e. computing the reflexive and transitive closure of $T^{\uparrow}$. For this, we follow another suggestion from Boissinot et al. consisting in using the following equation ([7, Equation (1)]): $T_v = \{v\} \cup \left[\bigcup_{w \in T_v^{\uparrow}} T_w\right]$, that we also call Equation (1). They note that, given a node $t$, all nodes $t'$ in $T_t^{\uparrow}$ have a DFS preorder number[4] smaller than that of $t$. This means that if we treat the back edge targets by growing DFS preorder number, we can use this equation to compute $T$ for all the back edge targets.

In our Coq development, this step is performed by $\texttt{precompute\_t\_from\_t\_up\_1}$. It takes as arguments $\texttt{dom\_pre}$, the preorder numbering on the dominance tree, $\texttt{pre}$, the DFS preorder number, and $\texttt{t\_up}$ and $\texttt{back}$, returned by the previous step. It extracts the back edge targets from $\texttt{back}$, sorts them according to $\texttt{pre}$, and uses Equation (1) to compute $T$ for the back edge targets. It returns a map $\texttt{t'}$ which is $\texttt{t\_up}$ updated with the new values for the back edge targets. We are careful to preserve in $\texttt{t'}$ the sorting of the values of $\texttt{t\_up}$ according to $\texttt{dom\_pre}$.

Boissinot et al. also suggest computing $T$ for the rest of the nodes by traversing the reduced graph in a second phase. Instead, we choose to use the same

---

[4] This numbering must not be confused with $\texttt{dom\_pre}$, the preorder numbering on the dominance tree.

equation. This is the role of function `precompute_t_from_t_up_2`. It takes as an argument `dom_pre` and the map `t'` returned by `precompute_t_from_t_up_1`, and applies Equation (1) to every node in any arbitrary order. This means that we also apply it to back edge targets, though they already have the right value, but this is correct and probably not costly. As before, we take care to ensure that the values of the returned map, `t`, are sorted according to `dom_pre`. However, we drop the preorder number component from the elements of `t`. They are no longer necessary, and, as mentioned in Sect. 3, were only there as an optimization.

Finally, function `precompute_t_from_t_up` assembles both previous functions to compute $T$ from $T^\uparrow$.

```
Definition precompute_t_from_t_up dom_pre pre t_up back :=
  let t' := precompute_t_from_t_up_1 dom_pre pre t_up back in
  precompute_t_from_t_up_2 dom_pre t'.
```

**Assembling.** To obtain the full precomputation step, we just have to assemble the pieces introduced in the previous sections. This is the role of `precompute_r_t`.

```
Definition precompute_r_t (g:graph) (entry:node) (dom_pre:node→ Z) :=
  let '(r, c, t_up, back) := precompute_r_t_up g entry dom_pre in
  let pre u := match r ! u with | None ⇒ 1 | Some (n, _) ⇒ n end in
  let t := precompute_t_from_t_up dom_pre pre t_up back in
  (r, c, t, back).
```

It takes as arguments a graph `g`, an entry node `entry` and a preorder numbering on the dominance tree, `dom_pre`. It returns $R$ (encoded as `r` and `c`), $T$ (encoded as `t`) and the list of back edges, `back`. Note that `pre`, the DFS preorder number, is simply defined as a lookup in `r`.

## 3.2 Online Part

The implementation of the online part in Coq is faithful to Algorithm 1, but also takes advantage of optimizations discussed in [7]. More precisely, it is an adaptation of [7, Algorithm 3] that uses sorted lists instead of bitsets, and functional instead of imperative programming.

Indeed, Boissinot et al. suggest two optimizations to speed up Algorithm 1. The first one, that we call (**opt1**), consists in testing at the beginning whether $q$ is strictly dominated by the definition point of `a`. If that is not the case, as explained in Sect. 2.3, *false* can be returned immediately. The second one, denoted (**opt2**), uses dominance information more. The idea is that if we test a node $t$ in $T_{(q,a)}$ and that fails, then the test for any $t'$ dominated by $t$ will fail too, and thus we can skip all such nodes. For instance, in Fig. 2a, $T_{(5,x_0)} = \{3, 4, 5\}$, $R_3 \cap uses(x_0) = \emptyset$, and 3 dominates 4 and 5, thus we can return *false* without testing 4 and 5. Boissinot et al. suggest taking advantage of a preorder numbering on the dominance tree. This numbering can be used in two ways. It can be used to sort $T_q$, since the node with the lowest number is likely to dominate the other nodes to be tested (this is always the case if the CFG is reducible). It can also be used as described in Sect. 2.1, to build a dominance test in constant time.

Our implementation is parameterized by the following objects. `dom` : `map itv` associates to each node an interval based on its preorder number in the dominance tree (this numbering is actually used to implement `dom_pre` in the precomputation step, cf. Sect. 3.1); `def` : `reg` → `node` associates to each variable of type `reg` its definition point; `du_chain` : `map (list node)` connects each variable to the points where it is used; `r` : `map itv`, `c` : `map set` and `t` : `map (list node)` are the results of the precomputation part. Based on these objects, we implement function `is_live_in`, given in Fig. 5. `is_live_in x u` returns whether variable `x` is live at point `u`. It is a bit difficult to read due to Coq syntax and notations, but it is rather straightforward.

First, we get the definition point, `d`, of variable `x` (l. 1). Then we get the preorder intervals in the dominance tree of `d` and `u` (ll. 3–8). We check that the interval of `u` is strictly included in that of `d` (l. 9), meaning that `u` is strictly dominated by `d`, otherwise we directly return `false` (this is (**opt1**)). Then we get the list `uses` of program points where `u` is used (l. 10), and we read in `t` the list `l` of points to test to answer the liveness query (l. 11). Recall that `l` is sorted according to the preorder numbering on the dominance tree. Then we call `fold_t` that tests the nodes in `l` one after the other.

`fold_t` performs case analysis on `l`. If it is empty (l. 20), this means that we have tested all the nodes and none of them have revealed a path to a use of `x`, thus we return `false`. Else, we consider the first element `v` of `l` (l. 21) and its preorder interval `n_v` in the dominance tree (l. 22). If `n_v.(pre)`, the left bound of the interval `n_v`, is greater than `max` (l. 25), this means that `v` is not dominated by `d`, and neither are the other nodes in `l`, thus we can answer `false`. Otherwise, if `n_v.(pre)` is not larger than `min` (l. 26), this means that `v` is not strictly dominated by `d` or is dominated by a node that has been tested unsuccessfully in a previous iteration, thus we can skip `v`. Otherwise (l. 27), we test if a node in `uses` is reduced reachable from `v` thanks to function `is_cross_included`. If yes, we return `true`. Otherwise, we test the other nodes of `l` and update the minimal bound to `n_v.(post)`, the right bound of the interval `n_v`, so that nodes dominated by `v` are skipped in the next iterations.

## 4   Proof of Correctness

The functions described in Sect. 3 all come with proofs of their correctness. However, among the pieces of CompCertSSA on which we build our work, one, namely the formalization of the dominance test [4], turned out to be too weak for our purposes. Indeed, it is proved correct, but not complete, while its completeness is necessary to prove the correctness of our approach. There is an ongoing effort based on [10] to build a correct and complete dominance test, but for now, completeness is admitted.

Most of the proof effort lies in the precomputation part (`precompute_r_t`, 1700 lines of specification and 4000 lines of proof), and especially in the proof of `precompute_r_t_up` that required dozens of invariants. While this number could undoubtedly be decreased, it shows that the justification of the operations performed during the DFS is non-trivial.

```
1   Definition is_live_in (x : reg) (u : node) :=
2     let d := def x in
3     match dom ! d with
4     | None ⇒ false (* impossible *)
5     | Some n_d ⇒
6       match dom ! u with
7       | None ⇒ false (* impossible *)
8       | Some n_u ⇒
9         (n_d.(pre) <? n_u.(pre)) && (n_d.(post) <=? n_u.(post)) &&
10        let uses := du_chain ! x in
11        match t ! u with
12        | None ⇒ false (* impossible *)
13        | Some l ⇒ fold_t uses l n_d.(pre) n_d.(post)
14        end
15      end
16    end.

17  Definition fold_t (uses l : list node) (min max : itv) :=
18    let fix aux l min :=
19      match l with
20      | [] ⇒ false (* all nodes tested, not live *)
21      | v :: l ⇒
22        match dom ! v with
23        | None ⇒ false (* impossible *)
24        | Some n_v ⇒
25          if max <? n_v.(pre) then false
26          else if n_v.(pre) <=? min then aux l min
27          else existsb (is_cross_included r c v) uses || aux l num.(post)
28        end
29      end
30    in
31    aux l min.
```

**Fig. 5.** Function `is_live_in` implements the online part of the algorithm.

For lack of space, we do not detail the proofs of `precompute_r_t_up` and `precompute_t_from_t_up`. We just want to emphasize one point in the proof of the latter. `precompute_t_from_t_up` is written using a `fold_left` operation on the list of back-edge targets, and the validity of this computation is really subtle. Indeed, it relies on Equation (1) and the fact that nodes are considered in the right order, i.e. in increasing DFS preorder number. To ease the definition of complex invariants, we reuse the architecture of `precompute_r_t_up` (cf. Fig. 4), but this time only on the proof side. This form allows to express more easily properties involving the nodes that have already been processed or those that are to be processed. We then show the equivalence of this form with the `fold_left`-based version, and we conclude about the correctness of `precompute_t_from_t_up`.

To state the correctness theorems of `precompute_r_t`, we assume we are given a graph g, a node `entry` in g, and a labeling function `dom_pre`. We make two reasonable assumptions about `dom_pre` and g.

```
Hypothesis dom_pre_inj : forall u v, dom_pre u = dom_pre v →
  reachable g entry u → reachable g entry v → u = v.
Hypothesis g_closed : forall u, reachable g entry u → g ! u <> None.
```

`dom_pre_inj` ensures that the preorder numbering on the dominance tree modeled by `dom_pre` is injective. `g_closed` ensures a kind of well-formedness of g, namely that all nodes reachable from `entry` must be in g.

We can note that both hypotheses take as preconditions that the considered nodes are reachable from the entry node of the CFG. Actually, most of the results have this kind of hypothesis, since the DFS from node `entry` can only discover nodes reachable from `entry`. In this section, such hypotheses will appear in the formal statements, but we will ignore them in the discussion.

Under these hypotheses, we can state the two main correctness theorems of `precompute_r_t`. They state that it computes correctly relations $R$ and $T$.

```
Theorem precompute_r_t_r_c_correct :
  let '(r, c, t, back) := precompute_r_t g root dom_pre in
  forall u v, cross_included r c u v
    ↔ (reachable g root u ∧ g ! u <> None ∧ reduced_reachable g back u v).
```

`precompute_r_t_r_c_correct` states that a node v is reduced reachable from a node u if and only if u and v are related by predicate `cross_included`, meaning that v is tree reachable from one node in $C_u$.

```
Theorem precompute_r_t_t_correct :
  let '(r, c, t, back) := precompute_r_t g root dom_pre in
  forall u v, reachable g root u ∧ is_in_t g back u v ↔ t_linked t u v.
```

`precompute_r_t_t_correct` states that a node v is in $T_u$ (modeled by `is_in_t`) if and only if v is in the list associated to u in t (specified by `t_linked`).

The proof of correctness of the online part is much smaller (230 lines of specification, 1000 lines of proof). One big fragment of it is the proof of the link between $T$ and the existence of strictly dominated paths, that justifies the use of $T$ in the liveness analysis. `is_in_t_sdom_1` is a lemma from this fragment. It states that if p is a strictly d-dominated path between u and v, then there exists a node w in $T_u$, strictly d-dominated and from which $v$ is reduced reachable.

```
Lemma is_in_t_sdom_1 :
  let '(r, c, t, back) := precompute_r_t (succs f) f.(entry) dom_pre in
  forall d u p v, SSApath f (PState u) p (PState v) →
  Forall (sdom f d) (p ++ [v]) →
  exists w, is_in_t (successors f) back u w ∧ sdom f d w
    ∧ reduced_reachable back w v.
```

The proof of this lemma is interesting, because the proof given by Boissinot et al. in [7] was not easily translatable in Coq. Indeed, their proof consists in considering a path with a minimal number of back edges among the strictly d-dominated paths from $u$ to $v$. Such a property is not easy to express in Coq. We proved this result in another manner, by induction on the path.

Finally, theorem `analyze_correct` states the correctness of the liveness analysis, namely that if the analysis succeeds, a liveness query is answered *true* if and only if the considered variable is live at the considered program point.

`wf_ssa_function` is a predicate guaranteeing that function `f` is well-formed. It allows to prove the hypotheses of the lemma described above (e.g. `g_closed`).

```
Theorem analyze_correct :
  forall (f : function), wf_ssa_function f →
  let live := analyze f in
  forall a q, live a q = true ↔ live_spec f a q.
```

## 5   Experiments

To evaluate the efficiency of the liveness checking approach, we compare it experimentally w.r.t. a standard liveness analysis.

More precisely, our reference implementation, called **(impl1)**, is a standard analysis based on data-flow equations. As already mentioned, CompCertSSA contains several liveness analyses, but actually none of them are defined on SSA, so we adapted one of them to SSA. Like the existing ones, this analysis uses the data-flow solver provided by CompCert in module `Kildall`, but takes into account the particularities of SSA, especially the $\phi$-nodes.

The two other implementations, called **(impl2)** and **(impl3)**, are variations of the implementation presented in Sect. 3. They both implement **(opt1)** mentioned in Sect. 3.2. However, **(impl2)** implements **(opt2)** only partially, it only sorts the nodes in $T_q$ by their preorder number in the dominance tree, while **(impl3)** implements it fully, since it can also skip a subtree of the dominance tree when a test fails.

We ran the three implementations on a set of programs taken from CompCert's benchmarks. These programs cover a wide range of size. Most of these programs are one or a few hundred lines long, some of them (e.g. bzip2 and raytracer) are a few thousand lines long, and one of them (spass) contains more than 50,000 lines. Experiments were conducted on a Dell Latitude 7490 with an Intel Core i7-8650U processor at 1.90 GHz and 16 GB of memory.

To perform the comparison, we need a set of liveness queries. To generate these, the best option would be to use a real compiler pass relying on liveness. However, CompCertSSA does not include such a pass at the level of SSA. We came up with the following, admittedly contrived, solution. We generate one query per variable and per natural loop header (a node dominating one of its predecessors). We do not know whether this kind of query is representative of actual queries. However, we can verify that the number of queries is reasonable. In particular, we have two programs in common with Boissinot et al.'s benchmarks: bzip2 and mcf. On both programs, we ask more queries (bzip2: 275071 vs. 10100, mcf: 3748 vs 2369). As doing too many queries penalizes us, the results we give underestimate the benefits of our implementation. Yet, this way of generating queries is fundamentally biased, since depending on the number of loops in a function, the number of queries varies widely. In particular, the functions with no loops are not tested. One program (fib) even has no loop, thus no query. We thus removed it from the experiments.

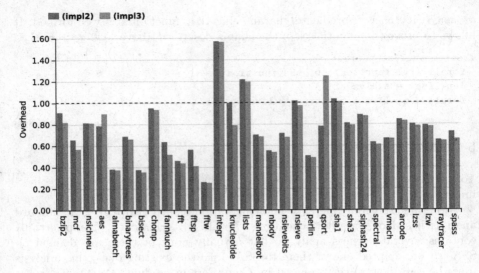

**Fig. 6.** Total overhead of (impl2) and (impl3) w.r.t. (impl1)

We first compared separately the precomputation and online parts of (impl2) and (impl3) w.r.t. (impl1). The results, not included in the paper for lack of space, but available in [1], confirm the expected trends: (impl1) is significantly slower than (impl2) and (impl3) in the precomputation part, and significantly faster in the online part. Then, we compared the total time taken by both parts performed successively in (impl2) and. (impl3) w.r.t. the time they take in (impl1) (see Fig. 6). We can observe that (impl2) and (impl3) are in nearly all the cases faster than (impl1). With the set of queries considered, liveness checking is thus a better trade-off than standard liveness analysis in terms of efficiency. If we compare our results to those obtained by Boissinot et al. [7], we observe a better average speedup (1.48, with (impl3)) of liveness checking w.r.t. standard liveness than them (1.16). But there are many differences in terms of implementation and testing process between Boissinot et al.'s work and ours, thus the comparison of these numbers is of limited value. On the comparison of (impl2) and (impl3), we can notice that (impl3) is in almost all cases faster than (impl2), although moderately, showing that the added complexity of (impl3) is worthwhile. There are two exceptions, `aes` and `qsort`, but with no clear explanation.

## 6   Conclusion and Perspectives

We have described the formalization and implementation in the CompCertSSA verified compiler of the liveness analysis described in [7]. This analysis belongs to the "liveness checking" category, i.e. it is designed to answer liveness queries of the form "is variable `a` live at point $q$?". Its proof of correctness involves the combination of non-trivial arguments about liveness, SSA form, dominance and depth first search. Limited experiments show that, as expected, this algorithm outperforms the classic data-flow based approach if the number of queries is low.

Boissinot et al.'s work is not the only alternative to the data-flow based technique. Appel [2] describes how to propagate liveness information backwards from uses to definitions in programs in SSA form. Boissinot et al. [5] extended the ideas of [7] in 2011, still for SSA-form programs, by taking advantage of an auxiliary structure called a loop-nesting forest. They also propose two variants of Appel's approach, and experimentally compare the three algorithms. Das et al. suggest DJ-graphs rather than loop-nesting forests as auxiliary structures. Among all these works, only [7] and Das et al. [8] embrace the "liveness checking" approach.

One limitation of this work is that it has not been used in a real pass of CompCertSSA yet. This is the reason why we came up with an artificial criterion to evaluate our approach. One pass where it could be used is SSA destruction. Indeed, Boissinot et al. detail in yet another work [6] an SSA destruction pass that uses liveness checking. We could take advantage of [9] that already formalized most of [6] in CompCertSSA, but used a traditional data-flow-based liveness analysis. However, [6] describes an approach with a linear number of queries, while, for the sake of simplicity, [9] makes a quadratic number of them. As the "liveness checking" approach is interesting only if the number of queries is low, we would need to implement the clever approach of [6] first.

A natural extension of this work is the mechanization of Boissinot et al.'s algorithm based on loop-nesting forests [5]. The formalization of a such a complex structure would certainly add a level of difficulty to the correctness proof, but this structure is generic enough to serve as a basis for other program analyses and transformations (e.g. [14]), thus formalizing it could turn out to be profitable.

**Acknowledgments.** This work is supported by a European Research Council (ERC) Consolidator Grant for the project "VESTA", funded under the European Union's Horizon 2020 Framework Programme (grant agreement no. 772568).

# References

1. Companion website. http://www.irisa.fr/celtique/ext/fast_liveness/
2. Appel, A.W., Palsberg, J.: Modern Compiler Implementation in Java, 2nd edn. Cambridge University Press, Cambridge (2002)
3. Barthe, G., Demange, D., Pichardie, D.: Formal verification of an SSA-based middle-end for CompCert. ACM Trans. Program. Lang. Syst. **36**(1), 4:1–4:35 (2014). https://doi.org/10.1145/2579080
4. Blazy, S., Demange, D., Pichardie, D.: Validating dominator trees for a fast, verified dominance test. In: Urban, C., Zhang, X. (eds.) ITP 2015. LNCS, vol. 9236, pp. 84–99. Springer, Cham (2015). https://doi.org/10.1007/978-3-319-22102-1_6
5. Boissinot, B., Brandner, F., Darte, A., de Dinechin, B.D., Rastello, F.: A non-iterative data-flow algorithm for computing liveness sets in strict SSA programs. In: Yang, H. (ed.) APLAS 2011. LNCS, vol. 7078, pp. 137–154. Springer, Heidelberg (2011). https://doi.org/10.1007/978-3-642-25318-8_13

6. Boissinot, B., Darte, A., Rastello, F., de Dinechin, B.D., Guillon, C.: Revisiting out-of-SSA translation for correctness, code quality and efficiency. In: Proceedings of the CGO 2009, The Seventh International Symposium on Code Generation and Optimization, Seattle, Washington, USA, 22–25 March 2009, pp. 114–125. IEEE Computer Society (2009). https://doi.org/10.1109/CGO.2009.19
7. Boissinot, B., Hack, S., Grund, D., de Dinechin, B.D., Rastello, F.: Fast liveness checking for SSA-form programs. In: Soffa, M.L., Duesterwald, E. (eds.) Sixth International Symposium on Code Generation and Optimization (CGO 2008), 5–9 April 2008, Boston, MA, USA, pp. 35–44. ACM (2008). https://doi.org/10.1145/1356058.1356064
8. Das, D., de Dinechin, B.D., Upadrasta, R.: Efficient liveness computation using merge sets and DJ-graphs. TACO 8(4), 27:1–27:18 (2012). https://doi.org/10.1145/2086696.2086706
9. Demange, D., de Retana, Y.F.: Mechanizing conventional SSA for a verified destruction with coalescing. In: Zaks, A., Hermenegildo, M.V. (eds.) Proceedings of the 25th International Conference on Compiler Construction, CC 2016, Barcelona, Spain, 12–18 March 2016, pp. 77–87. ACM (2016). https://doi.org/10.1145/2892208.2892222
10. Georgiadis, L., Tarjan, R.E.: Dominator tree certification and divergent spanning trees. ACM Trans. Algorithms 12(1), 11:1–11:42 (2016). https://doi.org/10.1145/2764913
11. Lattner, C., Adve, V.S.: LLVM: a compilation framework for lifelong program analysis & transformation. In: 2nd IEEE/ACM International Symposium on Code Generation and Optimization (CGO 2004), 20–24 March 2004, San Jose, CA, USA, pp. 75–88. IEEE Computer Society (2004). https://doi.org/10.1109/CGO.2004.1281665
12. Leroy, X.: A formally verified compiler back-end. J. Autom. Reasoning 43(4), 363–446 (2009). https://doi.org/10.1007/s10817-009-9155-4
13. Leroy, X., Blazy, S., Kästner, D., Schommer, B., Pister, M., Ferdinand, C.: CompCert - a formally verified optimizing compiler. In: ERTS 2016: Embedded Real Time Software and Systems. SEE (2016)
14. Ramalingam, G.: On loops, dominators, and dominance frontiers. ACM Trans. Program. Lang. Syst. 24(5), 455–490 (2002). https://doi.org/10.1145/570886.570887
15. Tan, Y.K., Myreen, M.O., Kumar, R., Fox, A.C.J., Owens, S., Norrish, M.: The verified CakeML compiler backend. J. Funct. Program. 29, e2 (2019). https://doi.org/10.1017/S0956796818000229

# Verification of Closest Pair of Points Algorithms

Martin Rau and Tobias Nipkow[✉] [iD]

Fakultät für Informatik, Technische Universität München, Munich, Germany

**Abstract.** We verify two related divide-and-conquer algorithms solving one of the fundamental problems in Computational Geometry, the *Closest Pair of Points* problem. Using the interactive theorem prover Isabelle/HOL, we prove functional correctness and the optimal running time of $\mathcal{O}(n \log n)$ of the algorithms. We generate executable code which is empirically competitive with handwritten reference implementations.

## 1 Introduction

The *Closest Pair of Points* or *Closest Pair* problem is one of the fundamental problems in Computational Geometry: Given a set $P$ of $n \geq 2$ points in $\mathbb{R}^d$, find the closest pair of $P$, i.e. two points $p_0 \in P$ and $p_1 \in P$ ($p_0 \neq p_1$) such that the distance between $p_0$ and $p_1$ is less than or equal to the distance of any distinct pair of points of $P$.

Shamos and Hoey [25] are one of the first to mention this problem and introduce an algorithm based on *Voronoi* diagrams for the planar case, improving the running time of the best known algorithms at the time from $\mathcal{O}(n^2)$ to $\mathcal{O}(n \log n)$. They also prove that this running time is optimal for a deterministic computation model. One year later, in 1976, Bentley and Shamos [2] publish a, also optimal, divide-and-conquer algorithm to solve the Closest Pair problem that can be non-trivially extended to work in arbitrary dimensions. Since then the problem has been the focus of extensive research and a multitude of optimal algorithms have been published. Smid [24] provides a comprehensive overview over the available algorithms, including randomized approaches which improve the running time even further to $\mathcal{O}(n)$.

The main contribution of this paper is the first verification of two related functional implementations of the divide-and-conquer algorithm solving the Closest Pair problem for the two-dimensional Euclidean plane with the optimal running time of $\mathcal{O}(n \log n)$. We use the interactive theorem prover Isabelle/HOL [17,18] to prove functional correctness as well as the running time of the algorithms. In contrast to many publications and implementations we do not assume all points of $P$ to have unique $x$-coordinates which causes some tricky complications. Empirical testing shows that our verified algorithms are competitive with handwritten reference implementations. Our formalizations are available online [23] in the Archive of Formal Proofs.

© Springer Nature Switzerland AG 2020
N. Peltier and V. Sofronie-Stokkermans (Eds.): IJCAR 2020, LNAI 12167, pp. 341–357, 2020.
https://doi.org/10.1007/978-3-030-51054-1_20

This paper is structured as follows: Sect. 2 familiarizes the reader with the algorithm by presenting a high-level description that covers both implementations. Section 3 presents the first implementation and its functional correctness proof. Section 4 proves the running time of $\mathcal{O}(n \log n)$ of the implementation of the previous section. Section 5 describes our second implementation and illustrates how the proofs of Sects. 3 and 4 need to be adjusted. We also give an overview over further implementation approaches. Section 6 describes final adjustments to obtain executable versions of the algorithms in target languages such as OCaml and SML and evaluates them against handwritten imperative and functional implementations. Section 7 concludes.

## 1.1  Related Verification Work

Computational geometry is a vast area but only a few algorithms and theorems seem to have been verified formally. We are aware of a number of verifications of convex hull algorithms [4,14,20] (and a similar algorithm for the intersection of zonotopes [12]) and algorithms for triangulation [3,7]. Geometric models based on maps and hypermaps [6,22] are frequently used.

Work on theorem proving in geometry (see [15] for an overview) is also related but considers fixed geometric constructions rather than algorithms.

## 1.2  Isabelle/HOL and Notation

The notation $t :: \tau$ means that term $t$ has type $\tau$. Basic types include *bool*, *nat*, *int* and *real*; type variables are written $'a$, $'b$ etc; the function space arrow is $\Rightarrow$. Functions *fst* and *snd* return the first and second component of a pair.

We suppress numeric conversion functions, e.g. *real* :: *nat* $\Rightarrow$ *real*, except where that would result in ambiguities for the reader.

Most type constructors are written postfix, e.g. $'a$ *set* and $'a$ *list*. Sets follow standard mathematical notation. Lists are constructed from the empty list [] via the infix cons-operator (#). Functions *hd* and *tl* return head and tail, function *set* converts a list into a set.

## 2  Closest Pair Algorithm

In this section we provide a high-level overview of the *Closest Pair* algorithm and give the reader a first intuition without delving into implementation details, functional correctness or running time proofs.

Let $P$ denote a set of $n \geq 2$ points. If $n \leq 3$ we solve the problem naively using the brute force approach of examining all $\cdot\binom{n}{2}$ possible closest pair combinations. Otherwise we apply the divide-and-conquer tactic.

We divide $P$ into two sets $P_L$ and $P_R$ along a vertical line $l$ such that the sizes of $P_L$ and $P_R$ differ by at most 1 and the $x$-coordinate of all points $p_L \in P_L$ ($p_R \in P_R$) is $\leq l$ ($\geq l$).

We then conquer the left and right subproblems by applying the algorithm recursively, obtaining $(p_{L0}, p_{L1})$ and $(p_{R0}, p_{R1})$, the respective closest pairs of $P_L$ and $P_R$. Let $\delta_L$ and $\delta_R$ denote the distance of the left and right closest pairs and let $\delta = min\ \delta_L\ \delta_R$. At this point the closest pair of $P$ is either $(p_{L0}, p_{L1})$, $(p_{R0}, p_{R1})$ or a pair of points $p_0 \in P_L$ and $p_1 \in P_R$ with a distance strictly less than $\delta$. In the first two cases we have already found our closest pair.

Now we assume the third case and have reached the most interesting part of divide-and-conquer algorithms, the *combine* step. It is not hard to see that both points of the closest pair must be within a $2\delta$ wide vertical strip centered around $l$. Let $ps$ be a list of all points in $P$ that are contained within this $2\delta$ wide strip, sorted in ascending order by $y$-coordinate. We can find our closest pair by iterating over $ps$ and computing for each point its closest neighbor. But in the worst case $ps$ contains all points of $P$, and we might think our only option is to again check all $\binom{n}{2}$ point combinations. This is not the case. Let $p$ denote an arbitrary point of $ps$, depicted as the square point in Fig. 1. If $p$ is one of the points of the closest pair, then the distance to its closest neighbor is strictly less than $\delta$ and we only have to check all points $q \in set\ ps$ that are contained within the $2\delta$ wide horizontal strip centered around the $y$-coordinate of $p$.

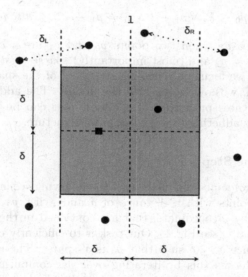

**Fig. 1.** The *combine* step

In Sect. 4 we prove that, for each $p \in set\ ps$, it suffices to check only a constant number of closest point candidates. This fact allows for an implementation of the *combine* step that runs in linear time and ultimately lets us achieve the familiar recurrence of $T(n) = T(\lceil n/2 \rceil) + T(\lfloor n/2 \rfloor) + \mathcal{O}(n)$, which results in the running time of $\mathcal{O}(n \log n)$.

We glossed over some implementation details to achieve this time complexity. In Sect. 3 we refine the algorithm and in Sect. 4 we prove the $\mathcal{O}(n \log n)$ running time.

## 3  Implementation and Functional Correctness Proof

We present the implementation of the divide-and-conquer algorithm and the corresponding correctness proofs using a bottom-up approach, starting with the *combine* step. The basis for both implementation and proof is the version presented by Cormen *et al.* [5]. But first we need to define the closest pair problem formally.

A point in the two-dimensional Euclidean plane is represented as a pair of (unbounded) integers[1]. The library HOL-Analysis provides a generic distance function *dist* applicable to our point definition. The definition of this specific *dist* instance corresponds to the familiar Euclidean distance measure.

The closest pair problem can be stated formally as follows: A set of points $P$ is $\delta$-**sparse** iff $\delta$ is a lower bound for the distance of all distinct pairs of points of $P$.

$$sparse\ \delta\ P = (\forall\, p_0 \in P.\ \forall\, p_1 \in P.\ p_0 \neq p_1 \longrightarrow \delta \leq dist\ p_0\ p_1)$$

We can now state easily when two points $p_0$ and $p_1$ are a *closest pair* of $P$: $p_0 \in P$, $p_1 \in P$, $p_0 \neq p_1$ and (most importantly) *sparse* $(dist\ p_0\ p_1)\ P$.

In the following we focus on outlining the proof of the sparsity property of our implementation, without going into the details. The additional set membership and distinctness properties of a closest pair can be proved relatively straightforwardly by adhering to a similar proof structure.

### 3.1  The Combine Step

The essence of the *combine* step deals with the following scenario: We are given an initial pair of points with a distance of $\delta$ and a list *ps* of points, sorted in ascending order by $y$-coordinate, that are contained in the $2\delta$ wide vertical strip centered around $l$ (see Fig. 1). Our task is to efficiently compute a pair of points with a distance $\delta' \leq \delta$ such that *ps* is $\delta'$-sparse. The recursive function *find_closest_pair* achieves this by iterating over *ps*, computing for each point its closest neighbor by calling the recursive function *find_closest* that considers only the points within the shaded square of Fig. 1, and updating the current pair of closest points if the newly found pair is closer together. We omit the implementation of the trivial base cases.

---

[1] We choose integers over reals because be we cannot implement mathematical reals. See Sect. 6. Alternatively we could have chosen rationals.

$find\_closest\_pair :: point \times point \Rightarrow point\ list \Rightarrow point \times point$
$find\_closest\_pair\ (c_0,\ c_1)\ (p_0\ \#\ ps) =$
$(let\ p_1 = find\_closest\ p_0\ (dist\ c_0\ c_1)\ ps$
$\quad in\ if\ dist\ c_0\ c_1 \leq dist\ p_0\ p_1\ then\ find\_closest\_pair\ (c_0,\ c_1)\ ps$
$\quad\quad else\ find\_closest\_pair\ (p_0,\ p_1)\ ps)$

$find\_closest :: point \Rightarrow real \Rightarrow point\ list \Rightarrow point$
$find\_closest\ p\ \delta\ (p_0\ \#\ ps) =$
$(if\ \delta \leq snd\ p_0 - snd\ p\ then\ p_0$
$\quad else\ let\ p_1 = find\_closest\ p\ (min\ \delta\ (dist\ p\ p_0))\ ps$
$\quad\quad in\ if\ dist\ p\ p_0 \leq dist\ p\ p_1\ then\ p_0\ else\ p_1)$

There are several noteworthy aspects of this implementation. The recursive search for the closest neighbor of a given point $p$ of *find_closest* starts at the first point spatially above $p$, continues upwards and is stopped early at the first point whose vertical distance to $p$ is equal to or exceeds the given $\delta$. Thus the function considers, in contrast to Fig. 1, only the upper half of the shaded square during this search. This is sufficient for the computation of a closest pair because for each possible point $q$ preceding $p$ in $ps$ we already considered the pair $(q, p)$, if needed, and do not have to re-check $(p, q)$ due to the commutative property of our closest pair definition. Note also that $\delta$ is updated, if possible, during the computation and consequently the search space for each point is limited to a $2\delta \times \delta'$ rectangle with $\delta' \leq \delta$. Lastly we intentionally do not minimize the number of distance computations. In Sect. 6 we make this optimization for the final executable code.

The following lemma captures the desired sparsity property of our implementation of the *combine* step so far. It is proved by induction on the computation.

**Lemma 1.** $sorted\_snd\ ps \wedge (p_0,\ p_1) = find\_closest\_pair\ (c_0,\ c_1)\ ps$
$\quad \Longrightarrow sparse\ (dist\ p_0\ p_1)\ (set\ ps)$

where $sorted\_snd\ ps$ means that $ps$ is sorted in ascending order by $y$-coordinate.

We wrap up the *combine* step by limiting our search for the closest pair to only the points contained within the $2\delta$ wide vertical strip and choosing as argument for the initial pair of points of *find_closest_pair* the closest pair of the two recursive invocations of our divide-and-conquer algorithm with the smaller distance $\delta$.

$combine :: point \times point \Rightarrow point \times point \Rightarrow int \Rightarrow point\ list \Rightarrow point \times point$
$combine\ (p_{0L},\ p_{1L})\ (p_{0R},\ p_{1R})\ l\ ps =$
$(let\ (c_0,\ c_1) =$
$\quad\quad if\ dist\ p_{0L}\ p_{1L} < dist\ p_{0R}\ p_{1R}\ then\ (p_{0L},\ p_{1L})\ else\ (p_{0R},\ p_{1R})$
$\quad in\ find\_closest\_pair\ (c_0,\ c_1)$
$\quad\quad (filter\ (\lambda p.\ dist\ p\ (l,\ snd\ p) < dist\ c_0\ c_1)\ ps))$

Lemma 2 shows that if there exists a pair $(p_0,\ p_1)$ of distinct points with a distance $< \delta$, then both its points are contained in the mentioned vertical strip, otherwise we have already found our closest pair $(c_0,\ c_1)$, and the pair returned by *find_closest_pair* is its initial argument.

**Lemma 2.** $p_0 \in set\ ps \wedge p_1 \in set\ ps \wedge p_0 \neq p_1 \wedge dist\ p_0\ p_1 < \delta \wedge$
$(\forall\, p \in P_L.\ fst\ p \leq l) \wedge sparse\ \delta\ P_L \wedge$
$(\forall\, p \in P_R.\ l \leq fst\ p) \wedge sparse\ \delta\ P_R \wedge$
$set\ ps = P_L \cup P_R \wedge ps' = filter\ (\lambda p.\ dist\ p\ (l\ ,\ snd\ p) < \delta)\ ps$
$\implies p_0 \in set\ ps' \wedge p_1 \in set\ ps'$

We then can prove, additionally using Lemma 1, that *combine* computes indeed a pair of points $(p_0,\ p_1)$ such that our given list of points *ps* is (*dist* $p_0\ p_1$)-sparse.

**Lemma 3.** $sorted\_snd\ ps \wedge set\ ps = P_L \cup P_R \wedge$
$(\forall\, p \in P_L.\ fst\ p \leq l) \wedge sparse\ (dist\ p_{0L}\ p_{1L})\ P_L \wedge$
$(\forall\, p \in P_R.\ l \leq fst\ p) \wedge sparse\ (dist\ p_{0R}\ p_{1R})\ P_R \wedge$
$(p_0,\ p_1) = combine\ (p_{0L},\ p_{1L})\ (p_{0R},\ p_{1R})\ l\ ps$
$\implies sparse\ (dist\ p_0\ p_1)\ (set\ ps)$

One can also show that $p_0$ and $p_1$ are in *ps* and distinct (and thus a closest pair of *set ps*), if $p_{0L}$ ($p_{0R}$) and $p_{1L}$ ($p_{1R}$) are in $P_L$ ($P_R$) and distinct.

## 3.2   The Divide & Conquer Algorithm

In Sect. 2 we glossed over some implementation detail which is necessary to achieve to running time of $\mathcal{O}(n \log n)$. In particular we need to partition the given list[2] of points *ps* along a vertical line $l$ into two lists of nearly equal length during the divide step and obtain a list *ys* of the same points, sorted in ascending order by $y$-coordinate, for the *combine* step in *linear* time at each level of recursion.

Cormen *et al.* propose the following top-down approach: Their algorithm takes three arguments: the set of points $P$ and lists *xs* and *ys* which contain the same set of points $P$ but are respectively sorted by $x$ and $y$-coordinate. The algorithm first splits *xs* at *length xs div* 2 into two still sorted lists $xs_L$ and $xs_R$ and chooses $l$ as either the $x$-coordinate of the last element of $xs_L$ or the $x$-coordinate of the first element of $xs_R$. It then constructs the sets $P_L$ and $P_R$ respectively consisting of the points of $xs_L$ and $xs_R$. For the recursive invocations it needs to obtain in addition lists $ys_L$ and $ys_R$ that are still sorted by $y$-coordinate and again respectively refer to the same points as $xs_L$ and $xs_R$. It achieves this by iterating once through *ys* and checking for each point if it is contained in $P_L$ or not, constructing $ys_L$ and $ys_R$ along the way.

But this approach requires an implementation of sets. In fact, if we want to achieve the overall worst case running time of $\mathcal{O}(n \log n)$ it requires an implementation of sets with linear time construction and constant time membership test, which is nontrivial, in particular in a functional setting. To avoid sets many publications and implementations either assume all points have unique $x$-coordinates or preprocess the points by applying for example a rotation such that the input

---

[2] Our implementation deals with concrete lists in contrast to the abstract sets used in Sect. 2.

fulfills this condition. For distinct $x$-coordinates one can then compute $ys_L$ and $ys_R$ by simply filtering $ys$ depending on the $x$-coordinate of the points relative to $l$ and eliminate the usage of sets entirely.

But there exists a third option which we have found only in Cormen *et al.* where it is merely hinted at in an exercise left to the reader. The approach is the following. Looking at the overall structure of the closest pair algorithm we recognize that it closely resembles the structure of a standard mergesort implementation and that we only need $ys$ for the *combine* step *after* the two recursive invocations of the algorithm. Thus we can obtain $ys$ by merging 'along the way' using a bottom-up approach. This is the actual code:

```
closest_pair_rec :: point list ⇒ point list × point × point
closest_pair_rec xs =
(let n = length xs
 in if n ≤ 3 then (mergesort snd xs, closest_pair_bf xs)
    else let (xs_L, xs_R) = split_at (n div 2) xs;
             (ys_L, p_0L, p_1L) = closest_pair_rec xs_L;
             (ys_R, p_0R, p_1R) = closest_pair_rec xs_R;
             ys = merge snd ys_L ys_R
         in (ys, combine (p_0L, p_1L) (p_0R, p_1R) (fst (hd xs_R)) ys))

closest_pair :: point list ⇒ point × point
closest_pair ps =
(let (ys, c_0, c_1) = closest_pair_rec (mergesort fst ps) in (c_0, c_1))
```

Function *closest_pair_rec* expects a list of points $xs$ sorted by $x$-coordinate. The construction of $xs_L$, $xs_R$ and $l$ is analogous to Cormen *et al.* In the base case we then sort $xs$ by $y$-coordinate and compute the closest pair using the brute-force approach (*closest_pair_bf*). The recursive call of the algorithm on $xs_L$ returns in addition to the closest pair of $xs_L$ the list $ys_L$, containing all points of $xs_L$ but now sorted by $y$-coordinate. Analogously for $xs_R$ and $ys_R$. Furthermore, we reuse function *merge* from our *mergesort* implementation, which we utilize to presort the points by $x$-coordinate, to obtain $ys$ from $ys_L$ and $ys_R$ in linear time at each level of recursion.

Splitting of $xs$ is performed by the function *split_at* via a simple linear pass over $xs$. Our implementation of *mergesort* sorts a list of points depending on a given projection function, *fst* for 'by $x$-coordinate' and *snd* for 'by $y$-coordinate'.

Using Lemma 3, the functional correctness proofs of our mergesort implementation and several auxiliary lemmas proving that *closest_pair_rec* also sorts the points by $y$-coordinate, we arrive at the correctness proof of the desired sparsity property of the algorithm.

**Theorem 1.** $1 < length\ xs \land sorted\_fst\ xs \land (ys, p_0, p_1) = closest\_pair\_rec\ xs$
  $\implies sparse\ (dist\ p_0\ p_1)\ xs$

Corollary 1 together with Theorems 2 and 3 then show that the pair $(p_0, p_1)$ is indeed a closest pair of $ps$.

**Corollary 1.** $1 < length\ ps \land (p_0,\ p_1) = closest\_pair\ ps$
$\implies sparse\ (dist\ p_0\ p_1)\ ps$

**Theorem 2.** $1 < length\ ps \land (p_0,\ p_1) = closest\_pair\ ps$
$\implies p_0 \in set\ ps \land p_1 \in set\ ps$

**Theorem 3.** $1 < length\ ps \land distinct\ ps \land (p_0,\ p_1) = closest\_pair\ ps$
$\implies p_0 \neq p_1$

# 4   Time Complexity Proof

To formally verify the running time we follow the approach in [16]. For each function $f$ we define a function $t\_f$ that takes the same arguments as $f$ but computes the number of function calls the computation of $f$ needs, the 'time'. Function $t\_f$ follows the same recursion structure as $f$ and can be seen as an abstract interpretation of $f$. To ensure the absence of errors we derive $f$ and $t\_f$ from a monadic function that computes both the value and the time but for simplicity of presentation we present only $f$ and $t\_f$. We also simplify matters a bit: we count only expensive operations where the running time increases with the size of the input; in particular we assume constant time arithmetic and ignore small additive constants. Due to reasons of space we only show one example of such a 'timing' function, $t\_find\_closest$, which is crucial to our time complexity proof.

$t\_find\_closest :: point \Rightarrow real \Rightarrow point\ list \Rightarrow nat$
$t\_find\_closest\ p\ \delta\ [] = 0$
$t\_find\_closest\ p\ \delta\ [p_0] = 1$
$t\_find\_closest\ p\ \delta\ (p_0\ \#\ ps) = 1\ +$
$(if\ \delta \leq snd\ p_0 - snd\ p\ then\ 0$
$\ else\ let\ p_1 = find\_closest\ p\ (min\ \delta\ (dist\ p\ p_0))\ ps$
$\quad in\ t\_find\_closest\ p\ (min\ \delta\ (dist\ p\ p_0))\ ps\ +$
$\qquad (if\ dist\ p\ p_0 \leq dist\ p\ p_1\ then\ 0\ else\ 0))$

We set the time to execute $dist$ computations to 0 since it is a combination of cheap operations. For the base cases of recursive functions we fix the computation time to be equivalent to the size of the input. This choice of constants is arbitrary and has no impact on the overall running time analysis but leads in general to 'cleaner' arithmetic bounds.

## 4.1   Time Analysis of the Combine Step

In Sect. 2 we claimed that the running time of the algorithm is captured by the recurrence $T(n) = T(\lceil n/2 \rceil) + T(\lfloor n/2 \rfloor) + O(n)$, where $n$ is the length of the given list of points. This claim implies an at most linear overhead at each level of recursion. Splitting of the list $xs$, merging $ys_L$ and $ys_R$ and the filtering operation of the *combine* step run in linear time. But it is non-trivial that the function *find_closest_pair*, central to the *combine* step, also exhibits a linear time complexity. It is applied to an argument list of, in the worst case, length $n$, iterates once through the list and calls *find_closest* for each element. Consequently our proof obligation is the constant running time of *find_closest* or, considering our timing function, that there exists some constant $c$ such that *t_find_closest* $p$ $\delta$ $ps \leq c$ holds in the context of the *combine* step.

Looking at the definition of *find_closest* we see that the function terminates as soon as it encounters the first point within the given list $ps$ that does not fulfill the predicate $(\lambda q.\ \delta \leq snd\ q - snd\ p)$, the point $p$ being an argument to *find_closest*, or if $ps$ is a list of length $\leq 1$. The corresponding timing function *t_find_closest* computes the number of recursive function calls, which is, in this case, synonymous with the number of examined points. For our time complexity proof it suffices to show the following bound on the result of *t_find_closest*. The proof is by induction on the computation of *t_find_closest*. The function *count f* is an abbreviation for *length* $\circ$ *filter f*.

**Lemma 4.** *t_find_closest* $p$ $\delta$ $ps \leq 1 + count\ (\lambda q.\ snd\ q - snd\ p \leq \delta)\ ps$

Therefore we need to prove that the term *count* $(\lambda q.\ snd\ q - snd\ p \leq \delta)\ ps$ does not depend on the length of $ps$. Looking back at Fig. 1, the square point representing $p$, we can assume that the list $p\ \#\ ps$ is distinct and sorted in ascending order by $y$-coordinate. From the precomputing effort of the *combine* step we know that its points are contained within the $2\delta$ wide vertical strip centered around $l$ and can be split into two sets $P_L$ $(P_R)$ consisting of all points which lie to the left (right) of or on the line $l$. Due to the two recursive invocations of the algorithm during the conquer step we can additionally assume that both $P_L$ and $P_R$ are $\delta$-sparse, suggesting the following lemma which implies *t_find_closest* $p$ $\delta$ $ps \leq 8$ and thus the constant running time of *find_closest*.

**Lemma 5.** *distinct* $(p\ \#\ ps) \land$ *sorted_snd* $(p\ \#\ ps) \land 0 \leq \delta \land$
$(\forall q \in set\ (p\ \#\ ps).\ l - \delta < fst\ q \land fst\ q < l + \delta) \land$
*set* $(p\ \#\ ps) = P_L \cup P_R \land$
$(\forall q \in P_L\ .\ fst\ q \leq l) \land$ *sparse* $\delta$ $P_L \land$
$(\forall q \in P_R\ .\ l \leq fst\ q) \land$ *sparse* $\delta$ $P_R$
$\implies count\ (\lambda q.\ snd\ q - snd\ p \leq \delta)\ ps \leq 7$

*Proof.* The library HOL-Analysis defines some basic geometric building blocks needed for the proof. A *closed box* describes points contained within rectangular shapes in Euclidean space. For our purposes the planar definition is sufficient.

$$cbox\ (x_0,\ y_0)\ (x_1,\ y_1) = \{(x,\ y) \mid x_0 \leq x \land x \leq x_1 \land y_0 \leq y \land y \leq y_1\}$$

The box is 'closed' since it includes points located on the border of the box. We then introduce some useful abbreviations:

- The rectangle $R$ is the upper half of the shaded square of Fig. 1:
  $R = cbox\ (l - \delta,\ snd\ p)\ (l + \delta,\ snd\ p + \delta)$

- The set $R_{ps}$ consists of all points of $p \# ps$ that are encompassed by $R$:
  $R_{ps} = R \cap set\ (p \# ps)$
- The squares $S_L$ and $S_R$ denote the left and right halves of $R$:
  $S_L = cbox\ (l - \delta,\ snd\ p)\ (l,\ snd\ p + \delta)$
  $S_R = cbox\ (l,\ snd\ p)\ (l + \delta,\ snd\ p + \delta)$
- The set $S_{PL}$ holds all points of $P_L$ that are contained within the square $S_L$.
  The definition of $S_{PR}$ is analogous:
  $S_{PL} = P_L \cap S_L,\ S_{PR} = P_R \cap S_R$

Let additionally $ps_f$ abbreviate the term *filter* $(\lambda q.\ snd\ q - snd\ p \le \delta)$ $ps$. First we prove *length* $(p \# ps_f) \le |R_{ps}|$: Let $q$ denote an arbitrary point of $p \# ps_f$. We know $snd\ p \le snd\ q$ because the list $p \# ps$ and therefore $p \# ps_f$ is sorted in ascending order by $y$-coordinate and $snd\ q \le snd\ p + \delta$ due to the filter predicate $(\lambda q.\ snd\ q - snd\ p \le \delta)$. Using the additional facts $l - \delta \le fst\ q$ and $fst\ q \le l + \delta$ (derived from our assumption that all points of $p \# ps$ are contained within the $2\delta$ strip) and the definitions of $R_{ps}$, $R$ and *cbox* we know $q \in R_{ps}$ and thus $set\ (p \# ps_f) \subseteq R_{ps}$. Since the list $p \# ps_f$ maintains the distinctness property of $p \# ps$ we additionally have *length* $(p \# ps_f) = |set\ (p \# ps_f)|$. Our intermediate goal immediately follows because $|set\ (p \# ps_f)| \le |R_{ps}|$ holds for the finite set $R_{ps}$.

But how many points can there be in $R_{ps}$? Lets first try to determine an upper bound for the number of points of $S_{PL}$. All its points are contained within the square $S_L$ whose side length is $\delta$. Moreover, since $P_L$ is $\delta$-sparse and $S_{PL} \subseteq P_L$, $S_{PL}$ is also $\delta$-sparse, or the distance between each distinct pair of points of $S_{PL}$ is at least $\delta$. Therefore the cardinality of $S_{PL}$ is bounded by the number of points we can maximally fit into $S_L$, maintaining a pairwise minimum distance of $\delta$. As the left-hand side of Fig. 2 depicts, we can arrange at most four points adhering to these restrictions, and consequently have $|S_{PL}| \le$ 4. An analogous argument holds for the number of points of $S_{PR}$. Furthermore we know $R_{ps} = S_{PL} \cup S_{PR}$ due to our assumption $set\ (p \# ps) = P_L \cup P_R$ and the fact $R = S_L \cup S_R$ and can conclude $|R_{ps}| \le 8$. Our original statement then follows from *length* $(p \# ps_f) \le |R_{ps}|$.     □

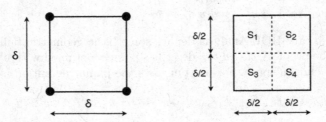

**Fig. 2.** Core argument.

Note that the intermediate proof for the bound on $|R_{ps}|$ relies on basic human geometric intuition. Indeed Cormen *et al.* [5] and most of the proofs in the literature do. But for a formal proof we have to be rigorous. First we show two auxiliary lemmas: The maximum distance between two points in a square $S$ with side length $\delta$ is less than or equal to $\sqrt{2}\delta$.

**Lemma 6.** $p_0 = (x, y) \wedge p_1 = (x + \delta, y + \delta) \wedge 0 \leq \delta \wedge$
$(x_0, y_0) \in \text{cbox } p_0 \, p_1 \wedge (x_1, y_1) \in \text{cbox } p_0 \, p_1$
$\implies \text{dist } (x_0, y_0) \, (x_1, y_1) \leq \sqrt{2} * \delta$

The proof is straightforward. Both points are contained within the square $S$, the difference between their $x$ and $y$ coordinates is hence bounded by $\delta$ and we finish the proof using the definition of the Euclidean distance. Below we employ the following variation of the *pigeonhole* principle:

**Lemma 7.** *finite* $B \wedge A \subseteq \bigcup B \wedge |B| < |A|$
$\implies \exists x \in A. \, \exists y \in A. \, \exists S \in B. \, x \neq y \wedge x \in S \wedge y \in S$

Finally we replace human intuition with formal proof:

**Lemma 8.** $(\forall p \in P. \, p \in \text{cbox } (x, y) \, (x + \delta, y + \delta)) \wedge \text{sparse } \delta \, P \wedge 0 \leq \delta$
$\implies |P| \leq 4$

*Proof.* Let $S$ denote the square with a side length of $\delta$ and suppose, for the sake of contradiction, that $4 < |P|$. Then $S$ can be split into the four congruent squares $S_1, S_2, S_3, S_4$ along the common point $(x + \delta/2, y + \delta/2)$ as depicted by the right-hand side of Fig. 2. Since all points of $P$ are contained within $S$ and $S = \bigcup \{S_1, S_2, S_3, S_4\}$ we have $P \subseteq \bigcup \{S_1, S_2, S_3, S_4\}$. Using Lemma 7 and our assumption $4 < |P|$ we know there exists a square $S_i \in \{S_1, S_2, S_3, S_4\}$ and a pair of distinct points $p_0 \in S_i$ and $p_1 \in S_i$. Lemma 6 and the fact that all four sub-squares have the same side length $\delta / 2$ shows that the distance between $p_0$ and $p_1$ must be less than or equal to $\sqrt{2} / 2 * \delta$ and hence strictly less than $\delta$. But we also know that $\delta$ is a lower bound for this distance because $p_0 \in P$, $p_0 \in P$, $p_0 \neq p_1$ and our premise that $P$ is $\delta$-sparse, a contradiction. $\square$

## 4.2 Time Analysis of the Divide & Conquer Algorithm

In summary, the time to evaluate *find_closest* $p$ $\delta$ $ps$ is constant in the context of the *combine* step and thus evaluating *find_closest_pair* $(p_0, p_1)$ $ps$ as well as *combine* $(p_{0L}, p_{1L})$ $(p_{0R}, p_{1R})$ $l$ $ps$ takes time linear in *length* $ps$.

Next we turn our attention to the timing of *closest_pair_rec* and derive (but do not show) the corresponding function *t_closest_pair_rec*. At this point we could prove a concrete bound on *t_closest_pair_rec*. But since we are dealing with a divide-and-conquer algorithm we should, in theory, be able to determine its running time using the 'master theorem' [5]. This is, in practice, also possible in Isabelle/HOL. Eberl [8] has formalized the Akra-Bazzi theorem [1], a generalization of the master theorem. Using this formalization we can derive the running time of our divide-and-conquer algorithm without a direct proof for *t_closest_pair_rec*. First we capture the essence of *t_closest_pair_rec* as a recurrence on natural numbers representing the length of the list argument of $(t_-)$*closest_pair_rec*:

*closest_pair_recurrence* :: *nat* $\Rightarrow$ *real*
*closest_pair_recurrence n* =
(*if* $n \leq 3$ *then* $n + mergesort\_recurrence\ n + n * n$
 *else* $13 * n + closest\_pair\_recurrence\ \lfloor n\ /\ 2 \rfloor\ +$
    $closest\_pair\_recurrence\ \lceil n\ /\ 2 \rceil$)

The time complexity of this recurrence is proved completely automatically:

**Lemma 9.** *closest_pair_recurrence* $\in \Theta(\lambda n.\ n * \ln n)$

Next we need to connect this bound with our timing function. Lemma 10 below expresses a procedure for deriving complexity properties of the form

$$t \in O[m\ going\_to\ at\_top\ within\ A](f \circ m)$$

where $t$ is a timing function on the data domain, in our case lists. The function $m$ is a measure on that data domain, $r$ is a recurrence or any other function of type *nat* $\Rightarrow$ *real* and $A$ is the set of valid inputs. The term '*m going_to at_top within A*' should be read as 'if the measured size of valid inputs is sufficiently large' and utilizes Eberls formalization of Landau Notation [9] and the "filter" machinery of asymptotics in Isabelle/HOL [11]. For readability we omit stating the filter and $m$ explicitly in the following and just state the conditions required of the input $A$. The measure $m$ always corresponds to the *length* function.

**Lemma 10.** $(\forall x \in A.\ t\ x \leq (r \circ m)\ x) \wedge r \in O(f) \wedge (\forall x \in A.\ 0 \leq t\ x)$
    $\implies t \in O[m\ going\_to\ at\_top\ within\ A](f \circ m)$

**Lemma 11.** *distinct ps* $\wedge$ *sorted_fst ps*
    $\implies t\_closest\_pair\_rec\ ps \leq (closest\_pair\_recurrence \circ length)\ ps$

Using Lemmas 9, 10 and 11 we arrive at Theorem 4, expressing our main claim: the running time of the divide-and-conquer algorithm.

**Theorem 4.** *For inputs that are distinct and sorted by x-coordinate:*
    $t\_closest\_pair\_rec \in O(\lambda n.\ n * \ln n)$

Since the function *closest_pair* only presorts the given list of points using our mergesort implementation and then calls *closest_pair_rec* we obtain Corollary 2 and finish the time complexity proof.

**Corollary 2.** *For distinct inputs:* $t\_closest\_pair \in O(\lambda n.\ n * \ln n)$

## 5    Alternative Implementations

In the literature there exist various other algorithmic approaches to solve the closest pair problem. Most of them are closely related to our implementation of Sect. 3, deviating primarily in two aspects: the exact implementation of the *combine* step and the approach to sorting the points by $y$-coordinate we already discussed in Subsect. 3.2. We present a short overview, concentrating on the *combine* step and the second implementation we verified.

## 5.1   A Second Verified Implementation

Although the algorithm described by Cormen *et al.* is the basis for our implementation of Sect. 3, we took the liberty to optimize it. During execution of *find_closest p δ ps* our algorithm searches for the closest neighbor of $p$ within the rectangle $R$, the upper half of the shaded square $S$ of Fig. 1, and terminates the search if it examines points on or beyond the upper border of $R$. Cormen *et al.* originally follow a slightly different approach. They search for a closest neighbor of $p$ by examining a constant number of points of *ps*, the first 7 to be exact. This is valid because there are at most 7 points within $R$, not counting $p$, and the 8th point of *ps* would again lie on or beyond the upper border of $R$. This slightly easier implementation comes at the cost of being less efficient in practice. Cormen *et al.* are always assuming the worst case by checking all 7 points following $p$. But it is unlikely that the algorithm needs to examine even close to 7 points, except for specifically constructed inputs. They furthermore state that the bound of 7 is an over-approximation and dare the reader to lower it to 5 as an exercise. We refrain from doing so since a bound of 7 suffices for the time complexity proof of our, inherently faster, implementation. At this point we should also mention that the specific optimization of Sect. 3 is not our idea but rather an algorithmic detail which is unfortunately rarely mentioned in the literature.

Nonetheless we can adapt the implementation of Sect. 3 and the proofs of Sect. 4 to verify the original implementation of Cormen *et al.* as follows: We replace each call of *find_closest p δ ps* by a call to *find_closest_bf p (take 7 ps)* where *find_closest_bf* iterates in brute-force fashion through its argument list to find the closest neighbor of $p$. To verify this implementation we then reuse most of the elementary lemmas and proof structure of Sects. 3 and 4, only a slightly adapted version of Lemma 5 is necessary. Note that this lemma was previously *solely* required for the time complexity proof of the algorithm. Now it is already necessary during the functional correctness proof since we need to argue that examining only a constant number of points of *ps* is sufficient. The time analysis is overall greatly simplified: A call of the form *find_closest_bf p (take 7 ps)* runs in constant time and we again are able to reuse the remaining time analysis proof structure of Sect. 4. For the exact differences between both formalizations we encourage the reader the consult our entry in the Archive of Formal Proofs [23].

## 5.2   Related Work

Over the years a considerable amount of effort has been made to further optimize the *combine* step. Central to these improvements is the 'complexity of computing distances', abbreviated CCP in the following, a term introduced by Zhou *et al.* [26] which measures the number of Euclidean distances computed by a closest pair algorithm. The core idea being, since computing the Euclidean distance is more expensive than other primitive operations, it might be possible to improve overall algorithmic running time by reducing this complexity measure. In the same paper they introduce an optimized version of the closest pair algorithm with a CCP of $2n \log n$, in contrast to $7n \log n$ which will be the worst case CCP of the algorithm of Sect. 3 after we minimize the number of distance computations in Sect. 6. They improve upon the algorithm presented by Preparata and Shamos [21] which achieves a CCP of $3n \log n$. Ge *et al.* [10] base their, quite sophisticated, algorithm on the version of Zhou *et al.* and prove an even lower CCP of $\frac{3}{2}n \log n$ for their implementation. The race for the lowest number of distance computations culminates so far with the work of Jiang and Gillespie [13] who present their algorithms 'Basic-2'[3] and '2-Pass' with a respective CCP of $2n \log n$ and (for the first time linear) $\frac{7}{2}n$.

# 6   Executable Code

Before we explore how our algorithm stacks up against Basic-2 (which is surprisingly the fastest of the CCP minimizing algorithms according to Jiang and Gillespie) we have to make some final adjustments to generate executable code from our formalization.

In Sect. 3 we fixed the data representation of a point to be a pair of mathematical ints rather than mathematical reals. During code export Isabelle then maps, correctly and automatically, its abstract data type *int* to a suitable concrete implementation of (arbitrary-sized) integers; for our target language OCaml using the library 'zarith'. For the data type *real* this is not possible since we cannot implement mathematical reals. We would instead have to resort to an approximation (e.g. floats) losing all proved guarantees in the process. But currently our algorithm still uses the standard Euclidean distance and hence mathematical reals due to the *sqrt* function. For the executable code we have to replace this distance measure by the squared Euclidean distance. To prove that we preserve the correctness of our implementation several small variations of the following lemma suffice:

$$dist\ p_0\ p_1 \leq dist\ p_2\ p_3 \longleftrightarrow (dist\ p_0\ p_1)^2 \leq (dist\ p_2\ p_3)^2$$

We apply two further code transformations. To minimize the number of distance computations we introduce auxiliary variables which capture and then replace repeated computations. For all of the shown functions that return a point or a pair of points this entails returning the corresponding computed distance as well. Furthermore we replace recursive auxiliary functions such as *filter* by corresponding tail-recursive implementations to allow the OCaml compiler to optimize the generated code and prevent stackoverflows. To make sure these transformations are correct we prove lemmas expressing the equivalence of old and new implementations for each function. Isabelles code export machinery can then apply these transformations automatically.

Now it is time to evaluate the performance of our verified code. Figure 3 depicts the running time ratios of several implementations of the algorithm of

---

[3] Pereira and Lobo [19] later independently developed the same algorithm and additionally present extensive functional correctness proofs for all Minkowski distances.

Sect. 3 (called Basic-$\delta$) and Basic-7 (the original approach of Cormen *et al.*) over Basic-2. Basis-$\delta$ is tested in three variations: the exported (purely functional) Isabelle code and equivalent handwritten functional and imperative implementations to gauge the overhead of the machine generated code. All algorithms are implemented in OCaml, use our bottom-up approach to sorting (imperative implementations sort in place) of Subsect. 3.2 and for each input of uniformly distributed points 50 independent executions were performed. Remarkably the exported code is only about 2.28[4] times slower than Basic-2 and furthermore most of the difference is caused by the inefficiencies inherent to machine generated code since its equivalent functional implementation is only 11% slower than Basic-2. Basic-7 is 2.26 times slower than the imperative Basic-$\delta$ which demonstrates the huge impact the small optimization of Subsect. 5.1 can have in practice.

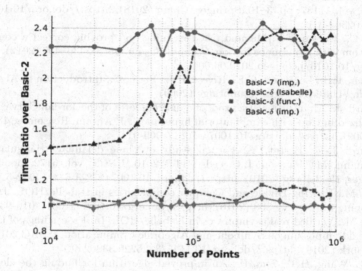

**Fig. 3.** Benchmarks.

# 7    Conclusion

We have presented the first verification (both functional correctness and running time) of two related closest pair of points algorithms in the plane, without assuming the $x$ coordinates of all points to be distinct. The executable code generated from the formalization is competitive with existing reference implementations. A challenging and rewarding next step would be to formalize and

---

[4] We measure differences between running times as the average over all data points weighted by the size of the input.

verify a closest pair of points algorithm in arbitrary dimensions. This case is treated rather sketchily in the literature.

**Acknowledgements.** Research supported by DFG grants NI 491/16-1 and 18-1.

# References

1. Akra, M., Bazzi, L.: On the solution of linear recurrence equations. Comput. Optim. Appl. **10**(2), 195–210 (1998). https://doi.org/10.1023/A:1018373005182
2. Bentley, J.L., Shamos, M.I.: Divide-and-Conquer in multidimensional space. In: Proceedings of Eighth Annual ACM Symposium on Theory of Computing, STOC 1976, pp. 220–230. ACM (1976). https://doi.org/10.1145/800113.803652
3. Bertot, Y.: Formal verification of a geometry algorithm: a quest for abstract views and symmetry in Coq proofs. In: Fischer, B., Uustalu, T. (eds.) ICTAC 2018. LNCS, vol. 11187, pp. 3–10. Springer, Cham (2018). https://doi.org/10.1007/978-3-030-02508-3_1
4. Brun, C., Dufourd, J., Magaud, N.: Designing and proving correct a convex hull algorithm with hypermaps in Coq. Comput. Geom. **45**(8), 436–457 (2012). https://doi.org/10.1016/j.comgeo.2010.06.006
5. Cormen, T.H., Leiserson, C.E., Rivest, R.L., Stein, C.: Introduction to Algorithms, 3rd edn. The MIT Press, Cambridge (2009)
6. Dufourd, J.: An intuitionistic proof of a discrete form of the Jordan Curve Theorem formalized in Coq with combinatorial hypermaps. J. Autom. Reasoning **43**(1), 19–51 (2009). https://doi.org/10.1007/s10817-009-9117-x
7. Dufourd, J.-F., Bertot, Y.: Formal study of plane delaunay triangulation. In: Kaufmann, M., Paulson, L.C. (eds.) ITP 2010. LNCS, vol. 6172, pp. 211–226. Springer, Heidelberg (2010). https://doi.org/10.1007/978-3-642-14052-5_16
8. Eberl, M.: Proving Divide and Conquer complexities in Isabelle/HOL. J. Autom. Reasoning **58**(4), 483–508 (2016). https://doi.org/10.1007/s10817-016-9378-0
9. Eberl, M.: Verified real asymptotics in Isabelle/HOL. In: Proceedings of the International Symposium on Symbolic and Algebraic Computation, ISSAC 2019. ACM, New York (2019). https://doi.org/10.1145/3326229.3326240
10. Ge, Q., Wang, H.T., Zhu, H.: An improved algorithm for finding the closest pair of points. J. Comput. Sci. Technol. **21**(1), 27–31 (2006). https://doi.org/10.1007/s11390-006-0027-7
11. Hölzl, J., Immler, F., Huffman, B.: Type classes and filters for mathematical analysis in Isabelle/HOL. In: Blazy, S., Paulin-Mohring, C., Pichardie, D. (eds.) ITP 2013. LNCS, vol. 7998, pp. 279–294. Springer, Heidelberg (2013). https://doi.org/10.1007/978-3-642-39634-2_21
12. Immler, F.: A verified algorithm for geometric zonotope/hyperplane intersection. In: Certified Programs and Proofs, CPP 2015, pp. 129–136. ACM (2015). https://doi.org/10.1145/2676724.2693164
13. Jiang, M., Gillespie, J.: Engineering the Divide-and-Conquer closest pair algorithm. J. Comput. Sci. Technol. **22**(4), 532–540 (2007)
14. Meikle, L.I., Fleuriot, J.D.: Mechanical theorem proving in computational geometry. In: Hong, H., Wang, D. (eds.) ADG 2004. LNCS (LNAI), vol. 3763, pp. 1–18. Springer, Heidelberg (2006). https://doi.org/10.1007/11615798_1

15. Narboux, J., Janicic, P., Fleuriot, J.: Computer-assisted theorem proving in synthetic geometry. In: Sitharam, M., John, A.S., Sidman, J. (eds.) Handbook of Geometric Constraint Systems Principles. Discrete Mathematics and Its Applications, Chapman and Hall/CRC (2018)
16. Nipkow, T.: Verified root-balanced trees. In: Chang, B.-Y.E. (ed.) APLAS 2017. LNCS, vol. 10695, pp. 255–272. Springer, Cham (2017). https://doi.org/10.1007/978-3-319-71237-6_13
17. Nipkow, T., Klein, G.: Concrete Semantics with Isabelle/HOL. Springer, Heidelberg (2014). http://concrete-semantics.org
18. Nipkow, T., Paulson, L., Wenzel, M.: Isabelle/HOL – A Proof Assistant for Higher-Order Logic. LNCS, vol. 2283. Springer, Heidelberg (2002). https://doi.org/10.1007/3-540-45949-9
19. Pereira, J.C., Lobo, F.G.: An optimized Divide-and-Conquer algorithm for the closest-pair problem in the planar case. J. Comput. Sci. Technol. **27**(4), 891–896 (2012)
20. Pichardie, D., Bertot, Y.: Formalizing convex hull algorithms. In: Boulton, R.J., Jackson, P.B. (eds.) TPHOLs 2001. LNCS, vol. 2152, pp. 346–361. Springer, Heidelberg (2001). https://doi.org/10.1007/3-540-44755-5_24
21. Preparata, F.P., Shamos, M.I.: Computational Geometry: An Introduction. Springer, Heidelberg (1985). https://doi.org/10.1007/978-1-4612-1098-6
22. Puitg, F., Dufourd, J.: Formalizing mathematics in higher-order logic: a case study in geometric modelling. Theor. Comput. Sci. **234**(1–2), 1–57 (2000). https://doi.org/10.1016/S0304-3975(98)00228-X
23. Rau, M., Nipkow, T.: Closest pair of points algorithms. Archive of Formal Proofs, Formal proof development, January 2020. http://isa-afp.org/entries/Closest_Pair_Points.html
24. Sack, J.R., Urrutia, J. (eds.): Handbook of Computational Geometry. North-Holland Publishing Co., Amsterdam (2000)
25. Shamos, M.I., Hoey, D.: Closest-point problems. In: 16th Annual Symposium on Foundations of Computer Science (SFCS 1975), pp. 151–162, October 1975. https://doi.org/10.1109/SFCS.1975.8
26. Zhou, Y., Xiong, P., Zhu, H.: An improved algorithm about the closest pair of points on plane set. Comput. Res. Dev. **35**(10), 957–960 (1998)

# Reasoning Systems and Tools

# A Polymorphic Vampire
## (Short Paper)

Ahmed Bhayat$^{(\boxtimes)}$ (iD) and Giles Reger$^{(\boxtimes)}$ (iD)

University of Manchester, Manchester, UK
ahmed_bhayat@hotmail.com, giles.reger@manchester.ac.uk

**Abstract.** We have modified the Vampire theorem prover to support rank-1 polymorphism. In this paper we discuss the changes required to enable this and compare the performance of polymorphic Vampire against other polymorphic provers. We also compare its performance on monomorphic problems against standard Vampire. Finally, we discuss how polymorphism can be used to support theory reasoning and present results related to this.

## 1 Introduction

Vampire is a well known automated theorem prover for first-order logic with equality [14]. For a long period, Vampire supported only untyped first-order logic. Around 2011 it was extended to support monomorphic first-order logic (FOL). As part of recent work on supporting higher-order logic reported on elsewhere [1], Vampire has been extended to support rank-1, polymorphic, first-order logic[1].

Polymorphic types have a number of advantages over their monomorphic counterparts. Firstly, they provide the user with a more succinct language for describing their problem. Secondly, they provide an elegant solution to dealing with theories. For example, when dealing with the theory of arrays, rather than having to provide separate sets of axioms for arrays of different sorts, polymorphism allows us to provide a single set of axioms. Thirdly, polymorphism also permits higher-order logic to be finitely axiomatizable in first-order logic by introducing polymorphic axioms for the **SK**-combinators.

There are several ways of encoding polymorphism. However, many of these are cumbersome and some even unsound [8]. Blanchette et al. [2] list a number of common translation methods including the use of type tags, type guards and type arguments. Of these, the last is unsound and the first two cumbersome. As the type tags and guards are ubiquitous in the literature, we provide a comparison between native handling of polymorphism and the use of these encodings in Sect. 5. Given issues with encodings, it makes sense to deal natively with polymorphism if possible. We are certainly not the first to attempt to do so. Bobot et

---

[1] At the time of publication this extension exists in a separate branch from the main Vampire development. See https://vprover.github.io/download.html.

© Springer Nature Switzerland AG 2020
N. Peltier and V. Sofronie-Stokkermans (Eds.): IJCAR 2020, LNAI 12167, pp. 361–368, 2020.
https://doi.org/10.1007/978-3-030-51054-1_21

**Problem 1** Sample TF1 problem (truncated)

```
tff(map, type, map : ($tType * $tType) > $tType).
tff(lookup, type,
    lookup : !>[A : $tType, B : $tType]: ((map(A, B) * A) > B)).
tff(update, type,
    update : !>[A : $tType, B : $tType]:
        ((map(A, B) * A * B) > map(A, B))).
tff(lookup_update_same, axiom,
    ![A : $tType, B : $tType, M : map(A, B), K : A, V : B]:
        lookup(A, B, update(A, B, M, K, V), K) = V).
tff(update_idem, conjecture,
    ![A : $tType, B : $tType, M : map(A, B), K : A, V : B]:
        update(A, B, update(A, B, M, K, V), K, V) =
        update(A, B, M, K, V)).
```

al. [5] have introduced polymorphism into their SMT solver Alt-Ergo. Similarly, the first-order provers ZenonModulo [11] and Zipperposition [9] support some form of polymorphism. However, it remains the case that few first-order provers can handle polymorphism.

In this short paper we begin by describing the relatively modest changes that had to be made to Vampire to support polymorphism (Sect. 2). We then present experimental results demonstrating that these changes are useful (Sect. 3). Finally, we discuss work-in-progress to use these polymorphic extensions to improve theory reasoning in Vampire both in terms of proof search and implementation (Sect. 4).

Before this, we give a brief (and informal) reminder of what rank-1 polymorphism is. A polymorphic type is a type variable, or n-ary type constructor applied to $n$ types. The type of all types is represented as $tType in TPTP syntax [17] which is used throughout this paper. Terms, in polymorphic FOL, are either a variable or a function symbol applied to $m$ type arguments and $n$ term arguments. Rank-1 polymorphism allows type and term variables to be quantified with the rule that an existentially quantified type may not occur underneath a universal term quantifier. On skolemisation such a construct would become a dependent type and require superposition into types.

## 2    Implementation

To support polymorphism modifications had to be carried out in three main areas. Firstly, changes had to be made to the concept of types in Vampire. Secondly, some inferences had to be modified slightly and finally preprocessing required consideration. We describe the work undertaken in this order.

Problem 1 is an example of a problem in TPTP TF1 syntax [3,17]. We use this problem to illustrate our implementation. The major change undertaken was to replace types with terms. In monomorphic Vampire each type in the input problem is stored as an unsigned integer. Function symbols are then assigned a type signature which is merely a list of unsigned integers representing the argument and return types. In polymorphic first-order logic, types have all the structure of terms. Therefore, it made sense to replace the types with terms. Type signatures then become a list of terms.

The type of a term of the form `func_sym(arg_1 ... arg_n)` can be calculated by substituting the type arguments that the head symbol is applied to into its result type. For example, the type of `update($int, $i, map, 1, X)` would be `map($int, $i)`. For a term `func_sym(arg_1 ... arg_n)`, the type of the ith term argument can be calculated in the same way. The one problem that arises is with two variable literals such as `X = Y`. In this case, the type of the terms `X` and `Y` have to be stored as a separate field in the literal.

The elegance of treating types as terms can be gauged when attention is turned to unification. Had types and terms been kept separate, unifying terms would have become an involved process requiring the unification of term and type arguments separately. As it is, type unification comes for 'free' with one caveat as shall be seen. Consider unifying the terms `update($int, $i, map, 1, X)` and `update(Y, Z, map, Z', a)`. The existing unification procedure in Vampire can handle this and return the type and term unifier $\{Y \rightarrow \$int, Z \rightarrow \$i, Z' \rightarrow 1, X \rightarrow a\}$. The one hitch occurs when unifying a term with a variable. As variables carry no type information, a second call must be made to the unification procedure to ensure that the type of the variable and the type of the term are unifiable.

As far as changes to inferences are concerned, no updates were required for inferences that do not work at subterms such as resolution or equality factoring. For inferences that work at subterms such as superposition and demodulation, we modified the iterators that return candidate subterms so that they do not return type arguments as superposition into types is unnecessary. We mentioned that the modifications required to support polymorphism were light. They also (in theory, see later experiment) add no overhead when dealing with monomorphic problems. In this case all types are constants and unifiability checking of types in the variable case degenerates to a syntactic equality check.

Finally, regarding preprocessing, implementing skolemisation posed a subtle issue. A skolem function must be applied to the free term and *type* variables in its context. For example, the skolemisation of `![X: $int, Y: $tType] : ?[Z : $i] : (func_sym(Y, X, Z))` would be `![X: $int, Y: $tType] : ?[Z : $i] : (func_sym(Y, X, sk(Y, Z)))`. This required us to update the notion of free variable within the code (e.g., when iterating over the free variables of a formula).

## 3   Results

To test our implementation we ran two experiments. All experiments were carried out with a CPU time limit of 300 s on StarExec [16] nodes equipped with four

**Table 1.** Problems proved theorem or unsat

|                   | TF1 problems | | TF0 problems | |
|-------------------|--------|---------|--------|---------|
|                   | Solved | Uniques | Solved | Uniques |
| Leo-III 1.4       | 224    | 10      | 8,665  | 100     |
| Vampire 4.4       | –      | –       | **11,338** | **476** |
| Vampire-poly      | **239** | **21** | 10,641 | 88      |
| ZenonModulo 0.4.2 | 80     | 1       | 2,926  | 14      |

2.40 GHz Intel Xeon CPUs. Our experimental results are publicly available.[2] All solver configurations used where taken from the last CASC in which that solver was entered. The portfolio of strategies for the two Vampire variants were the same.

*Experiment 1.* Firstly we ran Vampire on the set of 539 TF1 (rank-1 polymorphic) problems in the TPTP library. We compared the results against those of two other provers able to parse TPTP syntax and handle polymorphism that we are aware of, Leo-III [15] and ZenonModulo [11].[3] Vampire solved 15 more problems than Leo-III and 21 problems that neither Leo-III nor ZenonModulo could solve (see Table 1), although both solvers also solved problems Vampire was unable to solve. Vampire solves 7 previously unsolved rating 1.00 problems.

*Experiment 2.* We also wanted to ascertain how much overhead had been added for non-polymorphic problems, so we tested the polymorphic version of Vampire, Vampire-poly, against the previous version on the set of all 33,843 monomorphic or untyped first order problems in the TPTP library not containing arithmetic. Note that this simply tests whether we go from solving a problem to not solving it (or vice versa) and not the time taken to find a solution, i.e., we test the impact on proof search and whether any time overhead takes us past the given time limit.

The results (see Table 1) are interesting. For TF0 problems Vampire 4.4 does indeed outperform its polymorphic sibling. However, at the time of writing, there is a bug in the polymorphic parser that resulted in 324 problems being incorrectly rejected. Even taking this into account, the performance of Vampire-poly lags behind and the cause of this remains to be fully investigated, although is likely to be due to the fragile nature of proof search in Vampire. Note that Vampire-poly solves 88 problems unsolved by Vampire 4.4.

---

[2] https://github.com/vprover/vampire_publications/tree/master/experimental_data /IJCAR-2020-POLY-VAMP - this contains the results themselves and a link to the Vampire executable that produced them. Polymorphism is not yet supported in the main branch of Vampire but is available in the `polymorphic_vampire` branch, which may be merged in the future.

[3] At a late stage, we realised that Zipperposition [10] can also parse TF1 syntax. Unfortunately, it was too late to incorporate it into the results.

# 4   Polymorphism and Theory Reasoning

Vampire has built-in support for the polymorphic theory of arrays [12] and the polymorphic theory of first-class tuples [13]. Here we briefly discuss work-in-progress to improve the implementation of these theories (and future similar theories) using polymorphism.

Both theories are supported by detecting instances of the polymorphic theory and adding the relevant instances of that theory's axioms to the input problem. For example, for the polymorphic theory of arrays, for each array sort `array(t1, t2)` detected in the input problem, we add instances of the axioms

```
![V:array(t1, t2), X: t1, Y: t2, Z:t2] : (Y = Z
    => select(store(V, Y, X), Z) = X)
![V:array(t1, t2), X: t1, Y: t2, Z:t2] : (Y != Z
    => select(store(V, Y, X), Z) = select(V, Z))
![V:array(t1, t2), X: array(t1, t2), Y: t1] :
    (select(V, Y) = select(X, Y) => V = X)
```

With support for polymorphism, as soon as we detect that arrays of any kind are used we can simply add the three polymorphic axioms

```
!>[T1: $tType, T2: $tType]:
  ( ![V:array(T1, T2), X: T1, Y: T2, Z:T2] :
    (Y = Z => select(store(V, Y, X), Z) = X))
!>[T1: $tType, T2: $tType]:
  ( ![V:array(T1, T2), X: T1, Y: T2, Z: T2] :
    (Y != Z  => select(store(V, Y, X), Z) = select(V, Z)))
!>[T1: $tType, T2: $tType]:
  ( ![V:array(T1, T2), X: array(T1, T2), Y: T1] :
    ((select(V, Y) = select(X, Y) => V = X))
```

This has a minor impact on proof search. Instead of adding $3n$ clauses when we have $n$ different instances of the polymorphic theory, we only add 3 clauses. As $n$ is usually small, this is unlikely to have a significant impact. At the same time, we should not see any negative impact, these polymorphic axioms will act in the same way as the $3n$ instances did.

The main impact is on the implementation of theories within Vampire. A non-trivial amount of complexity is required within the Vampire codebase to support the current mechanisms for the two supported polymorphic theories. Adding a new polymorphic theory of this kind involves a lot of boilerplate code and updating of various definitions. Replacing this with polymorphic theory axioms will simplify the code significantly. For example, if the SMT-LIB language is extended to support polymorphism in the future (this has been discussed, e.g., [6] but not implemented), internal support for polymorphism would make supporting the polymorphic theory of term algebras straightforward.

Moreover, not all polymorphic theories are supported by the mechanism described above; our current approach of adding instantiated axioms based on the input is complete for the theory of arrays, but cannot be complete in general as shown by Bobot and Paskevich [4]. For the theory of combinatory logic for example, no decision procedure can exist for selecting a set of monomorphic combinator axioms to add to a problem and ensure completeness (even though such a set must exist).

## 5   Related Work

The polymorphism of TPTP's TF1 language is inspired by ML-style polymorphism but differs in the use of type quantifiers. As pointed out by Blånchette et al. [3], ML-style polymorphism avoids explicit type quantifiers, choosing to determine type signatures by the types of arguments, results or additional annotations (which are sometimes needed to guide Hindley-Milner type inference). Comparatively, type checking is more straightforward in TF1 due to an explicit signature and explicit type quantifiers.

As mentioned earlier, there are two main methods for reasoning in polymorphic logic: natively or via translations. We discuss related work for each direction.

Zipperposition [9] was built using *explicit polymorphism* – types are explicitly represented in terms and inferences perform unification on both terms and types. The main difference with our approach is that we are 'retro-fitting' polymorphism into a monomorphic theorem prover. Additionally, our 'types as terms' internal representation (mostly) removes the additional book-keeping required when performing separate term and type unification.

There are three main approaches to translation - type tags, type guards, and type arguments. The purpose of both the type tag and type guard encoding is to ensure that unsound inferences that violate typing constraints cannot occur in the untyped problem. We do not provide details of the encodings here, but refer readers to [2] with a further example given by Brown et al. [7] in their work translating between different TPTP formalisms. Consider the following satisfiable polymorphic problem with a polymorphic predicate p:

```
tff(a,type, p : !> [X : $tType] : X > $o).
tff(b, conjecture, ?[X:$i, Y:$int] : p($i,X) => p($int,Y)).
```

The negated conjecture becomes the two clauses ~p($i,X) and p($int,Y). Clearly, if we drop the types (i.e. via type erasure) then this satisfiable problem becomes unsatisfiable as we can no longer differentiate between the two versions of p. Using type tags we would get ~p(ti(X, $i)) and p(ti(Y, $int)) and with type guards we would get ~isi(X) | ~p (X) and ~isint(Y) | p(Y) – both prevent the unsatisfiability from type erasure at the expense of introducing extra functions or predicates. We achieve the same through type inference and unification. The type argument translation looks similar to our internal representation of types, e.g. types are encoded as terms. However, without being aware of the type of equalities where at least one side is a variable (as we are in our translation) this encoding can be unsound as equalities can capture cardinality constraints between types.

# 6   Conclusion

We have successfully extended a state-of-the-art first-order prover to support prenex polymorphism and shown that the difficulty in doing so is not as great as may be expected. We hope to encourage other researchers to do the same.

Theoretically, extending Vampire to polymorphic FOL should be graceful in the sense that no degradation of performance should be seen on non-polymorphic problems. Our experimental results do not bear this out. In future work, we hope to achieve two objectives. Firstly, to fix and refine our implementation of polymorphism such that no degradation on monomorphic or untyped problems is experienced. Secondly, as outlined above, to utilise polymorphism to simplify and extend theory reasoning in Vampire for polymorphic theories such as arrays.

# References

1. Bhayat, A., Reger, G.: A combinator-based superposition calculus for higher-order logic. In: Peltier, N., Sofronie-Stokkermans, V. (eds.) IJCAR 2020. LNCS, vol. 12166, pp. 278–296. Springer, Heidelberg (2020)
2. Blanchette, J.C., Böhme, S., Popescu, A., Smallbone, N.: Encoding monomorphic and polymorphic types. In: Piterman, N., Smolka, S.A. (eds.) TACAS 2013. LNCS, vol. 7795, pp. 493–507. Springer, Heidelberg (2013). https://doi.org/10.1007/978-3-642-36742-7_34
3. Blanchette, J.C., Paskevich, A.: TFF1: the TPTP typed first-order form with rank-1 polymorphism. In: Bonacina, M.P. (ed.) CADE 2013. LNCS (LNAI), vol. 7898, pp. 414–420. Springer, Heidelberg (2013). https://doi.org/10.1007/978-3-642-38574-2_29
4. Bobot, F., Paskevich, A.: Expressing polymorphic types in a many-sorted language. In: Tinelli, C., Sofronie-Stokkermans, V. (eds.) FroCoS 2011. LNCS (LNAI), vol. 6989, pp. 87–102. Springer, Heidelberg (2011). https://doi.org/10.1007/978-3-642-24364-6_7
5. Bobot, F., Conchon, S., Contejean, E., Lescuyer, S.: Implementing polymorphism in SMT solvers. In: ACM International Conference Proceeding Series, pp. 1–5 (2008)
6. Bonichon, R., Déharbe, D., Tavares, C.: Extending SMT-LIB v2 with λ-terms and polymorphism. In: SMT, pp. 53–62. Citeseer (2014)
7. Brown, C.E., Gauthier, T., Kaliszyk, C., Sutcliffe, G., Urban, J.: GRUNGE: a grand unified ATP challenge. In: Fontaine, P. (ed.) CADE 2019. LNCS (LNAI), vol. 11716, pp. 123–141. Springer, Cham (2019). https://doi.org/10.1007/978-3-030-29436-6_8
8. Couchot, J.-F., Lescuyer, S.: Handling polymorphism in automated deduction. In: Pfenning, F. (ed.) CADE 2007. LNCS (LNAI), vol. 4603, pp. 263–278. Springer, Heidelberg (2007). https://doi.org/10.1007/978-3-540-73595-3_18
9. Cruanes, S.: Extending superposition with integer arithmetic, structural induction, and beyond. Ph.D. thesis, INRIA (2015)
10. Cruanes, S.: Superposition with structural induction. In: Dixon, C., Finger, M. (eds.) FroCoS 2017. LNCS (LNAI), vol. 10483, pp. 172–188. Springer, Cham (2017). https://doi.org/10.1007/978-3-319-66167-4_10

11. Delahaye, D., Doligez, D., Gilbert, F., Halmagrand, P., Hermant, O.: Zenon Modulo: when achilles outruns the tortoise using deduction modulo. In: McMillan, K., Middeldorp, A., Voronkov, A. (eds.) LPAR 2013. LNCS, vol. 8312, pp. 274–290. Springer, Heidelberg (2013). https://doi.org/10.1007/978-3-642-45221-5_20

12. Kotelnikov, E., Kovács, L., Reger, G., Voronkov, A.: The vampire and the FOOL. In: Proceedings of the 5th ACM SIGPLAN Conference on Certified Programs and Proofs, pp. 37–48 (2016)

13. Kotelnikov, E., Kovács, L., Voronkov, A.: A FOOLish encoding of the next state relations of imperative programs. In: Galmiche, D., Schulz, S., Sebastiani, R. (eds.) IJCAR 2018. LNCS (LNAI), vol. 10900, pp. 405–421. Springer, Cham (2018). https://doi.org/10.1007/978-3-319-94205-6_27

14. Kovács, L., Voronkov, A.: First-order theorem proving and VAMPIRE. In: Sharygina, N., Veith, H. (eds.) CAV 2013. LNCS, vol. 8044, pp. 1–35. Springer, Heidelberg (2013). https://doi.org/10.1007/978-3-642-39799-8_1

15. Steen, A., Benzmüller, C.: The higher-order prover Leo-III. In: Galmiche, D., Schulz, S., Sebastiani, R. (eds.) IJCAR 2018. LNCS (LNAI), vol. 10900, pp. 108–116. Springer, Cham (2018). https://doi.org/10.1007/978-3-319-94205-6_8

16. Stump, A., Sutcliffe, G., Tinelli, C.: StarExec: a cross-community infrastructure for logic solving. In: Demri, S., Kapur, D., Weidenbach, C. (eds.) IJCAR 2014. LNCS (LNAI), vol. 8562, pp. 367–373. Springer, Cham (2014). https://doi.org/10.1007/978-3-319-08587-6_28

17. Sutcliffe, G.: The TPTP problem library and associated infrastructure. J. Autom. Reasoning **43**(4), 337–362 (2009). https://doi.org/10.1007/s10817-009-9143-8

# N-PAT: A Nested Model-Checker
## (System Description)

Hadrien Bride[1]([✉]), Cheng-Hao Cai[2], Jin Song Dong[1,3], Rajeev Gore[4],
Zhé Hóu[1], Brendan Mahony[5], and Jim McCarthy[5]

[1] Institute for Integrated and Intelligent Systems, Griffith University,
Brisbane, Australia
h.bride@griffith.edu.au
[2] School of Computer Science, University of Auckland, Auckland, New Zealand
[3] School of Computing, National University of Singapore, Singapore, Singapore
[4] Research School of Computer Science, The Australian National University,
Canberra, Australia
[5] Defence Science and Technology, Canberra, Australia

**Abstract.** N-PAT is a new model-checking tool that supports the verification of nested-models, i.e. models whose behaviour depends on the results of verification tasks. In this paper, we describe its operation and discuss mechanisms that are tailored to the efficient verification of nested-models. Further, we motivate the advantages of N-PAT over traditional model-checking tools through a network security case study.

## 1  Introduction

Model-checking is the problem of formally verifying that a model of a system meets a given specification. Automated model-checking techniques have been successfully applied to find subtle errors in complex industrial designs of e.g., hardware circuits, software controllers, and communication protocols [1]. However, the adoption rate of model-checking remains low in software engineering because of the computational complexity of model-checking algorithms. The state-space explosion problem [2] makes the verification of large models intractable unless high-level abstractions are used in the development and leveraged during verification.

Nowadays, complex systems are often designed in a modular and hierarchical fashion. Hierarchical models, also called multilevel models, are abstract representations of systems that span multiple levels of abstraction. They encode the hierarchical structure of systems explicitly and therefore enable reasoning about how properties of one level reflect across multiple levels of the model [5].

In this paper, we introduce the notion of nested model and nested model-checking. The main idea is to break up a large model-checking task into a hierarchy of smaller model-checking tasks. A Nested model is a high-level model which may contain several child models nested inside; its behaviour depends on the verification results of its child models. Note that the properties to be verified in child tasks may be different from the properties to be verified in parent tasks.

© Springer Nature Switzerland AG 2020
N. Peltier and V. Sofronie-Stokkermans (Eds.): IJCAR 2020, LNAI 12167, pp. 369–377, 2020.
https://doi.org/10.1007/978-3-030-51054-1_22

We present N-PAT – a nested model-checker suited to the verification of hierarchical systems and designed to perform nested model-checking tasks. In hierarchical modelling, some verification tasks may be used to determine and lift the properties of underlying child models to parent models. This structural abstraction provides modellers with the ability to structure the verification and guide the state-space exploration of model-checking methods, it also provides significant benefits in term of scalability for verification when compared to the traditional approach to modelling. We implement several optimisations leveraging the hierarchical structure of nested models. Also, since the time and space complexity of model-checking algorithms with respect to the size of models is super-linear, the divide-and-conquer approach employed by N-PAT significantly reduces the overall verification time. What sets N-PAT apart from existing model checkers (e.g., [4,6,7]) is the abstraction level of the modelling language. In our work, the modelling language of nested models has high-level primitives such as model checking and nested model instantiation.

## 2   Nested Model-Checking

*Standard* model checking is the problem of verifying whether a *standard* model complies with a given property. A standard model is a static and finite-state representation of a system, which may exhibit non-deterministic and probabilistic behaviours. The semantics of a standard model can be specified as a labelled transition system or a Markov decision process. Properties that can be verified include reachability, deadlock-freeness, divergence-freeness, reachability, and LTL formulae. The result of a model checking task depends on the type of the model. When checking a property over a non-probabilistic model, the result is 0 (not satisfied) or 1 (satisfied). When checking a property over a probabilistic model, the result is the min (alt. max) probability that the property is satisfied. Note that we only consider results in natural numbers, and a probability is represented in e.g., per thousand. Formally, let $M^s$ be the set of standard models and $\Phi$ be the set of properties. We denote by $\mathtt{mc} : M^s \times \Phi \to \mathbb{N}$ the model checking function that returns the results of checking a property over a standard model.

A *meta model*, also commonly referred to as a *template*, is a model of standard models. It can be viewed as a function that has a finite number of arguments and returns a standard model. Formally, a meta model is a function of the form $A_1 \times ... \times A_n \to M^s$ where $n \in \mathbb{N}$ and $A_1, ..., A_n \in \mathbb{N}$. We denote by $M^m$ the set of meta models. In order to instantiate a meta model, every argument must be known. An instantiated meta model is a standard model and can be verified using standard model checking.

Traditionally, meta models are instantiated from values specified by the modeller. In our work, we consider the verification of meta models instantiated from values that are the result of model checking tasks. Such a meta model is called a *nested* model and denoted as $M^n$. Figure 1 illustrates the structure and components of a nested model. Each diamond represents a standard model. Each box represents a meta model. Verification tasks are symbolised by circles. The text

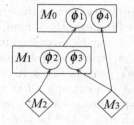

**Fig. 1.** Illustration of the structure of a nested model.

*binding* ::= *const* = *expr*          *mc* ::= **mc**(*model*, *Φ*)
*model* ::= $M^s$ | $M^m$(*binding* [, *binding*]*)     *op* ::= *expr* (+ | − | × | /) *expr*
*def* ::= **let** *binding* [, *binding*]* **in** *expr*     *expr* ::= ℕ | *op* | *mc* | *const* | *def*

**Fig. 2.** The BNF grammar of nested model checking problems.

within each circle is the property to be checked in the corresponding verification task. The arrows symbolise dependencies among verification tasks. In Fig. 1, there are two standard models: $M_2$ and $M_3$, and two meta models: $M_0$ and $M_1$. For instance, $M_1$ requires the verification results from $\mathrm{mc}(M_2, \phi_2)$ and $\mathrm{mc}(M_3, \phi_3)$ to be instantiated. After instantiation, $M_1$ will become a standard model. To verify the property $\phi_0$ over the nested-model $M_0$, we need to evaluate the following expression: $\mathrm{mc}(M_0(\mathrm{mc}(M_1(\mathrm{mc}(M_2, \phi_2), \mathrm{mc}(M_3, \phi_3)), \phi_1), \mathrm{mc}(M_3, \phi_4)), \phi_0)$.

A nested model checking problem is an expression that can be evaluated to an integer; it is formulated in a language that has two main primitives: meta model instantiation and standard model checking. For convenience, the language of nested model checking problems is extended with integers, basic arithmetic operations and restricted scope constant definitions. Let *const* be the set of constant identifiers, the syntax of nested model checking problems is given by the grammar in Fig. 2, where [$\alpha$]* denotes repeating $\alpha$ zero or more times.

eval($n, \Gamma$) = $n$          eval($m_s, \Gamma$) = $m_s$          eval($c, \Gamma$) = $v$ where $\langle c, v \rangle \in \Gamma$

eval($e_1$ *op* $e_2, \Gamma$) = eval($e_1, \Gamma$) *op* eval($e_2, \Gamma$)     eval(mc($m, \phi$), $\Gamma$) = mc(eval($m, \Gamma$), $\phi$)
eval($m_m(\{\langle c_1, e_1 \rangle, ..., \langle c_n, e_n \rangle\}), \Gamma$) = m($\{\langle c_1, v_1 \rangle, ..., \langle c_n, v_n \rangle\}$) where

$v_1$ = eval($e_1, \Gamma$), ..., $v_n$ = eval($e_n, \Gamma$)
eval(let $\{\langle c_1, e_1 \rangle, ..., \langle c_n, e_n \rangle\}$ in $e$) = eval($e, \Gamma'$) where

$\Gamma' = \Gamma \cup \{\langle c_1, \text{eval}(e_1, \Gamma) \rangle, ..., \langle c_n, \text{eval}(e_n, \Gamma) \rangle\}$

**Fig. 3.** The semantics of nested model checking problems.

We assume that the set of verification tasks related to a nested model form a directed acyclic graph (as in Fig. 1), which defines the dependencies of tasks, and that the tasks and the graph are known before the verification stage. When a `let` expression introduces multiple bindings, these bindings must be independent of one another, non-recursive, and may be evaluated in parallel.

The semantics of nested model checking problems in a given *context* is defined by the *evaluation function* `eval`, given in Fig. 3. A *valued* binding is a tuple $\langle c, v \rangle$ where $c \in const$ and $v \in \mathbb{N}$. A context is a set of valued bindings. Let $\Gamma$ be a context, $e, e_1, \dots \in expr$ be expressions, $c, c_1, \dots \in const$ be constant identifiers, $m_s \in M^s$ be a standard model, $m_m \in M^m$ be a meta model, $m \in M^s \cup M^m$ be a model, $\phi \in \Phi$ be a property, $n, v, v_1, \dots \in \mathbb{N}$ be numbers, and $op \in \{+, -, *, /\}$ be an integer operator.

# 3    N-PAT: Implementation

N-PAT is built on top of Process Analysis Toolkit [7] (PAT) – an industrial scale model-checker which employs an expressive modelling language called Communicating Sequential Processes with C# (CSP#) developed by Hoare [3] and others [9]. PAT features a model editor and an animated simulator using a mature and IDE-style user interface. Further, PAT facilitates new language and algorithm design and implementation as extended modules. Over the past 10 years, we have extended PAT with new verification modules for timed automata [9], real-time systems [8], and probabilistic systems [10]. We implemented N-PAT in C# as an extension of PAT. N-PAT is open-source and freely available online.[1]

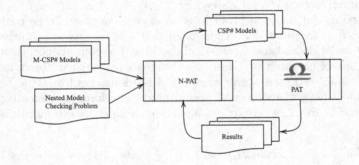

**Fig. 4.** N-PAT data-flow overview.

Standard models are specified by (probabilistic) CSP# models and properties are specified by CSP# assertions [7]. Meta models are specified by a *meta-level* CSP# language. This language, called meta-CSP#, introduces labelled placeholders, of the form $[id]$ where $id \in const$ is a label. These labelled place-holders extend the CSP# language and can be used in place of integer constants (e.g., variable initial values, choice probabilities). Let $m$ be a meta-CSP# model and $id_1, \dots, id_n$ where $n \in \mathbb{N}$ be the set of placeholders that appears in its definition.

---

Let $v = \{\langle id_1, v_1\rangle, ..., \langle id_n, v_n\rangle\}$ where $v_1, ..., v_n \in \mathbb{N}$ be a set of *valued* bindings. The meta-CSP# model $m$ can be instantiated using $v$ into a CSP# model by substituting the occurrences of $[id_i]$ by $v_i$ for all $i \in \{1, ..., n\}$. Nested model checking problems are specified using the language described in Sect. 2. Model checking is performed by N-PAT through the orchestration of calls to PAT.

Figure 4 depicts the overall data-flow of N-PAT. The input of N-PAT is a set of standard CSP# and meta-CSP# models and a nested model checking problem. N-PAT evaluates the result of the nested model checking problem similarly to how a dynamic interpreter evaluates an expression. The nested model checking problem is first parsed, and a corresponding abstract syntactic tree is built. This step is implemented using parser combinators (i.e., using recursive descent parsing). The resulting abstract syntactic tree is then recursively evaluated in a bottom-up fashion.

N-PAT exploits the hierarchical nature of nested models and provides improved verification scalability when compared to traditional model checkers that operate on flattened models. First, since CSP# models are static (i.e., they are not modified during execution), N-PAT applies stage-wise partial evaluation of verification sub-tasks to optimise the verification phase of nested CSP# models. Second, given a nested CSP# model, we assume that its verification sub-tasks are independent and can be computed concurrently. Thus N-PAT uses parallelism to speed up verification on modern architectures with multiple cores. This parallelism manifests itself in three places: bindings, operands, and the evaluation of meta model arguments.

## 4   Case Study and Experiment

We introduce a network security case study to illustrate the modelling and scalability advantages of nested model-checking. The case study is concerned with computing the probabilistic security level of a network. It is a simplified version of a real-life example which is studied by Australian Defence. The problem is hierarchical by nature, which illustrates code-reuse and modularisation of the proposed modelling approach. Another nice property of this example is that we can create models of different sizes to test the scalability of the verification, as will be shown in the experiment.

The details of the example are as follows: suppose there is a cluster of computation nodes, and each node can be in one of the three states: *safe*, *compromised*, and *isolated*. Initially, each node is safe, but it has a chance to be *hacked*, which changes the state of the node to compromised. When a node is compromised, it can either be *patched*, which will make the node safe again, or be *isolated*, which will disconnect the node from the cluster. If a node is isolated, then it loses the connection to other nodes and thus cannot contribute to the computational power of the cluster. When the node is isolated, it has a chance to be *recovered*, which will lead the node to the safe state again. Otherwise, the node stays isolated, in which case we increase down_nodes_counter, i.e., the number of nodes that are offline, by 1. For simplicity, we assume that the hacking of each node

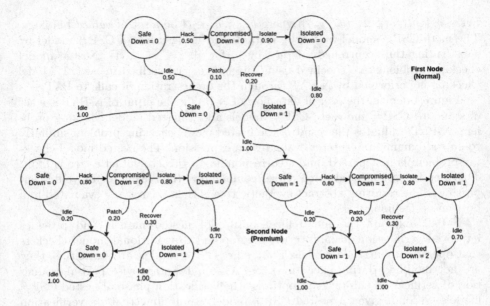

**Fig. 5.** The Markov chain of the traditional model, exemplified with two nodes.

is independent. We model two types of nodes: *normal* node and *premium* node, where the latter has a higher chance to be patched when it is compromised and a higher chance to be recovered when it is isolated.

*Traditional Method:* We shall formalise the above as a Markov chain in CSP#, and we have to define the state transitions for both types of nodes. Next, we model a `cluster manager`, which iterates through each node in the cluster and checks whether the node is offline. The manager will report that the cluster is in critical condition if at least half of the nodes in the cluster are offline. Consider an example where there are only two nodes in the network: the first node is a normal node, and the second node is a premium node. We show the overall Markov chain in Fig. 5. We annotate the event and its probability on the arrows. The upper part of the figure describes the process of checking the normal node, which can be either safe or isolated. The lower part of the figure splits into two cases when checking the premium node. The `Down = x` line in each circle indicates the `down_nodes_counter`. The premium node in Fig. 5 has a higher chance to be hacked than because in our hypothetical scenario, the hacker is more likely to target a premium node.

*Nested Model Checking Method:* We break the model in Fig. 5 down into two levels of abstraction: the node level and the cluster level. The idea is to create more modular models so that the model for a node can be used for both normal nodes and for premium nodes, and potentially can be used in future developments of other models. At the node level, we study the common properties of the

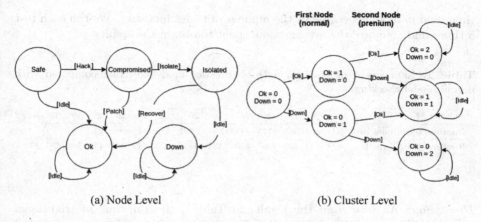

(a) Node Level                     (b) Cluster Level

**Fig. 6.** Modular models of Fig. 5 for nested model checking. a) Node Level b) Cluster Level

two types of nodes, and try to generalise the model so that the model-checking result is exactly what we need at the cluster level. As the cluster manager, we only need to know whether a node is offline or not. Therefore, besides safe, isolated, and compromised, we give a node two additional states: ok and down. Like in the traditional model, each node is initialised to be in the safe state. Since we do not consider cluster manager operations at the node level, we do not use the counter for offline nodes. Instead, when the node stays isolated, we change its state to down, and when it is not hacked/patched/recovered, we change its state to ok. The state transition diagram is shown in Fig. 6a.

At the cluster level, we abstract the notion of a node into only two states: ok and down. The probability of going to these two states will be given by the model-checking results from the node level. If a node is down, then we increment down_nodes_counter. We then model the cluster manager similarly as in the traditional model. The (partial) Markov chain for checking nodes is illustrated in Fig. 6b, where we only show two nodes. Note that places holders for probabilities in Fig. 6a, such as [Isolated], will be instantiated with specific values, depending on the type of the node, at run-time. Place holders in Fig. 6b will be instantiated with the results of node-level verification at run-time. Compared to the traditional model (cf. Fig. 5), the nested model is much simpler, and we will show that this leads to significant improvement in scalability.

*Experimental Comparison:* We compare the performance of both modelling approaches by evaluating the overall security of network of different sizes. In each case, the number of normal nodes are 4/5 of the total number. All experiments were carried on a desktop with Core i7-7700 quad-core processor at 3.6 GHz and 32 GB RAM. We verify multiple instances of the above models, starting with 8 nodes in a cluster, and increase the number of nodes by 2 at a time, and maintain the number of normal nodes at $(4 \times num\_of\_nodes)/5$. We then observe the time used in model checking as the number of nodes increases. We run each test 5 times and compute the average time spent to obtain the results.

**Table 1.** Experiment of traditional (probabilistic) model-checking compared with nested model checking.

| Number of nodes | 8 | 10 | 12 | 14 | 16 | 18 | 20 | 22 | 24 | 26 | 28 | 30 | 32 | 34 |
|---|---|---|---|---|---|---|---|---|---|---|---|---|---|---|
| Runtime traditional (ms) | 248 | 306 | 569 | 1771 | 7411 | 35K | 279K | Out of memory | | | | | | |
| Runtime N-PAT (ms) | 427 | 430 | 430 | 430 | 430 | 431 | 445 | 438 | 442 | 461 | 458 | 469 | 465 | 476 |

*Discussion:* As seen from the results in Table 1, the run-time of traditional model-checking grows rapidly as the size of the model (the number of nodes) increases. On the other hand, the run-time growth of nested model checking is moderate, and it solves instances up to 34 nodes less than 0.5 s. N-PAT also uses very little memory compared to PAT which uses up to 26.6 GB memory when running the 20 nodes instance. For small examples, the traditional modelling approach may be faster because the verification of nested models involves several calls to PAT which incur marginal overhead. However, the nested model checking approach scales better. The source code of this experiment, i.e. both the traditional model and the hierarchical model, can be found online.[2]

## 5   Conclusion and Future Work

We presented N-PAT – a high-level model checker that enables the verification of models that relies on the results of other verification tasks. We demonstrated in a case study in network security that this tool permits the use of high-level abstraction mechanisms and can therefore significantly improve the time and memory efficiency of verification tasks. These results indicate that nested model checking provides a novel modelling approach that can in some cases scale better than traditional model-checking.

In future work, we intend to provide more modelling flexibility by allowing dynamic calls to verification tasks that are not known a priori. We also planned to apply dynamic language optimisation techniques such as memoisation to speed up verification. Finally, we planned on supporting a fully reflective modelling language that permits inspection and modification of the behaviour and structure of models at verification-time.

## References

1. Clarke, E.M., Henzinger, T.A., Veith, H., Bloem, R. (eds.): Handbook of Model Checking. Springer, Heidelberg (2018). https://doi.org/10.1007/978-3-319-10575-8

---

[2] https://formal-analysis.com/research/npat/examples.html.

2. Clarke, E.M., Klieber, W., Nováček, M., Zuliani, P.: Model checking and the state explosion problem. In: Meyer, B., Nordio, M. (eds.) LASER 2011. LNCS, vol. 7682, pp. 1–30. Springer, Heidelberg (2012). https://doi.org/10.1007/978-3-642-35746-6_1

3. Hoare, C.A.R.: Communicating sequential processes. Commun. ACM **21**(8), 666–677 (1978)

4. López-Fernández, J., Guerra, E., De Lara, J.: Meta-model validation and verification with MetaBest. In: Proceedings of the 29th ACM/IEEE International Conference on Automated Software Engineering, pp. 831–834. ACM (2014)

5. Pârvu, O., Gilbert, D.: A novel method to verify multilevel computational models of biological systems using multiscale spatio-temporal meta model checking. PLoS One **11**(5), e0154847 (2016)

6. Steffen, B., Murtovi, A.: $M3C$: modal meta model checking. In: Howar, F., Barnat, J. (eds.) FMICS 2018. LNCS, vol. 11119, pp. 223–241. Springer, Cham (2018). https://doi.org/10.1007/978-3-030-00244-2_15

7. Sun, J., Liu, Y., Dong, J.S.: Model checking CSP revisited: introducing a process analysis toolkit. In: Margaria, T., Steffen, B. (eds.) ISoLA 2008. CCIS, vol. 17, pp. 307–322. Springer, Heidelberg (2008). https://doi.org/10.1007/978-3-540-88479-8_22

8. Sun, J., Liu, Y., Dong, J.S., Liu, Y., Shi, L., André, É.: Modeling and verifying hierarchical real-time systems using stateful timed CSP. ACM Trans. Softw. Eng. Methodol. **22**(1), 3:1–3:29 (2013)

9. Sun, J., Liu, Y., Dong, J.S., Zhang, X.: Verifying stateful timed CSP using implicit clocks and zone abstraction. In: Breitman, K., Cavalcanti, A. (eds.) ICFEM 2009. LNCS, vol. 5885, pp. 581–600. Springer, Heidelberg (2009). https://doi.org/10.1007/978-3-642-10373-5_30

10. Sun, J., Song, S., Liu, Y.: Model checking hierarchical probabilistic systems. In: Dong, J.S., Zhu, H. (eds.) ICFEM 2010. LNCS, vol. 6447, pp. 388–403. Springer, Heidelberg (2010). https://doi.org/10.1007/978-3-642-16901-4_26

# HYPNO: Theorem Proving with Hypersequent Calculi for Non-normal Modal Logics (System Description)

Tiziano Dalmonte[1], Nicola Olivetti[1], and Gian Luca Pozzato[2]

[1] Aix Marseille Univ, Université de Toulon, CNRS, LIS, Marseille, France
{tiziano.dalmonte,nicola.olivetti}@lis-lab.fr
[2] Dipartimento di Informatica, Universitá degli Studi di Torino, Turin, Italy
gianluca.pozzato@unito.it

**Abstract.** We present HYPNO (HYpersequent Prover for NOn-normal modal logics), a Prolog-based theorem prover and countermodel generator for non-normal modal logics. HYPNO implements some hypersequent calculi recently introduced for the basic system **E** and its extensions with axioms M, N, and C. It is inspired by the methodology of lean $T^A P$, so that it does not make use of any ad-hoc control mechanism. Given a formula, HYPNO provides either a proof in the calculus or a countermodel, directly built from an open saturated hypersequent. Preliminary experimental results show that the performances of HYPNO are very promising with respect to other theorem provers for the same class of logics.

**Keywords:** Non-normal modal logics · Hypersequent calculi · Prolog

## 1 Introduction

Non-Normal Modal Logics (NNMLs for short) are a generalization of ordinary modal logics that do not satisfy some axioms or rules of minimal normal modal logic **K**. They have been studied since the seminal works by C.I. Lewis, Scott, Lemmon, and Chellas (for an introduction see [3]), and along the years have gained interest in several areas such as epistemic, deontic, and agent reasoning among others [1,7,12–14]. NNMLs are characterised by the neighbourhood semantics. In [4,6], a variant of it is presented, called bi-neighbourhood semantics, this variant is more suitable for logics lacking the monotonicity property, although equivalent to the standard one.

This work has been partially supported by the ANR project TICAMORE ANR-16-CE91-0002-01 and by the INdAM project GNCS 2019 "METALLIC #2".

N. Peltier and V. Sofronie-Stokkermans (Eds.): IJCAR 2020, LNAI 12167, pp. 378–387, 2020.
https://doi.org/10.1007/978-3-030-51054-1_23

Not many theorem provers for NNMLs have been developed so far.[1] In [8] optimal decision procedures are presented for the whole cube of NNMLs; these procedures reduce a validity/satisfiability checking in NNMLs to a set of SAT problems and then call an efficient SAT solver. Despite undoubtably efficient, they do not provide explicitly "proofs", nor countermodels. A theorem prover for logic **EM** based on a tableaux calculus very similar to the one in [10], is presented in [9]: the system, is implemented in ELAN and handles also more complex Coalition Logic and Alternating Time Temporal logic. In [11] it is presented a Prolog implementation of a NNML containing both the $[\forall\forall]$ and the $[\exists\forall]$ modality; its $[\exists\forall]$ fragment coincides with the logic **EM**, also covered by HYPNO. Finally in [5] it is presented PRONOM, a theorem prover for the whole range of NNMLs, which implements *labelled* sequent calculi in [6]; PRONOM provides both proofs and countermodels in the mentioned bi-neighbourhood semantics.

In this paper we describe HYPNO (HYpersequent Prover for NOn-normal modal logics) a Prolog theorem prover for the whole cube of NNMLs. The prover HYPNO implements the optimal calculi for NNMLs recently introduced in [4]. These calculi handle hypersequents, where a hypersequent represents intuitively a metalogical disjunction of sequents; sequents in themselves can be interpreted as formulas of the language. While the hypersequent structure is not strictly needed for proving formulas, it ensures a direct computation of a countermodel from *one* failed proof-branch. Consequently, the prover takes as input a formula and either returns a proof or a countermodel of it in the bi-neighbourhood semantics mentioned above. The Prolog implementation closely corresponds to the calculi: each rule is encoded by a Prolog clause of a `prove` predicate. This correspondence ensures in principle both the soundness and completeness of the theorem prover. Termination of proof search is obtained by preventing redundant application of rules. Although there are no benchmarks in the literature, the performance seems promising, in particular it outperforms the theorem prover PRONOM based on labelled calculi.

The program HYPNO as well as all the Prolog source files, including those used for the performance evaluation, are available for free usage and download at http://193.51.60.97:8000/HYPNO/.

## 2   Axioms, Semantics, and Hypersequent Calculi

We present first the axiomatization and semantics of NNMLs of the classical cube and then the hypersequent calculi implemented by HYPNO.

Given a countable set of propositional variables Atm, the formulas of the language $\mathcal{L}$ of NNMLs are built as follows: $A ::= p \mid \bot \mid \top \mid A \vee A \mid A \wedge A \mid A \rightarrow A \mid \Box A$, where $p \in$ Atm. The minimal NNML **E** is defined in $\mathcal{L}$ by extending classical propositional logic with the rule RE below. The systems of the classical cube are then obtained by adding to **E** any combination of axioms M, C, and N. We obtain in this way eight distinct logics (see the *classical cube*, below on the right), where the top system **EMCN** coincides with normal modal logic **K**.

---

[1] We only mention here *implemented* systems, for a discussion on proof systems for NNMLs we refer to [4,6] and references therein.

$$\text{RE } \dfrac{A \leftrightarrow B}{\Box A \leftrightarrow \Box B} \qquad \begin{array}{ll} \text{M} & \Box(A \wedge B) \to \Box A \\[4pt] \text{C} & \Box A \wedge \Box B \to \Box(A \wedge B) \\[4pt] \text{N} & \Box \top \end{array}$$

Coming to the semantics, we consider the bi-neighbourhood models [6]. As a difference with standard neighbourhood semantics, in the bi-neighbourhood one, worlds are equipped with sets of *pairs* of neighbours which can be thought as lower and upper approximations of neighbourhood in the standard semantics.

**Definition 1.** *A* bi-neighbourhood model *is a tuple* $\mathcal{M} = \langle \mathcal{W}, \mathcal{N}_b, \mathcal{V} \rangle$, *where* $\mathcal{W}$ *is a non-empty set of worlds,* $\mathcal{V}$ *is a valuation function, and* $\mathcal{N}_b$ *is a bi-neighbourhood function* $\mathcal{W} \longrightarrow \mathcal{P}(\mathcal{P}(\mathcal{W}) \times \mathcal{P}(\mathcal{W}))$. *We say that* $\mathcal{M}$ *is a M-model if* $(\alpha, \beta) \in \mathcal{N}_b(w)$ *implies* $\beta = \emptyset$, *it is a N-model if for all* $w \in \mathcal{W}$ *there is* $\alpha \subseteq \mathcal{W}$ *such that* $(\alpha, \emptyset) \in \mathcal{N}_b(w)$, *and it is a C-model if* $(\alpha_1, \beta_1), (\alpha_2, \beta_2) \in \mathcal{N}_b(w)$ *implies* $(\alpha_1 \cap \alpha_2, \beta_1 \cup \beta_2) \in \mathcal{N}_b(w)$. *The forcing relation for boxed formulas is as follows:* $\mathcal{M}, w \Vdash \Box A$ *if and only if there is* $(\alpha, \beta) \in \mathcal{N}_b(w)$ *s.t.* $\alpha \subseteq [A] \subseteq \mathcal{W} \setminus \beta$, *where* $[A]$ *denotes the truth set of* $A$ *in* $\mathcal{M}$.

Bi-neighbourhood models can be easily transformed into equivalent standard neighbourhood models, and vice versa. Moreover, bi-neighbourhood semantics characterises the whole cube of NNMLs [6], in the sense that a formula $A$ is derivable in $\mathbf{E(M/C/N)}$ if and only if it is valid in all bi-neighbourhood (M/N/C)-models of the corresponding class.

The hypersequent calculi for NNMLs implemented by HYPNO are introduced in [4]. Their syntax is as follows: a *block* is a structure $\langle \Sigma \rangle$, where $\Sigma$ is a multiset of formulas of $\mathcal{L}$. A *sequent* is a pair $\Gamma \Rightarrow \Delta$, where $\Gamma$ is a multiset of formulas and blocks, and $\Delta$ is a multiset of formulas. A *hypersequent* is a multiset $S_1 \mid \dots \mid S_n$, where $S_1, \dots, S_n$ are sequents. Single sequents can be interpreted into the language as: $i(A_1, \dots, A_n, \langle \Sigma_1 \rangle, \dots, \langle \Sigma_m \rangle \Rightarrow B_1, \dots, B_k) = \bigwedge_{i \le n} A_i \wedge \bigwedge_{j \le m} \Box \bigwedge \Sigma_j \to \bigvee_{\ell \le k} B_\ell$. We say that a sequent $S$ is *valid* in a bi-neighbourhood model $\mathcal{M}$ (written $\mathcal{M} \models S$) if for all $w \in \mathcal{M}$, $\mathcal{M}, w \Vdash i(S)$. Moreover, a hypersequent $H$ is *valid* in $\mathcal{M}$ ($\mathcal{M} \models H$) if $\mathcal{M} \models S$ for some $S \in H$, and it is valid in (M/C/N-)models if it is valid in all models of that kind.

The calculi implemented by HYPNO are a minor variant of the ones in [4]: they contain an additional arrow $\Rightarrow$ used to represent that the formulas on the left of $\Rightarrow$ entails the *conjunction* (rather than their disjunction) of the formulas on its right. By this modification, all rules of the calculi are at most binary; the equivalence of the modified calculi with the original ones in [4] is straightforward.

The hypersequent calculi are defined by the rules in Fig. 1 (for propositional rules we only show the initial sequents and the rules for implication). In particular: $\mathcal{H}_E :=$ propositional rules $+ \Box_L + \Box_R + \Rightarrow_1 + \Rightarrow_2$; $\mathcal{H}_{EN} := \mathcal{H}_E + N$; $\mathcal{H}_{EC} := \mathcal{H}_E + C$; $\mathcal{H}_{ECN} := \mathcal{H}_E + C + N$; $\mathcal{H}_M :=$ propositional rules $+ \Box_L + {}_M\Box_R$;

$$\text{init} \dfrac{}{G \mid p, \Gamma \Rightarrow \Delta, p} \qquad \perp_L \dfrac{}{G \mid \perp, \Gamma \Rightarrow \Delta} \qquad \mathsf{T}_R \dfrac{}{G \mid \Gamma \Rightarrow \Delta, \mathsf{T}}$$

$$\rightarrow_L \dfrac{G \mid A \rightarrow B, \Gamma \Rightarrow \Delta, A \qquad G \mid B, A \rightarrow B, \Gamma \Rightarrow \Delta}{G \mid A \rightarrow B, \Gamma \Rightarrow \Delta} \qquad \rightarrow_R \dfrac{G \mid A, \Gamma \Rightarrow \Delta, A \rightarrow B, B}{G \mid \Gamma \Rightarrow \Delta, A \rightarrow B}$$

$$\square_L \dfrac{G \mid \langle A \rangle, \square A, \Gamma \Rightarrow \Delta}{G \mid \square A, \Gamma \Rightarrow \Delta} \qquad \mathsf{M}\square_R \dfrac{G \mid \langle \Sigma \rangle, \Gamma \Rightarrow \Delta, \square B \mid \Sigma \Rightarrow B}{G \mid \langle \Sigma \rangle, \Gamma \Rightarrow \Delta, \square B}$$

$$\square_R \dfrac{G \mid \langle \Sigma \rangle, \Gamma \Rightarrow \Delta, \square B \mid \Sigma \Rightarrow B \qquad G \mid \langle \Sigma \rangle, \Gamma \Rightarrow \Delta, \square B \mid B \Rightarrow \Sigma}{G \mid \langle \Sigma \rangle, \Gamma \Rightarrow \Delta, \square B}$$

$$\Rightarrow_1 \dfrac{G \mid A \Rightarrow B}{G \mid A \Rrightarrow B} \qquad \Rightarrow_2 \dfrac{G \mid A \Rightarrow B \qquad G \mid A \Rrightarrow \Sigma}{G \mid A \Rrightarrow B, \Sigma} \; |\Sigma| \geq 1$$

$$\mathsf{N} \dfrac{G \mid \langle \mathsf{T} \rangle, \Gamma \Rightarrow \Delta}{G \mid \Gamma \Rightarrow \Delta} \qquad \mathsf{C} \dfrac{G \mid \langle \Sigma, \Pi \rangle, \langle \Sigma \rangle, \langle \Pi \rangle, \Gamma \Rightarrow \Delta}{G \mid \langle \Sigma \rangle, \langle \Pi \rangle, \Gamma \Rightarrow \Delta}$$

**Fig. 1.** Rules of $\mathcal{H}_{\mathsf{E}*}$.

$\mathcal{H}_{\mathsf{MN}} := \mathcal{H}_{\mathsf{M}} + \mathsf{N}$; $\mathcal{H}_{\mathsf{MC}} := \mathcal{H}_{\mathsf{M}} + \mathsf{C}$; and $\mathcal{H}_{\mathsf{MCN}} := \mathcal{H}_{\mathsf{M}} + \mathsf{C} + \mathsf{N}$. In the following, we denote by $\mathcal{H}_{\mathsf{E}*}$ any extension of $\mathcal{H}_{\mathsf{E}}$.

## 3   Design of HYPNO

The prover HYPNO implements the calculi of Fig. 1. It is inspired by the "lean" methodology of lean$T^AP$ [2], even if it does not follow its style in a rigorous manner. The program comprises a set of clauses, each one of them implementing a rule or an axiom of the mentioned calculi. The proof search is provided for free by the mere depth-first search mechanism of Prolog, without any additional ad hoc mechanism. Before presenting in details the code of the theorem prover, let us discuss a general design choice.

As mentioned, HYPNO searches for a derivation of an input formula and in case of failure, on demand, it produces a countermodel of it. The proof search procedure is implemented by a predicate `terminating_proof_search` which tries to generate a derivation of the given input formula. In case it fails, on demand by the user, another predicate `build_saturated_branch` is invoked that computes an open saturated branch from which a countermodel is extracted. The predicate `build_saturated_branch` is in some sense "dual" of the proof search one. We have chosen to implement a two-phase computation, instead of a single one taking care of both tasks, for the following reason: a single-phase procedure would need to carry over all information for extracting a countermodel anyway; this information would be completely useless in case of a successful derivation and would unacceptably overload proof-search. As matter of fact, the time spent to "recompute" the saturated branch is negligible with respect to the overload of a proof-search procedure storing also information for countermodel extraction. By this choice we get a simpler and more readable code, and of course, more suited for countermodel generation only "on demand".

HYPNO represents an hypersequent with a Prolog list whose elements are Prolog terms of the form `singleSeq([Gamma,Delta],Additional)`, each one representing a sequent in the hypersequent. `Gamma`, `Delta`, and `Additional` are in turn Prolog lists: `Gamma` and `Delta` represent the left side and the right side of the single sequent, respectively, whereas `Additional` keeps track of the rules already applied to each sequent in order to ensure termination by avoiding multiple redundant applications of the same rule to a given hypersequent. Elements of `Gamma` and `Delta` are either formulas or Prolog lists representing blocks. Symbols $\top$ and $\bot$ are represented by constants `true` and `false`, respectively, whereas connectives $\neg$, $\wedge$, $\vee$, $\rightarrow$, and $\square$ are represented by `-`, `^`, `?`, `->`, and `box`. The symbol of provability $\Rightarrow$ in systems with axiom C is represented by `=>`. Propositional variables are represented by Prolog atoms. As an example, the Prolog list

```
[singleSeq( [[box (a ^ c), [true], [a,c]], [a, b, a -> b, box b]],
  [n, right(a -> b), apdR([a,c],b)]), singleSeq ([[P], [P]] ,[ ])]
```

is used to represent the hypersequent $\square(A \wedge C), \langle \top \rangle, \langle A, C \rangle \Rightarrow A, B, A \vee B, \square B \mid P \Rightarrow P$, to which the rules N, $\vee_R$ and $\square_R$ have been already applied, the last one by using the block $\langle A, C \rangle$ and the formula $\square B$ as the principal formulas. In turn, no rule has been applied to $P \Rightarrow P$ (the list `Additional` is empty).

Given a NNML formula $F$ represented by the Prolog term `f`, HYPNO executes the main predicate of the prover, called **prove**[2], whose only two clauses implement the functioning of HYPNO: the first clause checks whether $F$ is valid and, in case of a failure, the second one enables the graphical interface to invoke a predicate called `counter` to compute a model falsifying $F$. In detail, the predicate `prove` first checks whether the formula is valid by executing the predicate:

$$\texttt{terminating\_proof\_search(Hyper, ProofTree).}$$

This predicate succeeds if and only if the hypersequent represented by the list `Hyper` is derivable in $\mathcal{H}_{E^*}$. When it succeeds, the output term `ProofTree` matches with a representation of the derivation found by the prover. As an example, in order to prove that the sequent $\square(A \wedge (B \vee C)) \Rightarrow \square((A \wedge B) \vee (A \wedge C))$ is valid in **E**, one queries HYPNO with the goal:

```
terminating_proof_search([singleSeq([[box (a ^ (b ? c))], [box ((a ^ b) ?
                    (a ^ c))]], [ ]), ProofTree).
```

Each clause of `terminating_proof_search` implements an axiom or rule of the sequent calculi $\mathcal{H}_{E^*}$. To search for a derivation of a sequent $\Gamma \Rightarrow \Delta$, HYPNO proceeds as follows. First of all, if $\Gamma \Rightarrow \Delta$ is an instance of an axiom, then the goal will succeed immediately by using one of the clauses implementing the axioms. As an example, the clause implementing init is as follows:

```
terminating_proof_search(Hyper,tree(axiom,PrintableHyper,no,no)):-
    member(singleSeq([Gamma,Delta],_),Hyper),
    member(P,Gamma), member(P,Delta),!,
    extractPrintableSequents(Hyper,PrintableHyper).
```

---

[2] The user can run HYPNO without using the interface of the web application. To this aim, he just needs to invoke the goal `prove(f)`.

The auxiliary predicate `extractPrintableSequents` is used just for a graphical rendering of the hypersequent. If $\Gamma \Rightarrow \Delta$ is not an instance of the axioms, then the first applicable rule will be chosen, e.g. if `Gamma` contains a list `Sigma` representing a block $\langle \Sigma \rangle \in \Gamma$, and `Delta` contains box `b` representing that $\Box B \in \Delta$, then the clause for $\Box_\mathsf{R}$ will be chosen, and HYPNO will be recursively invoked on its premises. HYPNO proceeds in a similar way for the other rules. The ordering of the clauses is such that the application of branching rules is postponed as much as possible. As an example, here is the clause implementing $\Box_\mathsf{R}$:

```
1.  terminating_proof_search(Hyper,tree(rbox,PrintableHyper,Sub1,Sub2)):-
2.    select(singleSeq([Gamma,Delta],Additional),Hyper,NewHyper),
3.    member(Sigma,Gamma), is_list(Sigma),member(box B,Delta),
4.    list_to_ord_set(Sigma,SigmaOrd), \+member(apdR(SigmaOrd,B),Additional),!,
5.    terminating_proof_search([singleSeq([Sigma,[B]],[])|
        [singleSeq([Gamma,Delta],[apdR(SigmaOrd,B)|Additional])|NewHyper]],Sub1),
6.    terminating_proof_search([singleSeq([[],[B => Sigma]],[])|
        [singleSeq([Gamma,Delta],[apdR(SigmaOrd,B)|Additional])|NewHyper]],Sub2),
7.    extractPrintableSequents(Hyper,PrintableHyper).
```

Line 3 checks whether `Gamma` contains an item `Sigma` which is a list representing a block and if a box formula `box B` belongs to the list `Delta`. Line 4 implements the restriction on the application of the rule used in order to ensure a terminating proof search: if the `Additional` list contains the Prolog term `apdR(SigmaOrd,B)`[3], this means that the rule $\Box_\mathsf{R}$ has been already applied on that sequent by using $\Box B$ and the block $\Sigma$, and HYPNO does no longer apply it. Otherwise, the predicate `terminating_proof_search` is recursively invoked on the two premises of the rule (lines 5 and 6), by introducing $\Sigma \Rightarrow B$ and $B \Rightarrow \Sigma$ respectively. Since the rule is invertible, Prolog cut ! is used in line 4 to eventually block backtracking.

When the predicate `terminating_proof_search` fails, then the initial formula is not valid. On user demand, as recalled at the beginning of this section, HYPNO extracts a model falsifying such a formula from an open saturated branch, following the model extraction method described in [4]. The model is computed by executing the predicate:

<code>build_saturated_branch(Hyper, Model).</code>

When this predicate succeeds, the variable `Model` matches a description of an open saturated branch obtained by applying the rules of $\mathcal{H}_{\mathsf{E}^*}$ to the initial formula. Since the very objective of this predicate is to build an open saturated hypersequent in the sequent calculus, its clauses are essentially the same as the ones for the predicate `terminating_proof_search`, however rules introducing a branching in a backward proof search are implemented by *pairs* of (disjoint) clauses, each one attempting to build an open saturated hypersequent from the corresponding premise. As an example, the following clauses implement the saturation in presence of a block $\Sigma$ in the left hand side and of a boxed formula $\Box B$ in the right hand side of a sequent:

---

[3] The predicate `list_to_ord_set` is used in order to check the applicability of the rule by ignoring the order of the formulas in the block.

```
build_saturated_branch(Hyper,Model):-
  select(singleSeq([Gamma,Delta],Additional),Hyper,NewHyper),
  member(Sigma,Gamma),is_list(Sigma),member(box B,Delta),
  list_to_ord_set(Sigma,SigmaOrd),\+member(apdR(SigmaOrd,B),Additional),
  build_saturated_branch([singleSeq([Sigma,[B]],[])|
   [singleSeq([Gamma,Delta],[apdR(SigmaOrd,B)|Additional])|NewHyper]],Model).
build_saturated_branch(Hyper,Model):-
  select(singleSeq([Gamma,Delta],Additional),Hyper,NewHyper),
  member(Sigma,Gamma),is_list(Sigma),member(box B,Delta),
  list_to_ord_set(Sigma,SigmaOrd),\+member(apdR(SigmaOrd,B),Additional),
  build_saturated_branch([singleSeq([[],[B => Sigma]],[])|
   [singleSeq([Gamma,Delta],[apdR(SigmaOrd,B)|Additional])|NewHyper]],Model).
```

HYPNO will first try to build a countermodel by considering the left premise of $\Box_R$, whence recursively invoking the predicate build_saturated_branch on the premise with the sequent $\Sigma \Rightarrow B$. In case of a failure, it will carry on the saturation process by using the right premise of $\Box_R$ with the sequent $B \Rrightarrow \Sigma$.

Clauses implementing axioms for the predicate terminating_proof_search are replaced by the last clause, checking whether the current sequent represents an open and saturated hypersequent:

```
build_saturated_branch(Hyper,model(Hyper)):-\+instanceOfAnAxiom(Hyper).
```

Since this is the very last clause of build_saturated_branch, it is considered by HYPNO only if no other clause is applicable, then the hypersequent is saturated. The auxiliary predicate instanceOfAnAxiom checks whether the hypersequent is open by proving that it is not an instance of the axioms. The second argument matches a term model representing the countermodel extracted from Hyper.

The implementation of the calculi for extensions of **E** is very similar: given the modularity of the calculi $\mathcal{H}_{E*}$, each system is obtained by just adding clauses for both the predicates terminating_proof_search and build_saturated_branch corresponding to the specific axioms/rules. However, we provide a different Prolog file for each system of the cube. This choice is justified by two reasons: first of all readiness of the code: one may be interested only in one specific system, wishing to have all the rules in a stand-alone file. Second and more important, an implementation of calculi for a family of logic cannot be completely modular: the computation (both proof-search and countermodel extraction) is sensitive to the order of application of the rules, so that the insertion of different rules may result in different orders of application of the whole set of rules.

HYPNO can be used on any computer or mobile device through a web interface implemented in php, which allows the user to choose the modal logic. When a formula is valid, HYPNO builds a pdf file showing a derivation in the corresponding calculus, as well as the LATEX source file. Otherwise, a countermodel falsifying the initial formula is displayed. Prolog source codes are also available.

## 4   Performance of **HYPNO**

We have compared the performance of HYPNO with those of the prover PRONOM [5], which deals with the same set of logics, obtaining promising results. We have tested it by running SWI-Prolog, version 7.6.4, on an Apple MacBook Pro, 2.7 GHz Intel Core i7, 8 GB RAM machine. First, we have tested HYPNO over hundred valid formulas in **E** and considered extensions obtained by generalizing schemas of valid formulas by varying some crucial parameters, like the modal degree (the level of nesting of the $\Box$ connective), already used for testing PRONOM. For instance, we have considered the schemas (valid in all systems):

$$(\Box(\Box(A_1 \wedge (B_1 \vee C_1)) \wedge \cdots \wedge \Box(A_n \wedge (B_n \vee C_n)))) \rightarrow (\Box(\Box((A_1 \wedge B_1) \vee (A_1 \wedge C_1)) \wedge \cdots \wedge \Box((A_n \wedge B_n) \vee (A_n \wedge C_n)))$$

$$(\Box^n C_1 \wedge \cdots \wedge \Box^n C_j \wedge \Box^n A) \rightarrow (\Box^n A \vee \Box^n D_1 \vee \cdots \vee \Box^n D_k)$$

obtaining encouraging results: Table 1 reports the number of timeouts of HYPNO and PRONOM over a set of valid formulas in system **E**.

**Table 1.** Percentage of timeouts over valid formulas in **E**.

| System | 0,1 ms | 1 ms | 100 ms | 1 s | 5 s |
|--------|--------|------|--------|-----|-----|
| HYPNO | 91,50% | 78,91% | 28,23% | 9,52% | 5,78% |
| PRONOM | 85,71% | 77,55% | 57,82% | 31,16% | 19,80% |

HYPNO is able to answer in less than one second on more than the 90% of the tests, whereas PRONOM is not even if we extend the time limit to 5 s.

We have also tested HYPNO on randomly generated formulas, fixing different time limits, numbers of propositional variables, and levels of nesting of connectives. We have compared the performances of HYPNO with those of PRONOM, obtaining the results in Table 2: in each pair, the first number is the percentage of timeouts of HYPNO, the second number is the percentage of timeouts of PRONOM given the fixed time limit.

**Table 2.** Percentage of timeouts in 5000 random tests (system **E**).

| Vars/Depth | 1 ms | 10 ms | 1 s | 10 s |
|------------|------|-------|-----|------|
| 3 vars - depth 5 | 4–5,58% | 0,78–1,76% | 0,02–0,48% | 0–0,22% |
| 3 vars - depth 7 | 23,78–25,18% | 10,86–20,16% | 3,16–14,40% | 2,02–12% |
| 7 vars - depth 10 | 45,22–44,94% | 34,36–42,36% | 19,06–30,30% | 16,06–20,34% |

Also in case of formulas generated from 3 different atomic variables and with a higher level of nesting (7), HYPNO is able to answer in more than 96% of the cases within 1 s, against the 85% of PRONOM. We have repeated the experiments also for all the extensions of **E** considered by HYPNO: complete results can be found at http://193.51.60.97:8000/HYPNO/#experiments. Moreover, we are planning to perform more accurate tests following the approach of [8], where randomly generated formulas can be obtained by selecting different degrees of probability about their validity.

## 5   Conclusions

We have presented HYPNO, a prover for the cube of NNMLs based on some hypersequent calculi for these logics recently introduced. HYPNO produces both proofs and countermodels in the bi-neighbourhood semantics. Although no specific optimisation has been implemented, the performances of HYPNO are promising. In the future we intend to extend possible optimisation, in particular to minimize the size of countermodels. Moreover we intend to extend it to other non-normal modal logics in the realm of deontic and agent-ability logics.

**Acknowledgements.** We thank the reviewers for their careful reading that helped us to improve this paper. We are currently developing a new version of HYPNO taking into account all the suggestions about its implementation.

## References

1. Askounis, D., Koutras, C.D., Zikos, Y.: Knowledge means 'all', belief means 'most'. J. Appl. Non-Class. Log. **26**(3), 173–192 (2016). https://doi.org/10.1080/11663081.2016.1214804
2. Beckert, B., Posegga, J.: LeanTAP: lean tableau-based deduction. J. Autom. Reasoning **15**(3), 339–358 (1995). https://doi.org/10.1007/BF00881804
3. Chellas, B.F.: Modal Logic. Cambridge University Press, Cambridge (1980)
4. Dalmonte, T., Lellmann, B., Olivetti, N., Pimentel, E.: Countermodel construction via optimal hypersequent calculi for non-normal modal logics. In: Artemov, S., Nerode, A. (eds.) LFCS 2020. LNCS, vol. 11972, pp. 27–46. Springer, Cham (2020). https://doi.org/10.1007/978-3-030-36755-8_3
5. Dalmonte, T., Negri, S., Olivetti, N., Pozzato, G.L.: PRONOM: proof-search and countermodel generation for non-normal modal logics. In: Alviano, M., Greco, G., Scarcello, F. (eds.) AI*IA 2019. LNCS (LNAI), vol. 11946, pp. 165–179. Springer, Cham (2019). https://doi.org/10.1007/978-3-030-35166-3_12
6. Dalmonte, T., Olivetti, N., Negri, S.: Non-normal modal logics: Bi-neighbourhood semantics and its labelled calculi. In: Proceedings of AiML 12, pp. 159–178. College Publications (2018)
7. Elgesem, D.: The modal logic of agency. Nord. J. Philos. Log. **2**(2), 1–46 (1997)
8. Giunchiglia, E., Tacchella, A., Giunchiglia, F.: SAT-based decision procedures for classical modal logics. J. Autom. Reasoning **28**(2), 143–171 (2002). https://doi.org/10.1023/A:1015071400913
9. Hansen, H.: Tableau games for coalition logic and alternating-time temporal logic-theory and implementation. Master's thesis, University of Amsterdam (2004)
10. Lavendhomme, R., Lucas, T.: Sequent calculi and decision procedures for weak modal systems. Studia Logica **66**, 121–145 (2000). https://doi.org/10.1023/A:1026753129680
11. Lellmann, B.: Combining monotone and normal modal logic in nested sequents – with countermodels. In: Cerrito, S., Popescu, A. (eds.) TABLEAUX 2019. LNCS (LNAI), vol. 11714, pp. 203–220. Springer, Cham (2019). https://doi.org/10.1007/978-3-030-29026-9_12
12. Pacuit, E.: Neighborhood Semantics for Modal Logic. Springer, Heidelberg (2017). https://doi.org/10.1007/978-3-319-67149-9

13. Pauly, M.: A modal logic for coalitional power in games. J. Log. Comput. **12**(1), 149–166 (2002)
14. Vardi, M.Y.: On epistemic logic and logical omniscience. In: Theoretical Aspects of Reasoning About Knowledge, pp. 293–305. Elsevier (1986)

# Implementing Superposition in iProver
# (System Description)

André Duarte$^{(\boxtimes)}$ and Konstantin Korovin$^{(\boxtimes)}$

The University of Manchester, Manchester, UK
{andre.duarte,konstantin.korovin}@manchester.ac.uk

**Abstract.** iProver is a saturation theorem prover for first-order logic with equality, which is originally based on an instantiation calculus Inst-Gen. In this paper we describe an extension of iProver with the superposition calculus. We have developed a flexible simplification setup that subsumes and generalises common architectures such as DISCOUNT or Otter. This also includes the concept of "immediate simplification", wherein newly derived clauses are more aggressively simplified among themselves, and the concept of "light normalisation", wherein ground unit equations are stored in an interreduced TRS which is then used to simplify new clauses. We have also added support for associative-commutative theories (AC), namely by deletion of AC-joinable clauses, semantic detection of AC axioms, and preprocessing normalisation.

iProver[1] [10] is an automated theorem prover for first-order logic. It is a saturation prover, and is based primarily on the Inst-Gen calculus [7], but also implements resolution and supports running them in combination in an abstraction-refinement framework [9,12]. In this work we detail how iProver was extended with support for the superposition calculus.

Currently, iProver deals with equality axiomatically, which can be inefficient for problems heavy on equality. At the same time, the superposition calculus is a set of complete inference rules specialised for first-order logic with equality. It complements the instantiation calculus since it is effective on problems where instantiation struggles, and vice-versa. We show that running the two calculi in combination yields better results than either plain instantiation or plain superposition.

Rules in a calculus can be classified as "generating", if they derive new clauses, or "simplifying", if some premise gets deleted. While generating rules are the ones necessary for completeness of a calculus, simplification rules are crucial for practical performance. Intuitively, simplification rules are beneficial to taming the growth of the search space as more clauses get generated. However, the computation of those simplifications itself takes time, so being too eager in applying them will also grind the prover to a halt. It is an open problem, what the optimal strategy to balance these conflicting requirements is, and although there is

---

[1] Available at http://www.cs.man.ac.uk/~korovink/iprover.

© Springer Nature Switzerland AG 2020
N. Peltier and V. Sofronie-Stokkermans (Eds.): IJCAR 2020, LNAI 12167, pp. 388–397, 2020.
https://doi.org/10.1007/978-3-030-51054-1_24

a huge amount of flexibility in how to perform simplifications (see e.g., [21]), most provers are rather restrictive about this. In iProver we developed a flexible simplification setup that subsumes and generalises most common architectures.

Finally, we have also implemented specialised techniques to deal with associative-commutative (AC) theories. These are theories of great interest which arise in several domains [20], and are traditionally problematic for theorem provers to deal with, due to combinatorial explosion and the non-orientability of the AC axioms.

The paper is structured as follows: first, we give a quick overview of the architecture of iProver. Then, we describe the implementation of superposition in iProver, its modifications, the simplification architecture and given clause loop, and AC reasoning rules. For a more in-depth description of basic features of iProver, see [9].

# 1   Overview

iProver is based on the Inst-Gen calculus, which is based on the following idea. We approximate the first-order problem to a propositional problem, and submit it to a black box SAT solver. It either finds an inconsistency, which is also an inconsistency at the first-order level, or else it returns a model, which guides the instantiation of new clauses whose abstraction witnesses some inconsistency at the ground level. If no such instantiation exists, then the problem is satisfiable.

The SAT solver is also used to implement "global" simplification rules (in the sense that they involve reasoning with the clause set which is shared between different calculi), such as global propositional subsumption [9]. In a nutshell, we submit ground abstractions of clauses in $S$ to a set $S_{gr}$. If the SAT solver finds that $S_{gr}$ propositionally implies $D\gamma$, with $D$ a strict subset of $C$ and $\gamma$ an injective substitution of variables to fresh constants, then we can replace $C\theta$, in $S$, by $D$.

$$\text{Glob. prop. subs.} \qquad \frac{C\theta}{D}, \quad \begin{array}{l} \text{where } D \subsetneq C \\ \text{and } C \in S, \ S \models S_{gr}, \ S_{gr} \models D\gamma \end{array} \qquad (1)$$

As mentioned before, iProver can also run other calculi. This is beneficial because (i) some problems are solved easily by one strategy and not by others, and (ii) clauses derived in e.g. resolution can be passed to the instantiation solver to participate in simplifications. For example, clauses derived in all calculi are submitted to a shared global propositional subsumption set, which is in turn used by all calculi to simplify its clauses.

Schematically, the high-level architecture of iProver is summarised in Fig. 1. We can view it as a modular architecture where each calculus (Inst-Gen, resolution, superposition) runs its own saturation loop, and can (i) query external SAT and SMT solvers, and (ii) submit clauses to, and retrieve clauses from, the 'Exchange' module. The instantiation and resolution modules are discussed in [9]. Here we will focus on the superposition calculus.

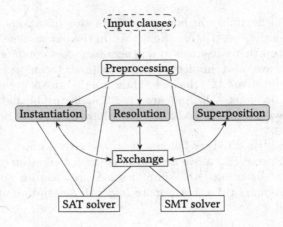

**Fig. 1.** Architecture of iProver.

## 2    Extension with Superposition

The superposition inference system consists of the following rules [15]:[2]

$$\text{Superposition} \qquad \frac{l \approx r \vee C \quad t[s] \mathbin{\dot{\approx}} u \vee D}{(t[s \mapsto r] \mathbin{\dot{\approx}} u \vee C \vee D)\theta} \tag{2}$$

where $\theta = \text{mgu}(l, s)$, $l\theta \not\preceq r\theta$, $t\theta \not\preceq u\theta$, and $s$ not a variable,

$$\text{Eq. Resolution} \qquad \frac{l \not\approx r \vee C}{C\theta} \quad \text{where } \theta = \text{mgu}(l, r), \tag{3}$$

$$\text{Eq. Factoring} \qquad \frac{l \approx r \vee l' \approx r' \vee C}{(l \approx r \vee r \not\approx r' \vee C)\theta} \quad \begin{array}{l} \text{where } \theta = \text{mgu}(l, l'), \\ l\theta \not\preceq r\theta \text{ and } r\theta \not\preceq r'\theta. \end{array} \tag{4}$$

We assume that $\prec$ is a simplification ordering. Non-equality predicates $P(t)$ are encoded as $P(t) \approx \top$. The calculus can easily be generalised to the many-sorted case, which iProver uses even in untyped problems, since it can perform sub-type inference during preprocessing. Superposition is sound and refutationally complete for first-order logic with equality (see [2,15]) and implemented in a number of state-of-the-art theorem provers: Vampire [11], E [19] and SPASS [22]. Currently, iProver uses non-perfect discrimination trees to find unification candidates efficiently [8,16]. For the literal selection, to ensure completeness, we must select either a negative literal, or all maximal literals. In iProver we use a variant of the Knuth-Bendix ordering which prioritises non-equational literals.

---

[2] '$\mathbin{\dot{\approx}}$' means '$\approx$' or '$\not\approx$'; also rules are to be read modulo flipping the equalities.

**Simultaneous Superposition.** In (2), by $t[s]$ and $t[s \mapsto r]$ we can mean resp. "a distinguished occurrence of $s$ as a subterm of $t$" and "replacing that subterm at that position by $r$. We call the variant *simultaneous superposition* where we mean instead "replacing all occurrences of $s$ in $t$ by $r$". This variant is still refutationally complete [3]. In cases where there are several occurrences of the same term, as in $f(\underbrace{s, s, \ldots, s}_{n \text{ times}})$, this avoids producing $2^n - 1$ intermediate clauses with $f(r, s, \ldots, s)$, $f(s, r, \ldots, s)$, $f(r, r, \ldots, s)$, etc., instead producing only $f(r, r, r, \ldots, r)$. In iProver, we implement this variant of superposition.

## 2.1 Simplifications

Apart from the generating inferences, necessary for completeness, we can add *simplification inferences*. In our implementation, we use the following rules: tautology deletion, syntactic equality resolution, subsumption, subset subsumption, subsumption resolution, demodulation and global subsumption [9, 11, 17, 22].

**Light Normalisation.** In addition, we introduce the following rule:

$$\text{Light Normalisation} \qquad \frac{R \quad C[l]}{C[l \mapsto r]}, \quad \text{where } l \rightarrow r \in R \qquad (5)$$

where $R$ is a set of interreduced wrt. (5) oriented rewrite rules. It can be seen as a restricted case of the demodulation rule, but it is advantageous to formulate this separately because it may be implemented much more efficiently than demodulation, by simply looking up terms in a hashtable, rather than having to do matching with variable instantiations.

A *light normalisation index* consists of (i) a hashtable that indexes rewrite rules in $R$ by their left-hand sides for forward light normalisations and (ii) a map of all subterms in $R$ for keeping $R$ interreduced. When we derive a unit equality, we first normalise it wrt. $R$ by recursively replacing each subterm by its normal form wrt. $R$. Then, if the simplified equality is orientable (wrt. $\prec$) we use it to normalise rules in $R$ add it to $R$. If there is a conflict between two rules $t \rightarrow s$ and $t \rightarrow u$ where $t \succ s \succ u$, we keep the rules $t \rightarrow u$ and $s \rightarrow u$. If $s$ and $u$ are incomparable wrt. $\succ$ we keep one of the rules in $R$. Since $R$ is only used for simplifications this choice does not affect the completeness. In general, we can restrict which orientable equations we add to $R$ (e.g. only ground ones, or small in size).

## 2.2 Given Clause Algorithm

In a standard given clause loop, the clause set is split into an active set, where inferences among the clauses have been performed, and a passive set, of clauses waiting to participate in inferences. Clauses are initially added to the passive, then in each iteration one given clause is picked from the passive set, added to the active set, and all inferences between given and active are performed. Newly derived clauses are pushed into the passive. The loop finishes when all clauses have been moved to the active set, meaning the initial set is satisfiable, or when a contradiction is derived.

**Immediate Simplification Set.** Next, we introduce the idea of *immediate simplification*. The intuition is as follows. Clauses that are derived in each loop are "related" to each other. It may be beneficial to keep the set of immediate conclusions inter-simplified. Also, throughout the execution of the program the set of generated clauses in each loop remains small compared to the set of passive or active clauses. Therefore, we can get away with applying more expensive rules that we do not necessarily want to apply on the set of all clauses (e.g. only "light" simplifications between newly derived clauses and passive clauses, but more expensive "full" simplifications among newly derived clauses). Finally, during this process, it is possible that the given clause itself becomes redundant (e.g. subsumed by one of its children). If this happens, we can add *only* the clauses responsible for making it redundant to the passive set, then remove the given clause from the active set, and throw away the rest of the iteration's newly generated clauses, abort the rest of the iteration, and proceed to the next given clause. In some problems, a significant number of iterations may be aborted, which means that fewer new clauses are added to the passive queue, and that we avoid the work of computing those inferences. This can be seen as a variant of orphan elimination [17]. Even when the given clause is not eliminated it is often beneficial to extensively inter-simplify immediate descendants of the given clause.

**Simplification Setup.** How all these simplifications are performed can greatly impact the performance of the solver, so care is needed, and tuning this part of the solver can pay off significantly. There is a significant amount of choice in how to perform simplifications. We can choose which simplifications to perform, and at what times, and with respect to which clauses. Additionally, some of these simplifications require auxiliary data structures (here referred to generally as "indices") to be done efficiently, and some indices support several simplification rules. Therefore we also need to choose which clauses to add to which indices at which stages.

For example, Otter-style loops [14] perform simplifications on clauses before adding them to the passive set. The problem with this is that the passive set is often orders of magnitude larger than the active set, therefore performance will degrade significantly as this set grows, and the system will spend most of its time performing simplifications on clauses that may not even end up being used. On the other hand, DISCOUNT-style loops [4] perform simplifications only with clauses that have been added to the active set. This has the benefit of reducing the time spent in simplifications, at the cost of potentially missing many valuable simplifications wrt. passive clauses. It is not clear where the "sweet spot" is, in terms of these setups.

It is possible, for example, to choose to apply only "cheap" simplifications to the full active + passive set (e.g. subset subsumption, and light normalisation), and use more expensive ones only on the small active set (e.g. full subsumption and demodulation). In Listing 1.1 we describe the 'iProver-Sup' given-clause saturation loop for superposition. A *simplification set* consists of a collection of indices, each of which supports one or more simplification rules. In our given clause loop we have four such sets: $S_{\text{passive}}$, to which we add the clauses added to the passive set, $S_{\text{active}}$, for the clauses in the active set, $S_{\text{immed}}$ for newly derived clauses (this set is cleared at the end of every given-clause iteration, and the non-redundant clauses added to the passive queue), and $S_{\text{input}}$ for preprocessing input clauses. Each set supports the following operations: add, which adds a clause to all indices in a set $S$, and simplify, which simplifies via some rules $R$ wrt. a set $S$. These are called at several points in the loop (see Listing 1.1 ), and the user can configure which indices/rules are involved in each operation. When simplifying, some rules *forward simplify* the clause wrt. the existing set, and others *backward simplify* the clauses in the set wrt. the new clause.

The simplification setup is specified by the rules to apply at each stage ($R_x$). and the indices to which to add at each stage ($S_x$). These can be specified by the user via command-line options. $S_x$ are lists of indices from {Subsumption, SubSetSubsumption, FwDemod, BwDemod, LightNorm, PropSubsSet}. $R_x$ are lists of rules from {EqResSimp, TautologyElim, EqTautologyElim, TrivRules, FwPropSubs, FwSubsumption, FwSubsumptionStrict, FwSubsumptionRes, FwDemod, FwLightNorm, FwLightNormDemodLoop, ACJoinability, BwSubsumption, BwSubsumptionRes, BwDemod}. Their usage is documented in the command-line help. The default options are presented in Table 1.

Currently iProver uses non-perfect discrimination trees for implementing backward and forward demodulation [8, 16], feature vector indices for subsumption [18], tries for subset subsumption [8], and MiniSat [6] for global subsumption.

Generally, when during immediate simplification a parent clause of a newly derived clause is made redundant, we can remove all the children of that clause from the immediate set (and thus avoid adding them to the passive queues), except for the ones which caused it to be redundant. Currently, we restrict this feature to the given clause rather than to all the parent clauses, therefore, this simplifies to checking whether the given clause is made redundant in $S_{\text{immed}}$, and if so abort the loop, add only the clauses that made it redundant to the passive, and remove the given clause.

## 2.3   AC Reasoning

If a problem contains associativity and commutativity axioms,

$$f(x, f(y, z)) = f(f(x, y), z), \qquad\qquad f(x, y) = f(y, x), \qquad (6)$$

then $f$ is said to be AC.

**Table 1.** Default simplification options

| $S_{passive}$ | SubsetSubsumption, PropSubs |
|---|---|
| $S_{active}$ | Subsumption, LightNorm, FwDemod, BwDemod |
| $S_{immed}$ | SubsetSubsumption, Subsumption, LightNorm, FwDemod, BwDemod |
| $S_{input}$ | SubsetSubsumption, Subsumption, LightNorm, FwDemod, BwDemod |
| $R_{passive}$ | TrivRules, ACJoinability, FwLightNormDemod, FwSubsumption |
| $R_{active}$ | TrivRules, FwPropSubs, FwLightNormDemod, FwSubsumption, FwSubsumptionRes, BwDemod |
| $R_{immed}$ | TrivRules, FwLightNormDemod, FwSubsumption, FwSubsumptionRes, BwDemod, BwSubsumption |
| $R_{input}$ | TrivRules, FwLightNormDemod, FwSubsumption, FwSubsumptionRes, BwDemod, BwSubsumption, BwSubsumptionRes |

AC axioms are particularly problematic in theorem proving, because they are non-orientable, which means they can generate permutations of arguments of AC functions. This leads to combinatorial explosion in the number of clauses. In particular, they will combine with each other to produce an exponential number of instances.

**Listing 1.1: iProver-Sup given-clause loop algorithm**

```
S_input = ∅
for c in input_clauses:
  simplify(c wrt S_input via R_input)
  add(c to S_input)
add_to_passive_queue(S_input)


S_active = S_passive = ∅
loop:
  S_immed = ∅
  given = pop_from_passive_queue()
  simplify(given wrt S_active ∪ S_passive via R_active)
  add_to_active_set(given)
  add(given to S_active)
  add(given to S_immed)
  for c in generating inferences between given and active:
    simplify(c wrt S_immed via R_immed)
    if given was eliminated in S_immed by clauses U:
      add(U to S_passive)
      remove(given from S_active)
      continue
    simplify(c wrt S_active ∪ S_passive via R_passive)
    add(c to S_immed)
  add_to_passive_queue(S_immed)
  add(S_immed to S_passive)
```

AC problems are ubiquitous and appear in a variety of domains [20]. Although theoretical developments behind AC reasoning have a long history, AC support in most theorem provers is limited due to implementation complexity and is mainly restricted to unit equality problems. We extended some of the techniques to be applicable to the general clausal case, see Theorem 1 below, and implemented them in iProver.

**AC Preprocessing.** During preprocessing we can transform the input problem into any equisatisfiable form. We can normalise AC terms, by e.g. collecting nested AC subterms into a flat list, sorting wrt. some total extension of the term ordering, and making them right-associative.

**Deletion of Joinable Equations.** A *rewrite system* is a set of rules $l \to r$, such that, if $l \to r$, then for any substitution $\sigma$, $l\sigma \to r\sigma$, and for any term $u$, $u[l] \to u[l \mapsto r]$. By abuse of notation we can also use unorientable equalities $l \leftrightarrow r$, in which case they stand for the set of its orientable instances, $\{l\sigma \to r\sigma \mid l\sigma \succ r\sigma\} \cup \{r\sigma \to l\sigma \mid r\sigma \succ l\sigma\}$. Two terms $s$ and $t$ are *joinable* wrt. a rewrite system $R$ (written $s \downarrow_R t$) if $s \xrightarrow{*} c \xleftarrow{*} t$, where '$\to$' denotes a rewrite step with a rule in $R$ and '$\xrightarrow{*}$' its reflexive-transitive closure. Two terms are *ground joinable* ($s \Downarrow_R t$) if all its ground instances are joinable. Two terms are *strongly ground joinable* (written $s \Downarrow_{\triangleright R} t$) if, for all $s' = s\sigma$, $t' = t\sigma$ ground instances of $s, t$ resp., with $s' \succeq t'$, either $s'$ is $t'$ or else $s' \xrightarrow{l \to r \in R} u' \downarrow_R t'$ where either $l \prec s'$ or $l$ is $s'$ but not $u' \succ t'$ (see [1,13]).

**Theorem 1.** *If $s \Downarrow_{\triangleright R} t$, then $s \approx t \lor C$ is redundant wrt. $R$. If $s \Downarrow_R t$ then $s \not\approx t \lor C$ is redundant wrt. $R \cup \{C\}$.*

Theorem 1 was shown in the context of unit equality reasoning in [1]; we extended this theorem to general clauses and provided a different proof [5].

This abstract theorem can be used for AC reasoning, provided we have a criterion to test $l \Downarrow_{\triangleright} r$. We use the following criterion [1]. Let $R_{\mathrm{AC}}$ be

$$f(x, y) \leftrightarrow f(y, x), \tag{7a}$$
$$f(f(x, y), z) \to f(x, f(y, z)), \tag{7b}$$
$$f(x, f(y, z)) \leftrightarrow f(y, f(x, z)), \tag{7c}$$

Unless $l \approx r$ is an instance of $R_{\mathrm{AC}}$ or can be simplified by an equation in $R_{\mathrm{AC}}$, $l =_{\mathrm{AC}} r$ implies $l \Downarrow_{\triangleright R_{\mathrm{AC}}} r$, which means we can use Theorem 1 to simplify/delete clauses wrt. $R_{\mathrm{AC}}$. This is a cheap test for strong ground joinability to apply in practice, since in order to check whether $s =_{\mathrm{AC}} t$ we can simply treat nested applications of $f$ as a flat-list, and then sort wrt. some total order on terms (see above discussion on AC normalisation).

**Semantic Detection of Axioms.** Some problems are AC even though the input does not contain the axioms explicitly. We say that a problem $S$ is AC if $S \models \text{AC}$. The usual syntactic detection checks if $\text{AC} \in S$. But we wish also to detect AC problems even when this is not the case.

During preprocessing, we query an SMT solver to find out whether $S \models \text{AC}$. Since SMT solvers only accept ground problems, we need to use a sound approximation of the entailment relation. We do this using an injective substitution mapping variables to fresh constants similar as it is done for global subsumption [9]. This is a sound approximation, since $\phi(\bar{c}) \models \psi(\bar{c}) \Rightarrow \forall x \phi(\bar{x}) \models \forall x \psi(\bar{x})$. In order to make SMT reasoning more efficient we can further restrict reasoning to fast rules like unit propagation or place a limit on the number of backtracks. Apart from this, we also check if the AC axioms (6) get produced at some point during saturation, among binary symbols of sort $\alpha \times \alpha \rightarrow \alpha$.

# 3    Implementation and Experimental Results

We integrated the simultaneous superposition calculus, with the iProver-Sup saturation loop, into iProver and evaluated it over 15 168 first-order problems in TPTP-v7.2.0. The superposition loop can solve 7375 (49%), the instantiation loop (on the previous version of iProver) can solve 7884 (52%), and their combination can solve 8708 (57%). We can see that the combination with superposition and the iProver-Sup simplification setup improved the performance of iProver over the whole TPTP library.

Among problems that were solved by superposition, (excluding trivial problems solved by preprocessing), immediate simplification was used in 71.7 % of problems and light normalisation was used in 64.5 % of problems. AC axioms were detected in 1903 problems.

# References

1. Avenhaus, J., Hillenbrand, T., Löchner, B.: On using ground joinable equations in equational theorem proving. J. Symb. Comput. **36**(1–2), 217–233 (2003). https://doi.org/10.1016/S0747-7171(03)00024-5
2. Bachmair, L., Ganzinger, H.: Rewrite-based equational theorem proving with selection and simplification. J. Log. Comput. **4**(3), 217–247 (1994). https://doi.org/10.1093/logcom/4.3.217
3. Benanav, D.: Simultaneous paramodulation. In: Stickel, M.E. (ed.) CADE 1990. LNCS, vol. 449, pp. 442–455. Springer, Heidelberg (1990). https://doi.org/10.1007/3-540-52885-7_106
4. Denzinger, J., Kronenburg, M., Schulz, S.: DISCOUNT - a distributed and learning equational prover. J. Autom. Reasoning **18**(2), 189–198 (1997). https://doi.org/10.1023/A:1005879229581
5. Duarte, A., Korovin, K.: AC Reasoning Revisited (2020, to appear)
6. Eén, N., Sörensson, N.: An extensible SAT-solver. In: Giunchiglia, E., Tacchella, A. (eds.) SAT 2003. LNCS, vol. 2919, pp. 502–518. Springer, Heidelberg (2004). https://doi.org/10.1007/978-3-540-24605-3_37

7. Ganzinger, H., Korovin, K.: New directions in instantiation-based theorem proving. In: Proceedings of the 18th IEEE Symposium on Logic in Computer Science (LICS 2003), pp. 55–64. IEEE Computer Society Press (2003)

8. Graf, P. (ed.): Term Indexing. LNCS, vol. 1053. Springer, Heidelberg (1995). https://doi.org/10.1007/3-540-61040-5. 284 p. ISBN 978-3-540-61040-3

9. Korovin, K.: Inst-Gen – a modular approach to instantiation-based automated reasoning. In: Voronkov, A., Weidenbach, C. (eds.) Programming Logics. LNCS, vol. 7797, pp. 239–270. Springer, Heidelberg (2013). https://doi.org/10.1007/978-3-642-37651-1_10

10. Korovin, K.: iProver – an instantiation-based theorem prover for first-order logic (system description). In: Armando, A., Baumgartner, P., Dowek, G. (eds.) IJCAR 2008. LNCS (LNAI), vol. 5195, pp. 292–298. Springer, Heidelberg (2008). https://doi.org/10.1007/978-3-540-71070-7_24

11. Kovács, L., Voronkov, A.: First-order theorem proving and VAMPIRE. In: Sharygina, N., Veith, H. (eds.) CAV 2013. LNCS, vol. 8044, pp. 1–35. Springer, Heidelberg (2013). https://doi.org/10.1007/978-3-642-39799-8_1

12. Lopez Hernandez, J.C., Korovin, K.: An abstraction-refinement framework for reasoning with large theories. In: Galmiche, D., Schulz, S., Sebastiani, R. (eds.) IJCAR 2018. LNCS (LNAI), vol. 10900, pp. 663–679. Springer, Cham (2018). https://doi.org/10.1007/978-3-319-94205-6_43

13. Martin, U., Nipkow, T.: Ordered rewriting and confluence. In: Stickel, M.E. (ed.) CADE 1990. LNCS, vol. 449, pp. 366–380. Springer, Heidelberg (1990). https://doi.org/10.1007/3-540-52885-7_100

14. McCune, W.: OTTER 3.3 Reference Manual. CoRR cs.SC/0310056 (2003). arXiv: cs.SC/0310056

15. Nieuwenhuis, R., Rubio, A.: Paramodulation-based theorem proving. In: Robinson, J.A., Voronkov, A. (eds.) Handbook of Automated Reasoning, vol. 2, pp. 371–443. Elsevier and MIT Press, Cambridge (2001). ISBN 0-444-50813-9

16. Robinson, J.A., Voronkov, A. (eds.): Handbook of Automated Reasoning, vol. 2. Elsevier and MIT Press, Cambridge (2001). ISBN 0-444-50813-9

17. Schulz, S.: E - a brainiac theorem prover. J. AI Commun. **15**(2/3), 111–126 (2002)

18. Schulz, S.: Simple and efficient clause subsumption with feature vector indexing. In: Bonacina, M.P., Stickel, M.E. (eds.) Automated Reasoning and Mathematics. LNCS (LNAI), vol. 7788, pp. 45–67. Springer, Heidelberg (2013). https://doi.org/10.1007/978-3-642-36675-8_3

19. Schulz, S.: System description: E 1.8. In: McMillan, K., Middeldorp, A., Voronkov, A. (eds.) LPAR 2013. LNCS, vol. 8312, pp. 735–743. Springer, Heidelberg (2013). https://doi.org/10.1007/978-3-642-45221-5_49

20. Sutcliffe, G.: The TPTP problem library and associated infrastructure. from CNF to TH0, TPTP v6.4.0. J. Autom. Reasoning **59**(4), 483–502 (2017). https://doi.org/10.1007/s10817-017-9407-7

21. Waldmann, U., Tourret, S., Robillard, S., Blanchette, J.: A comprehensive framework for saturation theorem proving. In: Peltier, N., Sofronie-Stokkermans, V. (eds.) IJCAR 2020. LNCS, vol. 12166, pp. xx–yy. Springer, Heidelberg (2020)

22. Weidenbach, C., Schmidt, R.A., Hillenbrand, T., Rusev, R., Topic, D.: System description: SPASS version 3.0. In: Pfenning, F. (ed.) CADE 2007. LNCS (LNAI), vol. 4603, pp. 514–520. Springer, Heidelberg (2007). https://doi.org/10.1007/978-3-540-73595-3_38

# MOIN: A Nested Sequent Theorem Prover for Intuitionistic Modal Logics (System Description)

Marianna Girlando[✉] and Lutz Straßburger

Inria, Equipe Partout, Centre Inria Saclay & École Polytechnique, LIX,
Palaiseau, France
marianna.girlando@inria.fr

**Abstract.** We present a simple Prolog prover for intuitionistic modal logics based on nested sequent proof systems. We have implemented single-conclusion systems (Gentzen-style) and multi-conclusion systems (Maehara-style) for all logics in the intuitionistic modal IS5-cube. While the single-conclusion system are better investigated and have an internal cut-elimination, the multi-conclusion systems can provide a countermodel in case the proof search fails. To our knowledge this is the first automated theorem prover for intuitionistic modal logics. For wider usability, we also implemented classical normal modal logics in the S5-cube.

**Keywords:** Intuitionistic modal logic · Nested sequents · Prolog

## 1    Introduction

In the last decade, nested sequent calculi have been successfully used to define cut-free deductive systems for various normal modal logics (classical [3], intuitionistic [17,23], constructive [1] and tense logics [10]) and non-normal modal logics [14]. For some variants there also exist focused calculi [5,6].

Even though many of these calculi yield terminating decision procedures, most of them have not (yet) been implemented in automated theorem provers. In fact, the only implementations of nested sequent systems that we are aware of are for non-normal logics [13,14] and normal conditional logics [18]. But for the more widely used classical and intuitionistic normal modal logics in the S5-cube no implementations using nested sequents exist. In fact, for intuitionistic modal logics we are not aware of any automated theorem provers.

For this reason we present here a modular Prolog implementation for nested sequent calculi for all intuitionistic normal modal logics in the IS5-cube (see Fig. 1). For the systems whose decidability is known, our prover terminates. For IS4, IK4, and ID4, for which decidability is still an open problem, we cannot ensure termination. For the sake of higher usability, we included in the implementation also all classical normal modal logics in the S5-cube, but they are not discussed here, as their implementation is rather straightforward [3].

© Springer Nature Switzerland AG 2020
N. Peltier and V. Sofronie-Stokkermans (Eds.): IJCAR 2020, LNAI 12167, pp. 398–407, 2020.
https://doi.org/10.1007/978-3-030-51054-1_25

The prover, called MOIN[1] (for *MOdal Intuitionistic Nested sequents*), is inspired by the *lean deduction* methodology introduced in [2]: the program is constituted by a set of Prolog clauses, one for each inference rule of the calculus, and the proof search is provided by the Prolog depth-first mechanism[2]. The advantages of this approach, as stated in [2], are high degree of adaptability and safety (i.e. the Prolog clauses are easy to verify), and high efficiency.

This article is organized as follows. Section 2 introduces intuitionistic modal logics; Sect. 3 and 4 present nested sequents (single- and multi-conclusion) and the proof systems we use. Section 5 treats termination and countermodel construction, and Sect. 6 describes the main features of MOIN.

## 2    Intuitionistic Modal Logics

We consider here the variant of intuitionistic modal logic that has first been studied in [7] and [20], and then been investigated in detail by Simpson in his PhD-thesis [22]. It is obtained from intuitionistic propositional logic, by adding the two modalities $\Box$ and $\Diamond$, together with the necessitation rule (if $A$ is provable, then so is $\Box A$), and the following five axioms:

$$k_1 : \Box(A \to B) \to (\Box A \to \Box B) \qquad k_3 : \Diamond(A \lor B) \to (\Diamond A \lor \Diamond B) \qquad k_5 : \Diamond\bot \to \bot$$
$$k_2 : \Box(A \to B) \to (\Diamond A \to \Diamond B) \qquad k_4 : (\Diamond A \to \Box B) \to \Box(A \to B)$$

In classical modal logic, only $k_1$ is added, as the others follow via the De Morgan duality between $\Box$ and $\Diamond$, which is not present in the intuitionistic case.

But as in the classical case, we can extend the logic by additional axioms. We restrict ourselves here to the five axioms d, t, b, 4 and 5, which in the classical as well as in the intuitionistic case generate 15 different logics, which can be arranged in the so-called S5-cube (classical) or IS5-cube (intuitionistic). Figure 1 shows the intuitionistic variant of that cube, together with the five aforementioned axioms. The reason for the restriction to these logics in this presentation is that the nested sequent proof systems for them are particularly well-studied [3,5,6,8,17,23]. The Prolog code is written in a way that it can easily be extended to other logics. The semantics of the logics in the IS5-cube is defined in terms of *bi-relational models*, introduced in [7,20,22]. No complexity results are known for these logics; however, their complexity has to be at least PSPACE-hard, as intuitionistic propositional logic is PSPACE-complete.

## 3    Nested Sequents

Nested sequents have been introduced independently in [3,11], and [21]. Whereas an ordinary sequent is a list (or multiset or set) of formulas, a nested sequent is

---

[1] Available at www.lix.polytechnique.fr/Labo/Lutz.Strassburger/Software/Moin.
[2] An introduction on Prolog theorem provers design is Jens Otten's tutorial at TABLEAUX 2019 [19].

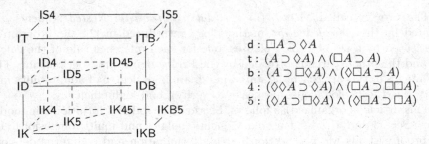

**Fig. 1.** The intuitionistic modal IS5-cube

a tree of lists (or multisets or sets) of formulas. In the classical setting a nested sequent $\Gamma$ is written as

$$\Gamma ::= A_1, \dots, A_n, [\Gamma_1], \dots, [\Gamma_m],$$

where $A_1, \dots, A_n$ are formulas and $\Gamma_1, \dots, \Gamma_m$ are nested sequents. This corresponds to a one-sided setting, and the formula interpretation $fm(\Gamma)$ of the sequent above is $fm(\Gamma) = A_1 \vee \cdots \vee A_n \vee \Box fm(\Gamma_1) \vee \cdots \vee \Box fm(\Gamma_m)$.

This setting is used by Brünnler in [3], to give cut-free systems for all logics in the classical S5-cube, including a decision procedure. He has also shown how a finite countermodel can be extracted from a failed proof search. We implemented his method in MOIN, but we will not go into details here because there are already many different provers for classical modal logic in the literature[3], and the details of our implementation for classical modal logics follow straightforwardly from our implementation for intuitionistic modal logics, that we discuss below.

The first observation is that for intuitionistic logic we need two-sided sequents, which are formally generated by

$$\Gamma ::= A_1^\bullet, \dots, A_n^\bullet, B_1^\circ, \dots, B_k^\circ, [\Gamma_1], \dots, [\Gamma_m],$$

where $A_1, \dots, A_n$ are the formulas that would occur on the left of the turnstile if there was a turnstile, $B_1, \dots, B_k$ are the formulas that would occur on the right of the turnstile if there was a turnstile, and $\Gamma_1, \dots, \Gamma_m$ are nested sequents. We use the •- and ∘-superscripts as polarity indication for formulas.

In this (multi-conclusion) intuitionistic two-sided setting, it is not possible to give a formula interpretation as in the one-side classical case. However, this is possible in the single-conclusion setting, where there is only one formula $B^\circ$ in the whole sequent. Single-conclusion sequents are formally generated by

$$\Gamma ::= \Lambda, B^\circ \mid \Lambda, [\Gamma] \qquad \Lambda ::= \emptyset \mid A^\bullet, \Lambda \mid [\Lambda_1], \Lambda_2$$

where $\Gamma$ stands for a (non-empty) sequent that contains exactly one formula with ∘-polarity, and $\Lambda$ for a (possibly empty) sequent in which all formulas have

---

[3] For an extensive list of provers for classical modal logic, see http://www.cs.man.ac.uk/~schmidt/tools/, maintained by Renate Schmidt.

•-polarity. Then the formula interpretation is:

$$fm(\emptyset) = \bot \qquad fm(A^\bullet, \Lambda) = A \wedge fm(\Lambda) \qquad fm([\Lambda_1], \Lambda_2) = \Diamond fm(\Lambda_1) \wedge fm(\Lambda_2)$$
$$fm(\Lambda, B^\circ) = fm(\Lambda) \supset B \qquad fm(\Lambda, [\Gamma]) = fm(\Lambda) \supset \Box fm(\Gamma)$$

In order to define inference rules on nested sequents, we need the notion of context, denoted as $\Gamma\{\cdot\}$, which is a nested sequent that contains exactly one occurrence of the hole $\{\cdot\}$ in the place of a formula (of either polarity). Then we write $\Gamma\{A\}$ (respectively $\Gamma\{\Delta\}$, respectively $\Gamma\{\emptyset\}$) for the sequent obtained from the context $\Gamma\{\cdot\}$ by replacing the hole $\{\cdot\}$ by the formula $A$ (respectively the sequent $\Delta$, respectively deleting the hole). Finally, we define $\Gamma^*\{\cdot\}$ to be the context that is obtained from $\Gamma\{\cdot\}$ by removing all ∘-formulas.

**Example 3.1** The nested sequent $\Gamma\{[C^\bullet]\} = A^\bullet, B^\bullet, [C^\bullet], [D^\bullet, E^\bullet, [F^\bullet, G^\circ]]$ is the result of filling context $\Gamma\{\cdot\} = A^\bullet, B^\bullet, \{\cdot\}, [D^\bullet, E^\bullet, [F^\bullet, G^\circ]]$ with $[C^\bullet]$. As $\Gamma\{[C^\bullet]\}$ is a single-conclusion sequent, we can give its formula interpretation: $(A \wedge B \wedge \Diamond C) \supset \Box((D \wedge E) \supset \Box(F \supset G))$. Then, for $\Gamma^*\{\cdot\} = A^\bullet, B^\bullet, \{\cdot\}, [D^\bullet, E^\bullet, [F^\bullet]]$, we have $\Gamma^*\{A^\circ\} = A^\bullet, B^\bullet, A^\circ, [D^\bullet, E^\bullet, [F^\bullet]]$, whose formula interpretation is $(A \wedge B \wedge \Diamond(D \wedge E \wedge \Diamond F) \supset A)$. Finally, $\Gamma\{\emptyset\} = A^\bullet, B^\bullet, [D^\bullet, E^\bullet, [F^\bullet, G^\circ]]$.

## 4 The Proof Systems

There are two kinds of nested sequent systems for intuitionistic logic: single-conclusion [6,17,23] in the style of Gentzen [9] and multi-conclusion [12], in the style of Maehara [15]. In MOIN, we implemented both. More precisely, we implemented the single-conclusion systems of [23] and (a minor variation of) the multi-conclusion systems presented in [12].

Figure 2 shows the basic single-conclusion system NIKs for the modal logic IK as presented in [23]. Formulas marked in gray are formally not part of the rules, but are part of the implementation in order to ease termination checks, and to reduce the search space. Gray formulas can never be principal in a rule application; they are kept in the premises for book-keeping. The use of $\Gamma^*$ in the $\supset^\bullet_s$-rule is due to the fact that every sequent occurring in a proof has to be a single-conclusion sequent.

Then, Fig. 3 shows the extension rules corresponding to the axioms in Fig. 1. They are taken from [23], and we refer the reader to [23] or [17] for a discussion on which set of rules gives which logic in the cube in Fig. 1. For a set $X \subseteq \{d, t, b, 4, 5\}$, we write NIKs + X for the system obtained from NIKs by adding the corresponding rules in Fig. 3. If $d \in X$ we need in some cases also the $d^{\|}_s$-rule shown in Fig. 4 (see [16] for more details).[4]

Finally, Fig. 4 shows the rules needed for the multi-conclusion system. For each $X \subseteq \{d, t, b, 4, 5\}$, we write NIKm + X for the system obtained from NIKs + X by removing the rules with an $s$ in the subscript and replacing them with the

---

[4] If $d^{\|}_s$ is in the system, the rules $d^\circ_s$ and $d^\bullet$ can be omitted, as they become admissible.

corresponding rules in Fig. 4 with an $m$ in the subscript. The $\mathsf{d}_s^{\parallel}$-rule is never needed in the multi-conclusion system. For more details see [12].

## 5    Termination and Countermodel Construction

All the classical modal logics in the S5-cube are decidable, and a simple loop check in the rules that create a new nesting is enough to ensure termination (see [3] for details). It is also explained in [3] how to obtain a finite countermodel from a failed proof search, and we implemented that method in MOIN.

In the intuitionistic case, not all logics in the IS5-cube are known to be decidable. For the logics ID4, IK4 and IS4, this is still an open problem. But, as already observed in [22], all other logics are decidable. Termination for logics that do not need the 4- and 5-axioms can be proved similarly as in the case of propositional intuitionistic logic: the only loop-check to be performed concerns application of the $\supset_s^\bullet$-rule, where it is needed to check whether the current sequent already occurs in the derivation branch.

For the logics containing the 5-axiom, it is enough to restrict the depth of a sequent to 1: when the $\Diamond^\bullet$-, $\Box^\circ$-, or one of the d-rules creates a new nesting at a depth $> 1$, the nesting is introduced at the root level of the sequent. Completeness of the resulting proof system has been proved in [8]. Thus, the restriction on the $\supset_s^\bullet$ rule mentioned above is the only loop-check needed for these systems.

For the remaining logics (ID4, IK4, IS4), decidability is not known, and we implement only two naive loop-checks: one for $\supset_s^\bullet$ mentioned above, and one that is similar to the loop-check for classical modal logics. Refer to [22] for an explanation on why this strategy does not suffice to ensure termination.

When using the single-conclusion systems, termination and completeness can easily be shown via cut-elimination [23]. However, it is not clear how to obtain a finite countermodel from a failed proof search. The reason is that there is not such a close correspondence between models and sequents as for classical modal logic. For this reason we also implemented the multi-conclusion systems of [12]. They are related to the intuitionistic (bi-relational) Kripke models [7, 20], as classical nested sequent systems to classical Kripke models. The tree-structure of the sequent tree corresponds to the $R$-relation (the accessibility relation for the modalities), and the tree-structure of the proof tree corresponds to the intuitionistic $\leq$-relation (often interpreted as future-relation). We follow the construction of [12] for obtaining a countermodel from a failed proof search, with certain simplifications, as in our case the model is always finite.

Termination for NIKm + X is ensured by the same arguments as for NIKs + X, with the difference that the loop-check (which is performed for $\supset_s^\bullet$ in NIKs) is performed for $\supset_m^\circ$ and $\Box_m^\circ$ in NIKm + X.

## 6    The Prolog Implementation

MOIN implements the nested sequents for classical modal logics from [3], and the single- and multi-conclusion calculi for intuitionistic modal logics shown in

$$\perp^{\bullet} \frac{}{\Gamma\{\perp^{\bullet}\}} \qquad\qquad \text{id} \frac{}{\Gamma\{a^{\bullet}, a^{\circ}\}}$$

$$\wedge^{\bullet} \frac{\Gamma\{A \wedge B^{\bullet}, A^{\bullet}, B^{\bullet}\}}{\Gamma\{A \wedge B^{\bullet}\}} \qquad \wedge^{\circ} \frac{\Gamma\{A \wedge B^{\circ}, A^{\circ}\} \quad \Gamma\{A \wedge B^{\circ}, B^{\circ}\}}{\Gamma\{A \wedge B^{\circ}\}}$$

$$\vee^{\bullet} \frac{\Gamma\{A \vee B^{\bullet}, A^{\bullet}\} \quad \Gamma\{A \vee B^{\bullet}, B^{\bullet}\}}{\Gamma\{A \vee B^{\bullet}\}} \qquad \vee_{s1}^{\circ} \frac{\Gamma\{A^{\circ}\}}{\Gamma\{A \vee B^{\circ}\}} \quad \vee_{s2}^{\circ} \frac{\Gamma\{B^{\circ}\}}{\Gamma\{A \vee B^{\circ}\}}$$

$$\supset_s^{\bullet} \frac{\Gamma^*\{A \supset B^{\bullet}, A^{\circ}\} \quad \Gamma\{A \supset B^{\bullet}, B^{\bullet}\}}{\Gamma\{A \supset B^{\bullet}\}} \qquad \supset_s^{\circ} \frac{\Gamma\{A^{\bullet}, B^{\circ}\}}{\Gamma\{A \supset B^{\circ}\}}$$

$$\Box^{\bullet} \frac{\Gamma\{\Box A^{\bullet}, [A^{\bullet}, \Delta]\}}{\Gamma\{\Box A^{\bullet}, [\Delta]\}} \quad \diamond^{\bullet} \frac{\Gamma\{\diamond A^{\bullet}, [A^{\bullet}]\}}{\Gamma\{\diamond A^{\bullet}\}} \quad \Box_s^{\circ} \frac{\Gamma\{[A^{\circ}]\}}{\Gamma\{\Box A^{\circ}\}} \quad \diamond^{\circ} \frac{\Gamma\{\diamond A^{\circ}, [A^{\circ}, \Delta]\}}{\Gamma\{\diamond A^{\circ}, [\Delta]\}}$$

**Fig. 2.** System NIKs

$$d_s^{\circ} \frac{\Gamma\{\diamond A^{\circ}, [A^{\circ}]\}}{\Gamma\{\diamond A^{\circ}\}} \quad t_s^{\circ} \frac{\Gamma\{\diamond A^{\circ}, A^{\circ}\}}{\Gamma\{\diamond A^{\circ}\}} \quad b_s^{\circ} \frac{\Gamma\{[\Delta, \diamond A^{\circ}], A^{\circ}\}}{\Gamma\{[\Delta, \diamond A^{\circ}]\}} \quad 4_s^{\circ} \frac{\Gamma\{\diamond A^{\circ}, [\diamond A^{\circ}, \Delta]\}}{\Gamma\{\diamond A^{\circ}, [\Delta]\}}$$

$$d^{\bullet} \frac{\Gamma\{\Box A^{\bullet}, [A^{\bullet}]\}}{\Gamma\{\Box A^{\bullet}\}} \quad t^{\bullet} \frac{\Gamma\{\Box A^{\bullet}, A^{\bullet}\}}{\Gamma\{\Box A^{\bullet}\}} \quad b^{\bullet} \frac{\Gamma\{[\Delta, \Box A^{\bullet}], A^{\bullet}\}}{\Gamma\{[\Delta, \Box A^{\bullet}]\}} \quad 4^{\bullet} \frac{\Gamma\{\Box A^{\bullet}, [\Box A^{\bullet}, \Delta]\}}{\Gamma\{\Box A^{\bullet}, [\Delta]\}}$$

$$5_{1s}^{\circ} \frac{\Gamma\{[\Delta, \diamond A^{\circ}], \diamond A^{\circ}\}}{\Gamma\{[\Delta, \diamond A^{\circ}]\}} \quad 5_{2s}^{\circ} \frac{\Gamma\{[\Delta, \diamond A^{\circ}], [\diamond A^{\circ}, \Sigma]\}}{\Gamma\{[\Delta, \diamond A^{\circ}], [\Sigma]\}} \quad 5_{3s}^{\circ} \frac{\Gamma\{[\Delta, \diamond A^{\circ}, [\diamond A^{\circ}, \Sigma]]\}}{\Gamma\{[\Delta, \diamond A^{\circ}, [\Sigma]]\}}$$

$$5_1^{\bullet} \frac{\Gamma\{[\Delta, \Box A^{\bullet}], \Box A^{\bullet}\}}{\Gamma\{[\Delta, \Box A^{\bullet}]\}} \quad 5_2^{\bullet} \frac{\Gamma\{[\Delta, \Box A^{\bullet}], [\Box A^{\bullet}, \Sigma]\}}{\Gamma\{[\Delta, \Box A^{\bullet}], [\Sigma]\}} \quad 5_3^{\bullet} \frac{\Gamma\{[\Delta, \Box A^{\bullet}, [\Box A^{\bullet}, \Sigma]]\}}{\Gamma\{[\Delta, \Box A^{\bullet}, [\Sigma]]\}}$$

**Fig. 3.** Intuitionistic $\diamond^{\circ}$- and $\Box^{\bullet}$-rules for the axioms d, t, b, 4, and 5.

$$d_s^{[]} \frac{\Gamma\{[\,]\}}{\Gamma\{\emptyset\}} \qquad\qquad \supset_m^{\bullet} \frac{\Gamma\{A \supset B^{\bullet}, A^{\circ}\} \quad \Gamma\{A \supset B^{\bullet}, B^{\bullet}\}}{\Gamma\{A \supset B^{\bullet}\}}$$

$$\supset_m^{\circ} \frac{\Gamma^*\{A^{\bullet}, B^{\circ}\}}{\Gamma\{A \supset B^{\circ}\}} \quad \Box_m^{\circ} \frac{\Gamma^*\{[A^{\circ}]\}}{\Gamma\{\Box A^{\circ}\}} \quad \vee_m^{\circ} \frac{\Gamma\{A \vee B^{\circ}, A^{\circ}, B^{\circ}\}}{\Gamma\{A \vee B^{\circ}\}}$$

$$d_m^{\circ} \frac{\Gamma\{\diamond A^{\circ}, [A^{\circ}]\}}{\Gamma\{\diamond A^{\circ}\}} \quad t_m^{\circ} \frac{\Gamma\{\diamond A^{\circ}, A^{\circ}\}}{\Gamma\{\diamond A^{\circ}\}} \quad b_m^{\circ} \frac{\Gamma\{[\Delta, \diamond A^{\circ}], A^{\circ}\}}{\Gamma\{[\Delta, \diamond A^{\circ}]\}} \quad 4_m^{\circ} \frac{\Gamma\{\diamond A^{\circ}, [\diamond A^{\circ}, \Delta]\}}{\Gamma\{\diamond A^{\circ}, [\Delta]\}}$$

$$5_{1m}^{\circ} \frac{\Gamma\{[\Delta, \diamond A^{\circ}], \diamond A^{\circ}\}}{\Gamma\{[\Delta, \diamond A^{\circ}]\}} \quad 5_{2m}^{\circ} \frac{\Gamma\{[\Delta, \diamond A^{\circ}], [\diamond A^{\circ}, \Sigma]\}}{\Gamma\{[\Delta, \diamond A^{\circ}], [\Sigma]\}} \quad 5_{3m}^{\circ} \frac{\Gamma\{[\Delta, \diamond A^{\circ}, [\diamond A^{\circ}, \Sigma]]\}}{\Gamma\{[\Delta, \diamond A^{\circ}, [\Sigma]]\}}$$

**Fig. 4.** The structural $d_s^{[]}$-rule, and the rules for the multi-conclusion system NIKm + X

Figs. 2, 3, 4. The prover is composed of a set clauses, each implementing a rule of the sequent calculus. The structure of MOIN reflects the modularity of the calculi: implementations of stronger systems are obtained by adding the clauses corresponding to the rules to the set of clauses for weaker systems.

The biggest difficulty in implementing nested sequents lies in the choice of the data structure. For ordinary sequents, it is straightforward to use the Prolog-lists. However, nested sequents are not lists of formulas, but trees of lists of formulas, and Prolog does not come with an efficient representation of trees, which would allow formulas occurring in lists inside a node of the tree to be easily accessed and replaced. There exist several approaches to overcome this difficulty. The implementation of conditional modal logics in [18] and of non-normal modal logics in [14] implement nested sequents as nested lists, while the implementation of non-normal modal logics in [13] uses a tree structure, in which nodes are annotated with sequents. We follow a different approach, representing nested sequents as Prolog lists with *annotations*. This can be compared to labelled calculi [22]: an annotation is an index labelling formulas of a nested sequent, with formulas occurring at the same node sharing the same annotation. This data structure allows for an easier countermodel extraction from a failed branch.

Propositional variables are represented in MOIN as Prolog atoms a,b,...; ⊥ and ⊤ are Prolog `false` and `true`, and the connectives ¬, ∧, ∨, ⊃, □ and ◊ are respectively represented by ~, v ^, ->, ! and ?. Nested sequents for classical modal logic are represented by means of two Prolog lists Rel,Seq. Seq is a list of triples (X,F,Sign), where F is a formula in MOIN syntax, X is the annotation of F, i.e., an integer keeping track of the component of the nested sequent at which the formula occurs, and Sign is either + or -. Rules can only be applied to formulas with a positive sign, while formulas with a negative sign are used for book-keeping (they are the gray formulas from Figs. 2, 3 and 4). Rel is a list of pairs (X,Y), representing the parent-child relation between nodes of a nested sequent. Single-conclusion nested sequents are represented as Rel,Lambda,Out, where Lambda is a list of elements (X,F,Sign) storing the •-formulas, and Out is a pair (X,F) storing the ∘-formula. Finally, multi-conclusion nested sequents are represented by means of three lists Rel,Gamma,Delta, where Gamma and Delta are lists of elements (X,F,Sign), respectively representing •- and ∘-formulas.

Proof search is invoked by the predicate derive(PS,Axioms,F), where F is the formula to be checked, PS selects the proof system (k for classical, i for intuitionistic single-conclusion, m for intuitionistic multi-conclusion nested sequents), and Axioms is a (possibly empty) list specifying the additional axioms to be used (d, t, b, 4 and 5). For instance, derive(i,[b,5], (?a -> !?a) ^ (?!a -> !a)) triggers the derivation of axiom 5 in NIKs + {b,5}. The predicate derive queries the corresponding predicate prove_k\4, prove_i\3 or prove_m\4. These predicates are recursively invoked and generate the proof-search tree (and the countermodel). To ensure termination, application of some rules needs to be restricted (see Sect. 5). The loop-checks are implemented by auxiliary predicates. The application of prove to a branch stops when an axiom clause is reached (success), or when no clause succeeds, producing a

$$A = \Box_1,\ldots,\Box_8(a \supset b) \supset (\Box_1,\ldots,\Box_8\, a \supset \Box_1,\ldots,\Box_8\, b)$$
$$B = (\Diamond_1,\ldots,\Diamond_5 a \supset \Box_1,\ldots,\Box_5 b) \supset \Box_1,\ldots,\Box_5(a \supset b)$$
$$C = (\Box a \wedge \neg\Box\Box a) \supset \Diamond(\Box\Box a \wedge \neg\Box\Box\Box a)$$
$$D = \Box(\Box(a \supset \Box a) \supset a) \supset (\Diamond\Box a \supset a)$$
$$E = (\Diamond(a \vee b) \wedge \Box(\Diamond(a \vee b))) \supset (\Diamond\Diamond\Diamond\Diamond a \vee \Diamond\Diamond\Diamond\Diamond\Diamond b)$$

**Fig. 5.** Formulas $A$ and $B$ are generated by adding modal operators to $k_1$ and $k_4$, and are both derivable in all systems. Formulas $C$ and $D$ are from [4]; $C$ is valid in all systems with 4 and $D$ in all systems with b. Formula $E$ is not derivable in any system.

| | IK | ID | IT | IKB | IK5 | IK45 | IKB5 | IDB | ID5 | ID45 | ITB | IS5 |
|---|---|---|---|---|---|---|---|---|---|---|---|---|
| $A$ | $0,1\cdot0,1$ | $0,1\cdot0,7$ | $0,2\cdot0,2$ | $0,1\cdot0,1$ | $19,6\cdot0,2$ | $113,9\cdot0,3$ | $16,1\cdot0,3$ | $0,2\cdot38,7$ | $*\cdot*$ | $*\cdot*$ | $0,2\cdot0,2$ | $32,8\cdot0,5$ |
| $B$ | $0,1\cdot0,1$ | $0,1\cdot0,2$ | $0,1\cdot0,1$ | $0,1\cdot0,1$ | $0,1\cdot0,1$ | $0,1\cdot0,2$ | $0,1\cdot0,2$ | $0,1\cdot0,5$ | $0,2\cdot40,2$ | $0,2\cdot46,2$ | $0,1\cdot0,1$ | $0,1\cdot0,2$ |
| $C$ | $0,1\cdot0,1$ | $0,3\cdot0,1$ | $0,1\cdot0,1$ | $0,6\cdot0,1$ | $0,2\cdot0,1$ | $0,1\cdot0,1$ | $0,2\cdot0,1$ | $3,0\cdot0,1$ | $8,1\cdot0,1$ | $0,1\cdot0,1$ | $1,2\cdot0,1$ | $0,1\cdot0,1$ |
| $D$ | $0,1\cdot0,2$ | $0,1\cdot0,1$ | $0,1\cdot0,1$ | $0,1\cdot0,1$ | $0,1\cdot0,1$ | $0,1\cdot0,1$ | $0,1\cdot0,1$ | $0,1\cdot0,1$ | $0,1\cdot0,1$ | $0,1\cdot0,1$ | $0,1\cdot0,1$ | $0,1\cdot0,1$ |
| $E$ | $0,1\cdot0,1$ | $0,1\cdot0,1$ | $0,1\cdot0,1$ | $0,1\cdot0,1$ | $0,1\cdot0,1$ | $0,1\cdot0,1$ | $0,1\cdot0,1$ | $0,1\cdot0,1$ | $0,1\cdot0,3$ | $0,1\cdot0,8$ | $0,1\cdot0,1$ | $0,1\cdot0,1$ |

**Fig. 6.** Results in seconds of tests in implementations for NIKs + X (red/left entry) and NIKm + X (blue/right entry). Symbol $*$ means execution time $> 150$ s. (Color figure online)

failed branch. In case of success of the proof-search, MOIN produces a LaTeX file containing the derivation. Otherwise, for classical and intuitionistic multi-conclusion sequents, MOIN prints out a countermodel in a LaTeX file. While in the classical system the information contained in the leaf of a failed branch is enough to extract a countermodel, this is not the case for NIKm + X, where all the failed branches need to be considered.

## 7  Performances

Due to the absence of benchmarks for intuitionistic modal logics, we have measured the performance MOIN with some adhoc (valid and non valid) formulas shown in Fig. 5. The construction of a set of benchmark formulas has to be postponed to future work, as the 15 different logics in the IS5-cube make this not an easy task.

We used SWI Prolog 8.0.3 on a Dell XPS 13 9370 laptop running with a 1.80 GHz Quad Core, Intel Core i7-8550U, 8 GB RAM, under Linux Mint 19.1. The numbers reported in the Fig. 6 are the results given in seconds of the user `time` command from bash script, approximated at 0,05 s. Numbers in red (left side) are the results of the single-conclusion calculus, numbers in blue (right side) of the multi-conclusion one. We removed the LaTeX-output part from MOIN before proceeding to the tests.

The results vary depending on the formula under scope. In some cases (formula $A$ tested in IK45 and IS5; formula $C$ in IDB, ID5) the implementation of

NIKm + X seems to have better performances than the one of NIKs + X, since the multi-conclusion prover has a smaller number of backtrack points (only the rules $\supset_m^\circ$ and $\Box_m^\circ$ are non-invertible). In some other cases (formula $A$ tested in IDB; formula $B$ tested in ID5 and ID45), the NIKm + X prover is slower than the NIKs + X one: this is due to the fact that in the multi-conclusion implementation the order in which the rules are applied is different from the single-conclusion one, and a larger number of sequents can be introduced in the proof search.

## 8    Conclusion and Future Work

The main purpose of MOIN is to be a tool for experimentation. The proof theory of intuitionistic modal logics has seen many advances in the last decade, and we felt the need for a tool that makes these advances accessible to a wider audience. We also hope to get through the use of MOIN some new insights in the decision problem for ID4, IK4 and IS4.

Further future work is to find more efficient implementations of the termination checks and to implement the focused systems of [5,6].

## References

1. Arisaka, R., Das, A., Straßburger, L.: On nested sequents for constructive modal logics. Log. Methods Comput. Sci. **11**(3) (2015). https://doi.org/10.2168/LMCS-11(3:7)2015
2. Beckert, B., Posegga, J.: leanTAP: lean tableau-based deduction. J. Autom. Reasoning **15**(3), 339–358 (1995)
3. Brünnler, K.: Deep sequent systems for modal logic. Arch. Math. Logic **48**(6), 551–577 (2009). https://doi.org/10.1007/s00153-009-0137-3
4. Catach, L.: TABLEAUX: a general theorem prover for modal logics. J. Autom. Reasoning **7**(4), 489–510 (1991). https://doi.org/10.1007/BF01880326
5. Chaudhuri, K., Marin, S., Straßburger, L.: Focused and synthetic nested sequents. In: Jacobs, B., Löding, C. (eds.) FoSSaCS 2016. LNCS, vol. 9634, pp. 390–407. Springer, Heidelberg (2016). https://doi.org/10.1007/978-3-662-49630-5_23
6. Chaudhuri, K., Marin, S., Straßburger, L.: Modular focused proof systems for intuitionistic modal logics. In: Kesner, D., Pientka, B. (eds.) 1st International Conference on Formal Structures for Computation and Deduction. FSCD 2016. LIPIcs, Porto, Portugal, 22–26 June 2016, vol. 52, pp. 16:1–16:18. Schloss Dagstuhl - Leibniz-Zentrum fuer Informatik (2016)
7. Fischer-Servi, G.: Axiomatizations for some intuitionistic modal logics. Rend. Sem. Mat. Univers. Politecn. Torino **42**(3), 179–194 (1984)
8. Galmiche, D., Salhi, Y.: Label-free proof systems for intuitionistic modal logic IS5. In: Clarke, E.M., Voronkov, A. (eds.) LPAR 2010. LNCS (LNAI), vol. 6355, pp. 255–271. Springer, Heidelberg (2010). https://doi.org/10.1007/978-3-642-17511-4_15
9. Gentzen, G.: Untersuchungen über das logische Schließen. I. Math. Z. **39**, 176–210 (1935). https://doi.org/10.1007/BF01201353
10. Goré, R., Postniece, L., Tiu, A.: Cut-elimination and proof search for bi-intuitionistic tense logic. In: 2010 Advances in Modal Logic, pp. 156–177 (2010)

11. Kashima, R.: Cut-free sequent calculi for some tense logics. Stud. Logica **53**(1), 119–136 (1994)
12. Kuznets, R., Straßburger, L.: Maehara-style modal nested calculi. Arch. Math. Logic **58**(3–4), 359–385 (2019). https://doi.org/10.1007/s00153-018-0636-1
13. Lellmann, B.: Combining monotone and normal modal logic in nested sequents – with countermodels. In: Cerrito, S., Popescu, A. (eds.) TABLEAUX 2019. LNCS (LNAI), vol. 11714, pp. 203–220. Springer, Cham (2019). https://doi.org/10.1007/978-3-030-29026-9_12
14. Lellmann, B., Pimentel, E.: Modularisation of sequent calculi for normal and non-normal modalities. ACM Trans. Comput. Logic (TOCL) **20**(2), 7 (2019)
15. Maehara, S.: Eine darstellung der intuitionistischen logik in der klassischen. Nagoya Math. J. **7**, 45–64 (1954)
16. Marin, S.: Modal proof theory through a focused telescope. Ph.D. thesis, Université Paris Saclay (2018)
17. Marin, S., Straßburger, L.: Label-free modular systems for classical and intuitionistic modal logics. In: Advances in Modal Logic 10, Groningen, Netherlands, August 2014
18. Olivetti, N., Pozzato, G.L.: NESCOND: an implementation of nested sequent calculi for conditional logics. In: Demri, S., Kapur, D., Weidenbach, C. (eds.) IJCAR 2014. LNCS (LNAI), vol. 8562, pp. 511–518. Springer, Cham (2014). https://doi.org/10.1007/978-3-319-08587-6_39
19. Otten, J.: How to build an automated theorem prover. Invited tutorial at TABLEAUX 2019 in London (2019). http://www.jens-otten.de/tutorial_tableaux19/
20. Plotkin, G.D., Stirling, C.P.: A framework for intuitionistic modal logic. In: Halpern, J.Y. (ed.) Theoretical Aspects of Reasoning About Knowledge (1986)
21. Poggiolesi, F.: The method of tree-hypersequents for modal propositional logic. In: Makinson, D., Malinowski, J., Wansing, H. (eds.) Towards Mathematical Philosophy. TL, vol. 28, pp. 31–51. Springer, Dordrecht (2009). https://doi.org/10.1007/978-1-4020-9084-4_3
22. Simpson, A.: The proof theory and semantics of intuitionistic modal logic. Ph.D. thesis, University of Edinburgh (1994)
23. Straßburger, L.: Cut elimination in nested sequents for intuitionistic modal logics. In: Pfenning, F. (ed.) FoSSaCS 2013. LNCS, vol. 7794, pp. 209–224. Springer, Heidelberg (2013). https://doi.org/10.1007/978-3-642-37075-5_14

# Make E Smart Again
# (Short Paper)

Zarathustra Amadeus Goertzel[(⊠)]

Czech Technical University in Prague, Prague, Czechia
zariuq@gmail.com

**Abstract.** In this work in progress, we demonstrate a new use-case for
the ENIGMA system. The ENIGMA system using the XGBoost implementation of gradient boosted decision trees has demonstrated high capability to learn to guide the E theorem prover's inferences in real-time.
Here, we strip E to the bare bones: we replace the KBO term ordering
with an identity relation as the minimal possible ordering, disable literal
selection, and replace evolved strategies with a simple combination of the
clause weight and FIFO (first in first out) clause evaluation functions.
We experimentally demonstrate that ENIGMA can learn to guide E as
well as the smart, evolved strategies even without these standard automated theorem prover functionalities. To this end, we experiment with
XGBoost's meta-parameters over a dozen loops.

## 1 Introduction: Making E Stupid and Then Smart Again

State-of-the-art saturation-based automated theorem provers (ATPs) for first-
order logic (FOL), such as E and Vampire [12], employ the *given clause algorithm* [13], translating the input FOL problem $T \cup \{\neg C\}$ (background theory and
negated conjecture) into a refutationally equivalent set of clauses. The search for
a contradiction is performed maintaining sets of *processed* ($P$) and *unprocessed*
($U$) clauses (the *proof state* $\Pi$). The algorithm repeatedly selects a *given clause*
$g$ from $U$, moves $g$ to $P$, and extends $U$ with all clauses inferred with $g$ and $P$.
This process continues until a contradiction is found, $U$ becomes empty, or a
resource limit is reached.

Historically, *term ordering*, together with *literal selection*, is used to guarantee
the completeness of the proof search [1] and to "tame the growth of the search
space and help steer proof search" [5]. Term ordering ensures that rewriting
happens in only one direction, toward smaller terms. Literal selection limits the
inferences done with each given clause $g$ to the selected literals, which slows
down the growth of the search space and reduces redundant inferences.

E includes a *strategy* language of *clause evaluation functions*, made up of
weight and priority functions, to heuristically guide the proof search. In this

Supported by the ERC Consolidator grant no. 649043 AI4REASON and by the
Czech project AI&Reasoning CZ.02.1.01/0.0/0.0/15_003/ 0000466 and the European
Regional Development Fund.

© Springer Nature Switzerland AG 2020
N. Peltier and V. Sofronie-Stokkermans (Eds.): IJCAR 2020, LNAI 12167, pp. 408–415, 2020.
https://doi.org/10.1007/978-3-030-51054-1_26

work, I use two algorithmically invented [6,7] strategies, E1 and E2[1], that use many sophisticated clause evaluation functions, the Knuth-Bendix ordering (KBO6), literal selection, and other E heuristics.

The ENIGMA [4,8–10] system with the XGBoost [2] implementation of gradient boosted decision trees has recently demonstrated high capability to learn to guide the E [14] theorem prover's inferences in real-time. ENIGMA uses the XGBoost model as a clause evaluation function to recommend clauses for selection based on clause and conjecture features. In particular, after several proving and learning iterations, its performance on the 57880 problems from the Mizar40 [11] benchmark improved by 70% (= 25397/14933) [10] over the strategy E1 used for the initial proving phase.

In this work, E is stripped to the bare bones by disabling term ordering and literal selection. KBO6 is replaced with an identity relation as the minimal possible ordering (called IDEN – an addition to E[2]). While this frees E to do inferences in any order, E can no longer perform rewriting inferences. The strategy E1 is replaced with the simple combination of the clause weight and FIFO (first in first out) evaluation functions. E is thus practically reduced to a basic superposition prover, without advanced heuristics, rewriting, or completeness guarantees. We call this strategy E0:

```
--definitional-cnf=24 --prefer-initial-clauses -tIDEN
--restrict-literal-comparisons -WNoSelection
-H'(5*Clauseweight(ConstPrio,1,1,1),1*FIFOWeight(ConstPrio))'
```

E0 solves only 3872 of the Mizar40 problems in 5 s compared to 14526 for E1. The first research question is the extent to which ENIGMA with this basic prover can learn ATP guidance completely on its own. The second is to what extent ENIGMA's learning can be boosted with data from strong strategies and models. That is, I explore how smart machine learning can become in this *zero-strategy* setting. The more general related question is to what extent can machine learning replace the sophisticated human-invented theorem-proving body of wisdom used in today's ATPs for restricting advanced proof calculi.

## 2   Experiments

We evaluate ENIGMA with the basic strategy, E0, in several scenarios and over two datasets of different sizes. All experiments are run with 5 s per problem[3] [4].

---

[1] Strategies E1 and E2 are displayed in the appendix.

[2] The E version used in this paper can be found at https://github.com/zariuq/eprover/tree/identity-order, and the library for running ENIGMA with E can be found at https://github.com/zariuq/enigmatic.

[3] As a rule of thumb, E solves most problems within a few seconds or *not for a very long time*.

[4] All the experiments are run on the same hardware unless otherwise specified: Intel(R) Xeon(R) Gold 6140 CPU @ 2.30GHz with 188GB RAM.

ENIGMA has so far been used in two ways: *coop* combines the learned model with some standard E strategy equally (50:50) while *solo* only uses the learned model for choosing the given clauses. The best results have been achieved by MaLARea-style [16] looping: that is, an ENIGMA model is trained and run with E (loop 0), then the resulting data are added to the initial training data and a new ENIGMA model is trained (loop 1).

In this work, ENIGMA trains with both *solo* and *coop* data. I present results from *solo* runs because they represent the most minimal setting.

## 2.1  Small Data (2000 Problems)

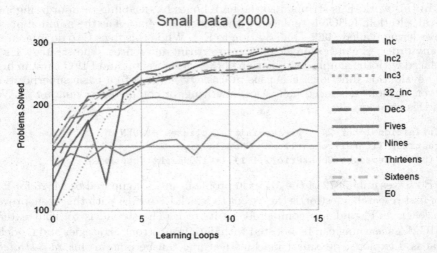

The E evaluations and XGBoost training can take a long time on the full Mizar40 dataset, so 2000 randomly sampled problems are used to test meta-parameters on. Each XGBoost model consists of $T$ decision trees of depth $D$, the most important training meta-parameters in addition to the learning rate ($\eta = 0.2$). In previous work with ENIGMA, $T$ and $D$ were fixed for all loops of learning. Here we try to vary the values of $T$ and $D$ during 16 loops. Let $S_{D,T}$ denote the experiment with specific $T$ and $D$. Of the many protocols tested, the following are included in the plot of solved problems (above): *Fives* ($S_{5,100}$), *Nines* ($S_{9,100}$), *Thirteens* ($S_{13,200}$), *Sixteens* ($S_{16,100}$).

We also experiment with adaptively setting the meta-parameters as the number of training examples increases according to the following protocols:

- *Inc* ($S_{[3,33],100}$) increases $D$ by 2 from 3 to 33 and keeps $T = 100$ fixed.
- *32_inc* ($S_{32,[50,250]}$) fixes $D = 32$ and gradually increases $T$ from 50 to 250.
- *Inc2* ($S_{[3,33],*}$) gradually decreases $T$ from 150 to 50, varying the value intuitively[5].

---

[5] Precise details of intuitively set parameters can be seen in the appendix.

- *Inc3* ($S_{[3,33],[50,250]}$) aims to be more systematic and steps $T$ from 50 to 250.
- *Dec3*($S_{[3,33],[250,50]}$) decreases $T$ from 250 to 50.

At the 16th loop *Inc*'s performance is best, solving 299 problems, doubling the performance of E0 (152). However *Inc2* and *Inc3* solve 298 problems and *32_inc* solves 291 problems. The conclusion is that simple protocols work well so long as $T$ or $D$ is incremented adaptively rather than fixed.

## 2.2  Big Data (57880 Problems)

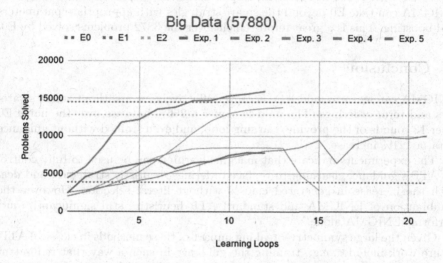

These experiments are done on the large benchmark of 57880 Mizar40 [11] problems from the MPTP dataset [15]. E1 and E2 are two strong E strategies that solve 14526 and 12788 problems.

- **Experiment 1** is done with $D = 9$ and $T = 200$ and uses our previously trained model that allowed us to solve 25562 problems when cooperating with E1 in our previous experiments [10]. This strong model, which hashes the features into 32768 ($2^{15}$) buckets [3, Sect. 3.4], is used with E0 now.
- **Experiment 2**'s parameters were intuitively toggled during the looping as in *Inc3*, and a feature size of $2^{16}$ is used. Exp. 2 uses training data from E1 and E2 for additional guidance up to the 4th loop (and then stops including them in the training data based on the assumption they may confuse learning).
- **Experiment 3** sets $T$ and $D$ according to protocol *Inc3*. Exp. 3 only learns from E run with E0 and trains on the GPU, which requires the feature size to be reduced to 256.
- **Experiment 4** mimics Exp. 3 but uses E1 and E2 data for training (up to the 4th loop).
- **Experiment 5** further tests boosting with data from an E0 ENIGMA model that proved 9759 problems and an E1 that proved 21542 problems.

Tree depth is intuitively varied among 32, 512, and 1000, the number of trees is varied among 2, 100, 200, and 32. The feature vector size starts at $2^{14}$ and is decreased to allow the data to fit on the RAM, down to 32 ($= 2^5$).

As seen in the figure, the strong model does not help much in guiding E without ordering or selection in Exp. 1. Exp. 2 learns gradually and catches up with Exp. 1, but seems to plateau around 10,000. Surprisingly the pure Exp. 3 learns fast with the small feature size, but plateaus and drops in performance (perhaps due to overfitting). Exp. 4 indicates that guidance is useful and surpasses E2 with 13805 in round 13. Exp. 5 solves 15990 problems, showing that ENIGMA can take E0 beyond the smart strategies with appropriate parameters and boosting. This is a great improvement over the 3872 problems solved by E0.

## 3   Conclusion

ENIGMA can learn to guide the E prover effectively even without smart strategies and term orderings. The models confer a 256% increase over the naive E0 after 13 rounds of the proving/learning loop, and even trained without guidance data, a 121% increase.

The experiments indicate that machine learning can be used to fully control an ATP's guidance, learning to replace orderings, heuristic strategies, and deal with the increase in generated clauses without literal selection. However the combination of ENIGMA and standard ATP heuristics still significantly outperforms ENIGMA alone.

Given the large symmetry-breaking impact of these methods in classical ATP, future work includes, e.g., training the guidance in such a way that redundant (symmetric) inferences are not done by the trained model once it has committed to a certain path. This probably means equipping the learning with more history and knowledge of the proof state in the saturation-style setting. ENIGMAWatch [4] may aid with symmetry breaking by focusing the proof search on particular proof paths. Additional work is needed to isolate the factors in Exp. 5's performance, and determine the most effective boosting methods in addition to increasing $D$ and $T$ with training loops. Ablation studies should be done to discover the impact of term ordering and literal selection individually on E and ENIGMA's performance. Perhaps term ordering alone is sufficient to train good ENIGMA models.

Running ENIGMA without term ordering and other restrictions is important because it may allow us to combine training data from different strategies, and it may allow ENIGMA to find novel proofs.

**Acknowledgments.** The research topic was proposed by Jan Jakubuv and Josef Urban, and further discussed with them, Martin Suda, and Thomas Tan. I also thank the AITP'20 anonymous referees for their comments on the first extended abstract of this work.

# A    Strategies

Strategy E1 is:

```
--definitional-cnf=24 --split-aggressive --simul-paramod
--forward-context-sr --destructive-er-aggressive --destructive-er
--prefer-initial-clauses -tKBO -winvfreqrank -c1 -Ginvfreq -F1
--delete-bad-limit=150000000 -WSelectMaxLComplexAvoidPosPred
-H'(1*ConjectureTermPrefixWeight(DeferSOS,1,3,0.1,5,0,0.1,1,4),
1*ConjectureTermPrefixWeight(DeferSOS,1,3,0.5,100,0,0.2,0.2,4),
1*Refinedweight(PreferWatchlist,4,300,4,4,0.7),
1*RelevanceLevelWeight2(PreferProcessed,0,1,2,1,1,1,200,200,2.5,9999.9,9999.9),
1*StaggeredWeight(DeferSOS,1),
1*SymbolTypeweight(DeferSOS,18,7,-2,5,9999.9,2,1.5),
2*Clauseweight(PreferWatchlist,20,9999,4),
2*ConjectureSymbolWeight(DeferSOS,9999,20,50,-1,50,3,3,0.5),
2*StaggeredWeight(DeferSOS,2))'
```

Strategy E2 is:

```
--definitional-cnf=24 --split-aggressive --split-reuse-defs
--simul-paramod --forward-context-sr --destructive-er-aggressive
--destructive-er --prefer-initial-clauses -tKBO -winvfreqrank
-c1 -Ginvfreq -F1 --delete-bad-limit=150000000
-WSelectMaxLComplexAvoidPosPred -H'(
3*ConjectureRelativeSymbolWeight(PreferUnitGroundGoals,0.1,100,100,50,100,0.3,1.5,1.5),
4*FIFOWeight(PreferNonGoals),
5*RelevanceLevelWeight2(ConstPrio,1,0,2,1,50,-2,-2,100,0.2,3,4))'
```

# B    Additional Protocol Details

In this section I include the details for *intuitively toggled* protocols.
Protocol *Inc2* is as follows:

| | 0 | 1 | 2 | 3 | 4 | 5 | 6 | 7 | 8 | 9 | 10 | 11 | 12 | 13 | 14 | 15 |
|---|---|---|---|---|---|---|---|---|---|---|---|---|---|---|---|---|
| Depth | 3 | 5 | 7 | 9 | 11 | 13 | 15 | 17 | 19 | 21 | 23 | 25 | 27 | 29 | 31 | 33 |
| Trees | 150 | 150 | 150 | 100 | 100 | 100 | 75 | 50 | 75 | 100 | 150 | 75 | 100 | 150 | 75 | 100 |

The protocol for Exp. 2 is as follows:

| | 0 | 1 | 2 | 3 | 4 | 5 | 6 | 7 | 8 | 9 | 10 | 11 | 12 | 13 | 14 | 15 | 16 | 17 | 18 | 19 | 20 | 21 |
|---|---|---|---|---|---|---|---|---|---|---|---|---|---|---|---|---|---|---|---|---|---|---|
| D | 4 | 5 | 6 | 7 | 8 | 9 | 10 | 11 | 12 | 13 | 14 | 15 | 16 | 16 | 32 | 9 | 16 | 32 | 64 | 24 | 25 | 32 |
| T | 50 | 150 | 160 | 170 | 180 | 190 | 200 | 200 | 200 | 200 | 210 | 220 | 225 | 225 | 225 | 300 | 300 | 225 | 150 | 250 | 250 | 250 |

The protocol for Exp. 5 requires some explanation. The motivation is to see how far E0 can be taken, even if the methods are too CPU-intensive for a thorough grid search.

Exp. 2 and Exp. 4 demonstrate the utility of boosting. Thus to create better boosting data I trained ENIGMA for 10 loops with strategies E1 through E12 and used this as boosting data for the first 4 of 10 loops of training. In addition to training E0, and in the spirit of ablation studies, I also trained ENIGMA models for E0 with KBO ordering (and no literal selection) and for E0 with KBO ordering and restricted literal comparisons. The motivation is that these versions may serve as a bridge between standard E and the basic E0.

Then I used these results to boost an ENIGMA model in loop 0, and trained based on this for 10 loops, proving 9759 problems.

Finally this data and the data from a loop 3 ENIGMA model trained with E1 is used to boost E0 with the following meta-parameters:

| | 0 | 1 | 2 | 3 | 4 | 5 | 6 | 7 | 8 | 9 | 10 | 11 |
|---|---|---|---|---|---|---|---|---|---|---|---|---|
| Depth | 512 | 512 | 32 | 1000 | 32 | 1000 | 32 | 1000 | 32 | 1000 | 1000 | 100 |
| Trees | 2 | 2 | 100 | 100 | 200 | 100 | 200 | 32 | 300 | 32 | 32 | 32 |
| Feature size | 16384 | 8192 | 4096 | 28 | 4096 | 28 | 4096 | 32 | 2048 | 64 | 32 | 128 |

# References

1. Bachmair, L., Ganzinger, H.: Rewrite-based equational theorem proving with selection and simplification. J. Log. Comput. **3**(4), 217–247 (1994)
2. Chen, T., Guestrin, C.: XGBoost: a scalable tree boosting system. In: Proceedings of the 22nd ACM SIGKDD International Conference on Knowledge Discovery and Data Mining (KDD 2016), pp. 785–794. ACM, New York (2016)
3. Chvalovský, K., Jakubův, J., Suda, M., Urban, J.: ENIGMA-NG: efficient neural and gradient-boosted inference guidance for E. In: Fontaine, P. (ed.) CADE 2019. LNCS (LNAI), vol. 11716, pp. 197–215. Springer, Cham (2019). https://doi.org/10.1007/978-3-030-29436-6_12
4. Goertzel, Z., Jakubův, J., Urban, J.: ENIGMAWatch: proofWatch meets ENIGMA. In: Cerrito, S., Popescu, A. (eds.) TABLEAUX 2019. LNCS (LNAI), vol. 11714, pp. 374–388. Springer, Cham (2019). https://doi.org/10.1007/978-3-030-29026-9_21
5. Hoder, K., Reger, G., Suda, M., Voronkov, A.: Selecting the selection. In: Olivetti, N., Tiwari, A. (eds.) IJCAR 2016. LNCS (LNAI), vol. 9706, pp. 313–329. Springer, Cham (2016). https://doi.org/10.1007/978-3-319-40229-1_22
6. Jakubův, J., Urban, J.: Hierarchical invention of theorem proving strategies. AI Commun. **31**(3), 237–250 (2018)
7. Jakubův, J., Urban, J.: Extending E prover with similarity based clause selection strategies. In: Kohlhase, M., Johansson, M., Miller, B., de Moura, L., Tompa, F. (eds.) CICM 2016. LNCS (LNAI), vol. 9791, pp. 151–156. Springer, Cham (2016). https://doi.org/10.1007/978-3-319-42547-4_11
8. Jakubův, J., Urban, J.: ENIGMA: efficient learning-based inference guiding machine. In: Geuvers, H., England, M., Hasan, O., Rabe, F., Teschke, O. (eds.) CICM 2017. LNCS (LNAI), vol. 10383, pp. 292–302. Springer, Cham (2017). https://doi.org/10.1007/978-3-319-62075-6_20

9. Jakubův, J., Urban, J.: Enhancing ENIGMA given clause guidance. In: Rabe, F., Farmer, W.M., Passmore, G.O., Youssef, A. (eds.) CICM 2018. LNCS (LNAI), vol. 11006, pp. 118–124. Springer, Cham (2018). https://doi.org/10.1007/978-3-319-96812-4_11

10. Jakubuv, J., Urban, J.: Hammering Mizar by learning clause guidance. In: Harrison, J., O'Leary, J., Tolmach, A. (eds.) 10th International Conference on Interactive Theorem Proving, (ITP 2019) of LIPIcs, 9–12 September 2019, Portland, OR, USA, vol. 141, pp. 34:1–34:8. Schloss Dagstuhl - Leibniz-Zentrum für Informatik (2019)

11. Kaliszyk, C., Urban, J.: MizAR 40 for Mizar 40. J. Autom. Reasoning 55(3), 245–256 (2015). https://doi.org/10.1007/s10817-015-9330-8

12. Kovács, L., Voronkov, A.: First-order theorem proving and VAMPIRE. In: Sharygina, N., Veith, H. (eds.) CAV 2013. LNCS, vol. 8044, pp. 1–35. Springer, Heidelberg (2013). https://doi.org/10.1007/978-3-642-39799-8_1

13. Overbeek, R.A.: A new class of automated theorem-proving algorithms. J. ACM 21(2), 191–200 (1974)

14. Schulz, S., Cruanes, S., Vukmirović, P.: Faster, higher, stronger: E 2.3. In: Fontaine, P. (ed.) CADE 2019. LNCS (LNAI), vol. 11716, pp. 495–507. Springer, Cham (2019). https://doi.org/10.1007/978-3-030-29436-6_29

15. Urban, J.: MPTP 0.2: Design, implementation, and initial experiments. J. Autom. Reasoning 37(1–2), 21–43 (2006)

16. Urban, J., Sutcliffe, G., Pudlák, P., Vyskočil, J.: MaLARea SG1 - machine learner for automated reasoning with semantic guidance. In: Armando, A., Baumgartner, P., Dowek, G. (eds.) IJCAR 2008. LNCS (LNAI), vol. 5195, pp. 441–456. Springer, Heidelberg (2008). https://doi.org/10.1007/978-3-540-71070-7_37

# Automatically Proving and Disproving Feasibility Conditions

Raúl Gutiérrez and Salvador Lucas

Valencian Research Institute for Artificial Intelligence,
Universitat Politècnica de València, Camino de Vera s/n, 46022 Valencia, Spain
{rgutierrez,slucas}@dsic.upv.es

**Abstract.** In the realm of term rewriting, given terms $s$ and $t$, a reachability condition $s \to^* t$ is called *feasible* if there is a substitution $\sigma$ such that $\sigma(s)$ rewrites into $\sigma(t)$ in zero or more steps; otherwise, it is called *infeasible*. Checking infeasibility of (sequences of) reachability conditions is important in the analysis of computational properties of rewrite systems like confluence or (operational) termination. In this paper, we generalize this notion of feasibility to arbitrary $n$-ary relations on terms defined by first-order theories. In this way, properties of computational systems whose operational semantics can be given as a first-order theory can be investigated. We introduce a framework for proving feasibility/infeasibility, and a new tool, in⁼Checker, which implements it.

**Keywords:** Conditional rewriting · Feasibility · Program analysis

## 1  Introduction

The *(in)feasibility* of sequences of goals $s \to^* t$ representing many step rewritings in Conditional Term Rewriting Systems (CTRSs, see [22, Section 7]) has been investigated by several authors. The word "feasibility" refers to the possibility of applying a substitution $\sigma$ as part of the desired test, i.e., checking whether $\sigma(s) \to^*_{\mathcal{R}} \sigma(t)$ holds for some substitution $\sigma$, rather than just checking $s \to^*_{\mathcal{R}} t$ (reachability test). The use of (in)feasibility tests in confluence and (operational) termination analysis of CTRSs has been investigated elsewhere (see, e.g., [13, 25] and the references therein). We generalize "feasibility of a reachability problem" by defining *feasibility conditions, sequences and goals* without any specific reference to rewriting systems or rewriting goals. Instead, we rely on first-order logic and use (two layered) sequences of *atoms* headed with a predicate $\bowtie$ as feasibility goals. The meaning of predicates $\bowtie$ is given by using first-order theories $\mathrm{Th}_{\bowtie}$ by provability of the corresponding atoms. New properties (also of CTRSs) can be investigated in this way.

Supported by EU (FEDER), and projects RTI2018-094403-B-C32, PROMETEO/2019/098, and SP20180225. Also by INCIBE program "Ayudas para la excelencia de los equipos de investigación avanzada en ciberseguridad" (Raúl Gutiérrez).

N. Peltier and V. Sofronie-Stokkermans (Eds.): IJCAR 2020, LNAI 12167, pp. 416–435, 2020.
https://doi.org/10.1007/978-3-030-51054-1_27

*Example 1.* Given a CTRS $\mathcal{R}$, a term $t$ *loops* if $t = t_1 \to_{\mathcal{R}} \cdots \to_{\mathcal{R}} t_n$ for some $n > 1$ such that $t$ is a subterm of $t_n$, written $t_n \trianglerighteq t$, cf. [2, Def. 3]. Provided that $\to$, $\to^*$ and $\trianglerighteq$ are given appropriate theories (see Example 3 below), *non-loopingness* of ground terms $t$ is the *infeasibility* of the sequence $t \to x, x \to^* y, y \trianglerighteq t$.

Now, looping CTRSs can be defined as those having looping terms. Thus, loopingness of CTRSs can be defined as the *feasibility* of $x \to y, y \to^* z, z \trianglerighteq x$.

In order to automatically analyze such (in)feasibility goals, we describe a framework similar to the Dependency Pair (DP) Framework for proving termination of TRSs [3]. After some preliminaries in Sect. 2, Sect. 3 presents the notion of feasibility goal. Section 4 describes the feasibility framework for proving and disproving feasibility goals. Section 5 describes our tool infChecker which provides a (partial) implementation of the framework introduced here. Section 6 provides an experimental evaluation and discusses some related work. Section 8 concludes.

## 2  Preliminaries

We use the standard notations in term rewriting (see, e.g., [22]). In this paper, $\mathcal{X}$ denotes a countable set of *variables* and $\mathcal{F}$ denotes a *signature*, i.e., a set of *function symbols* $\{f, g, \ldots\}$, each with a fixed *arity* given by a mapping $ar : \mathcal{F} \to \mathbb{N}$. The set of terms built from $\mathcal{F}$ and $\mathcal{X}$ is $\mathcal{T}(\mathcal{F}, \mathcal{X})$. The symbol labeling the root of $t$ is denoted as $root(t)$. The set of variables occurring in $t$ is $\mathcal{V}ar(t)$. Terms are viewed as labeled trees in the usual way. *Positions* $p, q, \ldots$ are represented by chains of positive natural numbers used to address subterms $t|_p$ of $t$. The *set of positions* of a term $t$ is $\mathcal{P}os(t)$. A substitution is a mapping from variables into terms which is homomorphically extended to a mapping from terms to terms. A conditional rule is written $\ell \to r \Leftarrow s_1 \approx t_1, \cdots, s_n \approx t_n$, where $\ell, r, s_1, t_1, \ldots, s_n, t_n \in \mathcal{T}(\mathcal{F}, \mathcal{X})$ and $\ell \notin \mathcal{X}$. As usual, $\ell$ and $r$ are called the left- and right-hand sides of the rule, and the sequence $s_1 \approx t_1, \cdots, s_n \approx t_n$ (often abbreviated to $c$) is the *conditional part* of the rule. We often write $s_i \approx t_i \in c$ to refer to the $i$-th atomic condition in $c$ or $s \to t \in c$ if the position of the atomic condition in $c$ does not matter. Rules $\ell \to r \Leftarrow c$ are classified according to the distribution of variables as follows: type 1 (or 1-rules), if $\mathcal{V}ar(r) \cup \mathcal{V}ar(c) \subseteq \mathcal{V}ar(\ell)$; type 2, if $\mathcal{V}ar(r) \subseteq \mathcal{V}ar(\ell)$; type 3, if $\mathcal{V}ar(r) \subseteq \mathcal{V}ar(\ell) \cup \mathcal{V}ar(c)$; and type 4, if no restriction is given. A CTRS $\mathcal{R}$ is a set of conditional rules; $\mathcal{R}$ is called an $n$-CTRS if it contains only $n$-rules; A 3-CTRS $\mathcal{R}$ is called *deterministic* if for each rule $\ell \to r \Leftarrow s_1 \approx t_1, \ldots, s_n \approx t_n$ in $\mathcal{R}$ and each $1 \leq i \leq n$, we have $\mathcal{V}ar(s_i) \subseteq \mathcal{V}ar(\ell) \cup \bigcup_{j=1}^{i-1} \mathcal{V}ar(t_j)$. *Oriented* CTRSs are those whose conditions $s \approx t$ are handled as *reachability* tests $\sigma(s) \to^* \sigma(t)$ for an appropriate substitution $\sigma$. For oriented CTRSs $\mathcal{R}$, an inference system $\mathcal{I}(\mathcal{R})$ is obtained from the following generic inference system $\mathfrak{I}_{\mathrm{CTRS}}$:

$$(\mathrm{Rf}) \quad \frac{}{x \to^* x} \qquad (\mathrm{C})_{f,i} \quad \frac{x_i \to y_i}{f(x_1, \ldots, x_i, \ldots, x_k) \to f(x_1, \ldots, y_i, \ldots, x_k)}$$
$$\text{for all } f \in \mathcal{F}^{(k)} \text{ and } 1 \leq i \leq k$$

$$(\mathrm{T}) \quad \frac{x \to y \quad y \to^* z}{x \to^* z} \qquad (\mathrm{Rl})_\alpha \quad \frac{s_1 \to^* t_1 \quad \cdots \quad s_n \to^* t_n}{\ell \to r}$$
$$\text{for } \alpha : \ell \to r \Leftarrow s_1 \approx t_1, \ldots, s_n \approx t_n \in \mathcal{R}$$

$$le(0, s(y)) \rightarrow true \tag{1}$$
$$le(s(x), s(y)) \rightarrow le(x, y) \tag{2}$$
$$le(x, 0) \rightarrow false \tag{3}$$
$$min(cons(x, nil)) \rightarrow x \tag{4}$$
$$min(cons(x, xs)) \rightarrow x \Leftarrow min(xs) \approx y, le(x, y) \approx true \tag{5}$$
$$min(cons(x, xs)) \rightarrow y \Leftarrow min(xs) \approx y, le(x, y) \approx false \tag{6}$$

**Fig. 1.** CTRS 551.trs in COPS database of confluence problems.

by *specializing* $(C)_{f,i}$ for each $k$-ary symbol $f$ in the signature $\mathcal{F}$ and $1 \leq i \leq k$ and $(Rl)_\alpha$ for all conditional rules $\alpha : \ell \rightarrow r \Leftarrow c$ in $\mathcal{R}$. Rules in $\mathcal{I}(\mathcal{R})$ are *schematic*: each inference rule $\frac{B_1 \cdots B_n}{A}$ can be used under any *instance* $\frac{\sigma(B_1) \cdots \sigma(B_n)}{\sigma(A)}$ of the rule by a substitution $\sigma$. We write $s \rightarrow_{\mathcal{R}} t$ (resp. $s \rightarrow_{\mathcal{R}}^* t$) iff there is a proof tree for $s \rightarrow t$ (resp. $s \rightarrow^* t$) using $\mathcal{I}(\mathcal{R})$. Operational termination of $\mathcal{R}$ is defined as the absence of infinite proof trees for goals $s \rightarrow t$ and $s \rightarrow^* t$ in $\mathcal{I}(\mathcal{R})$ [14].

A *structure* $\mathcal{A}$ for a first-order language is an interpretation of the function and predicate symbols ($f, g, \ldots$ and $P, Q, \ldots$, respectively) as mappings $f^{\mathcal{A}}, g^{\mathcal{A}}, \ldots$ and relations $P^{\mathcal{A}}, Q^{\mathcal{A}}, \ldots$ on a given set (carrier) also denoted $\mathcal{A}$. Then, the usual interpretation of first-order formulas with respect to $\mathcal{A}$ is considered. A model for a theory Th, i.e., a set of first-order sentences (formulas whose variables are all *quantified*), is just a structure $\mathcal{A}$ that makes them all true, written $\mathcal{A} \models$ Th. In the following, Th $\vdash \varphi$ means that formula $\varphi$ is a *logical consequence* of Th. We assume the use of a sound and complete proof method, in particular Gentzen's natural deduction, see [23]. In this setting, we often assume the use of the inference rules of natural deduction [23, p. 20] to deal with logical connectives and quantifiers when necessary.

## 3    Feasibility of Sequences and Goals

Consider a signature $\Sigma$ of function symbols and a set $\Pi$ of predicate symbols. As in [4], $(\Sigma, \Pi)$ is often called a *signature with predicates*. Let $\mathcal{F} \subseteq \Sigma$ be a signature and $\mathbb{P} \subseteq \Pi$ be a set of predicates (e.g., $\mathbb{P} = \{\rightarrow, \rightarrow^*, \downarrow, \leftrightarrow, \leftrightarrow^*, \trianglerighteq, \ldots\}$).[1] Let $\mathbb{T} = \{\mathsf{Th}_{\bowtie} \mid \bowtie \in \mathbb{P}\}$ be a $\mathbb{P}$-indexed set of first-order theories $\mathsf{Th}_{\bowtie}$ defining the predicates $\bowtie$ in $\mathbb{P}$, possibly involving predicate symbols which are *not* in $\mathbb{P}$.

*Example 2.* For the CTRS $\mathcal{R}$ in Fig. 1, we obtain a theory $\overline{\mathcal{R}}$ from $\mathcal{I}(\mathcal{R})$ as follows [11, Section 4.5]: the inference rules $(\rho) \frac{B_1 \cdots B_n}{A}$ in $\mathcal{I}(\mathcal{R})$ are considered as *sentences* $\overline{\rho}$ of the form $(\forall \boldsymbol{x}) B_1 \wedge \cdots \wedge B_n \Rightarrow A$, where $\boldsymbol{x}$ is the sequence of variables occurring in atoms $B_1, \ldots, B_n$ and $A$; if empty, we just write $B_1 \wedge \cdots \wedge B_n \Rightarrow A$ (see Fig. 2). For $\mathbb{P} = \{\rightarrow, \rightarrow^*\}$, we let $\mathsf{Th}_{\rightarrow} = \mathsf{Th}_{\rightarrow^*} = \overline{\mathcal{R}}$.

---

[1] For simplicity, in our exposition we restrict the attention to *binary* predicates, but the techniques and results in this paper easily generalize to $n$-ary predicates.

$$(\forall x)\ x \to^* x$$
$$(\forall x, y, z)\ x \to y \wedge y \to^* z \Rightarrow x \to^* z$$
$$(\forall x, y)\ x \to y \Rightarrow s(x) \to s(y)$$
$$(\forall x, y, z)\ x \to y \Rightarrow cons(x, z) \to cons(y, z)$$
$$(\forall x, y, z)\ x \to y \Rightarrow cons(z, x) \to cons(z, y)$$
$$(\forall x, y, z)\ x \to y \Rightarrow le(x, z) \to le(y, z)$$
$$(\forall x, y, z)\ x \to y \Rightarrow le(z, x) \to le(z, y)$$
$$(\forall x, y)\ x \to y \Rightarrow min(x) \to min(y)$$
$$(\forall y)\ le(0, s(y)) \to true$$
$$(\forall x, y)\ le(s(x), s(y)) \to le(x, y)$$
$$(\forall x)\ le(x, 0) \to false$$
$$(\forall x)\ min(cons(x, nil)) \to x$$
$$(\forall x, y, xs)\ min(xs) \to^* y \wedge le(x, y) \to^* true \Rightarrow min(cons(x, xs)) \to x$$
$$(\forall x, xs)\ min(xs) \to^* y \wedge le(x, y) \to^* false \Rightarrow min(cons(x, xs)) \to y$$

**Fig. 2.** Theory $\overline{\mathcal{R}}$ for the CTRS $\mathcal{R}$ in Example 2

Examples 6 and 7 illustrate and motivate the use of theories involving predicates not in $\mathbb{P}$.

An $(\mathcal{F}, \mathbb{P})$-*f-condition* $\gamma$ (or just *f-condition* if $\mathcal{F}$ and $\mathbb{P}$ are clear from the context) is an atom $s \bowtie t$ where $\bowtie \in \mathbb{P}$ and $s, t \in \mathcal{T}(\mathcal{F}, \mathcal{X})$. Sequences $\mathsf{F} = (\gamma_i)_{i=1}^n = (\gamma_1, \ldots, \gamma_n)$ of f-conditions are called *f-sequences*. A set $\mathcal{G} = \{\mathsf{F}_1; \ldots; \mathsf{F}_m\}$ of f-sequences is called an *f-goal*; we use ';' intead of ',' which is already considered in f-sequences. We often drop 'f-' when no confusion arises. Empty sequences and goals are written () and {}.

*Remark 1 (Notation).* In the following, we often use '$\in$' to denote membership of components in both sequences and goals.

**Definition 1 (Feasibility).** *A condition* $s \bowtie t$ *is* $(\mathbb{T}, \sigma)$-*feasible if* $\mathsf{Th}_\bowtie \vdash \sigma(s) \bowtie \sigma(t)$ *holds; otherwise, it is* $(\mathbb{T}, \sigma)$-*infeasible. We also say that* $s \bowtie t$ *is* $\mathbb{T}$-*feasible (or* $\mathsf{Th}_\bowtie$-*feasible, or just feasible if no confusion arises) if it is* $(\mathbb{T}, \sigma)$-*feasible for some substitution* $\sigma$; *otherwise, we call it infeasible.*

*A sequence* $\mathsf{F}$ *is* $\mathbb{T}$-*feasible (or just* feasible) *iff there is a substitution* $\sigma$ *such that, for all* $\gamma \in \mathsf{F}$, $\gamma$ *is* $(\mathbb{T}, \sigma)$-*feasible. Note that* () *is trivially feasible. A goal* $\mathcal{G}$ *is feasible iff it contains a feasible sequence* $\mathsf{F} \in \mathcal{G}$. *Now,* {} *is trivially infeasible.*

*Example 3.* (continuing Example 1) We can prove a ground term $t$ *non-looping* as the $\mathbb{T}$-infeasibility of $\mathcal{G} = \{(t \to y, y \to^* z, z \unrhd t)\}$, with $x$, $y$, and $z$ variables, and $\mathbb{T} = \{\mathsf{Th}_\to, \mathsf{Th}_{\to^*}, \mathsf{Th}_\unrhd\}$ such that $\mathsf{Th}_\to = \mathsf{Th}_{\to^*} = \overline{\mathcal{R}}$ and $\mathsf{Th}_\unrhd$ is given by:

$$(\forall x)\ x \unrhd x \quad (7) \qquad (\forall x_1, \ldots, x_k)\ f(x_1, \ldots, x_k) \unrhd x_i \quad (9)$$
$$(\forall x, y, z)\ x \unrhd y \wedge y \unrhd z \Rightarrow x \unrhd z \quad (8) \qquad \text{for each } f \in \mathcal{F} \text{ and } 1 \le i \le k$$

*Example 4.* A term $t$ is *root-stable* (with respect to a TRS $\mathcal{R}$) if $t$ cannot be reduced to a redex, i.e., there is no rule $\ell \to r \in \mathcal{R}$ such that $t \to^* \sigma(\ell)$ for some substitution $\sigma$. If $\mathcal{R}$ consists of rules $\ell_1 \to r_1, \ldots, \ell_p \to r_p$ (assume that different rules in $\mathcal{R}$ share no variable), we can prove a *ground* term $t$ root-stable by showing the $\{\mathsf{Th}_{\to^*}\}$-infeasibility of $\mathcal{G} = \{(t \to^* \ell_1); \cdots ; (t \to^* \ell_p)\}$ with $\mathsf{Th}_{\to^*} = \overline{\mathcal{R}}$.

More examples of theories $\mathsf{Th}_{\bowtie}$ can be found in [9, Sections 5.3 and 5.4] and [11, Sections 8.1 and 8.2].

Given theories $\mathsf{Th}, \mathsf{Th}'$ and a set of atoms $\mathbb{A}$, we write $\mathsf{Th} \equiv_{\mathbb{A}} \mathsf{Th}'$ if for all $A \in \mathbb{A}$, $\mathsf{Th} \vdash A$ if and only if $\mathsf{Th}' \vdash A$. Also, given a set of atoms $\mathbb{A}$ and a predicate symbol $\bowtie$, $\mathbb{A}_{\bowtie}$ is the subset of atoms in $\mathbb{A}$ with root $\bowtie$.

**Definition 2.** *Given a set of predicates $\mathbb{P}$ and a $\mathbb{P}$-indexed set of theories $\mathbb{T}$, we say that a theory $\mathsf{Th}$ preserves a feasibility sequence $\mathsf{F}$ (in $\mathbb{T}$) if for all predicates $\bowtie$ occurring in $\mathsf{F}$, $\mathsf{Th} \equiv_{\mathbb{A}_{\bowtie}} \mathsf{Th}_{\bowtie}$ holds. Thus, $\mathsf{Th}$ cannot prove more atoms rooted with $\bowtie$ than $\mathsf{Th}_{\bowtie}$. Similarly, $\mathsf{Th}$ preserves a goal $\mathcal{G} = \{\mathsf{F}_i\}_{i=1}^m$ if it preserves $\mathsf{F}_i$ for all $1 \le i \le m$.*

In the following, when no confusion arises, we do not explicitly mention the underlying set of theories $\mathbb{T}$. Given $\mathcal{G} = \{\mathsf{F}_i\}_{i=1}^m$ and $\mathsf{F}_i = (s_{ij} \bowtie_{ij} t_{ij})_{j=1}^{n_i}$ for $1 \le i \le m$, we let $\mathsf{Th}_{\mathsf{F}_i} = \bigcup_{j=1}^{n_i} \mathsf{Th}_{\bowtie_{ij}}$ and $\mathsf{Th}_{\mathcal{G}} = \bigcup_{i=1}^m \mathsf{Th}_{\mathsf{F}_i}$.

*Example 5.* It is not difficult to see that $\mathsf{Th}_{\mathcal{G}}$ preserves $\mathcal{G}$ in Example 3.

The following result provides a (first-order) provability perspective of feasibility.

**Theorem 1.** *1. A condition $\gamma = s \bowtie t$ is feasible iff $\mathsf{Th}_{\bowtie} \vdash (\exists \boldsymbol{x}) s \bowtie t$ holds.*
*2. If $\mathsf{F} = (s_i \bowtie_i t_i)_{i=1}^n$ is feasible, then $\mathsf{Th}_{\mathsf{F}} \vdash (\exists \boldsymbol{x}) \bigwedge_{i=1}^n s_i \bowtie_i t_i$ holds. If $\mathsf{Th}_{\mathsf{F}}$ preserves $\mathsf{F}$ and $\mathsf{Th}_{\mathsf{F}} \vdash (\exists \boldsymbol{x}) \bigwedge_{i=1}^n s_i \bowtie_i t_i$ holds, then $\mathsf{F}$ is feasible.*
*3. If $\mathcal{G} = \{\mathsf{F}_i\}_{i=1}^m$, where $\mathsf{F}_i = (s_{ij} \bowtie_{ij} t_{ij})_{j=1}^{n_i}$ for some $n_i$, is feasible, then we have that $\mathsf{Th}_{\mathcal{G}} \vdash (\exists \boldsymbol{x}) \bigvee_{i=1}^m \bigwedge_{j=1}^{n_i} s_{ij} \bowtie_{ij} t_{ij}$ holds. If $\mathsf{Th}_{\mathcal{G}} \vdash (\exists \boldsymbol{x}) \bigvee_{i=1}^m \bigwedge_{j=1}^{n_i} s_{ij} \bowtie_{ij} t_{ij}$ holds and $\mathsf{Th}_{\mathcal{G}}$ preserves $\mathcal{G}$, then $\mathcal{G}$ is feasible. Also, if there is $1 \le i \le m$ such that $\mathsf{Th}_{\mathsf{F}_i}$ preserves $\mathsf{F}_i$ and $\mathsf{Th}_{\mathsf{F}_i} \vdash (\exists \boldsymbol{x}) \bigwedge_{j=1}^{n_i} s_{ij} \bowtie_{ij} t_{ij}$ holds, then $\mathcal{G}$ is feasible.*

Sentences in Theorem 1 are *Existentially Closed Boolean Combinations of Atoms* (ECBCAs), i.e., formulas of the form $(\exists \boldsymbol{x}) \bigvee_{i=1}^m \bigwedge_{j=1}^{n_i} A_{ij}$, where $A_{ij}$ are atoms and $\boldsymbol{x}$ is the sequence of variables occurring in such atoms. We have investigated them in [9,11]. Requiring preservation is necessary for items (2) and (3) in Theorem 1.

*Example 6.* Let $\mathcal{R}_1 = \{a \to b \Leftarrow b \approx a\}$ and $\mathcal{R}_2 = \{b \to a\}$. With $\mathbb{P} = \{\to, \to^*\}$ and $\mathbb{T} = \{\mathsf{Th}_{\to}, \mathsf{Th}_{\to^*}\}$, where $\mathsf{Th}_{\to} = \overline{\mathcal{R}}_1 = \{(\forall x)x \to^* x, (\forall x,y,z)x \to y \wedge y \to^* z \Rightarrow x \to^* z, b \to^* a \Rightarrow a \to b\}$ and $\mathsf{Th}_{\to^*} = \overline{\mathcal{R}}_2 = \{(\forall x) x \to^* x, (\forall x,y,z) x \to y \wedge y \to^* z \Rightarrow x \to^* z, b \to a\}$, we have $\mathsf{Th} = \overline{\mathcal{R}}_1 \cup \overline{\mathcal{R}}_2$ and $\mathsf{Th} \vdash a \to b$. However, $a \to b$ is not $\mathbb{T}$-feasible because $\overline{\mathcal{R}}_1 \not\vdash a \to b$. Note that $\mathsf{Th}$ does not preserve $(a \to b)$.

Preservation is often achieved by distinguishing predicates describing different computations.

*Example 7.* Let $\overline{\mathcal{R}}_1'$ and $\overline{\mathcal{R}}_2'$ be theories for $\mathcal{R}_1$ and $\mathcal{R}_2$ in Example 6, where $\to_1^*$ is used instead of $\to^*$ in $\overline{\mathcal{R}}_1$ (but $\to$ remains as it is) to yield $\overline{\mathcal{R}}_1' = \{(\forall x)\, x \to_1^* x,\, (\forall x, y, z)\, x \to y \wedge y \to_1^* z \Rightarrow x \to_1^* z, b \to_1^* a \Rightarrow a \to b\}$ and $\to_2$ is used instead of $\to$ in $\overline{\mathcal{R}}_2$ (and $\to^*$ is still used) to yield $\overline{\mathcal{R}}_2' = \{(\forall x)\, x \to^* x,\, (\forall x, y, z)\, x \to_2 y \wedge y \to^* z \Rightarrow x \to^* z, b \to_2 a\}$. With $\mathsf{Th}_{\to}' = \overline{\mathcal{R}}_1'$ and $\mathsf{Th}_{\to^*}' = \overline{\mathcal{R}}_2'$ (and $\mathbb{T}' = \{\mathsf{Th}_{\to}', \mathsf{Th}_{\to^*}'\}$), we have that $\mathsf{Th}' = \overline{\mathcal{R}}_1' \cup \overline{\mathcal{R}}_2'$ preserves $\mathcal{G} = \{(a \to b)\}$ and $\mathsf{Th}' \not\vdash a \to b$. By Theorem 1 $a \to b$, is not $\mathbb{T}'$-feasible, as expected.

Theorem 1 characterizes feasibility of a goal $\mathcal{G}$ as provability of an ECBCA $\varphi_{\mathcal{G}}$. If $\varphi_{\mathcal{G}}$ is shown *unprovable*, we conclude infeasibility of $\mathcal{G}$. And, under appropriate preservation conditions, theorem proving can be used to conclude feasibility of a goal $\mathcal{G}$. However, (un)provability of atoms is often undecidable. For instance, for TRSs $\mathcal{R}$, it is well-known that given ground terms $s$ and $t$, it is in general undecidable whether $s$ rewrites into $t$; i.e., whether $\overline{\mathcal{R}} \vdash s \to^* t$ holds (as Post's correspondence problem is a particular case, see, e.g., [22, Section 4.1]). Hence, feasibility of conditions, sequences and goals remains, in general, undecidable. In order to obtain automatic proofs of feasibility, it is often useful to proceed using a 'divide-and-conquer' strategy. In the following, we exploit this idea to define a practical framework to prove/disprove feasibility of goals.

# 4  Feasibility Framework

In [3], proofs of termination of TRSs proceed by transforming the so-called *DP problems* $\tau$. A divide-and-conquer approach is applied by means of *processors* P mapping a DP problem $\tau$ into a (possibly empty) set $\mathsf{P}(\tau)$ of DP problems $\{\tau_1, \ldots, \tau_n\}$. DP problems $\tau_i$ returned by P can now be treated independently by using other processors. In this way, a *DP proof tree* is built.

In our setting, we first define notions of *f-problem* and *f-processor*, and then show how to use them to (dis)prove feasibility.

**Definition 3 (f-Problem and f-Processor).** *Given a set of predicates* $\mathbb{P}$, *a* $\mathbb{P}$-*indexed theory* $\mathbb{T}$, *and a goal* $\mathcal{G}$, *a pair* $\tau = (\mathbb{T}, \mathcal{G})$ *is called an* f-Problem. *We say that* $\tau$ *is feasible if* $\mathcal{G}$ *is* $\mathbb{T}$-*feasible; otherwise it is* infeasible.

*An* f-Processor P *is a partial function from f-Problems into sets of f-Problems. Alternatively, it can return* "yes". $\mathcal{D}om(\mathsf{P})$ *represents the domain of* P, *i.e., the set of f-Problems* $\tau$ *that* P *is defined for.*

**Definition 4 (Soundness and completeness).** *Let* P *be an f-Processor and* $\tau \in \mathcal{D}om(\mathsf{P})$. *We say that* P *is*

- sound *iff* $\tau$ *is feasible whenever either* $\mathsf{P}(\tau) =$ "yes" *or* $\exists \tau' \in \mathsf{P}(\tau)$, *such that* $\tau'$ *is feasible.*
- complete *iff* $\tau$ *is infeasible whenever* $\mathsf{P}(\tau) \neq$ "yes" *and* $\forall \tau' \in \mathsf{P}(\tau)$, $\tau'$ *is infeasible.*

Feasibility problems can be proved or disproved by using a proof tree as follows (where inner nodes include the root of the tree unless it consists of a single node).

**Definition 5 (Feasibility Proof Tree).** *Let* $\tau = (\mathbb{T}, \mathcal{G})$ *be an f-Problem. A feasibility proof tree (FP Tree)* $\mathcal{T}$ *for* $\tau$ *is a tree whose inner nodes are labeled with f-Problems and the leaves are labeled either with f-Problems, "yes" or "no". The root of* $\mathcal{T}$ *is labeled with* $\tau$ *and for every inner node* n *labeled with* $\tau'$, *there is an f-Processor* P *such that* $\tau' \in \mathcal{D}om(P)$ *and:*

1. *if* $P(\tau') = $ *"yes" then* n *has just one child, labeled with "yes".*
2. *if* $P(\tau') = \emptyset$ *then* n *has just one child, labeled with "no".*
3. *if* $P(\tau') = \{\tau_1, \dots, \tau_k\}$ *with* $k > 0$, *then* n *has* k *children labeled with the f-Problems* $\tau_1, \dots, \tau_k$.

**Theorem 2 (Feasibility Framework).** *Let* $\mathcal{T}$ *be a feasibility proof tree for* $\tau_I = (\mathbb{T}, \mathcal{G})$. *Then:*

1. *if all leaves in* $\mathcal{T}$ *are labeled with "no" and all involved f-Processors are complete for the f-Problems they are applied to, then* $\mathcal{G}$ *is* $\mathbb{T}$-*infeasible.*
2. *if* $\mathcal{T}$ *has a leaf labeled with "yes" and all f-Processors in the path from* $\tau_I$ *to the leaf are sound for the f-Problems they are applied to, then* $\mathcal{G}$ *is* $\mathbb{T}$-*feasible.*

In the following, we describe some sound and complete f-Processors. If no confusion arises, we use *processor* instead of *f-Processor*.

## 4.1   Splitting Processor

Our first processor decomposes a feasibility goal into its feasibility sequences.

**Definition 6 (Splitting Processor).** *Let* $\tau = (\mathbb{T}, \mathcal{G})$ *be an f-Problem. The processor* $P^{Spl}$ *is given by* $P^{Spl}(\tau) = \{(\mathbb{T}, \{F\}) \mid F \in \mathcal{G}\}$.

The proof of the following result is immediate by using Definitions 3 and 1.

**Theorem 3.** *Processor* $P^{Spl}$ *is sound and complete.*

*Example 8.* Consider the following TRS $\mathcal{R}$ [13, Example 9]:

$$a \to b \quad (10) \qquad\qquad f(x,x) \to c \quad (12)$$
$$b \to a \quad (11)$$

Following Example 4, we prove root-stability of $f(a,c)$ as the $\{\overline{\mathcal{R}}\}$-infeasibility of $\mathcal{G} = \{((f(a,c) \to^* a) ; (f(a,c) \to^* b); (f(a,c) \to^* f(x,x)))\}$. With $P^{Spl}$ we start the proof of infeasibility of $\tau = (\{\overline{\mathcal{R}}\}, \mathcal{G})$ as follows: $P^{Spl}(\tau) = \{\tau_1, \tau_2, \tau_3\}$, where $\tau_1 = (\{\overline{\mathcal{R}}\}, \{(f(a,c) \to^* a)\})$, $\tau_2 = (\{\overline{\mathcal{R}}\}, \{(f(a,c) \to^* b)\})$, and $\tau_3 = ((\{\overline{\mathcal{R}}\}, \{(f(a,c) \to^* f(x,x))\}))$.

## 4.2 Provability Processor

Our next processor exploits Theorem 1 to use theorem proving in proofs of feasibility.

**Definition 7 (Provability processor).** *Let* $\tau = (\mathbb{T}, \mathcal{G})$ *be an f-Problem with* $\mathcal{G} = \{F\} \uplus \mathcal{G}'$ *where* $F = (s_i \bowtie_i t_i)_{i=1}^n$. *Processor* $P^{Prov}$ *is given by*

$$P^{Prov}(\tau) = \text{``yes''} \; \textit{iff} \; \mathsf{Th}_F \vdash (\exists \boldsymbol{x}) \bigwedge_{i=1}^{n} s_i \bowtie_i t_i \; \textit{holds.}$$

Note that, whenever $n = 0$, i.e., $F = ()$, then $\bigwedge_{i=1}^{n} s_i \bowtie_i t_i$ is true and $P^{Prov}(\tau) =$ "yes".

**Theorem 4.** *Processor* $P^{Prov}$ *is complete. If* $\mathsf{Th}_F$ *preserves* F, *then it is sound.*

In infChecker, we use Prover9 [18] as a backend to implement $P^{Prov}$.

*Example 9.* For $\mathcal{R}$ in Fig. 1, the feasibility goal $\mathcal{G}$ (see file 903 in COPS):

$$\{(le(x, min(y)) \to^* false, min(y) \to^* x)\}$$

and the corresponding first-order formula:

$$(\exists x, y) \; le(x, min(y)) \to^* false \wedge min(y) \to^* x \tag{13}$$

with $\tau_I = (\{\overline{\mathcal{R}}\}, \mathcal{G})$, we have $P^{Prov}(\tau_I) = $ "yes" by using Prover9 to prove (13) by resolution as follows[2]:

```
(11) exists x y (le(x,min(y)) ->* false) & (min(y) ->* x) [goal]
(12) x ->* x [assumption]
(13) -(x -> y) | -(y ->* z) | (x ->* z) [assumption]
(15) -(x -> y) | (le(z,x) -> le(z,y)) [assumption]
(22) le(x,0) -> false [assumption]
(23) min(cons(x,nil)) -> x [assumption]
(27) -(le(x,min(y)) ->* false) | -(min(y) ->* x) [deny(11)]
(48) le(x,0) ->* false [ur(13,22,12)]
(59) -(le(min(x),min(x)) ->* false) [resolve(27,12)]
(67) -(le(min(x),min(x)) -> y) | -(y ->* false) [resolve(59,13)]
(69) -(le(min(x),y) ->* false) | -(min(x) -> y) [resolve(67,15)]
(76) -(le(min(cons(x,nil)),x) ->* false) [resolve(69,23)]
(77) $F [resolve(76,48)]
```

*Example 10.* Consider the two rules TRS $\mathcal{R} = \{a \to c(b), b \to c(b)\}$. For $\overline{\mathcal{R}} = \{(14) - (18)\}$ and $\mathsf{Th}_\unrhd = \{(19) - (21)\}$:

---

[2] For readability, the output is slightly pretty printed.

$$(\forall x)\ x \to^* x \qquad (14)$$
$$(\forall x, y, z)\ (x \to y \land y \to^* z \Rightarrow x \to^* z) \quad (15)$$
$$(\forall x, y)\ (x \to y \Rightarrow c(x) \to c(y)) \qquad (16)$$
$$a \to c(b) \qquad (17)$$
$$b \to c(b) \qquad (18)$$

$$(\forall x)\ x \trianglerighteq x \qquad (19)$$
$$(\forall x, y, z)\ x \trianglerighteq y \land y \trianglerighteq z \Rightarrow x \trianglerighteq z \qquad (20)$$
$$(\forall x)\ c(x) \trianglerighteq x \qquad (21)$$

infChecker can prove loopingness of $\mathcal{R}$ as the feasibility of $(\{\overline{\mathcal{R}}, \mathsf{Th}_{\trianglerighteq}\}, \mathcal{G})$ by relying on Prover9 with $\mathcal{G} = \{(x \to y, y \to^* z, z \trianglerighteq x)\}$ (see Example 1). Note that the union of $\overline{\mathcal{R}}$ and $\mathsf{Th}_{\trianglerighteq}$ preserves both $\overline{\mathcal{R}}$ and $\mathsf{Th}_{\trianglerighteq}$, as required for soundness of $\mathsf{P}^{\mathsf{Prov}}$.

If no proof of $\varphi_{\mathsf{F}} = (\exists \boldsymbol{x}) \bigwedge_{i=1}^n s_i \bowtie_i t_i$ is found, then $\mathsf{P}^{\mathsf{Prov}}$ does *not* apply. In this case, it is still possible that F is feasible, but the proof system failed to prove it. Also, it is possible that F is infeasible. In this case, our next processor, which tries to prove infeasibility as satisfiability [9], can be useful.

### 4.3   Satisfiability Processors

The next processor implements the satisfiability approach in [9].

**Definition 8 (Satisfiability Processor).** *Let $\tau = (\mathbb{T}, \mathcal{G})$ be an f-Problem with $\mathcal{G} = \{\mathsf{F}\} \uplus \mathcal{G}'$ for $\mathsf{F} = (s_i \bowtie_i t_i)_{i=1}^n$ and $\mathcal{A}$ be a structure. Processor $\mathsf{P}^{\mathsf{Sat}}$ is given by $\mathsf{P}^{\mathsf{Sat}}(\tau) = (\mathbb{T}, \mathcal{G}')$ iff $\mathcal{A} \models \mathsf{Th}_{\mathsf{F}} \cup \{\neg (\exists \boldsymbol{x}) \bigwedge_{i=1}^n s_i \bowtie_i t_i\}$.*

*Remark 2.* In the following, the soundness and completeness theorems given for the different introduced processors assume the notations previously introduced in the corresponding definitions.

In the following, we say that a theory Th is *stable* if for all terms $s, t$ and substitutions $\sigma$, if $\mathsf{Th} \vdash s \bowtie t$, then $\mathsf{Th} \vdash \sigma(s) \bowtie \sigma(t)$.

**Theorem 5.** *Processor $\mathsf{P}^{\mathsf{Sat}}$ is sound. If $\mathcal{T}(\mathcal{F}) \neq \emptyset$ and $\mathsf{Th}_{\mathsf{F}}$ is stable, then it is complete.*

In infChecker, we use the model generators AGES [5] and Mace4 [18] to find suitable structures $\mathcal{A}$ to be used in the implementation of $\mathsf{P}^{\mathsf{Sat}}$.

*Example 11.* For $\mathcal{R}$, $\overline{\mathcal{R}}$ and $\mathsf{Th}_{\trianglerighteq}$ as in Example 10, we can prove term $a$ non-looping. The following structure over $\mathbb{N} \cup \{-1\}$:

$$a^{\mathcal{A}} = -1 \qquad\qquad b^{\mathcal{A}} = 1 \qquad\qquad c^{\mathcal{A}}(x) = x$$
$$x \to^{\mathcal{A}} y \Leftrightarrow x \leq 1 \land y \geq 1 \qquad x (\to^*)^{\mathcal{A}} y \Leftrightarrow x \leq y \qquad x \trianglerighteq^{\mathcal{A}} y \Leftrightarrow x \leq y$$

satisfies $\overline{\mathcal{R}} \cup \mathsf{Th}_{\trianglerighteq} \cup \{\neg (\exists x, y)\ (a \to x \land x \to^* y \land y \trianglerighteq a)\}$. Thus, $a$ is non-looping.

The following version of $\mathsf{P}^{\mathsf{Sat}}$ often provides a direct answer about infeasibility of a goal.

**Definition 9 (One-Step Satisfiability Processor).** *Let* $\tau = (\mathbb{T}, \mathcal{G})$ *be an f-Problem with* $\mathcal{G} = \{F_1; \cdots; F_m\}$, *where, for all* $1 \leq i \leq m$, $F_i$ *is* $(s_{i1} \bowtie_{i1} t_{i1}, \ldots, s_{in_i} \bowtie_{in_i} t_{in_i})$ *for some* $n_i > 0$. *Let* $\mathcal{A}$ *be a structure. Processor* $\mathsf{P}^{SatAll}$ *is given by* $\mathsf{P}^{SatAll}(\tau) = \emptyset$ *iff* $\mathcal{A} \models \mathsf{Th}_{\mathcal{G}} \cup \{\neg(\exists \boldsymbol{x}) \bigvee_{i=1}^{m} \bigwedge_{j=1}^{n_i} s_{ij} \bowtie_{ij} t_{ij}\}$.

**Theorem 6.** *Processor* $\mathsf{P}^{SatAll}$ *is sound. If* $\mathcal{T}(\mathcal{F}) \neq \emptyset$ *and* $\mathsf{Th}_{\mathcal{G}}$ *is stable, then it is complete.*

*Example 12.* (continuing Example 8) With Mace4 we obtain a model of

$$\overline{\mathcal{R}} \cup \{\neg(\exists x) \ (f(a,c) \rightarrow^* a \vee f(a,c) \rightarrow^* b \vee f(a,c) \rightarrow^* f(x,x))\}$$

Since $\mathsf{P}^{SatAll}(\tau) = \emptyset$, $\mathcal{G}$ is $\mathbb{T}$-infeasible. This is an alternative proof without $\mathsf{P}^{Spl}$.

Although $\mathsf{P}^{SatAll}$ can be simulated by a single step of $\mathsf{P}^{Spl}$ followed by the application of $\mathsf{P}^{Sat}$ to the obtained f-Problems, from a practical point of view $\mathsf{P}^{SatAll}$ has the advantage of avoiding the overloading due to the split the initial goal into a set of sequences with the generation of several models by means of calls to (external) model generators. Instead, $\mathsf{P}^{SatAll}$ makes a single call to the model generator(s).

## 4.4 Usable Rules in CTRS Theories

As discussed in [13, Section 2], dealing with CTRSs $\mathcal{R} = (\mathcal{F}, R)$, we may often drop some rules in $R$ before establishing (in)feasibility of conditions $s \rightarrow^* t$. First, consider the following overapproximation of the set of rules that can be applied to a term $t$:

$$RULES(\mathcal{R}, t) = \{\ell \rightarrow r \Leftarrow c \in R \mid \exists p \in \mathcal{P}os(t), root(\ell) = root(t|_p)\}$$

The set of *usable rules* for $t$ is defined as follows:

$$\mathcal{U}(\mathcal{R}, t) = RULES(\mathcal{R}, t) \cup \bigcup_{l \rightarrow r \Leftarrow c \in RULES(\mathcal{R},t)} \left( \mathcal{U}(\mathcal{R}^{\sharp}, r) \cup \bigcup_{s \approx t \in c} \mathcal{U}(\mathcal{R}^{\sharp}, s) \right)$$

where $\mathcal{R}^{\sharp} = \mathcal{R} - RULES(\mathcal{R}, t)$.[3] Given a sequence F, we let

$$\mathcal{U}_{\rightarrow^*}(\mathcal{R}, \mathsf{F}) = \bigcup_{s \rightarrow^* t \in \mathsf{F}} \mathcal{U}(\mathcal{R}, s) \tag{22}$$

*Example 13.* For $\mathcal{R}$ in Fig. 1 and $\mathsf{F} = (le(0, s(0)) \rightarrow^* x)$, $\mathcal{U}_{\rightarrow^*}(\mathcal{R}, \mathsf{F}) = \{(1), (2), (3)\}$.

Let $\overline{\mathcal{R}}|_{\mathcal{U}_{\rightarrow^*}(\mathcal{R},\mathsf{F})}$ be the first-order theory for the CTRS $(\mathcal{F}, \mathcal{U}_{\rightarrow^*}(\mathcal{R}, \mathsf{F}))$, which keeps the original signature of $\mathcal{R}$.

---

[3] The use of $\mathcal{R}^{\sharp}$ instead of $\mathcal{R}$ is important for implementing the computation of usable rules. By decreasing (from $\mathcal{R}$ to $\mathcal{R}^{\sharp}$) the set of considered rules, the recursive definition is shown to be terminating.

**Definition 10.** *Let $\tau = (\mathbb{T}, \mathcal{G})$ be an f-Problem such that $\mathbb{T} = \{\mathsf{Th}_{\to^*}\} \uplus \mathbb{T}'$ with $\mathsf{Th}_{\to^*} = \overline{\mathcal{R}}$ for a CTRS $\mathcal{R}$ and $\mathcal{G} = \{\mathsf{F}\} \uplus \mathcal{G}'$. Let $\mathsf{Th}'_{\to^*} = \overline{\mathcal{R}}|_{\mathcal{U}_{\to^*}(\mathcal{R},\mathsf{F})}$. Processor $\mathsf{P}^{UR}$ is given by $\mathsf{P}^{UR}(\tau) = \{((\{\mathsf{Th}'_{\to^*}\} \uplus \mathbb{T}', \{\mathsf{F}\}), (\mathbb{T}, \mathcal{G}'))\}$.*

$\mathsf{P}^{UR}$ distributes the sequences in $\mathcal{G}$ in two new f-Problems: the first one consists of a goal with a single sequence $\mathsf{F}$ together with a refined version of $\mathbb{T}$ where $\overline{\mathcal{R}}$ is simplified into $\overline{\mathcal{R}}|_{\mathcal{U}_{\to^*}(\mathcal{R},\mathsf{F})}$; the second f-Problem consists of $\mathcal{G}'$ but keeps the original set of theories $\mathbb{T}$. By using [13, Proposition 4], we can see that the $(\{\overline{\mathcal{R}}|_{\mathcal{U}_{\to^*}(\mathcal{R},\mathcal{G})}\} \uplus \mathbb{T})$-feasibility of a sequence $\mathsf{F}$ implies its $(\{\overline{\mathcal{R}}\} \uplus \mathbb{T})$-feasibility. Similarly, $(\{\overline{\mathcal{R}}\} \uplus \mathbb{T})$-infeasibility of $\mathsf{F}$ can be proved as $(\{\overline{\mathcal{R}}|_{\mathcal{U}_{\to^*}(\mathcal{R},\mathsf{F})}\} \uplus \mathbb{T})$-infeasibility provided that all terms $s$ in feasibility conditions $s \to^* t$ in $\mathsf{F}$ are ground. Thus, we have the following:

**Theorem 7.** *$\mathsf{P}^{UR}$ is sound. If for all $s \to^* t \in \mathsf{F}$, $s$ is ground, then $\mathsf{P}^{UR}$ is complete.*

As discussed in the last paragraph of [13, Section 2], the groundness requirement cannot be dropped, in general (even for TRSs).

*Example 14.* For $\mathcal{R} = \{a \to b\}$, the sequence $\mathsf{F} = (x \to^* a, x \to^* b)$ is $\{\mathsf{Th}_{\to^*}\}$-feasible (just instantiate variable $x$ to $a$ and use the rule in $\mathcal{R}$). However, it is *not* $\{\mathsf{Th}'_{\to^*}\}$-feasible for $\mathsf{Th}'_{\to^*} = \overline{\mathcal{R}}|_{\mathcal{U}_{\to^*}(\mathcal{R},\mathsf{F})}$ because $\mathcal{U}(\mathcal{R}, x)$ is empty. Hence, $\mathcal{U}_{\to^*}(\mathcal{R}, \mathsf{F}) = \mathcal{U}(\mathcal{R}, x) \cup \mathcal{U}(\mathcal{R}, x)$ is also empty.

$\mathsf{P}^{UR}$ deals with many-step conditions $s \to^* t$ only. Furthermore, note that no change (simplification) in $\mathsf{Th}_{\to}$ (if used in $\mathbb{T}$) is introduced. For one-step conditions $s \to t$, we can use a similar (sound and complete) processor $\mathsf{P}^{UR1}$ as follows. Let $\mathcal{U}_{\to}(\mathcal{R}, \mathsf{F}) = \bigcup_{s \to t \in \mathsf{F}} \mathcal{U}(\mathcal{R}, s)$ and $\overline{\mathcal{R}}|_{\mathcal{U}_{\to}(\mathcal{R},\mathsf{F})}$ be the first-order theory for $(\mathcal{F}, \mathcal{U}_{\to}(\mathcal{R}, \mathsf{F}))$.

**Definition 11.** *Let $\tau = (\mathbb{T}, \mathcal{G})$ be an f-Problem such that $\mathbb{T} = \{\mathsf{Th}_{\to}\} \uplus \mathbb{T}'$ with $\mathsf{Th}_{\to} = \overline{\mathcal{R}}$ for a CTRS $\mathcal{R}$ and $\mathcal{G} = \{\mathsf{F}\} \uplus \mathcal{G}'$. Let $\mathsf{Th}'_{\to} = \overline{\mathcal{R}}|_{\mathcal{U}_{\to}(\mathcal{R},\mathsf{F})}$. Processor $\mathsf{P}^{UR1}$ is given by $\mathsf{P}^{UR1}(\tau) = \{((\{\mathsf{Th}'_{\to}\} \uplus \mathbb{T}', \{\mathsf{F}\}), (\mathbb{T}, \mathcal{G}'))\}$.*

**Theorem 8.** *$\mathsf{P}^{UR1}$ is sound. If for all $s \to t \in \mathsf{F}$, $s$ is ground, then $\mathsf{P}^{UR1}$ is complete.*

Starting from the f-Problem $(\mathbb{T}, \mathcal{G})$, where $\mathcal{G} = \{\mathsf{F}\} \uplus \mathcal{G}'$, both $\mathsf{P}^{UR}$ and $\mathsf{P}^{UR1}$ return two f-Problems $(\mathbb{T}_1, \mathcal{G}_1)$ and $(\mathbb{T}_2, \mathcal{G}_2)$. For both $\mathsf{P}^{UR}$ and $\mathsf{P}^{UR1}$, we have $\mathcal{G}_1 = \{\mathsf{F}\}$, $\mathcal{G}_2 = \mathcal{G}'$, and $\mathbb{T}_2 = \mathbb{T}$. As for $\mathbb{T}_1$, $\mathsf{P}^{UR}$ changes the component $\mathsf{Th}_{\to^*}$ of $\mathbb{T}$. On the other hand, $\mathsf{P}^{UR1}$ changes $\mathsf{Th}_{\to}$. With regard to the preservation property, which is relative to the goal and theory in a given f-Problem, whenever it holds for $(\mathbb{T}, \mathcal{G})$, it is not difficult to see (from the definition of preservation and usable rules) that it also remains true for $(\mathbb{T}_1, \mathcal{G}_1)$ and $(\mathbb{T}_2, \mathcal{G}_2)$.

## 4.5   Narrowing on Rewriting Conditions Processor

Reachability problems $\sigma(s) \to^* \sigma(t)$ are often investigated using *narrowing* and unification conditions directly over terms $s$ and $t$, thus avoiding the 'generation' of the required substitution $\sigma$. In this section, we use narrowing to simplify feasibility conditions in $\mathcal{G}$. Definition 12 describes how narrowing is defined in the context of CTRSs. In the following, we write $s =_\theta^? t$ if $s$ and $t$ unify with *mgu* $\theta$.

**Definition 12.** [16, Definition 79]  *Let $\mathcal{R}$ be a CTRS. A term $s$ narrows to a term $t$ (written $s \rightsquigarrow_{\mathcal{R},\theta,p} t$ or just $s \rightsquigarrow_{\mathcal{R},\theta} t$ or even $s \rightsquigarrow t$), iff there is a nonvariable position $p \in \mathcal{P}os_{\mathcal{F}}(s)$, a renamed rule $\ell \to r \Leftarrow s_1 \approx t_1, \ldots, s_n \approx t_n$ in $\mathcal{R}$, substitutions $\theta_0, \ldots, \theta_n, \tau_1, \ldots, \tau_n$, and terms $t_1', \ldots, t_n'$ such that:*

1. *$s|_p =_{\theta_0}^? \ell$,*
2. *for all $i$, $1 \le i \le n$, $\eta_{i-1}(s_i) \rightsquigarrow_{\mathcal{R},\theta_i}^* t_i'$ and $t_i' =_{\tau_i}^? \theta_i(\eta_{i-1}(t_i))$, where $\eta_0 = \theta_0$ and for all $i > 0$, $\eta_i = \tau_i \circ \theta_i \circ \eta_{i-1}$, and*
3. *$t = \theta(s[r]_p)$, where $\theta = \eta_n$.*

*We write $u \rightsquigarrow_{\mathcal{R},\beta}^* v$ for terms $u, v$ and substitution $\beta$ iff there are terms $u_1, \ldots, u_{m+1}$ and substitutions $\beta_1, \ldots, \beta_m$ for some $m \ge 0$ such that*

$$u = u_1 \rightsquigarrow_{\mathcal{R},\beta_1} u_2 \rightsquigarrow_{\mathcal{R},\beta_2} \cdots \rightsquigarrow_{\mathcal{R},\beta_m} u_{m+1} = v$$

*and $\beta = \beta_m \circ \cdots \circ \beta_1$ (or $\beta = \varepsilon$ if $m = 0$).*

Given a term $u$, the set $N_1(\mathcal{R}, u)$ represents the set of one-step $\mathcal{R}$-narrowings issued from $u$:

$$N_1(\mathcal{R}, u) = \{(v, \theta\!\downarrow_{\mathcal{V}ar(u)}) \mid u \rightsquigarrow_{\ell \to r \Leftarrow c, \theta} v, \ell \to r \Leftarrow c \in \mathcal{R}\} \tag{23}$$

where $\theta\!\downarrow_{\mathcal{V}ar(u)}$ is a substitution defined by $\theta\!\downarrow_{\mathcal{V}ar(u)}(x) = \theta(x)$ if $x \in \mathcal{V}ar(u)$ and $\theta\!\downarrow_{\mathcal{V}ar(u)}(x) = x$ otherwise.

As discussed in [16, Section 7.5], the set $N_1(\mathcal{R}, u)$ can be *infinite*. In [16, Proposition 87] some sufficient conditions for finiteness of $N_1(\mathcal{R}, u)$ are given. Knowing these restrictions, in Definition 13 we define a narrowing processor on feasibility conditions.

Given a sequence $\mathsf{F}_k = (s_j \bowtie_j t_j)_{j=1}^n$ in a goal $\mathcal{G}$, and $1 \le i \le n$ such that $\bowtie_i = \to^*$, $\overline{\mathcal{N}}(\mathcal{R}, \mathcal{G}, k, i)$ returns a new set of feasibility goals where each element of the set corresponds to a possible narrowing on the condition $i$:

$$\overline{\mathcal{N}}(\mathcal{R}, \mathcal{G}, k, i) = \{\mathcal{G}[\mathsf{F}_k[\boldsymbol{\theta}, w \to^* t_i]_i]_k \mid s_i \to^* t_i \in \mathsf{F}_k, (w, \theta) \in N_1(\mathcal{R}, s_i)\}$$

where $\boldsymbol{\theta}$ consists of new conditions $x_1 \to^* \theta(x_1), \ldots, x_m \to^* \theta(x_m)$ obtained from the bindings in $\theta$ for variables in $\mathcal{V}ar(s_i) = \{x_1, \ldots, x_m\}$.

**Definition 13 (Narrowing on f-Conditions Processor).** *Let $\tau = (\{\overline{\mathcal{R}}\}, \mathcal{G})$ be an f-Problem, $s_i \to^* t_i \in \mathsf{F}_k$ for some $\mathsf{F}_k$ in $\mathcal{G}$, and $\mathcal{N} \subseteq \overline{\mathcal{N}}(\mathcal{R}, \mathcal{G}, k, i)$ finite. $\mathsf{P}^{NC}$ is given by $\mathsf{P}^{NC}(\tau) = \{(\{\overline{\mathcal{R}}\}, \mathcal{G}') \mid \mathcal{G}' \in \mathcal{N}\}$.*

$$add(0, x) \rightarrow x$$
$$add(s(x), y) \rightarrow s(add(x, y))$$
$$div(0, s(x)) \rightarrow 0$$
$$div(s(x), s(y)) \rightarrow 0 \Leftarrow lte(s(x), y) \approx true$$
$$div(s(x), s(y)) \rightarrow s(q) \Leftarrow lte(s(x), y) \approx false, div(minus(x, y), s(y)) \approx q$$
$$lte(0, y) \rightarrow true$$
$$lte(s(x), 0) \rightarrow false$$
$$lte(s(x), s(y)) \rightarrow lte(x, y)$$
$$minus(0, s(y)) \rightarrow 0$$
$$minus(s(x), s(y)) \rightarrow minus(x, y)$$
$$minus(x, 0) \rightarrow x$$
$$mod(0, y) \rightarrow 0$$
$$mod(x, 0) \rightarrow x$$
$$mod(x, s(y)) \rightarrow mod(minus(x, s(y)), s(y)) \Leftarrow lte(s(y), x) \approx true$$
$$mod(x, s(y)) \rightarrow x \Leftarrow lte(s(y), x) \approx false$$
$$mult(0, y) \rightarrow 0$$
$$mult(s(x), y) \rightarrow add(mult(x, y), y)$$
$$power(x, 0) \rightarrow s(0)$$
$$power(x, n) \rightarrow mult(mult(y, y), s(0)) \Leftarrow n \approx s(n'),$$
$$mod(n, s(s(0))) \approx 0, power(x, div(n, s(s(0)))) \approx y$$
$$power(x, n) \rightarrow mult(mult(y, y), x) \Leftarrow n \approx s(n'),$$
$$mod(n, s(s(0))) \approx s(z), power(x, div(n, s(s(0)))) \approx y$$

**Fig. 3.** CTRS 529.trs in COPS

Given a term $s$, we let $NRules(\mathcal{R}, s)$ be the set of rules $\alpha : \ell \rightarrow r \Leftarrow c \in \mathcal{R}$ such that a nonvariable subterm $t$ of $s$ is a *narrex* of $\alpha$,[4] and, given a substitution $\theta$, we denote as $\theta\downarrow_{Var(s)}$ the substitution defined by $\theta\downarrow_{Var(s)}(x) = \theta(x)$ if $x \in Var(s)$ and $\theta\downarrow_{Var(s)}(x) = x$ otherwise.

**Theorem 9.** $P^{NC}$ *is sound. If* $\mathcal{N} = \overline{\mathcal{N}}(\mathcal{R}, \mathcal{G}, k, i)$ *and* $s_i \rightarrow^* t_i \in \mathsf{F}_k$ *is such that* $s_i$ *and* $t_i$ *do not unify and either* $s_i$ *is ground and* $\mathcal{R}$ *is a 2-CTRS or (1)* $NRules(\mathcal{R}, s_i)$ *is a TRS, (2)* $s_i$ *is linear, and (3)* $Var(s_i) \cap Var(t_i) = \emptyset$, *then* $P^{NC}$ *is complete.*[5]

Even with $\overline{\mathcal{N}}(\mathcal{R}, \mathcal{G}, k, i)$ infinite, a subset $\mathcal{N}$ of $\overline{\mathcal{N}}(\mathcal{R}, \mathcal{G}, k, i)$ can be sufficient to prove feasibility. However, to prove infeasibility we need to consider all possible narrowings.

---

[4] Given a CTRS $\mathcal{S}$, a non-variable term $t$ is a *narrowing redex* (or a *narrex*, for short) of a rule $\ell \rightarrow r \Leftarrow s_1 \approx t_1, \ldots, s_n \approx t_n \in \mathcal{S}$ if $t$ and $\ell$ unify with *mgu* $\theta$ (we assume $Var(t) \cap Var(\ell) = \emptyset$). However, if $(\theta(s_1) \approx \theta(t_1), \ldots, \theta(s_n) \approx \theta(t_n))$ can be proved $\{\overline{\mathcal{S}}\}$-infeasible, we can discard $t$, as no narrowing step is possible on it. In our current implementation, though, only the unification test is used.

[5] This processor is inspired by the processor defined in [17, Section 4.1]. A justification for the completeness conditions can be obtained from [17, Examples 18 and 19].

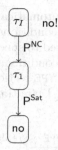

**Fig. 4.** Proof tree obtained from Example 15

*Example 15.* Consider the CTRS $\mathcal{R}$ in Fig. 3, $\mathcal{G} = \{(lte(s(x), 0) \rightarrow^* true)\}$, and $\tau_I = (\{\overline{\mathcal{R}}\}, \mathcal{G})$. Since $NRules(\mathcal{R}, lte(s(x), 0))$ contains the *lte* rules only, $\overline{\mathcal{N}}(\mathcal{R}, \mathcal{G}, 1, 1) = \{(x \rightarrow^* x, false \rightarrow^* true)\}$. Therefore, $\mathsf{P}^{\mathsf{NC}}(\tau_I) = \{\tau_1\}$ with $\tau_1 = (\{\overline{\mathcal{R}}\}, (x \rightarrow^* x, false \rightarrow^* true))$. Since $NRules(\mathcal{R}, lte(s(x), 0))$ is a TRS, $lte(s(x), 0)$ is linear, and $Var(lte(s(x), 0)) \cap Var(true) = \emptyset$, $\mathsf{P}^{\mathsf{NC}}$ is complete. Now, we apply $\mathsf{P}^{\mathsf{Sat}}$ to $\tau_1$ to obtain $\mathsf{P}^{\mathsf{Sat}}(\tau_1) = \emptyset$ by using Mace4. The obtained FP tree is in displayed in Fig. 4.

## 5   Implementation and Web Interface

infChecker 1.0 is written in Haskell and consists of 30 Haskell modules with more than 4500 lines of code. The tool can be used through its web interface here:

$$\text{http://zenon.dsic.upv.es/infChecker/}$$

The input format is an extended version of the Confluence Competition (CoCo) format [20], which is the official format used in the *infeasibility* (INF) category.[6] The input has two components:

1. A CTRS $\mathcal{R}$ in TPDB format[7] which can specify a *replacement map* $\mu$ for *context-sensitive rewriting* (*CSR* [10]) establishing the arguments $\mu(f) \subseteq \{1, \ldots, k\}$ which *can be rewritten* for each $k$-ary symbol $f$. This is top-down propagated to *positions* of terms, which are then called *active*. We write $s \hookrightarrow t$ if an *active* subterm of $s$ can be rewritten so that $s \rightarrow t$. Then, $\rightarrow^*$, $\downarrow$, $\leftrightarrow$, etc. are generalized to *CSR* as $\hookrightarrow^*$, $\mathord{\downarrow}$, $\overset{\hookleftarrow}{\hookrightarrow}$, etc., by using $\hookrightarrow$ instead of $\rightarrow$.
2. An f-goal built using the set

$$\mathbb{P}_{\mathsf{iCh}} = \{\,|>, \,|>=\} \cup \{\mathord{->}, \mathord{->}*, \mathord{->}*<\mathord{-}, <\mathord{--}>, <\mathord{--}>*\} \cup \{\backslash\mathord{->}, \backslash\mathord{->}*, \backslash\mathord{->}*<\mathord{-}/, <\mathord{-}/\backslash\mathord{->}, <\mathord{-}/\backslash\mathord{->}*\} \cup \{==\}$$

of (binary) predicates for (strict) subterm (|> and |>=), one or many rewriting steps (-> and ->*), joinability (->*<-), symmetric closure of -> (<-->), conversion (<-->*) and their context-sensitive versions \->, \->*, \->*<-/, <-/\->, and <-/\->*.

---

[6] See http://project-coco.uibk.ac.at/2019/categories/infeasibility.php.
[7] See http://zenon.dsic.upv.es/muterm/?page_id=31.

Theories Th$_{\bowtie}$ for each $\bowtie \in \mathbb{P}_{iCh}$ are automatically obtained from the components of $\mathcal{R}$ (signature, replacement map, conditional rules). Symbol == is borrowed from the COPS syntax, where it is used to specify the conditional part of rewrite rules (see $\approx$ in Fig. 1). Both in the conditional part of rules and in f-goals, its meaning depends on the CONDITIONTYPE section of the input specifying how the conditions of rules are evaluated [22, Definition 7.1.3] according to:

| CONDITIONTYPE | replace == by |
|---|---|
| ORIENTED | ->* |
| JOIN | -><- |
| SEMI-EQUATIONAL | <-->* |

In this respect, when using $\hookrightarrow$ or $\hookrightarrow^*$ (i.e., \-> or \->*) in f-goals, the associated theories Th$_{\hookrightarrow}$ and Th$_{\hookrightarrow^*}$ are those obtained by evaluating the conditions $s_i \approx t_i$ in rules using $\hookrightarrow^*$, $\int$, or $\leftrightarrow^*$, depending on the label ORIENTED, JOIN, or SEMI-EQUATIONAL specified in CONDITIONTYPE. If no replacement map has been specified (i.e., no STRATEGY section with CONTEXTSENSITIVE label is given), the *trivial* replacement map $\mu_\top(f) = \{1,\ldots,k\}$, establishing no replacement restrictions, is automatically assumed for each $k$-ary symbol $f$.

When the problem is introduced, a model generator for infeasibility (AGES, Mace4 or Automatic) can be selected. Then, pressing button Prove automatically initiates the procedure to check whether the problem is feasible or infeasible in the given timeout.

Currently, infChecker implements the construction of the FP tree in Definition 5 with the processors presented in Sect. 4 by depth-first generation of the nodes and orderly attempting the following sequence of processors to develop each node:[8] P$^{Spl}$, P$^{Prov}$, P$^{Sat}$, and P$^{NC}$. If the final answer is YES or NO, the tool displays a report in plain text. Otherwise, MAYBE is returned.

## 6   Experimental Evaluation

We participated in the INF category of the 2019 Confluence Competition (CoCo),[9] with a limit of 60s to return a proof of feasibility or infeasibility (or a *don't know* answer). infChecker obtained the following results:

---

[8] We use a Haskell library for parallelism. However, due to the *Breadth First Search* evaluation strategy of the library, in a parallel execution P$_1$ ∥ P$_2$ ∥ ⋯ ∥ P$_n$ of several processors we wait until the leftmost processor (P$_1$) is completely evaluated (returns a solution or reaches a timeout) before continuing with P$_2$ ∥ ⋯ ∥ P$_n$. Thus, there is a kind of 'restricted' parallelism in our implementation.

[9] http://project-coco.uibk.ac.at/2019/.

| INF Tool | Yes | No | Total |
|----------|-----|-----|-------|
| infChecker | 40 | 32 | 72 |
| ConCon | 31 | 0 | 31 |
| nonreach | 30 | 0 | 30 |
| Moca | 26 | 0 | 26 |
| maedmax | 15 | 0 | 15 |
| CO3 | 12 | 0 | 12 |

Answers *Yes/No* in the table refer to *infeasibility* (which is the focus of the competition). In our setting, given a CTRS $\mathcal{R}$ and an infeasibility problem given as a feasibility sequence $\mathcal{G}$, we just return "*Yes*" if $\tau_I$ is proved infeasible, and "*No*" if $\tau_I$ is proved feasible. Apart from the 32 "*No*" answers, there are 7 more examples that can be proved positively ("*Yes*") using infChecker only. There also are 10 examples that can be proved by other tools and cannot be proved by infChecker.

In the experiments $\mathsf{P^{UR}}$ was used 11 times and $\mathsf{P^{NC}}$ was used twice. We required a combination of processors in 13 examples: the sum of uses of $\mathsf{P^{UR}}$ and $\mathsf{P^{NC}}$. Being unable to provide a definite (YES/NO) answer, their use always requires another processor to finish the proof. According to the strategy described at the end of Sect. 5, such a combination is necessary. Thus, we need both $\mathsf{P^{UR}}$ and $\mathsf{P^{NC}}$ to solve the examples.

## 7  Related Work

The notion of (in)feasibility of a logic formula has been investigated in [12, Section 4.1], in the context of the analysis of operational termination of programs in general logics [15]. A satisfiability approach to prove infeasibility of first-order formulas with respect to an order-sorted first-order theory [4] is described in [12, Section 4.1.1]. No attempt to decompose such proofs by taking into account the structure of the logic formula (as done in our feasibility framework) is made. No technique for proving feasibility is proposed. Actually, our feasibility framework could be advantageously used to implement proofs of operational termination of programs in general logics.

Sternagel and Yamada [25] define a framework to prove reachability constraints $\phi$ for TRSs $\mathcal{R}$ as first-order formulas where only reachability atoms $s \twoheadrightarrow t$ (instead of $s \rightarrow^* t$) are allowed. As remarked in [25, footnote 1], negation and universal quantification are not considered, i.e., only ECBCAs with atoms $s \twoheadrightarrow t$ are (ultimately) considered. A constraint $s \twoheadrightarrow t$ is satisfied by a substitution $\sigma$ with respect to $\mathcal{R}$ if $\sigma(s) \rightarrow^*_{\mathcal{R}} \sigma(t)$. Reachability constraints $\phi$ are called satisfiable if there is a substitution $\sigma$ such that $\sigma(\phi)^{10}$ is satisfied in the usual

---

[10] Obtained by (i) *renaming* all bounded variables in $\phi$ using variables not occurring in bindings of $\sigma$ to obtain $\phi'$, and then (ii) *replacing* each free variable $x$ of $\phi'$ by $\sigma(x)$.

first-order logic sense. Our approach is more flexible as more predicates can be defined by appropriate theories (including CTRSs). For instance, *non-root reachability* constraints $s \xrightarrow{>\Lambda} t$ (given in [25, Section 4] in terms of reachability constraints) can be defined by $\mathsf{Th}_{\geq \Lambda_*}$ consisting of

$$(\forall x) \quad x \xrightarrow{>\Lambda}^* x \tag{24}$$

$$(\forall x)(\forall y)(\forall z) \quad x \xrightarrow{>\Lambda} y \wedge y \xrightarrow{>\Lambda}^* z \Rightarrow x \xrightarrow{>\Lambda}^* z \tag{25}$$

plus sentences $\overline{(\mathrm{Rl})_\alpha}$ for each rewrite rule $\alpha$, sentences $\overline{(\mathrm{C})_{f,i}}$ for each $k$-ary symbol $f$ and $1 \leq i \leq k$, and $(\forall \boldsymbol{x})(\forall y_i) x_i \rightarrow y_i \Rightarrow f(x_1, \ldots, x_i, \ldots, x_k) \xrightarrow{>\Lambda} f(x_1, \ldots, y_i, \ldots, x_k)$. We could also cover CTRSs by also adding $\overline{(\mathrm{T})}$, for the transitivity rule (T), necessary for the evaluation of the conditional part of conditional rules $\alpha$ (which may require root steps). Then, the non-reachability problems considered in [25, Section 4] for TRSs $\mathcal{R}$ could be treated in our framework using $\mathbb{P} = \{\rightarrow, \rightarrow^*, \xrightarrow{>\Lambda}^*\}$ and $\mathbb{T} = \{\mathsf{Th}_\rightarrow, \mathsf{Th}_{\rightarrow^*}, \mathsf{Th}_{\geq \Lambda_*}\}$, where $\mathsf{Th}_\rightarrow = \mathsf{Th}_{\rightarrow^*} = \overline{\mathcal{R}}$. Actually, we can 'import' the decomposition treatment for non-root reachability goals in [25, Definition 5] as a transformation processor (like $\mathsf{P}^{\mathsf{NC}}$) specific for non-root reachability conditions of TRSs as follows: let $\tau = (\mathbb{T}, \mathcal{G})$ and $\mathsf{F}_i \in \mathcal{G}$ be such that $\mathsf{F}_i = (\gamma_1, \ldots, \gamma_j, \ldots, \gamma_n)$ with $\gamma_j = f(s_1, \ldots, s_k) \xrightarrow{>\Lambda}^* f(t_1, \ldots, t_k) \in \mathsf{F}_i$ for some terms $s_i, t_i, 1 \leq i \leq k$. Then,

$$\mathsf{P}^{\mathsf{non\text{-}root\text{-}r}}(\tau) = \{(\mathbb{T}, \mathcal{G}[\mathsf{F}'_i]_i)\}$$

where $\mathsf{F}'_i = (\gamma_1, \ldots, \gamma_{j-1}, s_1 \rightarrow^* t_1, \ldots, s_k \rightarrow^* t_k, \gamma_{j+1}, \ldots, \gamma_n)$ and $\mathcal{G}[\mathsf{F}'_i]_i$ is the goal obtained by replacing the $i$-th sequence of $\mathcal{G}$ by $\mathsf{F}'_i$. This example also shows how techniques developed elsewhere could be smoothly integrated in our framework.

Decidability of reachability problems for *CSR* in TRSs (i.e., does $s \hookrightarrow^* t$ hold?) has been investigated using tree-automata techniques [1,6–8]. infChecker is able to (try to) prove and disprove reachability conditions as f-goals using predicate \->* without any specific restriction on the TRSs (e.g., left-linearity), as done in these papers.

Regarding the automation of proofs of infeasibility in (conditional) rewriting, 2019 was the first year the infeasibility category was included in the International Confluence Competition. The new category had a good reception, with 6 participating tools (summary of results in Sect. 6), and provided a good picture of the state of the art:

- CO3 tries to prove confluence and if it fails linearizes the condition and tries to compute a narrowing tree for the linearized condition. The applicability of narrowing trees in this context is restricted to *syntactically deterministic conditional* term rewriting systems (right-hand sides of conditions must be constructor terms or ground normal forms) that are constructor systems [21].

- ConCon uses a variety of methods to check for infeasibility of conditional critical pairs, ranging from a simple technique based on unification, via symbol transition graph analysis, reachability problem decomposition, the exploitation of certain equalities in the conditions, and tree automata completion to equational reasoning [24].
- Moca implements *maximal ordered completion* similar to maedmax [26] together with some *approximation techniques* not yet published.
- maedmax implements *maximal order completion* [26].
- nonreach uses two approaches: *transformations* based on decomposition and narrowing and *nonreachability checks* based on unification, symbol transition graphs, equational reasoning and tree automata completion [19, 25].

Thanks to representing CTRSs $\mathcal{R}$ as first-order theories $\overline{\mathcal{R}}$, infChecker was not only the most successful tool for checking infeasibility, but also the only tool currently able to *disprove* it (by proving feasibility). There also is room, however, for improvements, as witnessed by the 10 examples mentioned in the summary of experiments in Sect. 6. In order to deal with these examples (not handled by infChecker), nonreach uses *narrowing* and Moca uses satisfiability with LPO. Regarding ConCon, it is unclear from the report provided by the tool, which specific technique was used to solve the examples.

## 8    Conclusions and Future Work

We have extended and generalized the notion of *feasibility sequence* introduced in [13] by considering goals which are sets of sequences of conditions $s \bowtie t$ for arbitrary predicates $\bowtie$. Each predicate symbol $\bowtie$ is 'defined' by a first-order theory $\mathsf{Th}_\bowtie$. Such conditions, sequences, and goals have a precise logical characterization as ECBCAs, and its feasibility can be investigated as provability of such formulas (Theorem 1). We have shown some examples of properties (of CTRSs) which can be proved by using this approach. We have introduced a framework for proving and disproving feasibility of such goals. To the best of our knowledge, our logic-based notion of feasibility goal and the framework to prove and disprove them are new in the literature.

We have developed a new tool implementing our framework: infChecker. Currently, infChecker provides a first implementation of the framework introduced in this paper (restricted to CTRSs, but extended with context-sensitivity, subterm, etc.), and supports predicates like $\rightarrow$ (one-step rewriting), $\rightarrow^*$ (many-step rewriting), $\downarrow$ (joinability), $\leftrightarrow^*$ (conversion), and the analogous for context-sensitive rewriting. We also give support to $\unrhd$ (subterm) and $\rhd$ (strict subterm). As far as we know, infChecker is the first tool dealing with (in)feasibility problems supporting this set of predicates. Also, the use of provability/satisfiability techniques in proofs of (in)feasibility seems to be new in the literature. We participated in the 2019 Confluence Competition [20] in the INF (infeasibility) category, being the most powerful tool for checking both infeasibility and feasibility. In the near future, we plan to extend infChecker to provide full support to our framework, by allowing the explicit definition of (not necessarily binary) predicates and independent first-order theories associated to such predicates besides the built-in set of predicates $\mathbb{P}_{iCh}$ and associated theories which are available now.

**Acknowledgments.** We thank the anonymous referees for many remarks and suggestions that led to improve the paper.

# References

1. Andrianarivelo, N., Réty, P.: Over-approximating terms reachable by context-sensitive rewriting. In: Bojańczyk, M., Lasota, S., Potapov, I. (eds.) RP 2015. LNCS, vol. 9328, pp. 128–139. Springer, Cham (2015). https://doi.org/10.1007/978-3-319-24537-9_12
2. Dershowitz, N.: Termination of rewriting. J. Symb. Comput. **3**(1/2), 69–116 (1987). https://doi.org/10.1016/S0747-7171(87)80022-6
3. Giesl, J., Thiemann, R., Schneider-Kamp, P., Falke, S.: Mechanizing and improving dependency pairs. J. Autom. Reasoning **37**(3), 155–203 (2006). https://doi.org/10.1007/s10817-006-9057-7
4. Goguen, J.A., Meseguer, J.: Models and equality for logical programming. In: Ehrig, H., Kowalski, R., Levi, G., Montanari, U. (eds.) TAPSOFT 1987. LNCS, vol. 250, pp. 1–22. Springer, Heidelberg (1987). https://doi.org/10.1007/BFb0014969
5. Gutiérrez, R., Lucas, S.: Automatic generation of logical models with AGES. In: Fontaine, P. (ed.) CADE 2019. LNCS (LNAI), vol. 11716, pp. 287–299. Springer, Cham (2019). https://doi.org/10.1007/978-3-030-29436-6_17
6. Kojima, Y., Sakai, M.: Innermost reachability and context sensitive reachability properties are decidable for linear right-shallow term rewriting systems. In: Voronkov, A. (ed.) RTA 2008. LNCS, vol. 5117, pp. 187–201. Springer, Heidelberg (2008). https://doi.org/10.1007/978-3-540-70590-1_13
7. Kojima, Y., Sakai, M., Nishida, N., Kusakari, K., Sakabe, T.: Context-sensitive innermost reachability is decidable for linear right-shallow term rewriting systems. Inf. Media Technol. **4**(4), 802–814 (2009)
8. Kojima, Y., Sakai, M., Nishida, N., Kusakari, K., Sakabe, T.: Decidability of reachability for right-shallow context-sensitive term rewriting systems. IPSJ Online Trans. **4**, 192–216 (2011)
9. Lucas, S.: Analysis of rewriting-based systems as first-order theories. In: Fioravanti, F., Gallagher, J.P. (eds.) LOPSTR 2017. LNCS, vol. 10855, pp. 180–197. Springer, Cham (2018). https://doi.org/10.1007/978-3-319-94460-9_11
10. Lucas, S.: Context-sensitive computations in functional and functional logic programs. J. Funct. Logic Program. **1998**(1) (1998). http://danae.uni-muenster.de/lehre/kuchen/JFLP/articles/1998/A98-01/A98-01.html
11. Lucas, S.: Proving semantic properties as first-order satisfiability. Artif. Intell. **277** (2019). https://doi.org/10.1016/j.artint.2019.103174
12. Lucas, S.: Using well-founded relations for proving operational termination. J. Autom. Reasoning **64**(2), 167–195 (2019). https://doi.org/10.1007/s10817-019-09514-2
13. Lucas, S., Gutiérrez, R.: Use of logical models for proving infeasibility in term rewriting. Inf. Process. Lett. **136**, 90–95 (2018). https://doi.org/10.1016/j.ipl.2018.04.002

14. Lucas, S., Marché, C., Meseguer, J.: Operational termination of conditional term rewriting systems. Inf. Process. Lett. **95**(4), 446–453 (2005). https://doi.org/10.1016/j.ipl.2005.05.002

15. Lucas, S., Meseguer, J.: Proving operational termination of declarative programs in general logics. In: Chitil, O., King, A., Danvy, O. (eds.) Proceedings of the 16th International Symposium on Principles and Practice of Declarative Programming, Kent, Canterbury, United Kingdom, 8–10 September 2014, pp. 111–122. ACM (2014). https://doi.org/10.1145/2643135.2643152

16. Lucas, S., Meseguer, J., Gutiérrez, R.: The 2D dependency pair framework for conditional rewrite systems. Part I: definition and basic processors. J. Comput. Syst. Sci. **96**, 74–106 (2018). https://doi.org/10.1016/j.jcss.2018.04.002

17. Lucas, S., Meseguer, J., Gutiérrez, R.: The 2D dependency pair framework for conditional rewrite systems—Part II: advanced processors and implementation techniques. J. Autom. Reasoning (2020, in press)

18. McCune, W.: Prover9 and Mace4. https://www.cs.unm.edu/~mccune/mace4/

19. Meßner, F., Sternagel, C.: nonreach – a tool for nonreachability analysis. In: Vojnar, T., Zhang, L. (eds.) TACAS 2019. LNCS, vol. 11427, pp. 337–343. Springer, Cham (2019). https://doi.org/10.1007/978-3-030-17462-0_19

20. Middeldorp, A., Nagele, J., Shintani, K.: Confluence competition 2019. In: Beyer, D., Huisman, M., Kordon, F., Steffen, B. (eds.) TACAS 2019. LNCS, vol. 11429, pp. 25–40. Springer, Cham (2019). https://doi.org/10.1007/978-3-030-17502-3_2

21. Nishida, N., Maeda, Y.: Narrowing trees for syntactically deterministic conditional term rewriting systems. In: Kirchner, H. (ed.) Proceedings of the 3rd International Conference on Formal Structures for Computation and Deduction. FSCD 2018. Leibniz International Proceedings in Informatics (LIPIcs), vol. 108, pp. 26:1–26:20. Schloss Dagstuhl-Leibniz-Zentrum fuer Informatik (2018). https://doi.org/10.4230/LIPIcs.FSCD.2018.26

22. Ohlebusch, E.: Advanced Topics in Term Rewriting. Springer, Heidelberg (2002). http://www.springer.com/computer/swe/book/978-0-387-95250-5

23. Prawitz, D.: Natural Deduction: A Proof-Theoretical Study. Dover, New York (2006)

24. Sternagel, C., Sternagel, T., Middeldorp, A.: CoCo 2018 Participant: ConCon 1.5. In: Felgenhauer, B., Simonsen, J. (eds.) Proceedings of the 7th International Workshop on Confluence. IWC 2018, p. 66 (2018). http://cl-informatik.uibk.ac.at/events/iwc-2018/

25. Sternagel, C., Yamada, A.: Reachability analysis for termination and confluence of rewriting. In: Vojnar, T., Zhang, L. (eds.) TACAS 2019. LNCS, vol. 11427, pp. 262–278. Springer, Cham (2019). https://doi.org/10.1007/978-3-030-17462-0_15

26. Winkler, S., Moser, G.: MædMax: a maximal ordered completion tool. In: Galmiche, D., Schulz, S., Sebastiani, R. (eds.) IJCAR 2018. LNCS (LNAI), vol. 10900, pp. 472–480. Springer, Cham (2018). https://doi.org/10.1007/978-3-319-94205-6_31

# MU-TERM: Verify Termination Properties Automatically (System Description)

Raúl Gutiérrez[ID] and Salvador Lucas[(✉)][ID]

Valencian Research Institute for Artificial Intelligence,
Universitat Politècnica de València,
Camino de Vera s/n, 46022 Valencia, Spain
{rgutierrez,slucas}@dsic.upv.es

**Abstract.** We report on the new version of MU-TERM, a tool for proving termination properties of variants of rewrite systems, including conditional, context-sensitive, equational, and order-sorted rewrite systems. We follow a unified, logic-based approach to describe rewriting computations. The automatic generation of *logical models* for suitable first-order theories and formulas provides a common basis to implement the proofs.

## 1  Introduction

MU-TERM is a tool that can be used to automatically verify termination properties of variants of *Term Rewriting Systems* (TRSs): termination and innermost termination of TRSs using the DP Framework for TRSs [10] (this framework is also used to prove termination of *String Rewriting Systems*); termination and innermost termination of *context-sensitive rewriting* [16,17] using the *Context-Sensitive DP Framework* [2,13]; termination of *rewriting modulo associative/commutative theories* using the $A\lor C$-DP Framework [4]; termination of *order-sorted rewriting* using the *Order-Sorted DP Framework* [22]; and operational termination of Conditional TRSs (CTRSs) using the *2D DP Framework* [24,25]. In this setting, describing different kinds of rewriting computations as *proofs* of goals $s \to t$ and $s \to^* t$ with respect to an appropriate inference system is useful. Such an approach, exploiting the logic-based description of rewriting computations, involves the use of several techniques which have been recently investigated elsewhere: (i) the generation logical models and well-founded relations [19], (ii) modeling operational termination of CTRSs with conditional dependency pairs [23], (iii) the use of removal triples [24] generated by logical models [25], etc. Giving support to such techniques in termination proofs motivated the development of a new version of our tool, MU-TERM 6.0:

Supported by EU (FEDER), and projects RTI2018-094403-B-C32,PROMETEO/ 2019/098, and SP20180225. Also by INCIBE program "Ayudas para la excelencia de los equipos de investigación avanzada en ciberseguridad" (Raúl Gutiérrez).

N. Peltier and V. Sofronie-Stokkermans (Eds.): IJCAR 2020, LNAI 12167, pp. 436–447, 2020.
https://doi.org/10.1007/978-3-030-51054-1_28

http://zenon.dsic.upv.es/muterm

We report on the new logic-based approach followed by MU-TERM 6.0, and also on the new features included since the last description of the system in 2010 [3].

## 2   New Features of MU-TERM

In the following, we enumerate the new features of MU-TERM 6.0 and illustrate them with some examples. Examples are intended to provide a better understanding of the techniques and often display solutions not necessarily obtained by an automatic proof with the tool, where the use of a specific proof strategy combining a sequence of *several* techniques (see Sect. 3) may dismiss the focused technique. Although we use some examples from other papers, all proofs of (operational) termination displayed here are new. For instance, the CTRS in Example 1 was proved operationally terminating in [25, Example 33], but the use of models in Example 3 below to show the absence of a link in the dependency graph is new. Examples 6 and 7 (of Order-Sorted TRSs) are discussed and proved here for the first time.

### 2.1   Logic-Based Representation of CTRSs

Given an oriented CTRS $\mathcal{R}$, with rules $\ell \to r \Leftarrow s_1 \approx t_1, \ldots, s_n \approx t_n$,[1] an inference system $\mathcal{I}(\mathcal{R})$ is obtained from the following generic inference system $\mathfrak{I}_{\mathrm{CTRS}}$

$$(\mathrm{Rf}) \quad \frac{}{x \to^* x} \qquad (\mathrm{C})_{f,i} \quad \frac{x_i \to y_i}{f(x_1, \ldots, x_i, \ldots, x_k) \to f(x_1, \ldots, y_i, \ldots, x_k)}$$
$$\text{for all } f \in \mathcal{F}^{(k)} \text{ and } 1 \le i \le k$$

$$(\mathrm{T}) \quad \frac{x \to y \quad y \to^* z}{x \to^* z} \qquad (\mathrm{Rl})_\alpha \quad \frac{s_1 \to^* t_1 \quad \cdots \quad s_n \to^* t_n}{\ell \to r}$$
$$\text{for } \alpha : \ell \to r \Leftarrow s_1 \approx t_1, \ldots, s_n \approx t_n \in \mathcal{R}$$

by *specializing* $(\mathrm{C})_{f,i}$ for each $k$-ary symbol $f$ in the signature $\mathcal{F}$ and $1 \le i \le k$, and $(\mathrm{Rl})_\alpha$ for all conditional rules $\alpha : \ell \to r \Leftarrow c$ in $\mathcal{R}$. Rules $\frac{B_1 \, \cdots \, B_n}{A}$ in $\mathcal{I}(\mathcal{R})$ are *schematic*: they can be used under any *instance* $\frac{\sigma(B_1) \, \cdots \, \sigma(B_n)}{\sigma(A)}$ by a substitution $\sigma$. We write $s \to_{\mathcal{R}} t$ $(s \to_{\mathcal{R}}^* t)$ iff there is a proof tree for $s \to t$ $(s \to^* t)$ using $\mathcal{I}(\mathcal{R})$. Operational termination of $\mathcal{R}$ is defined as the absence of infinite proof trees for goals $s \to t$ and $s \to^* t$ in $\mathcal{I}(\mathcal{R})$ [21]. In the analysis of computational properties of $\mathcal{R}$, we use the first-order *theory* $\overline{\mathcal{R}}$ obtained from $\mathcal{I}(\mathcal{R})$ by translating the inference rules $(\rho)\frac{B_1 \, \cdots \, B_n}{A}$ in $\mathcal{I}(\mathcal{R})$ into *sentences* $\overline{\rho}$ of the form $(\forall \boldsymbol{x}) \, B_1 \wedge \cdots \wedge B_n \Rightarrow A$, for $\boldsymbol{x}$ the sequence of variables occurring in $A, B_1, \ldots, B_n$ [18, Sect. 4.5].

---

[1] Oriented CTRSs treat conditions $s_i \approx t_i$ in rules as *rewriting* goals $\sigma(s_i) \to^* \sigma(t_i)$ for appropriate substitutions $\sigma$ [27, Definition 7.1.3].

$$x \to^* x$$

$$x_1 \to y_1 \Rightarrow \mathsf{f}(x_1, x_2, x_3) \to \mathsf{f}(y_1, x_2, x_3)$$

$$x_3 \to y_3 \Rightarrow \mathsf{f}(x_1, x_2, x_3) \to \mathsf{f}(x_1, x_2, y_3)$$

$$x \to y \Rightarrow \mathsf{g}(x) \to \mathsf{g}(y)$$

$$x \to y \Rightarrow \mathsf{k}(x) \to \mathsf{k}(y)$$

$$\mathsf{h}(\mathsf{d}) \to \mathsf{c}(\mathsf{b})$$

$$\mathsf{h}(x) \to^* \mathsf{d} \wedge \mathsf{h}(x) \to^* \mathsf{c}(y) \Rightarrow \mathsf{g}(x) \to \mathsf{k}(y)$$

$$x \to y \wedge y \to^* z \Rightarrow x \to^* z$$

$$x_2 \to y_2 \Rightarrow \mathsf{f}(x_1, x_2, x_3) \to \mathsf{f}(x_1, y_2, x_3)$$

$$x \to y \Rightarrow \mathsf{c}(x) \to \mathsf{c}(y)$$

$$x \to y \Rightarrow \mathsf{h}(x) \to \mathsf{h}(y)$$

$$\mathsf{h}(\mathsf{d}) \to \mathsf{c}(\mathsf{a})$$

$$\mathsf{f}(\mathsf{k}(\mathsf{a}), \mathsf{k}(\mathsf{b}), x) \to \mathsf{f}(x, x, x)$$

**Fig. 1.** First-order theory $\overline{\mathcal{R}}$ for $\mathcal{R}$ in Example 1 (all variables universally quantified)

*Example 1.* For the following CTRS $\mathcal{R}$ [27, Example 7.2.51]

$$\mathsf{h}(\mathsf{d}) \to \mathsf{c}(\mathsf{a}) \tag{1}$$

$$\mathsf{h}(\mathsf{d}) \to \mathsf{c}(\mathsf{b}) \tag{2}$$

$$\mathsf{f}(\mathsf{k}(\mathsf{a}), \mathsf{k}(\mathsf{b}), x) \to \mathsf{f}(x, x, x) \tag{3}$$

$$\mathsf{g}(x) \to \mathsf{k}(y) \Leftarrow \mathsf{h}(x) \approx \mathsf{d}, \mathsf{h}(x) \approx \mathsf{c}(y) \tag{4}$$

the theory $\overline{\mathcal{R}}$ is given in Fig. 1.

## 2.2 Operational Termination of Conditional Rewrite Systems

In [24,25] a framework for automatically proving operational termination of (oriented) CTRSs using appropriate notions of *dependency pairs* (adapting the original notion for TRSs [5]) has been introduced: the 2D DP Framework.

**Dependency Pairs for CTRSs.** Given a CTRS $\mathcal{R}$, two new CTRSs $\mathsf{DP}_H(\mathcal{R})$ and $\mathsf{DP}_V(\mathcal{R})$ are introduced to capture the two *horizontal* and *vertical* dimensions of operational termination of CTRSs [23]: the usual absence of infinite rewrite sequences (termination), and the absence of infinite 'climbs' on a proof tree when trying to prove a goal $s \to t$ or $s \to^* t$ (called $V$-termination). $\mathsf{DP}_H(\mathcal{R})$ consists of rules $u \to v \Leftarrow c$ whose terms $u$ and $v$ capture the progress of infinite rewrite sequences involving rules $\ell \to r \Leftarrow c$ with $u$ and $v$ marked versions of $\ell$ and a subterm of $r$ respectively (only the root symbol $f$ is marked as $f^\sharp$, or just capitalized: $F$). Similarly, $\mathsf{DP}_V(\mathcal{R})$ consists of rules $u \to v \Leftarrow d$ where $v$ is a marked subterm of $s_i$ for a condition $s_i \approx t_i$ in $c$ and $d$ is $s_1 \approx t_1, \ldots, s_{i-1} \approx t_{i-1}$.[2]

*Example 2.* For $\mathcal{R}$ in Example 1, we have $\mathsf{DP}_H(\mathcal{R}) = \{\mathsf{F}(\mathsf{k}(\mathsf{a}), \mathsf{k}(\mathsf{b}), x) \to \mathsf{F}(x, x, x)\}$ and $\mathsf{DP}_V(\mathcal{R}) = \{\mathsf{G}(x) \to \mathsf{H}(x), \mathsf{G}(x) \to \mathsf{H}(x) \Leftarrow \mathsf{h}(x) \approx \mathsf{d}\}$

---

[2] A third set of dependency pairs $\mathsf{DP}_{VH}(\mathcal{R}) \subseteq \mathsf{DP}_H(\mathcal{R})$ is used in [23]. For simplicity, in the examples of this paper, $\mathsf{DP}_{VH}(\mathcal{R})$ is empty and we pay no attention to it.

As in [7, Sect. 5], we use a set of sorts $S_{DP} = \{\mathsf{S}, \mathsf{P}\}$ so that symbols $f$ are (automatically) given a rank $f : \mathsf{S} \cdots \mathsf{S} \to \mathsf{S}$ and *marked* symbols are given rank $F : \mathsf{S} \cdots \mathsf{S} \to \mathsf{P}$ [25, Sect. 4.3]. Variables of formulas in $\overline{\mathcal{R}}$ (e.g., Fig. 1) are then assumed to be universally quantified on sort $\mathsf{S}$.

**The 2D DP Framework for CTRSs.** The absence of infinite *chains* of 2D DPs (i.e., sequences of 2D DPs which model infinite branches in the proof trees for goals $s \to t$ and $s \to^* t$) characterizes operational termination of CTRSs [23]. This is proved using a divide-and-conquer strategy which successively decomposes operational termination problems into smaller and simpler ones. Processors $\mathsf{P}$ are used for this purpose [24]. They simplify problems by decomposing or shrinking them. In particular, the appropriate *estimation* of graphs $\mathcal{G}$ whose nodes are dependency pairs is useful to analyze the existence of such infinite chains as *cycles* in the graph. The absence of cycles implies operational termination. The presence of conditional rules and pairs introduces some particular issues which we enumerate, and discuss below.

1. Some pairs could be *infeasible*, i.e., unable to be used in any of the afore-mentioned chains. Then, we could remove them [24, Sect. 4]. Also, arcs in $\mathcal{G}$ are defined by specific (often undecidable) *sequences* $s_1 \bowtie_1 t_1, \ldots, s_n \bowtie_n t_n$ (called f-sequences [15]) where $s_i$ and $t_i$ are terms and $\bowtie_i$ are predicates $\to_i^*$ that capture the possibility of having two nodes involved in a *chain* [20, Sect. 4.5] and must be proved feasible or infeasible (for some substitution $\sigma$ which applies to terms $s_i$ and $t_i$). A typical strategy is discarding arcs whose associated sequence is *infeasible*. We discuss this in paragraph *Infeasibility in Termination Proofs* below (see Examples 3 and 4).
2. Some pairs could be '*harmless*', i.e., unable to be persistently used in any infinite chain. This can be shown if we prove a 'decrease' when such pairs are used in a chain. Again, we can remove them to obtain a simplification [25, Sect. 4.3]. We discuss this in paragraph *Use of Well-Founded Relations* below (see Example 5).

**Infeasibility in Termination Proofs.** Given a (C)TRS $\mathcal{R}$ we say that a sequence $s_1 \to^* t_1, \ldots, s_n \to^* t_n$ is $\mathcal{R}$-infeasible if there is no substitution $\sigma$ such that $\sigma(s_i) \to_{\mathcal{R}}^* \sigma(t_i)$ holds for all $1 \leq i \leq n$. In [20] it is proved that a sequence $s_1 \to^* t_1, \ldots, s_n \to^* t_n$ is $\mathcal{R}$-infeasible if there is a model of $\overline{\mathcal{R}} \cup \{\neg(\exists \boldsymbol{x}) \, s_1 \to^* t_1, \ldots, s_n \to^* t_n\}$, where $\boldsymbol{x}$ contains the variables in $s_1, t_1, \ldots, s_n, t_n$. In termination proofs, proving infeasibility is useful at different levels. As remarked above, when the conditional part $c$ of a pair $u \to v \Leftarrow c$ is proved infeasible, we can remove it. Also, the absence of an arc between two nodes (pairs) $u \to v \Leftarrow c$ and $u' \to v' \Leftarrow c'$ in the graph $\mathcal{G}$ can be treated as the infeasibility of $v \to^* u'$ (where, as usual, we assume that $v$ and $u'$ share no variable). For instance, for $\mathcal{R}$ in Example 1, it is possible to prove that there is no arc in the 'horizontal' graph which consists of a single node $\mathsf{F}(\mathsf{k}(\mathsf{a}), \mathsf{k}(\mathsf{b}), x) \to \mathsf{F}(x, x, x)$ (the only dependency pair in $\mathsf{DP}_H(\mathcal{R})$) by just finding a model of

$$\overline{\mathcal{R}} \cup \{\neg(\exists x, y : \mathsf{S}) \, \mathsf{F}(x, x, x) \to^* \mathsf{F}(\mathsf{k}(\mathsf{a}), \mathsf{k}(\mathsf{b}), y)\} \tag{5}$$

For this purpose, model generators AGES [14] and Mace4 [26] are used by MU-TERM.

*Example 3.* We obtain a model $\mathcal{A}$ of (5) with Mace4. The domain is $\mathcal{A} = \{0,1\}$ (Mace4 does not support sorts; thus, both S and P are merged into a single sort). Function and predicate symbols are interpreted as follows:

$$\mathsf{a}^{\mathcal{A}} = \mathsf{d}^{\mathcal{A}} = 0 \qquad \mathsf{b}^{\mathcal{A}} = \mathsf{c}^{\mathcal{A}}(x) = 1 \quad \mathsf{f}^{\mathcal{A}}(x,y,z) = \mathsf{g}^{\mathcal{A}}(x) = 0$$
$$\mathsf{h}^{\mathcal{A}}(x) = 1 - x \qquad \mathsf{k}^{\mathcal{A}}(x) = x \quad \mathsf{F}^{\mathcal{A}}(x,y,z) = \begin{cases} 1 \text{ if } x = 0 \text{ and } y = 1 \\ 0 \text{ otherwise} \end{cases}$$
$$x(\to_{\mathcal{R}})^{\mathcal{A}} y \Leftrightarrow x = y \qquad x(\to_{\mathcal{R}}^{*})^{\mathcal{A}} y \Leftrightarrow x = y$$

Discarding the arc would not be possible by the usual unification-based technique in [9]. With regard to infeasibility of pairs, consider the following well-known example.

*Example 4.* Consider the following CTRS $\mathcal{R}$ [8, p. 46]:

$$\mathsf{a} \to \mathsf{b} \tag{6}$$
$$\mathsf{f}(\mathsf{a}) \to \mathsf{b} \tag{7}$$
$$\mathsf{g}(x) \to \mathsf{g}(\mathsf{a}) \Leftarrow \mathsf{f}(x) \approx x \tag{8}$$

where $\mathsf{DP}_H(\mathcal{R}) = \{\mathsf{G}(x) \to \mathsf{G}(\mathsf{a}) \Leftarrow \mathsf{f}(x) \approx x, \mathsf{G}(x) \to \mathsf{A} \Leftarrow \mathsf{f}(x) \approx x\}$. Both pairs in $\mathsf{DP}_H(\mathcal{R})$ are $\mathcal{R}$-infeasible: no substitution $\sigma$ makes $\sigma(\mathsf{f}(x)) \to^{*} \sigma(x)$ true. We can prove it if a model $\mathcal{A}$ of $\overline{\mathcal{R}} \cup \{\neg(\exists x)\,\mathsf{f}(x) \to^{*} x\}$ can be found. We obtain a model with AGES. The domain is $\mathcal{A} = \mathbb{N} - \{0\}$ (since no marked symbol is involved, we can use a single interpretation domain); for function and predicate symbols:

$$\mathsf{a}^{\mathcal{A}} = 1 \quad \mathsf{b}^{\mathcal{A}} = 2 \quad \mathsf{f}^{\mathcal{A}}(x) = x + 1 \quad \mathsf{g}^{\mathcal{A}}(x) = 1 \quad x(\to_{\mathcal{R}})_{\mathsf{s}}^{\mathcal{A}} y \Leftrightarrow x(\to_{\mathcal{R}}^{*})_{\mathsf{s}}^{\mathcal{A}} y \Leftrightarrow y \geq x$$

We can safely remove both pairs. Thus no infinite chain of pairs in $\mathsf{DP}_H(\mathcal{R})$ exists.

**Use of Well-Founded Relations.** The *removal triple processor* [24, Def. 70] implements the use of removal triples $(\gtrsim, \succeq, \sqsupset)$, including a well-founded relation $\sqsupset$ to remove pairs from, and hence simplify, termination problems. For instance, as shown in [25, Sect. 4.3], $\mathcal{R}$ in Example 1 is operationally terminating if we find a model $\mathcal{A}$ of

$$\mathcal{S}_{\mathcal{R}}^{RT} \cup \{(\forall x : \mathsf{S})\,\mathsf{F}(\mathsf{k}(\mathsf{a}), \mathsf{k}(\mathsf{b}), x)\,\pi_{\sqsupset}\,\mathsf{F}(x, x, x)\} \tag{9}$$

where $\pi_{\sqsupset}$ (a new predicate symbol representing $\sqsupset$) is interpreted as a *well-founded relation* $\pi_{\sqsupset}^{\mathcal{A}}$, and $\mathcal{S}_{\mathcal{R}}^{RT}$ extends $\overline{\mathcal{R}}$ with the following additional requirements to apply the processor [24, Definitions 68 and 69]:

$$(\forall x, y : \mathsf{P})\,x\,\pi_{\gtrsim}\,y \wedge y\,\pi_{\sqsupset}\,z \Rightarrow x\,\pi_{\sqsupset}\,z \tag{10}$$
$$(\forall x, y : \mathsf{P})\,x \to y \Rightarrow x\,\pi_{\gtrsim}\,y \tag{11}$$

No predicate $\pi_{\succeq}$ is necessary in this example (where a single pair is considered).

*Example 5.* We obtain a model $\mathcal{A}$ of (9) with AGES. Domains are $\mathcal{A}_P = \{-1, 0, 1\}$ and $\mathcal{A}_S = \{0, 1\}$. With regard to function and predicate symbols:

$$\mathsf{a}^{\mathcal{A}} = \mathsf{d}^{\mathcal{A}} = 0 \quad \mathsf{b}^{\mathcal{A}} = \mathsf{c}^{\mathcal{A}}(x) = 1 \qquad \mathsf{f}^{\mathcal{A}}(x, y, z) = \mathsf{g}^{\mathcal{A}}(x) = 0$$
$$\mathsf{h}^{\mathcal{A}}(x) = 1 \qquad \mathsf{k}^{\mathcal{A}}(x) = 1 - x \quad \mathsf{F}^{\mathcal{A}}(x, y, z) = x - y$$

$$x \, (\to)_P^{\mathcal{A}} \, y \Leftrightarrow x \geq y \qquad x \, (\to^*)_P^{\mathcal{A}} \, y \Leftrightarrow true \qquad x \, (\to)_S^{\mathcal{A}} \, y \Leftrightarrow x = y$$
$$x \, (\to^*)_S^{\mathcal{A}} \, y \Leftrightarrow y \geq x \qquad x \, \pi_{\gtrsim}^{\mathcal{A}} \, y \Leftrightarrow x \geq y \qquad x \, \pi_{\sqsupset}^{\mathcal{A}} \, y \Leftrightarrow 6x \geq 1 + 6y$$

where, as in the semantic approach in [25, Sect. 4.3], $\to$ and $\to^*$ are *overloaded* for sorts P and S; thus, $(\to)_P^{\mathcal{A}}$, $(\to^*)_P^{\mathcal{A}}$, $(\to)_S^{\mathcal{A}}$, and $(\to^*)_S^{\mathcal{A}}$ are the corresponding interpretations. Note that $\pi_{\sqsupset}^{\mathcal{A}} = \{(x, y) \mid x, y \in \mathcal{A}_P, 6x \geq 1 + 6y\} = \{(0, -1), (1, -1), (1, 0)\}$ is well-founded on $\mathcal{A}_P$. Thus, we conclude operational termination of $\mathcal{R}$.

## 2.3  Termination of Order-Sorted Rewriting

Sorts are often used to reinforce program termination. Order-sorted dependency pairs were introduced in [22] for proving termination of order-sorted TRSs.

*Example 6.* The following many-sorted TRS $\mathcal{R}$ in [29, Sect. 3.3] (in the hopefully self-explained Maude format [6]) is a terminating version of Toyama's example, which is nonterminating as a TRS (i.e., without sort information):

```
mod Toyama-MS is
 sorts S1 S2 .
 ops a b : -> S1 .    op f : S1 S1 S1 -> S1 .   op g : S2 S2 -> S2 .
 vars x : S1 .        vars y z : S2 .
 rl g(y,z) => y .     rl g(y,z) => z .          rl f(a,b,x) => f(x,x,x) .
endm
```

The 2010 version of MU-TERM could not prove it terminating.[3] According to [22], $\mathcal{R}$ has a single dependency pair:

$$\mathsf{F}(\mathsf{a}, \mathsf{b}, x) \to \mathsf{F}(x, x, x) \tag{12}$$

where $\mathsf{F}$ has rank $\mathsf{S1} \, \mathsf{S1} \, \mathsf{S1} \to \mathsf{P}$ for a new sort P [22, Sect. 3.2] and $x$ has sort S1. We can prove that the dependency graph consisting of this single pair has no cycle. With AGES we can compute a model of $\overline{\mathcal{R}} \cup \{\neg(\exists x, y : \mathsf{S1}) \, \mathsf{F}(x, x, x) \to^* \mathsf{F}(\mathsf{a}, \mathsf{b}, y)\}$ which is as follows: $\mathcal{A}_{S1} = \{0, 1\}$, $\mathcal{A}_{S2} = \{1\}$, $\mathcal{A}_P = -\mathbb{N}$ (i.e., the set of nonpositive integers), and functions and predicates interpreted as follows:

$$\mathsf{a}^{\mathcal{A}} = 1 \qquad \mathsf{b}^{\mathcal{A}} = 0 \qquad \mathsf{f}^{\mathcal{A}}(x, y, z) = 1 \qquad \mathsf{g}^{\mathcal{A}}(x) = 1 \qquad \mathsf{F}^{\mathcal{A}}(x, y, z) = x - y - 1$$
$$x(\to_{\mathcal{R}})_{S1}^{\mathcal{A}} y \Leftrightarrow x = y = 1 \qquad x(\to_{\mathcal{R}}^*)_{S1}^{\mathcal{A}} y \Leftrightarrow true \qquad x(\to_{\mathcal{R}})_{S2}^{\mathcal{A}} y \Leftrightarrow true$$
$$x(\to_{\mathcal{R}}^*)_{S2}^{\mathcal{A}} y \Leftrightarrow true \qquad x(\to_{\mathcal{R}})_{P}^{\mathcal{A}} y \Leftrightarrow x = y \qquad x(\to_{\mathcal{R}}^*)_{P}^{\mathcal{A}} y \Leftrightarrow x \geq y$$

---

[3] Benchmarks available here: http://zenon.dsic.upv.es/muterm/benchmarks/benchmarks-ostrs/benchmarks.html.

The crucial point to obtain the proof in Example 6 is the ability to provide different interpretations to different sorts. The following example from [28] could not be handled by the 2010 version of MU-TERM because orderings were generated without paying attention to sorts (see [22, Sect. 6]).

*Example 7.* The following OS-TRS $\mathcal{R}$ [28, Example 11] is nonterminating as a TRS:

```
mod Example11-OL96 is
  sorts S S1 S2 S3 S4 .    subsorts S1 S2 S3 S4 < S .
  ops f g : S -> S .       op g : S3 -> S1 .           op g : S4 -> S2 .
  op h : S1 -> S2 .        op a : -> S3 .              op b : -> S4 .
  var x : S1 .             rl f(x) => f(h(x)) .        rl a => b .
endm
```

There is a single OS-DP for $\mathcal{R}$: $\mathsf{F}(x) \to \mathsf{F}(\mathsf{h}(x))$, where $\mathsf{F}$ has rank $\mathsf{S} \to \mathsf{P}$ for a new sort $\mathsf{P}$ and $x$ has sort $\mathsf{S1}$. We can prove termination of $\mathcal{R}$ by finding a removal triple $(\gtrsim, \succeq, \sqsupset)$ such that the rules of $\mathcal{R}$ are compatible with $\gtrsim$, and $\mathsf{F}(x) \sqsupset \mathsf{F}(\mathsf{h}(x))$ holds whenever $x$ ranges on terms of sort $\mathsf{S1}$. With AGES we obtain an interpretation $\mathcal{A}$ as follows: sorts are interpreted as $\mathcal{A}_{\mathsf{S}} = \{-1, 0, 1\}$, $\mathcal{A}_{\mathsf{S1}} = \{-1\}$, $\mathcal{A}_{\mathsf{S2}} = \{-1, 0\}$, $\mathcal{A}_{\mathsf{S3}} = \{-1\}$, $\mathcal{A}_{\mathsf{S4}} = \{-1, 0\}$, and $\mathcal{A}_{\mathsf{P}} = \mathbb{N} \cup \{-1\}$. Functions and predicates are interpreted as follows:

$$\mathsf{a}^{\mathcal{A}} = -1 \qquad \mathsf{b}^{\mathcal{A}} = 0 \qquad \mathsf{f}^{\mathcal{A}}(x, y, z) = x \qquad \mathsf{g}_{\mathsf{S}}^{\mathcal{A}}(x) = x$$
$$\mathsf{g}_{\mathsf{S3}}^{\mathcal{A}}(x) = -1 \qquad \mathsf{g}_{\mathsf{S4}}^{\mathcal{A}}(x) = x \qquad \mathsf{h}^{\mathcal{A}}(x, y, z) = 0 \qquad \mathsf{F}^{\mathcal{A}}(x, y, z) = -x$$

(where different overloaded versions of $\mathsf{g}$ use the input sort as a subindex) and

$$x(\to_{\mathcal{R}})_{\mathsf{S}}^{\mathcal{A}} y \Leftrightarrow y = 0 \wedge x = -1 \qquad x(\to_{\mathcal{R}}^{*})_{\mathsf{S}}^{\mathcal{A}} y \Leftrightarrow true \qquad x(\to_{\mathcal{R}})_{\mathsf{P}}^{\mathcal{A}} y \Leftrightarrow x \geq y$$
$$x(\to_{\mathcal{R}}^{*})_{\mathsf{P}}^{\mathcal{A}} y \Leftrightarrow true \qquad x \gtrsim^{\mathcal{A}} y \Leftrightarrow x \geq y \qquad x \sqsupset^{\mathcal{A}} y \Leftrightarrow x > y$$

Note that the interpretation of the 'original' rewrite relation concerns sort $\mathsf{S}$ only because it is the top sort of the full sort hierarchy.

## 2.4   Termination of Context-Sensitive Rewriting

In *context-sensitive rewriting* (CSR [16]), a *replacement map* $\mu$ is used to restrict the arguments $\mu(f) \subseteq \{1, \ldots, k\}$ which can be rewritten for each $k$-ary symbol $f$. The restriction on arguments is top-down propagated to positions of terms $t$, which are called *active* positions of $t$. We write $s \hookrightarrow t$ if an *active* subterm of $s$ can be rewritten so that $s \to t$. In the *dependency pair approach* for proving termination of CSR [2], rules of the form $f(\ell_1, \ldots, \ell_k) \to r$ are given *dependency pairs* $f^{\sharp}(\ell_1, \ldots, \ell_k) \to g^{\sharp}(s_1, \ldots, s_m)$, for $s = g(s_1, \ldots, s_m)$ a *replacing* subterm of $r$ (i.e., a subterm $s = r|_p$ occurring at an active position $p$ of $r$) and $g$ a defined symbol. The notation $f^{\sharp}$ means that $f$ is *marked* (capital letters $F$ are often used instead of $f^{\sharp}$). However, due to rules $\ell \to r \in \mathcal{R}$ with *migrating* variables $x \in \mathcal{V}ar^{\mu}(r) \backslash \mathcal{V}ar^{\mu}(\ell)$ (that are frozen, i.e., not active, in $\ell$ but become active in $r$, possibly 'awaking' infinite rewrite sequences), we also need *collapsing dependency pairs* $\ell^{\sharp} \to x$ where $x$ is a *migrating variable* of the rule.

*Example 8.* For the following TRS $\mathcal{R}$ in [30, Introduction]

$$\mathsf{p}(\mathsf{s}(x)) \rightarrow x \qquad\qquad \mathsf{if}(\mathsf{true}, x, y) \rightarrow x$$
$$0 + x \rightarrow x \qquad\qquad \mathsf{if}(\mathsf{false}, x, y) \rightarrow y$$
$$\mathsf{s}(x) + y \rightarrow \mathsf{s}(x + y) \qquad\qquad \mathsf{zero}(0) \rightarrow \mathsf{true}$$
$$0 \times y \rightarrow 0 \qquad\qquad \mathsf{zero}(\mathsf{s}(x)) \rightarrow \mathsf{false}$$
$$\mathsf{s}(x) \times y \rightarrow y + (x \times y) \qquad\qquad \mathsf{fact}(x) \rightarrow \mathsf{if}(\mathsf{zero}(x), \mathsf{s}(0), \mathsf{fact}(\mathsf{p}(x)) \times x)$$

and $\mu$ given by $\mu(\mathsf{if}) = \{1\}$ and $\mu(f) = \{1, \ldots, k\}$ for any other $k$-ary symbol $f$ [12, Example 1]. $\mathsf{DP}(\mathcal{R}, \mu)$ consists of pairs

$$\mathsf{s}(x) +^{\sharp} y \rightarrow x +^{\sharp} y \qquad \mathsf{s}(x) \times^{\sharp} y \rightarrow y +^{\sharp} (x \times y) \qquad \mathsf{s}(x) \times^{\sharp} y \rightarrow x \times^{\sharp} y$$
$$\mathsf{FACT}(x) \rightarrow \mathsf{ZERO}(x) \qquad \mathsf{FACT}(x) \rightarrow \mathsf{IF}(\mathsf{zero}(x), \mathsf{s}(0), \mathsf{fact}(\mathsf{p}(x)) \times x)$$
$$\mathsf{IF}(\mathsf{true}, x, y) \rightarrow x \qquad \mathsf{IF}(\mathsf{false}, x, y) \rightarrow y$$

Collapsing pairs capture a kind of *recursion* which is *hidden* below frozen parts of the terms involved in infinite context-sensitive rewrite sequences until a *migrating* variable within a rule $\ell \rightarrow r$ shows them up. The *hidden terms* of a TRS $\mathcal{R}$ are defined subterms occurring at frozen positions in the *rhs* of some rule of $\mathcal{R}$ [2]. *Hiding contexts* are contexts where hidden terms can occur *at active positions* within a context-sensitive rewrite sequence [1,13]. There, hidden terms can *restart* a delayed recursive call after the application of a rule with migrating variables (see [12] for a detailed analysis). For $\mathcal{R}$ and $\mu$ in Example 8, the only rule with hidden terms is $\mathsf{fact}(x) \rightarrow \mathsf{if}(\mathsf{zero}(x), \mathsf{s}(0), \mathsf{fact}(\mathsf{p}(x)) \times x)$. Symbols $\mathsf{fact}$ and '$\times$' hide position 1 because $\mathsf{p}(x)$ is rooted by a defined symbol. Symbol '$\times$' does *not* hide position 2. Symbol $\mathsf{p}$ hides no position. The refinements introduced in [12] have led to a more precise notion of hidden terms and contexts, enabling a better analysis of the connections between them. This has greatly improved the ability of MU-TERM to prove termination of CSR. For instance, the proof of termination of $\mathcal{R}$ and $\mu$ in Example 8, which could not be obtained with the 2010 version of MU-TERM, is now possible with MU-TERM 6.0, see the proof of `CSR_04/ExIntrod_Zan97.xml` in the 2019 Termination Competition

http://group-mmm.org/termination/competitions/Y2019/caches/
termination_33019.html

or in our local benchmarks:

http://zenon.dsic.upv.es/muterm/benchmarks/ijcar20/TRS_
Contextsensitive/benchmarks.html

## 3   Termination Expert

The arbitrary application of processors can generate a huge search space. Furthermore, proofs usually proceed under some *timeout*. For this reason, we need to choose a fixed strategy where fast processors that reduce the number of rules are first used, and slow processors, or processors that increase the number of rules, are used when fast processors fail. Hence, the frequency of use for the different processors depends on the chosen strategy. With small differences depending on the particular kind of problem, we do the following:

1. If $\mathcal{R}$ is a TRS or a CS-TRS, we check whether the system is innermost equivalent [3, Sect. 2.2]. If it is true, then we transform the problem into an innermost one.
2. Then, we obtain the corresponding dependency pairs, obtaining a CTRS, OS, CS, or DP problem. Then we perform the following steps repeatedly
   (a) Decision point between processors for proving (operational) non-termination and the *strongly connected component* (SCC) processor.
   (b) Subterm criterion processor.
   (c) Removal triple processor generating models with AGES (we try different configurations, from simpler to more complex).
   (d) If $\mathcal{R}$ is a CTRS, we apply simplification and removal processors on the conditions (using AGES when a model is necessary).
   (e) Transformation processors on rules, pairs and conditions: instantiation, forward instantiation, and narrowing.

Full explanations of the processors can be found in [4, 12, 13, 19, 20, 24, 25]. The MU-TERM 6.0 logic-based approach has led to dramatic improvements, as reported here:

   http://zenon.dsic.upv.es/muterm/benchmarks/ijcar20/Comparison/
   benchmarks.html

where the use of logical models is compared with the exclusive use of *polynomial interpretations* (as in MU-TERM 5.0). Polynomial interpretations are strictly less powerful in terms of solved examples (as every proof using polynomial interpretations can be obtained using the new logic-based approach). However, we *keep* them in MU-TERM 6.0 as they lead to *faster* proofs. We use polynomial interpretations as part of MU-TERM 6.0 strategy (via the *removal triple* processor).

   MU-TERM 6.0 consists of more than 30000 lines of Haskell code. In the web-based interface, besides the fully automatic use of the termination expert, we can also use specific techniques like polynomial orderings, matrix interpretations, (context-sensitive) recursive path ordering, etc., which we have found useful for *teaching* purposes.

## 4    Experimental Evaluation

Since 2014, MU-TERM has proven to be the most powerful tool for proving operational termination of *conditional* rewriting and termination of *context-sensitive* rewriting, each year winning the corresponding subcategory of the annual International Competition of Termination Tools, see http://zenon.dsic. upv.es/muterm/?page_id=82 for an historical account. In the CSR subcategory, since 2014 MU-TERM is able to prove *all* the examples proved by any other participating tool (thanks to the results in [12]).

The benchmarks web page of MU-TERM reports on specific experiments comparing the 2010 and 2020 versions. First, the 2010 version did not support CTRSs. For CS-TRSs, three new examples can be proved now (and all the examples handled by the 2010 version are also handled now). As for OS-TRSs, MU-TERM 6.0 is able to prove or disprove termination of *all* the OS-TRSs in the 2010 benchmark suite (except a non-sort-decreasing OS-TRS, not covered by the theory in [22], where sort-decreasingness [11] is required). The 2010 version could not disprove termination of OS-TRSs.

**Acknowledgments.** We thank the anonymous referees for many remarks and suggestions that led to improve the paper.

# References

1. Alarcón, B., et al.: Improving context-sensitive dependency pairs. In: Cervesato, I., Veith, H., Voronkov, A. (eds.) LPAR 2008. LNCS (LNAI), vol. 5330, pp. 636–651. Springer, Heidelberg (2008). https://doi.org/10.1007/978-3-540-89439-1_44
2. Alarcón, B., Gutiérrez, R., Lucas, S.: Context-sensitive dependency pairs. Inf. Comput. **208**(8), 922–968 (2010). https://doi.org/10.1016/j.ic.2010.03.003
3. Alarcón, B., Gutiérrez, R., Lucas, S., Navarro-Marset, R.: Proving termination properties with MU-TERM. In: Johnson, M., Pavlovic, D. (eds.) AMAST 2010. LNCS, vol. 6486, pp. 201–208. Springer, Heidelberg (2011). https://doi.org/10.1007/978-3-642-17796-5_12
4. Alarcón, B., Lucas, S., Meseguer, J.: A dependency pair framework for $A \vee C$-termination. In: Ölveczky, P.C. (ed.) WRLA 2010. LNCS, vol. 6381, pp. 35–51. Springer, Heidelberg (2010). https://doi.org/10.1007/978-3-642-16310-4_4
5. Arts, T., Giesl, J.: Termination of term rewriting using dependency pairs. Theor. Comput. Sci. **236**(1–2), 133–178 (2000). https://doi.org/10.1016/S0304-3975(99)00207-8
6. Clavel, M., et al.: All About Maude - A High-Performance Logical Framework. LNCS, vol. 4350. Springer, Heidelberg (2007). https://doi.org/10.1007/978-3-540-71999-1
7. Endrullis, J., Waldmann, J., Zantema, H.: Matrix interpretations for proving termination of term rewriting. J. Autom. Reasoning **40**(2–3), 195–220 (2008). https://doi.org/10.1007/s10817-007-9087-9
8. Giesl, J., Arts, T.: Verification of erlang processes by dependency pairs. Appl. Algebra Eng. Commun. Comput. **12**(1/2), 39–72 (2001). https://doi.org/10.1007/s002000100063
9. Giesl, J., Thiemann, R., Schneider-Kamp, P.: Proving and disproving termination of higher-order functions. In: Gramlich, B. (ed.) FroCoS 2005. LNCS (LNAI), vol. 3717, pp. 216–231. Springer, Heidelberg (2005). https://doi.org/10.1007/11559306_12
10. Giesl, J., Thiemann, R., Schneider-Kamp, P., Falke, S.: Mechanizing and improving dependency pairs. J. Autom. Reasoning **37**(3), 155–203 (2006). https://doi.org/10.1007/s10817-006-9057-7
11. Goguen, J.A., Meseguer, J.: Order-sorted algebra I: equational deduction for multiple inheritance, overloading, exceptions and partial operations. Theor. Comput. Sci. **105**(2), 217–273 (1992). https://doi.org/10.1016/0304-3975(92)90302-V

12. Gutiérrez, R., Lucas, S.: Function calls at frozen positions in termination of context-sensitive rewriting. In: Martí-Oliet, N., Ölveczky, P.C., Talcott, C. (eds.) Logic, Rewriting, and Concurrency. LNCS, vol. 9200, pp. 311–330. Springer, Cham (2015). https://doi.org/10.1007/978-3-319-23165-5_15

13. Gutiérrez, R., Lucas, S.: Proving termination in the context-sensitive dependency pair framework. In: Ölveczky, P.C. (ed.) WRLA 2010. LNCS, vol. 6381, pp. 18–34. Springer, Heidelberg (2010). https://doi.org/10.1007/978-3-642-16310-4_3

14. Gutiérrez, R., Lucas, S.: Automatic generation of logical models with AGES. In: Fontaine, P. (ed.) CADE 2019. LNCS (LNAI), vol. 11716, pp. 287–299. Springer, Cham (2019). https://doi.org/10.1007/978-3-030-29436-6_17

15. Gutiérrez, R., Lucas, S.: Automatically proving and disproving feasibility conditions. In: Peltier, N., Sofronie-Stokkermans, V. (eds.) IJCAR 2020. LNAI, vol. 12167, pp. 416–435. Springer, Heidelberg (2020)

16. Lucas, S.: Context-sensitive computations in functional and functional logic programs. J. Funct. Log. Program. **1998**(1), 1–61 (1998). http://danae.uni-muenster.de/lehre/kuchen/JFLP/articles/1998/A98-01/A98-01.html

17. Lucas, S.: Context-sensitive rewriting strategies. Inf. Comput. **178**(1), 294–343 (2002). https://doi.org/10.1006/inco.2002.3176

18. Lucas, S.: Proving semantic properties as first-order satisfiability. Artif. Intell. **277** (2019). https://doi.org/10.1016/j.artint.2019.103174

19. Lucas, S., Gutiérrez, R.: Automatic synthesis of logical models for order-sorted first-order theories. J. Autom. Reasoning **60**(4), 465–501 (2017). https://doi.org/10.1007/s10817-017-9419-3

20. Lucas, S., Gutiérrez, R.: Use of logical models for proving infeasibility in term rewriting. Inf. Process. Lett. **136**, 90–95 (2018). https://doi.org/10.1016/j.ipl.2018.04.002

21. Lucas, S., Marché, C., Meseguer, J.: Operational termination of conditional term rewriting systems. Inf. Process. Lett. **95**(4), 446–453 (2005). https://doi.org/10.1016/j.ipl.2005.05.002

22. Lucas, S., Meseguer, J.: Order-sorted dependency pairs. In: Antoy, S., Albert, E. (eds.) Proceedings of the 10th International ACM SIGPLAN Conference on Principles and Practice of Declarative Programming, 15–17 July 2008, Valencia, Spain, pp. 108–119. ACM (2008). https://doi.org/10.1145/1389449.1389463

23. Lucas, S., Meseguer, J.: Dependency pairs for proving termination properties of conditional term rewriting systems. J. Log. Algebraic Methods Program. **86**(1), 236–268 (2017). https://doi.org/10.1016/j.jlamp.2016.03.003

24. Lucas, S., Meseguer, J., Gutiérrez, R.: The 2D dependency pair framework for conditional rewrite systems. Part I: Definition and basic processors. J. Comput. Syst. Sci. **96**, 74–106 (2018). https://doi.org/10.1016/j.jcss.2018.04.002

25. Lucas, S., Meseguer, J., Gutiérrez, R.: The 2D dependency pair framework for conditional rewrite systems—part II: advanced processors and implementation techniques. J. Autom. Reasoning (2020). https://doi.org/10.1007/s10817-020-09542-3

26. McCune, W.: Prover9 & Mace4. Technical report (2005–2010). http://www.cs.unm.edu/~mccune/prover9/

27. Ohlebusch, E.: Advanced Topics in Term Rewriting. Springer (2002). https://doi.org/10.1007/978-1-4757-3661-8. http://www.springer.com/computer/swe/book/978-0-387-95250-5

28. Ölveczky, P.C., Lysne, O.: Order-sorted termination: the unsorted way. In: Hanus, M., Rodríguez-Artalejo, M. (eds.) ALP 1996. LNCS, vol. 1139, pp. 92–106. Springer, Heidelberg (1996). https://doi.org/10.1007/3-540-61735-3_6

29. Zantema, H.: Termination of term rewriting: interpretation and type elimination. J. Symb. Comput. **17**(1), 23–50 (1994). https://doi.org/10.1006/jsco.1994.1003

30. Zantema, H.: Termination of context-sensitive rewriting. In: Comon, H. (ed.) RTA 1997. LNCS, vol. 1232, pp. 172–186. Springer, Heidelberg (1997). https://doi.org/10.1007/3-540-62950-5_69

# ENIGMA Anonymous:
# Symbol-Independent Inference
# Guiding Machine (System Description)

Jan Jakubův[1], Karel Chvalovský[1], Miroslav Olšák[2], Bartosz Piotrowski[1,3],
Martin Suda[1(✉)], and Josef Urban[1]

[1] Czech Technical University in Prague, Prague, Czechia
martin.suda@cvut.cz
[2] University of Innsbruck, Innsbruck, Austria
[3] University of Warsaw, Warsaw, Poland

**Abstract.** We describe an implementation of gradient boosting and
neural guidance of saturation-style automated theorem provers that
does not depend on consistent symbol names across problems. For
the gradient-boosting guidance, we manually create abstracted features
by considering arity-based encodings of formulas. For the neural guidance, we use symbol-independent graph neural networks (GNNs) and
their embedding of the terms and clauses. The two methods are efficiently implemented in the E prover and its ENIGMA learning-guided
framework.

To provide competitive real-time performance of the GNNs, we have
developed a new context-based approach to evaluation of generated
clauses in E. Clauses are evaluated jointly in larger batches and with
respect to a large number of already selected clauses (context) by the
GNN that estimates their collectively most useful subset in several
rounds of message passing. This means that approximative inference
rounds done by the GNN are efficiently interleaved with precise symbolic inference rounds done inside E. The methods are evaluated on the
MPTP large-theory benchmark and shown to achieve comparable real-
time performance to state-of-the-art symbol-based methods. The methods also show high complementarity, solving a large number of hard
Mizar problems.

**Keywords:** Automated theorem proving · Machine learning · Neural
networks · Decision trees · Saturation-style proving

Supported by the ERC Consolidator grant AI4REASON no. 649043 (JJ, BP, MS,
and JU), the Czech project AI&Reasoning CZ.02.1.01/0.0/0.0/15_003/0000466 and
the European Regional Development Fund (KC and JU), the ERC Project *SMART*
Starting Grant no. 714034 (MO), grant 2018/29/N/ST6/02903 of National Science
Center, Poland (BP), and the Czech Science Foundation project 20-06390Y (MS).

N. Peltier and V. Sofronie-Stokkermans (Eds.): IJCAR 2020, LNAI 12167, pp. 448–463, 2020.
https://doi.org/10.1007/978-3-030-51054-1_29

# 1    Introduction: Symbol Independent Inference Guidance

In this work, we develop two *symbol-independent* (anonymous) inference guiding methods for saturation-style automated theorem provers (ATPs) such as E [25] and Vampire [20]. Both methods are based on learning clause classifiers from previous proofs within the ENIGMA framework [5,13,14] implemented in E. By *symbol-independence* we mean that no information about the symbol names is used by the learned guidance. In particular, if all symbols in a particular ATP problem are consistently renamed to new symbols, the learned guidance will result in the same proof search and the same proof modulo the renaming.

Symbol-independent guidance is an important challenge for learning-guided ATP, addressed already in Schulz's early work on learning guidance in E [23]. With ATPs being increasingly used and trained on large ITP libraries [2,3,6,8, 16,18], it is more and more rewarding to develop methods that learn to reason without relying on the particular terminology adopted in a single project. Initial experiments in this direction using concept alignment [10] methods have already shown performance improvements by transferring knowledge between the HOL libraries [9]. Structural analogies (or even terminology duplications) are however common already in a single large ITP library [17] and their automated detection can lead to new proof ideas and a number of other interesting applications [11].

This system description first briefly introduces saturation-based ATP with learned guidance (Sect. 2). Then we discuss symbol-independent learning and guidance using abstract features and gradient boosting trees (Sect. 3) and graph neural networks (Sect. 4). The implementation details are explained in Sect. 5 and the methods are evaluated on the MPTP benchmark in Sect. 6.

# 2    Saturation Proving Guided by Machine Learning

**Saturation-Based Automated Theorem Provers** (ATPs) such as E and Vampire are used to prove goals $G$ using a set of axioms $A$. They clausify the formulas $A \cup \{\neg G\}$ and try to deduce contradiction using the *given clause loop* [22] as follows. The ATP maintains two sets of processed ($P$) and unprocessed ($U$) clauses. At each loop iteration, a given clause $g$ from $U$ is selected, moved to $P$, and $U$ is extended with new inferences from $g$ and $P$. This process continues until the contradiction is found, $U$ becomes empty, or a resource limit is reached. The search space grows quickly and selection of the right given clauses is critical.

**Learning Clause Selection** over a set of related problems is a general method how to guide the proof search. Given a set of FOL problems $\mathcal{P}$ and initial ATP strategy $\mathcal{S}$, we can evaluate $\mathcal{S}$ over $\mathcal{P}$ obtaining training samples $\mathcal{T}$. For each successful proof search, training samples $\mathcal{T}$ contain the set of clauses processed during the search. *Positive* clauses are those that were *useful* for the proof search (they appeared in the final proof), while the remaining clauses were *useless*, forming the *negative* examples. Given the samples $\mathcal{T}$, we can *train* a machine learning *classifier* $\mathcal{M}$ which predicts usefulness of clauses in future proof searches. Some clause classifiers are described in detail in Sects. 3, 4, and 5.

**ATP Guidance By a Trained Classifier:** Once a clause classifier $\mathcal{M}$ is trained, we can use it inside an ATP. An ATP strategy $\mathcal{S}$ is a collection of proof search parameters such as term ordering, literal selection, and also given clause selection mechanism. In E, the given clause selection is defined by a collection of clause *weight functions* which alternate to select the given clauses. Our ENIGMA framework uses two methods of plugging the trained classifier $\mathcal{M}$ into $\mathcal{S}$. Either (1) we use $\mathcal{M}$ to select all given clauses (*solo mode* denoted $\mathcal{S} \odot \mathcal{M}$), or (2) we combine predictions of $\mathcal{M}$ with clause selection mechanism from $\mathcal{S}$ so that roughly 50% of the clauses is selected by $\mathcal{M}$ (*cooperative mode* denoted $\mathcal{S} \oplus \mathcal{M}$). Proof search settings other than clause selection are inherited from $\mathcal{S}$ in both the cases. See [5] for details. The phases of learning and ATP guidance can be iterated in a *learning/evaluation loop* [29], yielding growing sets of proofs $\mathcal{T}_i$ and stronger classifiers $\mathcal{M}_i$ trained over them. See [15] for such large experiment.

# 3   Clause Classification by Decision Trees

**Clause Features** are used by ENIGMA to represent clauses as sparse vectors for machine learners. They are based mainly on vertical/horizontal cuts of the clause syntax tree. We use simple *feature hashing* to handle theories with large number of symbols. A clause $C$ is represented by the vector $\varphi_C$ whose $i$-th index stores the value of a feature with hash index $i$. Values of conflicting features (mapped to the same index) are summed. Additionally, we embed *conjecture features* into the clause representation and we work with vector pairs $(\varphi_C, \varphi_G)$ of size $2 * base$, where $\varphi_G$ is the feature vector of the current goal (conjecture). This allows us to provide goal-specific predictions. See [15] for more details.

   **Gradient Boosting Decision Trees (GBDTs)** implemented by the XGBoost library [4] currently provide the strongest ENIGMA classifiers. Their speed is comparable to the previously used [14] weaker linear logistic classifier, implemented by the LIBLINEAR library [7]. In this work, we newly employ the LightGBM [19] GBDT implementation. A *decision tree* is a binary tree whose nodes contain Boolean conditions on values of different features. Given a feature vector $\varphi_C$, the decision tree can be navigated from the root to the unique tree leaf which contains the classification of clause $C$. GBDTs combine predictions from a collection of follow-up decision trees. While inputs, outputs, and API of XGBoost and LightGBM are compatible, each employ a different method of tree construction. XGBoost constructs trees level-wise, while LightGBM leaf-wise. This implies that XGBoost trees are well-balanced. On the other hand, LightGBM can produce much deeper trees and the tree depth limit is indeed an important learning meta-parameter which must be additionally set.

**New Symbol-Independent Features:** We develop a feature anonymization method based on symbol arities. Each function symbol name $s$ with arity $n$ is substituted by a special name "$\mathbf{f}n$", while a predicate symbol name $q$ with arity $m$ is substituted by "$\mathbf{p}m$". Such features lose the ability to distinguish different symbol names, and many features are merged together. Vector representations of two clauses with renamed symbols are clearly equal. Hence the underlying machine learning method will provide equal predictions for such clauses. For more detailed discussion and comparison with related work see Appendix B.

**New Statistics and Problem Features:** To improve the ability to distinguish different anonymized clauses, we add the following features. *Variable statistics* of clause $C$ containing (1) the number of variables in $C$ without repetitions, (2) the number of variables with repetitions, (3) the number of variables with exactly one occurrence, (4) the number of variables with more than one occurrence, (5–10) the number of occurrences of the most/least (and second/third most/least) occurring variable. *Symbol statistics* do the same for symbols instead of variables. Recall that we embed conjecture features in clause vector pair $(\varphi_C, \varphi_G)$. As $G$ embeds information about the conjecture but not about the problem axioms, we propose to additionally embed some statistics of the problem $P$ that $C$ and $G$ come from. We use 22 problem features that E prover already computes for each input problem to choose a suitable strategy. These are (1) number of goals, (2) number of axioms, (3) number of unit goals, etc. See E's manual for more details. Hence we work with vector triples $(\varphi_C, \varphi_G, \varphi_P)$.

## 4    Clause Classification by Graph Neural Network

Another clause classifier newly added to ENIGMA is based on graph neural networks (GNNs). We use the symbol-independent network architecture developed in [21] for premise selection. As [21] contains all the details, we only briefly explain the basic ideas behind this architecture here.

**Hypergraph.** Given a set of clauses $C$ we create a directed hypergraph with three kinds of nodes that correspond to clauses, function and predicate symbols $\mathcal{N}$, and unique (sub)terms and literals $\mathcal{U}$ occurring in $C$, respectively. There are two kinds of hyperedges that describe the relations between nodes according to $C$. The first kind encodes literal occurrences in clauses by connecting the corresponding nodes. The second hyperedge kind encodes the relations between nodes from $\mathcal{N}$ and $\mathcal{U}$. For example, for $f(t_1, \ldots, t_k) \in \mathcal{U}$ we loosely speaking connect the nodes $f \in \mathcal{N}$ and $t_1, \ldots, t_k \in \mathcal{U}$ with the node $f(t_1, \ldots, t_k)$ and similarly for literals, where their polarity is also taken into account.

**Message-Passing.** The hypergraph describes the relation between various kinds of objects occurring in $C$. Every node in the hypergraph is initially assigned a constant vector, called the *embedding*, based only on its kind ($C$, $\mathcal{N}$, or $\mathcal{U}$). These node embeddings are updated in a fixed number of message-passing rounds, based on the embeddings of each node's neighbors. The underlying idea of such neural message-passing methods[1] is to make the node embeddings encode more and more precisely the information about the connections (and thus various

---

[1] Graph convolutions are a generalization of the sliding window convolutions used for aggregating neighborhood information in neural networks used for image recognition.

properties) of the nodes. For this to work, we have to learn initial embeddings for our three kinds of nodes and the update function.[2]

**Classification.** After the message-passing phase, the final clause embeddings are available in the corresponding clause nodes. The estimated probability of a clause being a good given clause is then computed by a neural network that takes the final embedding of this clause and also aggregated final embeddings of all clauses obtained from the negated conjecture.

## 5   Learning and Using the Classifiers, Implementation

In order to use either GBDTs (Sect. 3) or GNNs (Sect. 4), a prediction model must be learned. Learning starts with training samples $\mathcal{T}$, that is, a set of pairs $(\mathcal{C}^+, \mathcal{C}^-)$ of positive and negative clauses. For each training sample $T \in \mathcal{T}$, we additionally know the source problem $P$ and its conjecture $G$. Hence we can consider one sample $T \in \mathcal{T}$ as a quadruple $(\mathcal{C}^+, \mathcal{C}^-, P, G)$ for convenience.

**GBDT.** Given a training sample $T = (\mathcal{C}^+, \mathcal{C}^-, P, G) \in \mathcal{T}$, each clause $C \in \mathcal{C}^+ \cup \mathcal{C}^-$ is translated to the feature vector $(\varphi_C, \varphi_G, \varphi_P)$. Vectors where $C \in \mathcal{C}^+$ are labeled as positive, and otherwise as negative. All the labeled vectors are fed together to a GBDT trainer yielding model $\mathcal{D}_T$.

When predicting a generated clause, the feature vector is computed and $\mathcal{D}_T$ is asked for the prediction. GBDT's binary predictions (positive/negative) are turned into E's clause weight (positives have weight 1 and negatives 10).

**GNN.** Given $T = (\mathcal{C}^+, \mathcal{C}^-, P, G) \in \mathcal{T}$ as above we construct a hypergraph for the set of clauses $\mathcal{C}^+ \cup \mathcal{C}^- \cup G$. This hypergraph is translated to a tensor representation (vectors and matrices), marking clause nodes as positive, negative, or goal. These tensors are fed as input to our GNN training, yielding a GNN model $\mathcal{N}_T$. The training works in iterations, and $\mathcal{N}_T$ contains one GNN per iteration epoch. Only one GNN from a selected epoch is used for predictions during the evaluation.

In evaluation, it is more efficient to compute predictions for several clauses at once. This also improves prediction quality as the queried data resembles more the training hypergraphs where multiple clauses are encoded at once as well. During an ATP run on problem $P$ with the conjecture $G$, we postpone evaluation of newly inferred clauses until we reach a certain amount of clauses $\mathcal{Q}$ to *query*.[3] To resemble the training data even more, we add a fixed number of the given clauses processed so far. We call these *context* clauses ($\mathcal{X}$). To evaluate $\mathcal{Q}$, we construct the hypergraph for $\mathcal{Q} \cup \mathcal{X} \cup G$, and mark clauses from $G$ as goals. Then model $\mathcal{N}_T$ is asked for predictions on $\mathcal{Q}$ (predictions for $\mathcal{X}$ are dropped). The numeric predictions computed by $\mathcal{N}_T$ are directly used as E's weights.

**Implementation and Performance.** We use GBDTs implemented by the XGBoost [4] and LightGBM [19] libraries. For GNN we use Tensorflow [1]. All

---

[2] We learn individual components, which correspond to different kinds of hyperedges, from which the update function is efficiently constructed.

[3] We may evaluate less than $\mathcal{Q}$ if E runs out of unevaluated unprocessed clauses.

**Table 1.** Model training and evaluation for anonymous GBDTs $(D_i)$ and GNN $(\mathcal{N}_i)$.

| $\mathcal{M}$ | TPR | TNR | Training | | | Real time | | Abstract time | |
|---|---|---|---|---|---|---|---|---|---|
| | [%] | [%] | Size | Time | Params | $\mathcal{S} \oplus \mathcal{M}$ | +% | $\mathcal{S} \oplus \mathcal{M}$ | +% |
| $\emptyset$ | - | - | - | - | - | 14 966 | 0.0 | 10 679 | 0.0 |
| $\mathcal{D}_0$ | 84.9 | 68.4 | 14M | 2h29m | X,d12 | 20 679 | 38.1 | 17 917 | 67.8 |
| $\mathcal{D}_1$ | 79.0 | 79.5 | 29M | 4h33m | X,d12 | 23 280 | 58.2 | 20 760 | 94.4 |
| $\mathcal{D}_2$ | 80.5 | 79.2 | 47M | 40m | L,d30,l1800 | 24 347 | 62.7 | 22 661 | 112.2 |
| $\mathcal{N}_0$ | 92.1 | 77.1 | 14M | 17h | e20,q128,c512 | 20 912 | 39.7 | 19 755 | 84.9 |
| $\mathcal{N}_1$ | 90.0 | 78.6 | 31M | 1d19h | e10,q128,c512 | 23 156 | 54.7 | 21 737 | 103.5 |
| $\mathcal{N}_2$ | 91.3 | 79.6 | 50M | 1d 8h | e50,q256,c768 | 23 262 | 55.4 | 22 169 | 107.6 |

the libraries provide Python interfaces and C/C++ APIs. We use the Python interfaces for training and the C APIs for the evaluation in E. The Python interfaces for XGBoost and LightGBM include the C APIs, while for Tensorflow this must be manually compiled, which is further complicated by poor documentation.

The libraries support training both on CPUs and on GPUs. We train Light-GBM on CPUs, and XGBoost and Tensorflow on GPUs. However, we always evaluate on a single CPU as we aim at practical usability on standard hardware. This is non-trivial and it distinguishes this work from evaluations done with large numbers of GPUs or TPUs and/or in prohibitively high real times. The LightGBM training can be parallelized much better – with 60 CPUs it is much faster than XGBoost on 4 GPUs. Neither using GPUs for LightGBM nor many CPUs for XGBoost provided better training times. The GNN training is slower than GBDT training and it is not easy to make Tensorflow evaluate reasonably on a single CPU. It has to be compiled with all CPU optimizations and restricted to a single thread, using Tensorflow's poorly documented experimental C API.

## 6    Experimental Evaluation

**Setup.** We experimentally evaluate[4] our GBDT and GNN guidance[5] on a large benchmark of 57880 Mizar40 [18] problems[6] exported by MPTP [28]. Hence this evaluation is compatible with our previous symbol-dependent work [15]. We evaluate GBDT and GNN separately. We start with a good-performing E strategy $\mathcal{S}$ (see [5, Appendix A]) which solves 14 966 problems with a 10 s limit per problem. This gives us training data $\mathcal{T}_0 = \mathsf{eval}(\mathcal{S})$ (see Sect. 5), and we start three iterations of the learning/evaluation loop (see Sect. 2).

---

[4] On a server with 36 hyperthreading Intel(R) Xeon(R) Gold 6140 CPU @ 2.30 GHz cores, 755 GB of memory, and 4 NVIDIA GeForce GTX 1080 Ti GPUs.
[5] Available at https://github.com/ai4reason/eprover-data/tree/master/IJCAR-20.
[6] http://grid01.ciirc.cvut.cz/~mptp/1147/MPTP2/problems_small_consist.tar.gz.

For GBDT, we train several models (with hash base $2^{15}$) and conduct a small learning meta-parameters *grid search*. For XGBoost, we try different tree depths ($d \in \{9, 12, 16\}$), and for LightGBM various combinations of tree depths and leaves count ($(d, l) \in \{10, 20, 30, 40\} \times \{1200, 1500, 1800\}$). We evaluate all these models in a cooperative mode with $\mathcal{S}$ on a random (but fixed) 10% of all problems (Appendix A). The best performing model is evaluated on the whole benchmark in both cooperative ($\oplus$) and solo ($\odot$) runs. These give us the next samples $\mathcal{T}_{i+1}$. We perform three iterations and obtain models $\mathcal{D}_0$, $\mathcal{D}_1$, and $\mathcal{D}_2$.

For GNN, we train a model with 100 epochs, obtaining 100 different GNNs. We evaluate GNNs from selected epochs ($e \in \{10, 20, 50, 75, 100\}$) and we try different settings of *query* ($q$) and *context* ($c$) sizes (see Sect. 5). In particular, $q$ ranges over $\{64, 128, 192, 256, 512\}$ and $c$ over $\{512, 768, 1024, 1536\}$. All possible combinations of $(e, q, c)$ are again evaluated in a grid search on the small benchmark subset (Appendix A), and the best performing model is selected for the next iteration. We run three iterations and obtain models $\mathcal{N}_0$, $\mathcal{N}_1$, and $\mathcal{N}_2$.

**Results** are presented in Table 1. For each model $\mathcal{D}_i$ and $\mathcal{N}_i$ we show (1) true positive/negative rates, (2) training data sizes, (3) train times, and (4) the best performing parameters from the grid search. Furthermore, for each model $\mathcal{M}$ we show the performance of $\mathcal{S} \oplus \mathcal{M}$ in (5) real and (6) abstract time. Details follow. (1) Model accuracies are computed on samples extracted from problems newly solved by each model, that is, on testing data not known during the training. Columns TPR/TNR show accuracies on positive/negative testing samples. (2) Train sizes measure the training data in millions of clauses. (4) Letter "X" stands for XGBoost models, while "L" for LightGBM. (5) For real time we use 10 s limit per problem, and (6) in abstract time we limit the number of generated clauses to 5000. We show the number of problems solved and the gain (in %) on $\mathcal{S}$. The abstract time evaluation is useful to assess the methods modulo the speed of the implementation. The first row shows the performance of $\mathcal{S}$ without learning.

**Evaluation.** The GNN models start better, but the GBDT models catch up and beat GNN in later iterations. The GBDT models show a significant gain even in the 3rd iteration, while the GNN models start stagnating. The GNN models report better testing accuracy, but their ATP performance is not as good.

For GBDTs, we see that the first two best models ($\mathcal{D}_0$ and $\mathcal{D}_1$) were produced by XGBoost, while $\mathcal{D}_2$ by LightGBM. While both libraries can provide similar results, LightGBM is significantly faster. For comparison, the training time for XGBoost in the third iteration was 7 h, that is, LightGBM is 10 times faster. The higher speed of LightGBM can overcome the problems with more complicated parameter settings, as more models can be trained and evaluated.

For GNNs, we observe higher training times and better models coming from earlier epochs. The training in the 1st and 2nd iterations was done on 1 GPU, while in the 3rd on 4 GPUs. The good abstract time performance indicates that further gain could be obtained by a faster implementation. But note that this is the first time that NNs have been made comparable to GBDTs in real time.

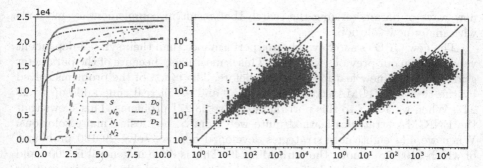

**Fig. 1.** Left: the number of problems solved in time; Right: the number of processed clauses (the $x$-axis for $\mathcal{S}$, and the $y$-axis for $\mathcal{S} \oplus \mathcal{D}_0$ and $\mathcal{S} \oplus \mathcal{N}_0$, respectively).

Figure 1 summarizes the results. On the left, we observe a slower start for GNNs caused by the initial model loading. On the right, we see a decrease in the number of processed clauses, which suggests that the guidance is effective.

**Complementarity.** The twelve (solo and cooperative) versions of the methods compared in Fig. 1 solve together 28271 problems, with the six GBDTs solving 25255 and the six GNNs solving 26571. All twenty methods tested by us solve 29118 problems, with the top-6 greedy cover solving (in 60 s) 28067 and the top-15 greedy cover solving (in 150 s) 29039. The GNNs show higher complementarity – varying the epoch as well as the size of the query and context produces many new solutions. For example, the most complementary GNN method adds to the best GNN method 1976 solutions. The GNNs are also quite complementary to the GBDTs. The second (GNN) strategy in the greedy cover adds 2045 solutions to the best (GBDT) strategy. Altogether, the twenty strategies solve (in 200 s) 2109 of the Mizar40 *hard* problems, i.e., the problems unsolved by any method developed previously in [18].

## 7   Conclusion

We have developed and evaluated symbol-independent GBDT and GNN ATP guidance. This is the first time symbol-independent features and GNNs are tightly integrated with E and provide good real-time results on a large corpus. Both the GBDT and GNN predictors display high ability to learn from previous proof searches even in the symbol-independent setting.

To provide competitive real-time performance of the GNNs, we have developed context-based evaluation of generated clauses in E. This introduces a new paradigm for clause ranking and selection in saturation-style proving. The generated clauses are not ranked immediatelly and independently of other clauses. Instead, they are judged in larger batches and with respect to a large number of already selected clauses (context) by a neural network that estimates their collectively most useful subset by several rounds of message passing. This also allows new ways of parameterizing the search that result in complementary methods with many new solutions.

The new GBDTs show even better performance than their symbol-dependent versions from our previous work [15]. This is most likely because of the parameter grid search and new features not used before. The union of the problems solved by the twelve ENIGMA strategies (both $\odot$ and $\oplus$) in real time adds up to 28 247. When we add $\mathcal{S}$ to this portfolio we solve 28 271 problems. This shows that the ENIGMA strategies learned quite well from $\mathcal{S}$, not losing many solutions. When we add eight more strategies developed here we solve 29 130 problems, of which 2109 are among the hard Mizar40. This is done in general in 200 s and without any additional help from premise selection methods. Vampire in 300 s solves 27 842 problems. Future work includes joint evaluation of the system on problems translated from different ITP libraries, similar to [9].

**Acknowledgments.** We thank Stephan Schulz and Thibault Gauthier for discussing with us their methods for symbol-independent term and formula matching.

## A    Additional Data From the Experiments

This appendix presents additional data from the experiments in Section 6. Figure 3 shows the results of the grid search for GNN models on one tenth of all benchmark problems done in order to find the best-performing parameters for *query* and *context* sizes. The $x$-axis plots the query size, the $y$-axis plots the context size, while the $z$-axis plots the ATP performance, that is, the number of solved problems. Recall that the grid search was performed on a randomly selected but fixed tenth of all benchmark problems with a 10 s real-time limit per problem. For $\mathcal{N}_0$ and $\mathcal{N}_1$, there is a separate graph for each iteration, showing only the best epochs. For $\mathcal{N}_2$, there are two graphs for models from epoch 20 and 50. Note how the later epoch 50 becomes more independent on the context size. The ranges of the grid search parameters were extended in later iterations when the best-performing value was at the graph edge.

Figure 4 shows the grid search results for the best LightGBM's GBDT models from iterations 1, 2, and 3 (denoted here $\mathcal{D}_0$, $\mathcal{D}_1$, and $\mathcal{D}_2$). The $x$-axis plots the number of tree leaves, the $y$-axis plots the tree depth, while the $z$-axis plots the number of solved problems. There are two models from the second iteration ($\mathcal{D}_1$), showing the effect of different learning rate ($\eta$). Again, the ranges of meta-parameters were updated in between the iterations by a human engineer.

Figure 5 shows the training accuracies and training loss for the LightGBM model $\mathcal{D}_2$. Accuracies (TPR and TNR) of the training data are computed from the first iteration ($\mathcal{T}_0$). The values for loss ($z$) are inverted ($1 - z$) so that higher values correspond to better models which makes a visual comparison easier. We can see a clear correlation between the accuracies and the loss, but not so clear correlation with the ATP performance. The ATP performance of $\mathcal{D}_2$ is the same as in Figure 4, repeated here for convenience.

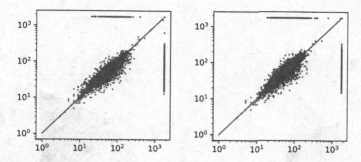

**Fig. 2.** Scatter plots for the lengths of the discovered proofs (the $x$-axis for $\mathcal{S}$, and the $y$-axis for $\mathcal{S} \oplus \mathcal{D}_2$ and $\mathcal{S} \oplus \mathcal{N}_2$, respectively).

Figure 2 compares the lengths of the discovered proofs. We can see that there is no systematic difference in this metric between the base strategy and the ENIGMA ones.

Finally, we have compared the feature vectors of the symbol-dependent and symbol-independent versions of the GBDTs. On the same data, we observe roughly 2x more collisions. The symbol-independent version has around 1% of colliding feature vectors, while the symbol-dependent version has 0.42%.

# B    Discussion of Anonymization

Our use of symbol-independent arity-based features for GBDTs differs from Schulz's anonymous *clause patterns* [23,24] (CPs) used in E for proof guidance and from Gauthier and Kaliszyk's (GK) anonymous abstractions used for their concept alignments between ITP libraries [10] in two ways:

1. In both CP and GK, serial (de Bruijn-style) numbering of abstracted symbols of the same arity is used. I.e., the term $h(g(a))$ will get abstracted to $F11(F12(F01))$. Our encoding is just $F1(F1(F0))$. It is even more lossy, because it is the same for $h(h(a))$.
2. ENIGMA with gradient boosting decision trees (GBDTs) can be (approximately) thought of as implementing weighted feature-based clause classification where the feature weights are learned. Whereas both in CP and GK, exact matching is used after the abstraction is done.[7] In CP, this is used for hint-style guidance of E. There, for clauses, such serial numbering however isn't stable under literal reordering and subsumption. Partial heuristics can be used, such as normalization based on a fixed global ordering done in both CP and GK.

Addressing the latter issue (stability under reordering of literals and subsumption) leads to the NP hardness of (hint) matching/subsumption. I.e.,

---

[7] We thank Stephan Schulz for pointing out that although CPs used exact matching by default, matching up to a certain depth was also implemented.

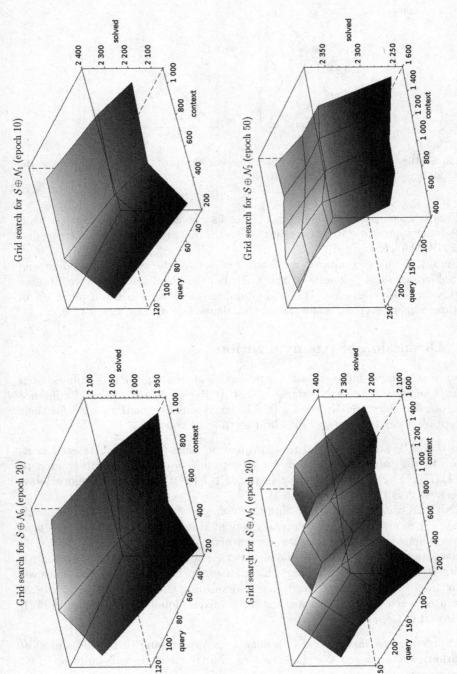

**Fig. 3.** Grid search results for GNN models ($\mathcal{N}_i$).

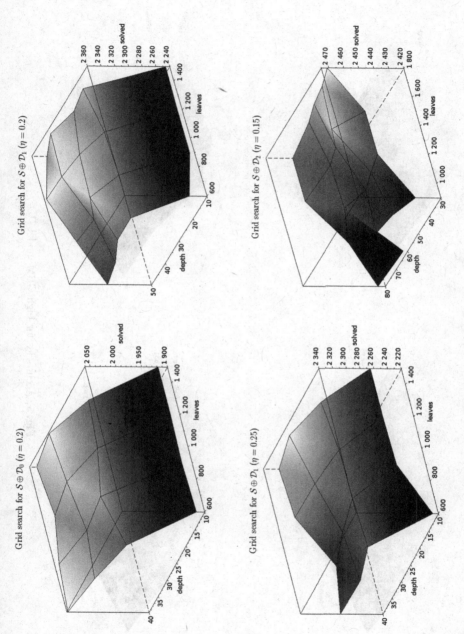

**Fig. 4.** Grid search results for LightGBM GBDT models ($\mathcal{D}_i$).

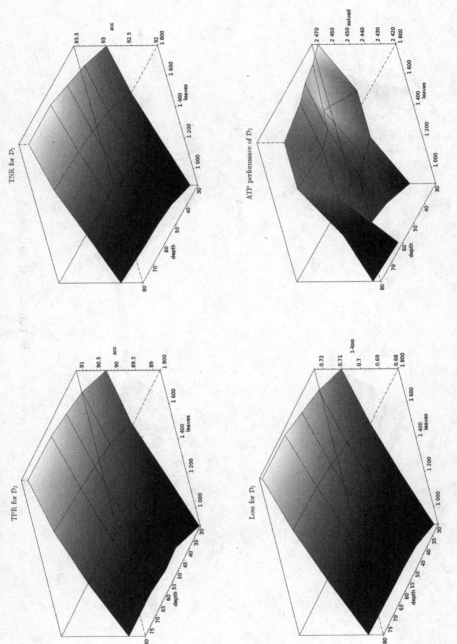

**Fig. 5.** Accuracies for LightGBM model $\mathcal{D}_2$.

the abstracted subsumption task can be encoded as standard first-order subsumption for clauses where terms like $F11(F12(F01))$ are encoded as $apply1(X1, apply1(X2, apply0(X3)))$. The NP hardness of subsumption is however here more serious in practice than in standard ATP because only applications behave as non-variable symbols during the matching.

Thus, the difference between our anonymous approach and CP is practically the same as between the standard symbol-based ENIGMA guidance and standard hint-based [30] guidance. In the former the matching (actually, clause classification) is approximate, weighted and learned, while with hints the clause matching/classification is crisp, logic-rooted and preprogrammed, sometimes running into the NP hardness issues. Our latest comparison [12] done over the Mizar/MPTP corpus in the symbol-based setting showed better performance of ENIGMA over using hints, most likely due to better generalization behavior of ENIGMA based on the statistical (GBDT) learning.

Note also that the variable and symbol statistics features to some extent alleviate the conflicts obtained with our encoding. E.g., $h(g(a))$ and $h(h(a))$ will have different *symbol statistics* (Section 3) features. To some extent, such features are similar to Schulz's feature vector and fingerprint indexing [26, 27].

# References

1. Abadi, M., et al.: TensorFlow: large-scale machine learning on heterogeneous systems (2015). http://tensorflow.org
2. Blanchette, J.C., Greenaway, D., Kaliszyk, C., Kühlwein, D., Urban, J.: A learning-based fact selector for Isabelle/HOL. J. Autom. Reasoning **57**(3), 219–244 (2016). https://doi.org/10.1007/s10817-016-9362-8
3. Blanchette, J.C., Kaliszyk, C., Paulson, L.C., Urban, J.: Hammering towards QED. J. Formalized Reasoning **9**(1), 101–148 (2016)
4. Chen, T., Guestrin, C.: XGBoost: a scalable tree boosting system. In: Proceedings of the 22nd ACM SIGKDD International Conference on Knowledge Discovery and Data Mining (KDD 2016), pp. 785–794. ACM, New York (2016)
5. Chvalovský, K., Jakubův, J., Suda, M., Urban, J.: ENIGMA-NG: efficient neural and gradient-boosted inference guidance for E. In: Fontaine, P. (ed.) CADE 2019. LNCS (LNAI), vol. 11716, pp. 197–215. Springer, Cham (2019). https://doi.org/10.1007/978-3-030-29436-6_12
6. Czajka, L., Kaliszyk, C.: Hammer for Coq: automation for dependent type theory. J. Autom. Reasoning **61**(1–4), 423–453 (2018). https://doi.org/10.1007/s10817-018-9458-4
7. Fan, R.-E., Chang, K.-W., Hsieh, C.-J., Wang, X.-R., Lin, C.-J.: Liblinear: a library for large linear classification. J. Mach. Learn. Res. **9**, 1871–1874 (2008)
8. Gauthier, T., Kaliszyk, C.: Premise selection and external provers for HOL4. In: Leroy, X., Tiu, A. (eds.) Proceedings of the 2015 Conference on Certified Programs and Proofs (CPP 2015), Mumbai, India, 15–17 January (2015), pp. 49–57. ACM (2015)
9. Gauthier, T., Kaliszyk, C.: Sharing HOL4 and HOL light proof knowledge. In: Davis, M., Fehnker, A., McIver, A., Voronkov, A. (eds.) LPAR 2015. LNCS, vol. 9450, pp. 372–386. Springer, Heidelberg (2015). https://doi.org/10.1007/978-3-662-48899-7_26

10. Gauthier, T., Kaliszyk, C.: Aligning concepts across proof assistant libraries. J. Symb. Comput. **90**, 89–123 (2019)
11. Gauthier, T., Kaliszyk, C., Urban, J.: Initial experiments with statistical conjecturing over large formal corpora. In: Kohlhase, A. (eds.) Joint Proceedings of the FM4M, MathUI, and ThEdu Workshops, Doctoral Program, and Work in Progress at the Conference on Intelligent Computer Mathematics 2016 co-located with the 9th Conference on Intelligent Computer Mathematics (CICM 2016) of CEUR Workshop Proceedings, Bialystok, Poland, 25–29 July 2016, vol. 1785, pp. 219–228. CEUR-WS.org (2016)
12. Goertzel, Z., Jakubův, J., Urban, J.: ENIGMAWatch: proofWatch meets ENIGMA. In: Cerrito, S., Popescu, A. (eds.) TABLEAUX 2019. LNCS (LNAI), vol. 11714, pp. 374–388. Springer, Cham (2019). https://doi.org/10.1007/978-3-030-29026-9_21
13. Jakubův, J., Urban, J.: ENIGMA: efficient learning-based inference guiding machine. In: Geuvers, H., England, M., Hasan, O., Rabe, F., Teschke, O. (eds.) CICM 2017. LNCS (LNAI), vol. 10383, pp. 292–302. Springer, Cham (2017). https://doi.org/10.1007/978-3-319-62075-6_20
14. Jakubův, J., Urban, J.: Enhancing ENIGMA given clause guidance. In: Rabe, F., Farmer, W.M., Passmore, G.O., Youssef, A. (eds.) CICM 2018. LNCS (LNAI), vol. 11006, pp. 118–124. Springer, Cham (2018). https://doi.org/10.1007/978-3-319-96812-4_11
15. Jakubuv, J., Urban, J.: Hammering Mizar by learning clause guidance. In: Harrison, J., O'Leary, J., Tolmach, A. (eds.) 10th International Conference on Interactive Theorem Proving (ITP 2019) of LIPIcs, 9–12 September 2019, Portland, OR, USA, vol. 141, pp. 34:1–34:8. Schloss Dagstuhl - Leibniz-Zentrum für Informatik (2019)
16. Kaliszyk, C., Urban, J.: Learning-assisted automated reasoning with Flyspeck. J. Autom. Reasoning **53**(2), 173–213 (2014). https://doi.org/10.1007/s10817-014-9303-3
17. Kaliszyk, C., Urban, J.: HOL(y)Hammer: online ATP service for HOL light. Math. Comput. Sci. **9**(1), 5–22 (2015). https://doi.org/10.1007/s11786-014-0182-0
18. Kaliszyk, C., Urban, J.: MizAR 40 for Mizar 40. J. Autom. Reasoning **55**(3), 245–256 (2015). https://doi.org/10.1007/s10817-015-9330-8
19. Ke, G., et al.: Lightgbm: a highly efficient gradient boosting decision tree. In: NIPS, pp. 3146–3154 (2017)
20. Kovács, L., Voronkov, A.: First-order theorem proving and VAMPIRE. In: Sharygina, N., Veith, H. (eds.) CAV 2013. LNCS, vol. 8044, pp. 1–35. Springer, Heidelberg (2013). https://doi.org/10.1007/978-3-642-39799-8_1
21. Olsák, M., Kaliszyk, C., Urban, J.: Property invariant embedding for automated reasoning. CoRR, abs/1911.12073 (2019)
22. Overbeek, R.A.: A new class of automated theorem-proving algorithms. J. ACM **21**(2), 191–200 (1974)
23. Schulz, S.: Learning Search Control Knowledge For Equational Deduction of DISKI, vol. 230. Infix Akademische Verlagsgesellschaft, Frankfurt (2000)
24. Schulz, S.: Learning search control knowledge for equational theorem proving. In: Baader, F., Brewka, G., Eiter, T. (eds.) KI 2001. LNCS (LNAI), vol. 2174, pp. 320–334. Springer, Heidelberg (2001). https://doi.org/10.1007/3-540-45422-5_23
25. Schulz, S.: E - a brainiac theorem prover. AI Commun. **15**(2–3), 111–126 (2002)
26. Schulz, S.: Fingerprint indexing for paramodulation and rewriting. In: Gramlich, B., Miller, D., Sattler, U. (eds.) IJCAR 2012. LNCS (LNAI), vol. 7364, pp. 477–483. Springer, Heidelberg (2012). https://doi.org/10.1007/978-3-642-31365-3_37

27. Schulz, S.: Simple and efficient clause subsumption with feature vector indexing. In: Bonacina, M.P., Stickel, M.E. (eds.) Automated Reasoning and Mathematics. LNCS (LNAI), vol. 7788, pp. 45–67. Springer, Heidelberg (2013). https://doi.org/10.1007/978-3-642-36675-8_3

28. Urban, J.: MPTP 0.2: design, implementation, and initial experiments. J. Autom. Reasoning **37**(1–2), 21–43 (2006)

29. Urban, J., Sutcliffe, G., Pudlák, P., Vyskočil, J.: MaLARea SG1 - machine learner for automated reasoning with semantic guidance. In: Armando, A., Baumgartner, P., Dowek, G. (eds.) IJCAR 2008. LNCS (LNAI), vol. 5195, pp. 441–456. Springer, Heidelberg (2008). https://doi.org/10.1007/978-3-540-71070-7_37

30. Veroff, R.: Using hints to increase the effectiveness of an automated reasoning program: case studies. J. Autom. Reasoning **16**(3), 223–239 (1996)

# The Imandra Automated Reasoning System (System Description)

Grant Passmore[✉], Simon Cruanes, Denis Ignatovich, Dave Aitken,
Matt Bray, Elijah Kagan, Kostya Kanishev, Ewen Maclean,
and Nicola Mometto

Imandra Inc., Austin, USA
grant@imandra.ai

**Abstract.** We describe Imandra, a modern computational logic theorem prover designed to bridge the gap between decision procedures such as SMT, semi-automatic inductive provers of the Boyer-Moore family like ACL2, and interactive proof assistants for typed higher-order logics. Imandra's logic is computational, based on a pure subset of OCaml in which all functions are terminating, with restrictions on types and higher-order functions that allow conjectures to be translated into multi-sorted first-order logic with theories, including arithmetic and datatypes. Imandra has novel features supporting large-scale industrial applications, including a seamless integration of bounded and unbounded verification, first-class computable counterexamples, efficiently executable models and a cloud-native architecture supporting live multiuser collaboration. The core reasoning mechanisms of Imandra are (i) a semi-complete procedure for finding models of formulas in the logic mentioned above, centered around the lazy expansion of recursive functions, (ii) an inductive waterfall and simplifier which "lifts" many Boyer-Moore ideas to our typed higher-order setting. These mechanisms are tightly integrated and subject to many forms of user control.

## 1  Introduction

Imandra is a modern computational logic theorem prover built around a pure, higher-order subset of OCaml. Mathematical models and conjectures are written as executable OCaml programs, and Imandra may be used to reason about them, combining models, proofs and counterexamples in a unified computational environment. Imandra is designed to bridge the gap between decision procedures such as SMT [2], semi-automatic inductive provers of the Boyer-Moore family like ACL2 [1,6], and interactive proof assistants for typed higher-order logics [4,5,7,8]. Our goal is to build a friendly, easy to use system by leveraging strong automation in proof search that can also robustly provide counterexamples for false conjectures. Imandra has novel features supporting large-scale industrial applications, including a seamless integration of bounded and unbounded verification, first-class computable counterexamples, efficiently executable models and a cloud-native architecture supporting live multiuser

© Springer Nature Switzerland AG 2020
N. Peltier and V. Sofronie-Stokkermans (Eds.): IJCAR 2020, LNAI 12167, pp. 464–471, 2020.
https://doi.org/10.1007/978-3-030-51054-1_30

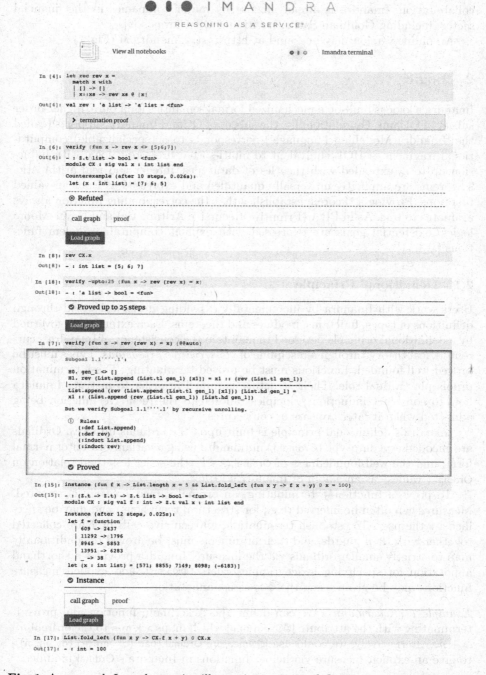

**Fig. 1.** An example Imandra session illustrating recursive definitions, computable counterexamples (**CX**), bounded verification (**verify** upto), unbounded verification with automated induction (**@@auto**), and higher-order instance synthesis.

collaboration. Imandra is already in use by major companies in the financial sector, including Goldman Sachs, Itiviti and OneChronos [9].

An online version may be found at https://try.imandra.ai (Fig. 1).

## 2    Logic

Imandra's logic is built on a mechanized formal semantics for a pure, higher-order subset of OCaml. Foundationally, the subset of OCaml Imandra supports (called the 'Imandra Modelling Language') corresponds to a (specializable) computational fragment of HOL equivalent to multi-sorted first-order logic with induction up to $\epsilon_0$ extended with theories of datatypes, integer and real arithmetic. Theorems are implicitly universally quantified and expressed as Boolean-valued functions. Proving a theorem establishes that the corresponding function always evaluates to **true**. As in PRA (Primitive Recursive Arithmetic) and Boyer-Moore logics, existential goals are expressed with explicit computable Skolem functions [1, 3, 11].

### 2.1    Definitional Principle

Users work with Imandra by incrementally extending its logical world through definitions of types, functions, modules and theorems. Each extension is governed by a definitional principle designed to maintain the consistency of Imandra's current logical theory through a discipline of *conservative extensions*. Types must be proved well-founded. Functions must be proved terminating. These termination proofs play a dual role: Their structure is mined in order to instruct Imandra how to construct induction principles tailored to the recursive function being admitted when it later appears in conjectures.

Imandra's definitional principle is built upon the ordinals up to $\epsilon_0$. Ordinals are encoded as a datatype (Ordinal.t) in Imandra using a variant of Cantor normal form, and the well-foundedness of Ordinal.(<<)—the strict less-than relation on Ordinal.t values—is an axiom of Imandra's logic.

To prove a function $f$ terminating, an ordinal-valued *measure* is required. Measures can often be inferred (e.g., for structural recursions) and may be specified by the user. To establish termination, all recursive calls of $f$ are collected together with their guards, and their arguments must be proved to conditionally map to strictly smaller ordinals via the measure. Imandra provides a shorthand annotation for specifying lexicographic orders (@@adm), and explicit measure functions may be given using the @@measure annotation.

*Example 1 (Ackermann).* We can define the Ackermann function and prove it terminating with the attribute [@@adm m,n] which maps ack m n to the ordinal $m \cdot \omega + n$. Alternatively, we could use [@@measure Ordinal.(pair (of_int m) (of_int n))] to give an explicit measure via helper functions in Imandra's Ordinal module.

```
let rec ack m n =
  if m <= 0 then n + 1 else if n <= 0 then ack (m−1) 1 else ack (m−1) (ack m (n−1))
[@@adm m,n]
```

*Example 2.* Here we have a naive version of the classic *left-pad* function [13], where termination depends on both arguments in a non-lexicographic manner:

```
let rec left_pad c n xs =
  if List.length xs >= n then xs else left_pad c n (c :: xs)
[@@measure Ordinal.of_int (n − List.length xs)]
```

## 2.2 Lifting, Specialization and Monomorphization

Imandra definitions may be polymorphic and higher-order. However, once Imandra is tasked with determining the truth value of a conjecture, the goal and its transitive dependencies are transformed into a family of ground, monomorphic first-order (recursive) definitions. These transformations include lambda lifting, specialization and monomorphization. Imandra's supported fragment of OCaml is designed so that all admitted definitions may be transformed in this way.

*Example 3.* To prove the following higher-order theorem

```
theorem same_len l =
  List.length (List.map (fun x −> x+1) l) = List.length l
```

we obtain a set of lower level definitions, where the anonymous function was lifted, the type list was monomorphised, and map and length were specialised:

```
type int_list = Nil_int | Cons_int of int * int_list
let rec length_int = function
  | Nil_int −> 0
  | Cons_int (_, tl) −> 1 + length_int tl
let map_lambda0 x = x+1
let rec map1 = function
  | Nil_int −> Nil_int
  | Cons_int (x, tl) −> Cons_int (map_lambda0 x, map1 tl)
theorem same_len (l:int_list) : bool =
  length_int (map1 l) = length_int l
```

# 3   Unrolling of Recursive Functions

A major feature of Imandra is its ability to automatically search for proofs and counterexamples in a logic with recursive functions. When a counterexample is found, it is reflected as a first-class value in Imandra's runtime and can be directly computed with and run through the model being analysed. In fact, the statement **verify** (fun x −> ...) does not try any inductive proving unless requested; the default strategy is recursive function *unrolling* for a fixed number of steps, a form of bounded symbolic model-checking.

Our core unrolling algorithm is similar in spirit to the work of Suter et al. [12] but with crucial strategic differences. In essence, Imandra uses the *assumption* mechanism of SMT to block all Boolean assignments that involve the evaluation of a (currently) uninterpreted ground instance of a recursive function. A refinement loop, based on extraction of unsat-cores from this set of assumptions, then expands (interprets) the function calls one by one until a model is found, an empty unsat-core is obtained, or a maximal number of steps is reached.

**Definition 1 (Function template).** *A function template for $f$ is a set of tuples $(g, t, p)$ such that the body of $f$ contains a call to $g(t)$ under the path $p$.*

*Example 4*

- $\mathsf{fact}(x) = \text{if } x > 1 \ (x * \mathsf{fact}(x-1)) \ 1$ has as template $\{(\mathsf{fact}, (\boldsymbol{x-1}), (x > 1))\}$
- $f(x) = 1 + \text{if } g(0) \ h(g(x)) \ h(42)$
  has as template
  $\{(g, (0), \top), (h, (\boldsymbol{g(x)}), (g(0) = \top)), (g, (\boldsymbol{x}), g(0) = \top), (h, (\boldsymbol{42}), (g(0) = \bot))\}$

We use what we call *reachability literals* to prevent the SMT solver from picking assignments that use function calls that are not expanded yet. A reachability literal is a Boolean atom that doesn't appear in the original problem, and that we associate to a given function call $f(t)$ regardless of where it occurs. This is to be contrasted with Suter et al.'s notion of *control literals* associated with individual occurrences of function calls within the expanded body of another function call. We denote by $b[f(t)]$ the unique reachability literal for $f(t)$.

```
1    def calls_of_term(t: Term):
2       return  {b[f(u⃗)] | f(u⃗) ◁ t}
3
4    def subcalls_of_call(f(t⃗): Term, expanded: Set[Term]):
5       return {(b[g(u⃗)], p) | (g, u⃗, ⋀ p) ∈ template(f)[t⃗/x⃗] ∧ g(u⃗) ∉ expanded}
6
7    def unroll(goal: Formula) −> SAT|UNSAT:
8       q = calls_of_term(goal), expanded = ∅
9       F = goal ∧ ⋀_{a∈q} a
10      while True:
11         is_sat, unsat_core = check_sat(F, assume={¬a | a ∈ q})
12         if is_sat == SAT: return SAT
13         else if is_sat == UNSAT:
14            if unsat_core == ∅: return UNSAT
15            b[f(t⃗)] = pick_from(unsat_core)  # next call to expand
16            expanded = {f(t⃗)} ∪ expanded
17            {(a_i, p_i)}_i = subcalls_of_call(f(t⃗), expanded)
18            q = q ∪ {a_i}_i \ b[f(t⃗)]
19            F = F ∧ b[f(t⃗)] ∧ f(t⃗) = body_f[t⃗/x⃗] ∧ ⋀_i (b[f(t⃗)] ∧ p_i ⇒ a_i)
```

**Fig. 2.** Unrolling algorithm

The main search loop is presented in Fig. 2, where $body_f$ is the body of $f$ (i.e. $f(x) \overset{\text{def}}{=} body_f$) and $t \lhd u$ means $t$ is a proper subterm of $u$. We start with $F$

initialized to the original goal, and the queue $q$ containing function calls in the goal (computed by calls_of_term). Each iteration of the loop starts by checking validity under the assumption that all reachability literals in $q$ are false (line 11). If no model is found, we pick an unexpanded function call $f(t)$ from the unsat core (line 15). Selection must be *fair*: all function calls must eventually be picked.

To expand $f(t)$, the corresponding reachability literal becomes true, we instantiate the body of $f$ on $t$, and use subcalls_of_call to compute the set of subcalls along with their control path within $f(t)$ (using $f$'s template). For each $b[g(u)]$ occurring under path $p$ inside $\text{body}_f(t)$, we need to block models that would make $p$ valid until $g(u)$ gets expanded. The assertions $\bigwedge_i (b[f(t)] \wedge p_i \Rightarrow a_i)$ delegate to SMT the work of tracking which paths are forbidden. This way, expanding one function call might lead to many paths becoming "unlocked" at once.

# 4 Induction

Imandra has extensive support for automated induction built principally around Imandra's *inductive waterfall*[1]. This combines techniques such as symbolic execution, lemma-based conditional rewriting, forward-chaining, generalization and the automatic synthesis of goal-specific induction principles. Induction principle synthesis depends upon data computed about a function's termination obtained when it was admitted via our definitional principle. Imandra's waterfall is deeply

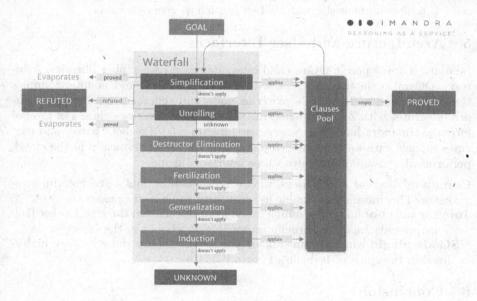

**Fig. 3.** Imandra's inductive waterfall

---

[1] More details about Imandra's waterfall and rule classes may be found in our online documentation at https://docs.imandra.ai.

inspired by the pioneering work of Boyer-Moore [1,6], and is in many ways a "lifting" of the Boyer-Moore waterfall to our typed, higher-order setting (Fig. 3).

Imandra's waterfall contains a simplifier which automatically makes use of previously proved lemmas. Once proved, lemmas may be installed as rewrite, forward-chaining, elimination or generalization rules. Imandra gives users feedback in order to help them design efficient collections of rules. With a good collection of rules (especially rewrite rules), it is hoped that "most" useful theorems over a given domain will be provable by simplification alone, and induction will only be applied as a last resort. In these cases, the subsequent waterfall moves are designed to prepare the simplified conjecture for induction (via, e.g., generalization) before goal-specific induction principles are synthesized.

Imandra's inductive waterfall plays an important role in what we believe to be a robust verification strategy for applying Imandra to real-world systems. Recall that all Imandra goals may be subjected to bounded verification via unrolling (cf. Sect. 3). In practice, we almost always attack a goal by unrolling first, attempting to verify it up to a bound before we consider trying to prove it by induction. Typically, for real-world systems, models and conjectures will have flaws, and unrolling will uncover many counterexamples, confusions and mistakes. As all models are executable and all counterexamples are reflected in Imandra's runtime, they can be directly run through models facilitating rapid investigation. It is typically only after iterating on models and conjectures until all (bounded) counterexamples have been eliminated that we consider trying to prove them by induction. Imandra's support for counterexamples also plays another important role: as a filter on heuristic waterfall steps such as generalization.

## 5    Architecture and User Interfaces

Imandra is developed in OCaml and integrates with its compiler libraries. Arbitrary OCaml code may interact with Imandra models and counterexamples through the use of Imandra's **program mode** and reflection machinery. Imandra integrates with Z3 [2] for checking satisfiability of various classes of ground formulas. Imandra has a client-server architecture: (i) the client parses and executes models with an integrated toplevel; (ii) the server, typically in the cloud, performs all reasoning. Imandra's user interfaces include:

**Command line** for power users, with tab-completion, hints, and colorful messages. This interface is similar in some ways to OCaml's utop.

**Jupyter notebooks** hosted online or via local installation through Docker [10]. This presents Imandra through interactive notebooks in the browser.

**VSCode plugin** where documents are checked on the fly and errors are underlined in the spirit of Isabelle's Prover IDE [14].

## 6    Conclusion

Imandra is an industrial-strength reasoning system combining ideas from SMT, Boyer-Moore inductive provers, and ITPs for typed higher-order logics. Imandra delivers an extremely high degree of automation and has novel techniques such as reflected computable counterexamples that we now believe are indispensible for the effective industrial application of automated reasoning. We are encouraged by Imandra's success in mainstream finance [9], and share a deep conviction that further advances in automation and UI—driven in large part by meeting the demands of industrial users—will lead to a (near-term) future in which automated reasoning is a widely adopted foundational technology.

# References

1. Boyer, R.S., Moore, J.S.: A Computational Logic. Academic Press Professional Inc, Cambridge (1979)
2. de Moura, L., Bjørner, N.: Z3: an efficient SMT solver. In: Ramakrishnan, C.R., Rehof, J. (eds.) TACAS 2008. LNCS, vol. 4963, pp. 337–340. Springer, Heidelberg (2008). https://doi.org/10.1007/978-3-540-78800-3_24
3. Goodstein, R.L.: Recursive Number Theory: A Development of Recursive Arithmetic in a Logic-Free Equation Calculus. North-Holland Pub. Co., Amsterdam (1957)
4. Gordon, M.J.C., Melham, T.F.: Introduction to HOL: A Theorem Proving Environment for Higher Order Logic. Cambridge University Press, Cambridge (1993)
5. Harrison, J.: HOL light: an overview. In: Berghofer, S., Nipkow, T., Urban, C., Wenzel, M. (eds.) TPHOLs 2009. LNCS, vol. 5674, pp. 60–66. Springer, Heidelberg (2009). https://doi.org/10.1007/978-3-642-03359-9_4
6. Kaufmann, M., Moore, J.S.: ACL2: an industrial strength version of Nqthm. In: Computer Assurance (COMPASS 1996), pp. 23–34. IEEE (1996)
7. Nipkow, T., Wenzel, M., Paulson, L.C. (eds.): Isabelle/HOL. LNCS, vol. 2283. Springer, Heidelberg (2002). https://doi.org/10.1007/3-540-45949-9
8. Owre, S., Rushby, J.M., Shankar, N.: PVS: a prototype verification system. In: Kapur, D. (ed.) CADE 1992. LNCS, vol. 607, pp. 748–752. Springer, Heidelberg (1992). https://doi.org/10.1007/3-540-55602-8_217
9. Passmore, G.O., Ignatovich, D.: Formal verification of financial algorithms. In: de Moura, L. (ed.) CADE 2017. LNCS (LNAI), vol. 10395, pp. 26–41. Springer, Cham (2017). https://doi.org/10.1007/978-3-319-63046-5_3
10. Pérez, F., Granger, B.E.: IPython: a system for interactive scientific computing. Comput. Sci. Eng. 9(3), 21–29 (2007)
11. Skolem, T.: The foundations of elementary arithmetic established by means of the recursive mode of thought, without the use of apparent variables ranging over infinite domains. In: van Heijenoort, J. (ed.) From Frege to Gödel. Harvard University Press, Cambridge (1967)
12. Suter, P., Köksal, A.S., Kuncak, V.: Satisfiability modulo recursive programs. In: Yahav, E. (ed.) SAS 2011. LNCS, vol. 6887, pp. 298–315. Springer, Heidelberg (2011). https://doi.org/10.1007/978-3-642-23702-7_23
13. Hillel, W.: The Great Theorem Prover Showdown (2018). https://www.hillelwayne.com/post/theorem-prover-showdown/
14. Wenzel, M.: Isabelle/jEdit – a prover IDE within the PIDE framework. In: Jeuring, J., et al. (eds.) CICM 2012. LNCS (LNAI), vol. 7362, pp. 468–471. Springer, Heidelberg (2012). https://doi.org/10.1007/978-3-642-31374-5_38

# A Programmer's Text Editor for a Logical Theory: The SUMOjEdit Editor (System Description)

Adam Pease[✉]

Articulate Software, San Jose, USA
apease@articulatesoftware.com

**Abstract.** SUMOjEdit is a programmer's text editor for the SUO-KIF language and SUMO http://www.ontologyportal.org theory. Modern procedural programming is done in a text editor with tool support. Development of ontologies and taxonomies has often been done in graphical editors, leading many developers to employ only logics of very limited expressiveness that can be manipulated visually. Developers in the theorem proving community typically work in text editors but often without the same degree of tool support that most programmers rely on. Beginners working with SUMO make some very predictable errors in syntax, logical formulation, and use of the library of theories. Many of these errors can be flagged during editing, resulting in reduced time to become a productive developer. An editor designed for working with SUMO has the potential to aid beginners and experienced SUMO developers.

## 1 Introduction

The Suggested Upper Merged Ontology (SUMO) [6,9] is a logical theory stated in a higher-order logic, in SUO-KIF syntax[1]. It has roughly 20,000 constant symbols and 80,000 statements that have been written by hand since the start of the effort in the year 2000. It has been mapped by hand [7] to all 100,000 word senses in the WordNet lexicon [4] and to the Open Multilingual WordNet [3] that includes some two dozen languages. SUMO has been used as a source theory in several CASC competitions [13]. The Sigma knowledge engineering environment (SigmaKEE) [11] is the tool set employed in development of SUMO. It includes translators that translate SUMO from SUO-KIF to TPTP FOF [12], TFF0 [10] and THF [2] and interfaces to provers including E [14] and Vampire [5].

## 2 Motivation: Support for New Ontologists

We recently employed SUMO on a project to improve a taxonomy of customer service requests for a consumer electronics company. They had thousands of tags

---

[1] https://github.com/ontologyportal/sigmakee/blob/master/suo-kif.pdf.

© Springer Nature Switzerland AG 2020
N. Peltier and V. Sofronie-Stokkermans (Eds.): IJCAR 2020, LNAI 12167, pp. 472–479, 2020.
https://doi.org/10.1007/978-3-030-51054-1_31

that were used to classify problems, products and solutions to problems. The tags were created by customer service people, essentially forming a "folksonomy" that had considerable overlap and duplication of concepts. Because of a lack of definition of the tags, they were often reused in an inconsistent manner, since users of the tags would often not have the same understanding of their intended meaning as their author.

The project started quickly, with very little time to recruit and train developers. Three recent college graduates with Bachelor's degrees in Computer Science or Physics were employed and given one week of training based on the SUMO textbook [9]. They were then expected to begin work and ramp up their rate of constant symbol and definition authorship over the course of a month to a target of 50 concepts with definitions per week. None had any prior experience writing logic expressions. The training consisted of exercises in the SUO-KIF language and the most commonly used terms in SUMO, and use of the Sigma environment to view and test new SUMO content. The training followed the textbook [9] on SUMO. It became clear over the course of the project that although developers could load their new theories into Sigma and get feedback on errors, and were required to do so each week, that getting more immediate feedback would be valuable. Getting feedback from Sigma is essentially a batch process. If a syntax error is found and the system can't recover and interpret the rest of the statements in a file, then that error has to be corrected and then the process must start over. If instead errors could be caught at editing time, it would improve productivity.

A retrospective analysis of coding errors was performed. Given the sample size of only three developers and that this was not a controlled study, it must be considered anecdotal, but may still be informative. Many corrections were provided in person to the developers by the author of SUMO, so only email comments could be analyzed. The result was a set of 104 corrections that have been grouped into 7 categories. Only data for the first month of work was sampled as after that time basic coding errors became less frequent. Figure 1 shows the results of the analysis. The seven categories of errors are:

- *syntax error* - This refers to any number of violations of the BNF grammar of SUO-KIF [8].
- *unused var & var name* - A variable may appear in a quantifier list and never get used or a variable might be used only once. In the first case that is logically harmless but may indicate a mistake. In the second case, it's allowed if the intent of the formula is to specify some restriction about an argument to a relation, without respect to the other arguments. However, it is also often indicative of a mistake.
- *term name* - The developer uses a name for a constant that is likely misspelled and not defined elsewhere.
- *arg order or num* - A common error is to reverse the arguments to case roles or other binary relations, or to forget an argument to a higher-arity relation.
- *class vs. instance* - It is a common error for beginners to confuse an instance with a type. An example of this error would be stating that the result of

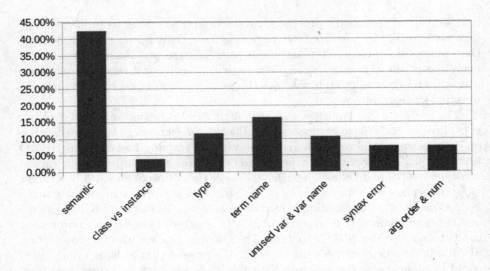

**Fig. 1.** Knowledge coding errors

a `Cooking` process is the class `PreparedFood` rather than an instance of `PreparedFood`.

- *type* - This refers to any error with types. All relations in SUMO have required types. Errors can include using an argument of the wrong type to a relation as well as using variables as arguments to relations that would require the variable to have two disjoint types.
- *semantic* - This refers to comments and corrections about whether the developer stated a fact that accurately reflects the domain. This sort of comment would be expected to be the most common correction. It's also one that automation cannot address. Semantic corrections become a much larger percentage as users gain facility in using the mechanics of SUMO and the SUO-KIF logical language.

## 3   SUMOjEdit Features and Functions

While theorem proving should in theory be able to find most contradictions, on large theories such as SUMO it may fail to find them in a reasonable amount of time. Theorem proving will also not be able to find a problem that doesn't result in a contradiction, like use of a constant symbol that only appears once. Much simpler inference procedures, coded in Java, can find many straightforward issues that would result in a logical contradiction, such as argument type violations, very quickly. In addition, such dedicated procedures can also find all such issues at once, and without the need for the user to diagnose the root cause by reading a potentially long or complicated proof. Apart from errors that result in logical contradictions, there are also many issues that can be classified as "warnings", that are indicative of possible errors, but are still logically consistent. Lastly,

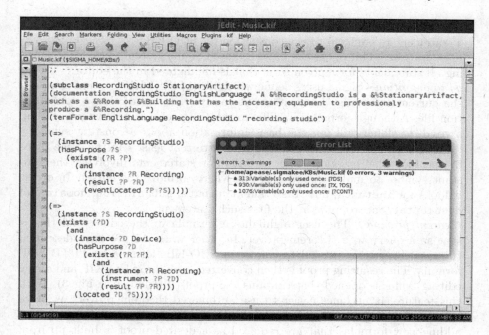

**Fig. 2.** An editor screen

some features may be viewed as helpful productivity enhancements, such as finding the likely definition of a constant symbol.

The features currently implemented in the editor are:

- *color coding* - there are only eight "reserved words" in the SUO-KIF language - and, or, not, =>, <=>, equal, forall, exists and they are given their own color. Comments, numbers and strings are uniquely color coded as shown in Fig. 2. The most fundamental defining relations in SUMO are given their own color. They are instance, subclass, domain, domainSubclass, range, rangeSubclass. Documentation predicates are also color-coded. These predicates are: documentation, termFormat and format. Lastly True and False are highlighted. Color coding often makes some errors obvious, such as a case of comment text not being preceded by a semicolon comment symbol.
- *check for errors and warnings* - There are many errors that can be checked for by the Sigma system and this feature applies all the checks to a file. A common error is for a developer to make an instance of an Object as the Agent of a Process. Agent(s) however must be sentient beings and not inanimate objects. Another common error is for a particular variable not to get used because one occurrence of it has a typo. Beginners often do copy-and-modify as an approach to writing rules and wind up with a conjunction with just one element, or a quantifier enclosing a set of literals, rather than a conjunction of literals. This function should catch all of the errors (other than "semantic") listed in Fig. 1.

- *formatting* - This is analogous to a Lisp "pretty-print" function and formats a statement to conform to the convention for SUO-KIF.
- *open browser* - This opens the browser-based Sigma system to the page showing all statements for a given term that is highlighted in the editor.
- *go to definition* - Sigma users define which files of SUO-KIF content comprise the current knowledge base either interactively or via editing a configuration file. Although formulas in different files can refer to a constant symbol, a good developer will put the basic information about a constant symbol - its class membership and documentation string at least - in the same place in one file. This function looks in order for statements involving the constant symbol to make a good guess at where to move the cursor. In order, it looks for `instance`, `subclass`, `subAttribute`, `subrelation`, `domain` and `documentation`, stopping at the first such statement.
- *theorem proving* - The user highlights a formula in the editor and this is sent as a query to a theorem prover. In other work [12] we have described the transformations needed to convert a SUO-KIF formula to TPTP FOF formula. The resulting proof is then converted back into SUO-KIF and a new editing buffer is opened that contains the proof (if found) (see Fig. 3). With this feature, it becomes easier to use automated theorem provers as proof assistants, trying to assemble lemmas that can be proven into a larger proof, adding new formulas that are required as a desired proof is built up from smaller proofs.

## 4   Related Work

Most prior work in this area is on editing tools for proofs using proof assistants. The focus of SUMO, and the SUMO editor, is on authoring logical theories, where the primary applications of the theories are as metadata for software engineering, including development of taxonomies and schemas.

**The Logic text editor** - [2] is designed for human authoring of statements and proofs. It is a set of extensions for the Markdown language. No automation support is provided for checking the statements or proofs that a user creates.

**ProofGeneral** [1] - [3] is an editor based on Emacs (and more recently, there is a proposed port to the Eclipse IDE) for proof assistants and supports many provers including Coq, EasyCrypt and PhoX. It includes color coding of proofs and interactive application of tactics.

**Isabelle/jEdit** [17][4] is an interface to the Isabelle interactive prover with many sophisticated functions. It includes color coding and the ability to display non-ASCII symbols in formulas. It does theory formatting. Many kinds of auto-completion of formulas and symbols can be performed. Proof checking can be done continuously as a background task.

---

[2] http://davidagler.com/teaching/logic/handouts/supplemental_material/ MarkdownForSymbolicLogic.html.
[3] https://proofgeneral.github.io/.
[4] https://isabelle.in.tum.de/dist/doc/jedit.pdf.

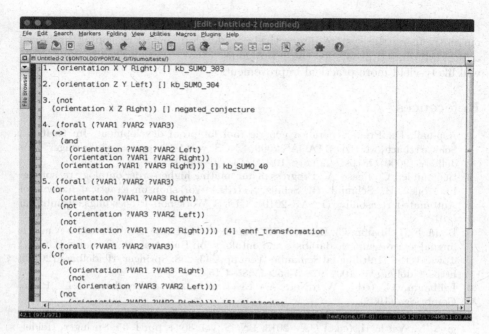

**Fig. 3.** A proof window

**MizarMode** [15,16] is focused on interactive theorem proving with the Mizar library. Like ProofGeneral, it runs on top of Emacs. It supports browsing semantically disambiguated queries, placing the cursor at the definition of a constant and sending a formula to the Mizar Proof Advisor.

Other systems include CoqIDE[5] and the Lean prover's UI[6].

## 5   Code and Future Work

SUMOjEdit is implemented in Java as a plugin for jEdit. It relies on the Sigma framework that is also written in Java, as well as the ErrorList jEdit plugin. It is open source and available at https://github.com/ontologyportal/SUMOjEdit.

Auto-complete would be a useful feature, especially given that there are 20,000 constant symbols in SUMO, and spelling unusual ones correctly can sometime be a challenge. A more thorough review of the proof assistant tools listed in the Related Work section will be helpful to see if the approach of proof assistants can be more thoroughly applied to work with automated provers. Finally, in this first version, much could be done to improve speed and efficiency. Load time is rather slow, and implementing jEdit's *listener* functionality would help to get the editor running while more complex operations proceed in the background and

---

[5] https://coq.inria.fr/refman/practical-tools/coqide.html.
[6] https://leanprover.github.io/.

then trigger the availability of the more sophisticated forms of analysis. We are currently using the system interactively to develop spatial reasoning problems and SUMO-based proofs for those problems. More experience with the system will likely yield more practical improvements.

# References

1. Aspinall, D.: Proof general: a generic tool for proof development. In: Graf, S., Schwartzbach, M. (eds.) TACAS 2000. LNCS, vol. 1785, pp. 38–43. Springer, Heidelberg (2000). https://doi.org/10.1007/3-540-46419-0_3
2. Benzmüller, C., Pease, A.: Progress in automating higher-order ontology reasoning. In: Konev, B., Schmidt, R., Schulz, S. (eds.) Workshop on Practical Aspects of Automated Reasoning (PAAR-2010). CEUR Workshop Proceedings, Edinburgh (2010)
3. Bond, F., Fellbaum, C., Hsieh, S.-K., Huang, C.-R., Pease, A., Vossen, P.: A multilingual lexico-semantic database and ontology. In: Buitelaar, P., Cimiano, P. (eds.) Towards the Multilingual Semantic Web, pp. 243–258. Springer, Heidelberg (2014). https://doi.org/10.1007/978-3-662-43585-4_15
4. Fellbaum, C. (ed.): WordNet: An Electronic Lexical Database. MIT Press, Cambridge (1998)
5. Kovács, L., Voronkov, A.: First-order theorem proving and VAMPIRE. In: Sharygina, N., Veith, H. (eds.) CAV 2013. LNCS, vol. 8044, pp. 1–35. Springer, Heidelberg (2013). https://doi.org/10.1007/978-3-642-39799-8_1
6. Niles, I., Pease, A.: Toward a standard upper ontology. In: Welty, C., Smith, B. (eds.) Proceedings of the 2nd International Conference on Formal Ontology in Information Systems (FOIS-2001), pp. 2–9 (2001)
7. Niles, I., Pease, A.: Linking lexicons and ontologies: mapping WordNet to the suggested upper merged ontology. In: Proceedings of the IEEE International Conference on Information and Knowledge Engineering, pp. 412–416 (2003)
8. Pease, A.: (2009). https://github.com/ontologyportal/sigmakee/blob/master/suo-kif.pdf
9. Pease, A.: Ontology: A Practical Guide. Articulate Software Press, Angwin (2011)
10. Pease, A.: Arithmetic and inference in a large theory. In: AI in Theorem Proving (2019)
11. Pease, A., Benzmüller, C.: Sigma: an integrated development environment for logical theories. In: Proceedings of the ECAI 2010 Workshop on Intelligent Engineering Techniques for Knowledge Bases (I-KBET-2010), Lisbon, Portugal (2010)
12. Pease, A., Schulz, S.: Knowledge engineering for large ontologies with Sigma KEE 3.0. In: Demri, S., Kapur, D., Weidenbach, C. (eds.) IJCAR 2014. LNCS (LNAI), vol. 8562, pp. 519–525. Springer, Cham (2014). https://doi.org/10.1007/978-3-319-08587-6_40
13. Pease, A., Sutcliffe, G., Siegel, N., Trac, S.: Large theory reasoning with SUMO at CASC. AI Commun. Spec. Issue Pract. Aspects Autom. Reasoning 23(2–3), 137–144 (2010)
14. Schulz, S., Cruanes, S., Vukmirović, P.: Faster, higher, stronger: E 2.3. In: Fontaine, P. (ed.) CADE 2019. LNCS (LNAI), vol. 11716, pp. 495–507. Springer, Cham (2019). https://doi.org/10.1007/978-3-030-29436-6_29
15. Urban, J.: Mizarmode - an integrated proof assistance tool for the mizar way of formalizing mathematics. J. Appl. Logic 4(4), 414–427 (2006). Towards Comput. Aided Math

16. Urban, J., Rudnicki, P., Sutcliffe, G.: ATP and presentation service for Mizar formalizations. J. Autom. Reasoning **50**(2), 229–241 (2013). https://doi.org/10.1007/s10817-012-9269-y
17. Wenzel, M.: Isabelle/jEDIT as IDE for domain-specific formal languages and informal text documents. In: Electronic Proceedings in Theoretical Computer Science, vol. 284, pp. 71–84, November 2018. https://doi.org/10.4204/EPTCS.284.6

# Sequoia: A Playground for Logicians
## (System Description)

Giselle Reis[(✉)] [iD], Zan Naeem, and Mohammed Hashim

Carnegie Mellon University, Doha, Qatar
giselle@cmu.edu, {znaeem,mqh}@andrew.cmu.edu

**Abstract.** Sequent calculus is a pervasive technique for studying logics and their properties due to the regularity of rules, proofs, and meta-property proofs across logics. However, even simple proofs can be large, and writing them by hand is often messy. Moreover, the combinatorial nature of the calculus makes it easy for humans to make mistakes or miss cases. Sequoia aims to alleviate these problems. Sequoia is a web-based application for specifying sequent calculi and performing basic reasoning about them. The goal is to be a user-friendly program, where logicians can specify and "play" with their calculi. For that purpose, we provide an intuitive interface where inference rules can be input in LaTeX and are immediately rendered with the corresponding symbols. Users can then build proof trees in a streamlined and minimal-effort way, in whichever calculus they defined. In addition to that, we provide checks for some of the most important meta-theoretical properties, such as weakening admissibility and identity expansion, given that they proceed by the usual structural induction. In this sense, the logician is only left with the tricky and most interesting cases of each analysis.

**Keywords:** Sequent calculus · Meta-properties · Web-based app

## 1 Introduction

Proof (and derivation) trees are a central structure for sequent calculus. They are used to check validity of formulas and sequents, as well as for checking meta-properties of the calculus (such as rule permutability, invertibility, and cut-elimination). In the latter, trees are generally *schematic*, using context variables to represent a family of trees with that shape.

When checking the validity of sequents, the required proof trees can become quite large (both in depth and breadth) to be written on paper. At the same time, the proofs of meta-properties may involve several (often small) proof trees to cover all the cases. Many times they are slight variations of each other, or largely the same across logics. Building and then verifying these proof trees is often a tedious task. Several tiresome issues arise: symbols can easily be misplaced, look unclear, or be confused by mistake; trees may have to be adjusted or re-sized to fit the writing space; and the proofs themselves may not look as elegant as their

© Springer Nature Switzerland AG 2020
N. Peltier and V. Sofronie-Stokkermans (Eds.): IJCAR 2020, LNAI 12167, pp. 480–488, 2020.
https://doi.org/10.1007/978-3-030-51054-1_32

typeset counterpart. However, a glance towards creating proof trees in a digital environment shows a separate set of issues. Currently there are very few tools for creating proof trees intuitively. The most common method is to write the proofs in LATEX, or a program that produces such proofs. Even then, the process can easily become too long and cumbersome.

Sequoia is a web application that makes sequent calculus tree building, whether schematic or not, simple and intuitive. Sequoia is aimed at students and academics who find the traditional methods too cumbersome, and provides a user-friendly means to create multiple calculi specified from user-defined symbols and inference rules, as well as a way to correctly build proof trees with their defined calculi. We use a sound and complete algorithm that computes all valid applications of a rule to a sequent. In addition, Sequoia features metatheoretical property checking for weakening admissibility, identity expansion, and permutability. It provides all the straightforward cases needed for the complete proofs of these properties, alleviating the user from the monotonous part, and allowing them to focus on the interesting cases. All these features are ready to be used now, and we are still improving Sequoia by adding more properties to check and more proof tree building tools.

Sequoia can be accessed at: https://logic.qatar.cmu.edu/sequoia/, and the source code is available at https://github.com/meta-logic/sequoia.

## 2   System Description

Sequoia consists of a front-end built in JavaScript and HTML, and a back-end built in MongoDB and Standard ML New Jersey. The application runs in a Node.js environment. The front-end is responsible for displaying the user defined rules and symbols, rendering the interactive proof tree, and presenting possible proof tree transformations for property testing, among other things. Aside from running the server, the back-end stores the user's defined calculi (including rules and symbols), computes all the possible valid proof trees when a rule is applied to a tree sequent, and constructs all possible tree derivations for a particular metaproperty. The following sections will progressively describe the design of our system by first providing the basic representations for the datatypes in SML, then describing the schematic tree building, and explaining our approach to automating the meta-property tests.

### 2.1   Datatypes

Currently, Sequoia supports sequent calculi with multiple contexts on the left and right. We restrict the rules in a calculus to operate on one connective at a time. We also require that rules have no restrictions on their contexts, such as:

$$\frac{\Box\Gamma \vdash A}{\Gamma, \Box\Gamma \vdash \Box A}$$

Note that such rules can often be rewritten using multiple contexts.

Our datatypes are defined as:

**Formula**   $F ::= p, q, ... \mid A, B, ... \mid \bullet_1(F_1, ..., F_{r_1}), \bullet_2(F_1, ..., F_{r_2}), ...$
**Context** $\Gamma/\Delta ::= \Theta_1, ..., \Theta_k, F_1, ...F_l$
**Sequent**   $S ::= \Gamma_1 /\!/_1 ... /\!/_{n-1} \Gamma_n \vdash \Delta_1 /\!/'_1 ... /\!/'_{m-1} \Delta_m$
**Rule**   $R ::= (\text{rule name}, S_i, [S_1, ...S_z])$
**Tree**   $T ::= (\text{rule name}, S_i, [T_1, ...T_y])$

Where $p, q, ...$ are atom variables, $A, B, ...$ are formula variables, $\bullet_1, \bullet_2, ...$ are connectives with arities $r_1, r_2, ...$ respectively, $\Theta_1, \Theta_2, ...$ are context variables, $/\!/$ is a context separator and $\vdash$ is a sequent sign. Context separators are symbols used to separate different contexts on either the left or right sides the sequent. For example, in the focused system for linear logic [3], three symbols ("$;$", "$\Downarrow$", "$\Uparrow$") are used to separate different parts of the right context. Note that all the mentioned symbols must be declared by the user and cannot contain superscripts.

Proof trees are recursive structures made up of a sequent, the rule name (if any) and a set of proof trees above the sequent; rules consist of a conclusion sequent and a set of premise sequents.

## 2.2   Core Operations

Rule application is the most important operation of Sequoia. It relies on three core operations: unification, substitution, and variable renaming. Rule application is a function applied to a rule and a sequent. We will use the following rule and sequent as our running example (assuming all the symbols have been declared by the user):

$$\frac{\Gamma_1 \vdash A \quad \Gamma_2 \vdash B}{\Gamma_1, \Gamma_2 \vdash A \wedge B} \wedge_r \qquad \Delta, F \vee G, H \vdash X \wedge Y$$

**Unification.** The first step for rule application is obtaining the valid unifiers between the sequent and the conclusion sequent of the rule. The reason we are not using pattern matching is because we may have to substitute context variables in the sequent as well as in the inference rule. For example, if the context is $\Gamma, r, p \wedge q$ and the conclusion of the rule is $\Gamma_1, \Gamma_2, A \wedge B$, then we need: one of $\Gamma_i$ substituted by a context variable and $r$, and $\Gamma$ substituted by two new context variables. Unification of sequents and contexts is defined as:

$$\frac{\vdash = \vdash' \quad \Gamma_1 /\!/_1 ... \Gamma_n \doteq \Gamma'_1 /\!/'_1 ... \Gamma'_n \mid \Sigma_1 \quad (\Delta_{1;1} ... \Delta_k)\sigma \doteq (\Delta'_{1;1} ... \Delta'_k)\sigma \mid \Sigma_2 \quad \forall \sigma \in \Sigma_1}{\Gamma_1 /\!/_1 ... \Gamma_n \vdash \Delta_1 /\!/_1 ... \Delta_k \doteq \Gamma'_1 /\!/'_1 ... \Gamma'_n \vdash' \Delta'_1 /\!/'_1 ... \Delta'_k \mid \Sigma_1 \circ \Sigma_2} \text{ seq}$$

$$\frac{}{\doteq \cdot \mid \{\}} \text{ ctx}_0 \qquad \frac{/\!/_1 = /\!/'_1 \quad \Gamma_1 \doteq \Gamma'_1 \mid \Sigma_1 \quad (\Gamma_2 /\!/_2 ... \Gamma_n)\sigma \doteq (\Gamma'_2 /\!/'_2 ... \Gamma'_n)\sigma \mid \Sigma_2 \quad \forall \sigma \in \Sigma_1}{\Gamma_1 /\!/_1 \Gamma_2 /\!/_2 ... \Gamma_n \doteq \Gamma'_1 /\!/'_1 \Gamma'_2 /\!/'_2 ... \Gamma'_n \mid \Sigma_1 \circ \Sigma_2} \text{ ctx}$$

Note that the symbols between contexts and the sequent sign must match for the unification to succeed, and that contexts are ordered. The unification of individual contexts is done through multiset unification using constraints [6].

Suppose that we want to apply the rule $\wedge_r$ to the sequent in the example above. Then, we need to get the unifiers between $\Gamma_1, \Gamma_2 \vdash A \wedge B$ ($\wedge_r$ conclusion) and $\Delta, F \vee G, H \vdash X \wedge Y$. This will produce a number of valid unifiers and a constraint theory alongside each unifier, one of them being:

$$\sigma = \{A \to X,\ B \to Y,\ \Gamma_1 \to [\Gamma_1', H],\ \Gamma_2 \to [\Gamma_2', F \vee G]\} \qquad (\Delta = \Gamma_1', \Gamma_2')$$

Constraint theories are used to maintain consistency between a conclusion and its premises in a proof tree. Its importance is discussed later in Sect. 2.3.

**Substitution.** Once unification is done, every valid unifier represents one possible way of applying the inference rule to the sequent. The premises are determined by applying the resulting unifier to the rule's premises. For example, the unifier above can be applied to the premises of $\wedge_r$, resulting in:

$$(\Gamma_1 \vdash A)\sigma = \Gamma_1', H \vdash X \qquad (\Gamma_2 \vdash Y)\sigma = \Gamma_2', F \vee G \vdash Y$$

Which is a correct set of premises when applying $\wedge_r$ to $\Delta, F \vee G, H \vdash X \wedge Y$, given the constraint $\Delta = \Gamma_1', \Gamma_2'$:

$$\frac{(\Gamma_1 \vdash A)\sigma \quad (\Gamma_2 \vdash B)\sigma}{(\Gamma_1, \Gamma_2 \vdash A \wedge B)\sigma} \wedge_r \quad = \quad \frac{\Gamma_1', H \vdash X \quad \Gamma_2', F \vee G \vdash Y}{\Delta, F \vee G, H \vdash X \wedge Y} \wedge_r$$

**Variable Renaming.** When applying a rule, we can assume that the sequent and the conclusion sequent of the rule have different variable names because of the symbol restrictions in the symbols tables. However, after applying a rule some problems might arise. For example, applying the $\wedge_r$ rule on $\Gamma \vdash (a \wedge b) \wedge c$ would yield the premises $\Gamma_1 \vdash a \wedge b$ and $\Gamma_2 \vdash c$ and the constraint $\Gamma = \Gamma_1, \Gamma_2$. However, applying $\wedge_r$ again on $\Gamma_1 \vdash a \wedge b$ would cause problems in unification as $\Gamma_1$ is used in both the rule and the sequent. To avoid this problem, all context variables are renamed after unification. To rename a context variable, we simply add a fresh superscript to the name of the variable, or update it if the name has one already.

## 2.3   Functionalities

The key features of Sequoia are that it allows the user to build ground and schematic proof trees, and to automate the process of testing for certain meta-properties. Currently, Sequoia is able to check rule permutability, weakening admissibility (for each context), and identity expansion.

**Tree Building.** Sequoia's tree building relies entirely on rule application. Given a tree, a constraint list, and a set of rules, a selected rule can be applied to an open sequent in the tree to produce a new tree and constraints with the appropriate updates. To do this, we first compute all possible unifiers of an open sequent and a rule. Then for each unifier, the empty premise set of the open sequent is replaced by the new premises obtained as explained above. The unifier is applied to each sequent in the tree, including the open sequent, and the constraint list is updated with the unifier's accompanying constraint theory. The constraint list is bound to the tree and accounts for the context variables changing at different levels in the tree as a result of multiple rule applications. The user can undo rule applications, as well as export the proof tree to LaTeX.

**Proof Transformations.** In some cases (such as checking for permutability), we need to decide whether a tree $T_1$ with end sequent $S$ can be transformed into another tree $T_2$ with the same end sequent. For that, we assume that $T_1$ is a closed tree, i.e., each premise $S_{1,i}$ in $T_1$ has a proof $\mathcal{D}_{1,i}$. Checking if $T_1$ can be transformed into $T_2$, amounts to checking that each open premise $S_{2,j}$ in $T_2$ can be proved using some $\mathcal{D}_{1,i}$. The proof $\mathcal{D}_{1,i}$ for $S_{1,i}$ can be used to prove $S_{2,j}$ if: (1) the two sequents are the same (modulo context variables), or (2) if weakening is admissible in some contexts, that $S_{1,i}$ can be obtained by weakening $S_{2,j}$. If the proof can be used, we add constraints specifying which multiset of context variables in $S_{2,j}$ is equal to the multiset of context variables in $S_{1,i}$. Given this set of constraints and the ones obtained from unification when applying the rules, which are equalities between multisets, we try to find an AC1[1] unifier [1, section 10.3] such that it does not map context variables of $T_1$ to empty or a multiset which contains more than one copy of each context. This approach to proof transformations has its limitations, since we do not take into consideration cases that succeed because of a rule's invertibility or the use of cut rules. Thus, the check is always sound, but not complete. The user needs to check by hand the cases that Sequoia cannot infer.

**Permutability.** Given two rules $R_1$ and $R_2$, the initial rules, and the weakening properties of the calculus, we say that $R_1$ *permutes up* $R_2$ if a proof tree $T$ ending with the rule $R_2$ applied over $R_1$ can be transformed into a proof tree $T'$ ending with $R_1$ applied over $R_2$. Sequoia performs this check by first generating all derivations $T$ where $R_2$ is applied over $R_1$, and all derivations $T'$ where $R_1$ is applied over $R_2$. Then, for each tree $T$, we try to find a tree $T'$ such that $T$ can be transformed into $T'$.

**Weakening Admissibility.** The admissibility of weakening for a calculus is checked for each context separately. Given a context $\Gamma$ in a sequent $S$, the theorem states: if a sequent $S[\Gamma]$ is provable, then so is $S[\Gamma, F]$. The usual proof

---

[1] Associative, commutative, with neutral element (the properties of multiset union).

proceeds by structural induction on the derivation of $S[\Gamma]$. Sequoia is able to check all "trivial" cases, i.e. the ones that require only the inductive hypothesis.

**Identity Expansion.** Identity expansion is the property that all identity rules can be applied on atoms. The usual proof proceeds by induction on the formula structure. Let $\bullet(F_1, ..., F_n)$ be a formula with main connective $\bullet$, $[\bullet_1], ..., [\bullet_k]$ be the rules for decomposing $\bullet$, and $[id]$ one identity rule. Sequoia checks if a proof ending with $[id]$ on $\bullet(F_1, ..., F_n)$ can be transformed into a proof using some of the rules $[\bullet_1], ..., [\bullet_k]$ and $[id]$ only on $F_1, ..., F_n$. This is done by applying left and right pairs of rules and trying to close the proof.

Once again, this check is sound, but not complete. For example, take the LJ calculus for intuitionistic logic with the following rules for conjunction left:

$$\frac{\Gamma, A_1 \vdash C}{\Gamma, A_1 \wedge A_2 \vdash C} \wedge_r^1 \quad \frac{\Gamma, A_2 \vdash C}{\Gamma, A_1 \wedge A_2 \vdash C} \wedge_r^2$$

Sequoia is not able to infer identity expansion because it will not apply contraction arbitrarily. Instead, if the following rule is used:

$$\frac{\Gamma, A_1, A_2 \vdash C}{\Gamma, A_1 \wedge A_2 \vdash C} \wedge_r$$

Then Sequoia succeeds in showing identity expansion for the case of $\wedge$.

## 3 Usage

Sequoia was made with the goal in mind that a calculus construction and tree building tool should have a nice design and an intuitive interface for students and academics. All input is compiled in LaTeX, as it is a familiar typesetting language with a vast access to symbols.

**Symbols Table.** Before creating rules and building trees, users must declare the symbols to be used and their types. A symbols table consists of the symbol (input in LaTeX) and its type (chosen from a drop-down menu). Symbols can be updated by changing their type or simply deleted. There are two symbols tables: a rule one (symbols used for the rules in a calculus) and an end-sequent one (symbols used on the end-sequent of a proof tree). The following restrictions apply to both tables: the same symbol cannot be assigned different types (per calculus), and symbols cannot contain superscripts. Moreover, the symbols used for context variables and formulas in rules and end-sequents must be disjoint.

**Calculus Specification.** The homepage displays a user's defined calculi, a form to create new calculi, and buttons to add sample calculi that are available by default. Each calculus will have a card. Clicking on a calculus card will direct the user to the main page for that calculus, which contains all its rules and the rule symbols table. "Add Rule" directs the user to the rule creation page, where they see the input fields for a rule: name, main connective, conclusion, and premise(s). There is also a drop-down menu for indicating on which side of the sequent the rule operates (left, right, or none). After filling in the information, clicking on "Preview" will show a rendering of the rule compiled in LATEX and the rule symbols table from the calculus page. When all symbols in the premises and conclusion are in the table, the new rules can be added to the calculus.

**Tree Building.** Clicking on "Proof Tree" (upper right corner) takes the user to a page where they can build proof trees using the calculus rules they have defined (listed on the right). After entering an end-sequent and clicking on "Preview", they will see a rendering of the sequent in LATEX and the end-sequent symbols table. Once all the context variable and formula symbols are declared in the table, the user can begin building the tree. The constraint list (initially empty) is shown on the left. To build on a proof tree, the user selects a leaf premise in the tree and a rule to apply on it. If the rule is not applicable, the user will be informed. Otherwise, the user is prompted with a selection of all the possible premise sets that result from applying the rule. Selecting a premise set renders the appropriate tree with these premises and the constraint list is updated with the associated constraint. In case the selected rule is *cut*, the user is prompted for the substitution to be used. They must type the cut-formula variable (used in the rule), and the cut-formula to be used for that variable.

**Properties Testing.** The properties page allows the user to test certain meta-properties for a sequent calculus system. Currently, the implemented meta-properties are: weakening admissibility, identity expansion, and permutability.

By clicking on "Weakening Admissibility" or "Identity Expansion", the user is presented with several cards representing contexts or connectives, respectively. Clicking on a card will show all its proof tree transformations for that property. By clicking on "Permutability", the user is shown all rules and must select two to perform the check. After clicking on "Permute Rules", Sequoia shows all the successful and failed proof transformations for permutability between them.

## 4    Related Work

There are several other tools for constructing and visualizing proofs. We will focus on the ones that are interactive and offer support for sequent calculi.

Closest to our approach is Carnap.io [5], a web-based tool built using proofJS and Haskell. Carnap supports different deductive systems and allows users to add their logic by implementing it in Haskell with the help of Carnap's type classes. Proof tree construction is done by typing the proof, while Carnap checks each step. The sequent calculus calculator [4] has an interface similar to Sequoia, and also allows the user to build proof trees in four different logics. The user needs to instantiate the rules before applying them. Axolotl [2] is a Java applet and mobile app for constructing proofs. It can handle proofs in sequent calculus, natural deduction, or Hilbert systems. Sequent calculus rules are displayed on a one-dimensional notation, and proof goals may not contain context variables.

Both Carnap.io and the sequent calculus calculator require manual input from the user when building proofs, which are checked. Sequoia instead computes all possibilities for a *correct* rule application, and prompts the user to choose one. This is less tedious for experienced users, and more user friendly for those not versed in sequent calculus. This is also the approach used in Axolotl.

Concerning different logics, while Carnap.io allows the user to add more calculi, this requires expertise with Haskell and type classes, which most undergraduate students lack. We believe that the approach of inputting sequent calculus via LATEX will be more appealing for those users. The other systems only work on a pre-determined set of calculi.

Different from all the aforementioned systems, Sequoia allows users to build *schematic* proofs, using context variables. The reason for including this feature is that, most of the time, logicians "play" with proofs using schemas as opposed to concrete formulas since they are trying to see patterns or investigate proof transformations regardless of concrete terms.

Tatu [7] and Quati [8] are web-based tools that allow users to check for certain meta-properties of sequent calculus systems. Tatu allows the user to check for identity expansion and cut admissibility, while Quati allows the user to check if rules permute over each other and shows the proof tree transformations rendered from LATEX. To use those tools, users have to define their sequent calculus system in linear logic with subexponentials, a non-trivial task that cannot be easily automated. Another approach for checking meta-properties is using rewrite logic. In [9] the authors used Maude to automate the checking of permutability, admissibility and invertibility of rules. Although there is no user interface, the technique seems powerful and could be used in Sequoia for other checks.

Sequoia improves on Tatu and Quati by facilitating considerably the input of systems. The main difference between Sequoia and the tool based on Maude is that Sequoia displays the proof transformations that were found, thus showing the user how they work and increasing the trust in the system.

## 5    Future Work

There are a number of features and improvements we plan to add to Sequoia.

We have recently finished the implementation of two new features: checking cut admissibility (using Gentzen-style proofs), and supporting rules with context restrictions (such as the one mentioned in Sect. 2.1). These will be added to the website soon. The next meta-property we would like to add support for is rule invertibility. It should not be hard to check the simplest cases, which use a short derivation with cut. Most, if not all, of the operations needed are already implemented. We also plan to add support for first-order systems, but this will be more challenging, since it requires changes to some of the core operations. It will also result in more prompts to the user.

To improve usability, we are investigating the possibility of inputting sequent calculi or proofs by taking pictures of hand-written objects. We believe this feature will make the system much more appealing, specially to undergrads who do their work by hand, and need to type it in LATEX afterwards. A simpler addition that increases usability is allowing that rules be reused between calculi.

Concerning the meta-property proofs, we want to give the user the ability to export (incomplete) proofs to LATEX. Given a stable framework for formalizing meta-properties, one could also think of exporting these proofs into partial proof scripts to be completed by the user.

# References

1. Baader, F., Nipkow, T.: Term Rewriting and All That. Cambridge University Press, New York (1998)
2. Cerna, D.M., Kiesel, R.P., Dzhiganskaya, A.: A mobile application for self-guided study of formal reasoning. In: Quaresma, P., Neuper, W., Marcos, J. (eds.) Proceedings of the 8th International Workshop on Theorem Proving Components for Educational Software. Electronic Proceedings in Theoretical Computer Science, vol. 313, pp. 35–53. Open Publishing Association (2020). https://doi.org/10.4204/EPTCS.313.3
3. Di Cosmo, R., Miller, D.: Linear logic. In: Zalta, E.N. (ed.) The Stanford Encyclopedia of Philosophy. Metaphysics Research Lab, Stanford University (2019)
4. Hauser, F.: Sequent Calculus Calculator. https://seqcalc.io/. Accessed Jan 2020
5. Leach-Krouse, G.: Carnap: an open framework for formal reasoning in the browser. In: Quaresma, P., Neuper, W. (eds.) Proceedings 6th International Workshop on Theorem proving components for Educational software. Electronic Proceedings in Theoretical Computer Science, vol. 267, pp. 70–88. Open Publishing Association (2018). https://doi.org/10.4204/EPTCS.267.5
6. Naeem, Z., Reis, G.: Unification of multisets with multiple labelled multiset variables. In: 33rd International Workshop on Unification (UNIF) (2019)
7. Nigam, V., Pimentel, E., Reis, G.: An extended framework for specifying and reasoning about proof systems. J. Logic Comput. **26**(2), 539–576 (2016). https://doi.org/10.1093/logcom/exu029
8. Nigam, V., Reis, G., Lima, L.: Quati: an automated tool for proving permutation lemmas. In: 7th International Joint Conference on Automated Reasoning (IJCAR 2014), pp. 255–261 (2014). https://doi.org/10.1007/978-3-319-08587-6_18
9. Olarte, C., Pimentel, E., Rocha, C.: Proving structural properties of sequent systems in rewriting logic. In: Rusu, V. (ed.) WRLA 2018. LNCS, vol. 11152, pp. 115–135. Springer, Cham (2018). https://doi.org/10.1007/978-3-319-99840-4_7

# Prolog Technology
# Reinforcement Learning Prover
## (System Description)

Zsolt Zombori[1,2]([✉]), Josef Urban[3], and Chad E. Brown[3]

[1] Alfréd Rényi Institute of Mathematics, Budapest, Hungary
zombori@renyi.hu
[2] Eötvös Loránd University, Budapest, Hungary
[3] Czech Technical University of Prague, Prague, Czechia

**Abstract.** We present a reinforcement learning toolkit for experiments with guiding automated theorem proving in the connection calculus. The core of the toolkit is a compact and easy to extend Prolog-based automated theorem prover called plCoP. plCoP builds on the leanCoP Prolog implementation and adds learning-guided Monte-Carlo Tree Search as done in the rlCoP system. Other components include a Python interface to plCoP and machine learners, and an external proof checker that verifies the validity of plCoP proofs. The toolkit is evaluated on two benchmarks and we demonstrate its extendability by two additions: (1) guidance is extended to reduction steps and (2) the standard leanCoP calculus is extended with rewrite steps and their learned guidance. We argue that the Prolog setting is suitable for combining statistical and symbolic learning methods. The complete toolkit is publicly released.

**Keywords:** Automated theorem proving · Reinforcement learning · Logic programming · Connection tableau calculus

## 1 Introduction

Reinforcement learning (RL) [36] is an area of Machine Learning (ML) that has been responsible for some of the largest recent AI breakthroughs [3,32–34]. RL develops methods that advise agents to choose from multiple actions in an environment with a delayed reward. This fits many settings in Automated Theorem Proving (ATP), where many inferences are often possible in a particular search state, but their relevance only becomes clear when a proof is found.

Several learning-guided ATP systems have been developed that interleave proving with supervised learning from proof searches [4,10–13,17,19,23,39]. In

ZZ was supported by the European Union, co-financed by the European Social Fund (EFOP-3.6.3-VEKOP-16-2017-00002) and the Hungarian National Excellence Grant 2018-1.2.1-NKP-00008. JU and CB were funded by the *AI4REASON* ERC Consolidator grant nr. 649043, the Czech project AI&Reasoning CZ.02.1.01/0.0/0.0/15_003/0000466 and the European Regional Development Fund.

N. Peltier and V. Sofronie-Stokkermans (Eds.): IJCAR 2020, LNAI 12167, pp. 489–507, 2020.
https://doi.org/10.1007/978-3-030-51054-1_33

the saturation-style setting used by ATP systems like E [31] and Vampire [21], direct learning-based selection of the most promising given clauses leads already to large improvements [14], without other changes to the proof search procedure.

The situation is different in the connection tableau [22] setting, where choices of actions rarely commute, and backtracking is very common. This setting resembles games like Go, where Monte-Carlo tree search [20] with reinforcement learning used for action selection (*policy*) and state evaluation (*value*) has recently achieved superhuman performance. First experiments with the rlCoP system in this setting have been encouraging [19], achieving more than 40% improvement on a test set after training on a large corpus of Mizar problems.

The connection tableau setting is attractive also because of its simplicity, leading to very compact Prolog implementations such as leanCoP [29]. Such implementations are easy to modify and extend in various ways [27,28]. This is particularly interesting for machine learning research over reasoning corpora, where automated learning and addition of new prover actions (tactics, inferences, symbolic decision procedures) based on previous proof traces seems to be a large upcoming topic. Finally, the proofs obtained in this setting are easy to verify, which is important whenever automated self-improvement is involved.

The goal of the work described here is to develop a reinforcement learning toolkit for experiments with guiding automated theorem proving in the connection calculus. The core of the toolkit (Sect. 2) is a compact and easy to extend Prolog-based automated theorem prover called plCoP. plCoP builds on the lean-CoP Prolog implementation and adds learning-guided Monte-Carlo Tree Search as done in the rlCoP [19] system. Other components include a Python interface to plCoP and state-of-the-art machine learners and an external proof checker that verifies the validity of the plCoP proofs. The proof checker has proven useful in discovering bugs during development. Furthermore, it is our long term goal to add new prover actions automatically, where proof checking becomes essential.

Prolog is traditionally associated with ATP research, and it has been used for a number of Prolog provers [5,24,29,35], as well as for rapid ATP prototyping, with core methods like unification for free. Also, Prolog is the basis for Inductive Logic Programming (ILP) [25] style systems and a natural choice for combining such symbolic learning methods with machine learning for ATP systems, which we are currently working on [41]. In more detail, the main contributions are:

1. We provide an open-source Prolog implementation of rlCoP, called plCoP, that uses the SWI-Prolog [40] environment.
2. We extend the guidance of leanCoP to reduction steps involving unification.
3. We extend leanCoP with rewrite steps while keeping the original equality axioms. This demonstrates the benefit of adding a useful but redundant inference rule, with its use controlled by the learned guidance.
4. We provide an external proof checker that certifies the validity of the proofs.
5. The policy model of rlCoP is trained using Monte Carlo search trees of all proof attempts. We show, however, that this introduces a lot of noise, and we get significant improvement by limiting policy training data to successful theorem proving attempts.

6. Policy and value models are trained in rlCoP using a subset of the Monte Carlo search nodes, called *bigstep nodes*. However, when a proof is found, not all nodes leading to the proof are necessarily bigstep nodes. We make training more efficient by explicitly ensuring that all nodes leading to proofs are included in the training dataset.
7. The system is evaluated in several iterations on two MPTP-based [37] benchmarks, showing large performance increases thanks to learning (Sect. 3). We also improve upon rlCoP with 12% and 7% on these benchmarks.

## 2   Prolog Technology Reinforcement Learning Prover

The toolkit is available at our repository.[1] Its core is our plCoP connection prover based on the leanCoP implementation and inspired by rlCoP. leanCoP [29] is a compact theorem prover for first-order logic, implementing connection tableau search. The proof search starts with a *start clause* as a *goal* and proceeds by building a connection tableau by applying *extension steps* and *reduction steps*. leanCoP uses iterative deepening to ensure completeness. This is removed in rlCoP and plCoP and learning-guided Monte-Carlo Tree Search (MCTS) [8] is used instead. Below, we explain the main ideas and parts of the system.

**Monte Carlo Tree Search (MCTS)** is a search algorithm for sequential decision processes. MCTS builds a tree whose nodes are states of the process, and edges represent sequential decisions. Each state (node) yields some reward. The aim of the search algorithm is to find trajectories (branches in the search tree) that yield high accumulated reward. The search starts from a single root node (*starting state*), and new nodes are added iteratively. In each node $i$, we maintain the number of visits $n_i$, the total reward $r_i$, and its prior probability $p_i$ given by a learned *policy* function. Each iteration, also called *playout*, starts with the addition of a new leaf node. This is done by recursively selecting a child that maximizes the standard UCT [20] formula (1), until a leaf is reached. In (1), $N$ is the number of visits of the parent, and $cp$ is a parameter that determines the balance between nodes with high value (exploitation) and rarely visited nodes (exploration). Each leaf is given an initial value, which is typically provided by a learned *value* function. Next, ancestors are updated: visit counts are increased by 1 and value estimates are increased by the value of the new node. The value and policy functions are learned in AlphaGo/Zero, rlCoP and plCoP.

$$\text{UCT}(i) = \frac{r_i}{n_i} + cp \cdot p_i \cdot \sqrt{\frac{lnN}{n_i}} \qquad (1)$$

**MCTS for Connection Tableau:** Both rlCoP and plCoP use the DAgger [30] meta-learning algorithm to learn the policy and value functions. DAgger interleaves ATP runs based on the current policy and value (*data collection phase*) with a *training phase*, in which these functions are updated to fit the collected

---

[1] https://github.com/zsoltzombori/plcop.

data. Such iterative interleaving of proving and learning has also been used successfully in ATP systems such as MaLARea [38] and ENIGMA [14]. During the proof search plCoP builds a Monte Carlo tree for each training problem. Its nodes are the proof states (partial tableaux), and edges represent inferences. A branch leading to a node with a closed tableau is a valid proof. Initially, plCoP uses simple heuristic value and policy functions, later to be replaced with learned guidance. To enforce deeper exploration, rlCoP and plCoP perform a *bigstep* after a fixed number of playouts: the starting node of exploration is moved one level down towards the child with the highest value (called *bigstep node*). Later MCTS steps thus only extend the subtree under the bigstep node.

Training data for policy and value learning is extracted from the tableau states of the bigstep nodes. The value model gets features from the current goal and path, while the policy model also receives features of the given action.

We use term walks of length up to 3 as main features. Both rlCoP and plCoP add also several more specific features.[2] The resulting sparse feature vectors are compressed to a fixed size (see Appendix E). For learning value, each bigstep node is assigned a label of $1^3$ if it leads to a proof and 0 otherwise. The policy model gets a target probability for each edge based on the relative frequency of the corresponding child. Both rlCoP and plCoP use gradient boosted trees (XGBoost [9]) for guidance. Training concludes one iteration of the DAgger method. See Appendix D for more details about policy and value functions.

**Prolog Implementation of plCoP:** To implement MCTS, we modify leanCoP so that the Prolog stack is explicitly maintained and saved in the Prolog database using assertions after each inference step. This is done in the `leancop_step.pl` submodule, described in Appendix C. This setup makes it possible to interrupt proof search and later continue at any previously visited state, required to interleave prover steps with Monte Carlo tree steps, also implemented in Prolog. The main MCTS code is explained in Appendix B. The MCTS search tree is stored in destructive Prolog hashtables.[4] These are necessary for efficient updates of the nodes statistics. The training data after each proof run is exported from the MCTS trees and saved for the XGBoost learning done in Python.

To guide the search, the trained XGBoost policy and value functions are accessed efficiently via the C foreign language interface of SWI-Prolog. This is done in 70 lines of C++ code, using the SWI C++ templates and the XGBoost C++ API. The main foreign predicate `xgb:predict` takes an XGBoost predictor and a feature list and returns the predicted value. A trained model performs 1000000 predictions in 19 s in SWI. To quantify the total slowdown due to the guidance, we ran plCoP with 200000 inference step limit and a large time limit (1000 s) on the M2k dataset (see Sect. 3) with and without guidance. The average execution time ratio on problems unsolved in both cases is 2.88, i.e., the XGBoost guidance roughly triples the execution time. Efficient feature collection in plCoP

---

[2] Number of open goals, number of symbols in them, their maximum size and depth, length of the current path, and two most frequent symbols in open goals.

[3] A discount factor of 0.99 is applied to positive rewards to favor shorter proofs.

[4] The `hashtbl` library of SWI by G. Barany – https://github.com/gergo-/hashtbl.

is implemented using declarative association lists (the `assoc` SWI library) implemented as AVL trees. Insertion, change, and retrieval is $O(log(N))$. We also use destructive hashtables for caching the features already computed. Compared to rlCoP, we can rely on SWI's fast internal hash function `term_hash/4` that only considers terms to a specified depth. This all contributes to plCoP's smaller size.

**Guiding Reduction Steps:** The connection tableau calculus has two steps: 1) extension that replaces the current goal with a new set of goals and 2) reduction that closes off the goal by unifying it with a literal on the path. rlCoP applies reduction steps eagerly, which can be harmful by triggering some unwanted unification. Instead, plCoP lets the guidance system learn when to apply reduction. Suppose the current goal is $G$. An input clause $\{H, B\}$, s.t. $H$ is a literal that unifies with $\neg G$ and $B$ is a clause, yields an extension step, represented as $ext(H, B)$, while a literal $P$ on the path that unifies with $\neg G$ yields a reduction step, represented as $red(P)$. The symbols $red$ and $ext$ are then part of the standard feature representation.

**Limited Policy Training:** rlCoP extracts policy training data from the child visit frequencies of the bigstep nodes. We argue, however, that node visit frequencies may not be useful when no proof was found, i.e., when no real reward was observed. A frequently selected action that did not lead to proof should not be reinforced. Hence, plCoP only extracts policy training data when a proof was found. Note that the same is not true for value data. If MCTS was not successful, then bigstep nodes are given a value of 0, which encourages exploring elsewhere.

**Training from all Proofsteps:** Policy and value models are trained in rlCoP using bigstep nodes. However, when a proof is found, not all nodes leading to the proof are necessarily bigstep nodes. We make training more efficient by explicitly ensuring that all nodes leading to proofs are included in the training dataset.

**(Conditional) Rewrite Steps:** plCoP extends leanCoP with rewrite steps that can handle equality predicates more efficiently. Let $t|_p$ denote the subterm of $t$ at position $p$ and $t[u]_p$ denote the term obtained after replacing in $t$ at position $p$ by term $u$. Given a goal $G$ and an input clause $\{X = Y, B\}$, s.t. for some position $p$ there is a substitution $\sigma$ such that $G|_p\sigma = X\sigma$, the rewrite step changes $G$ to $\{G[Y]_p\sigma, \neg B\sigma\}$. Rewriting is allowed in both directions, i.e., the roles of $X$ and $Y$ can be switched.[5] This is a valid and well-known inference step, which can make proofs much shorter. On the other hand, rewriting can be simulated by a sequence of extension steps. We add rewriting without removing the original congruence axioms, making the calculus redundant. We find, however, that the increased branching in the search space is compensated by learning since we only explore branches that are deemed "reasonable" by the guidance.

**Proof Checking:** After plCoP generates a proof, the `leancheck` program included in the toolkit does an independent verification. Proofs are first

---

[5] The rewrite step could probably be made more powerful by ordering equalities via a term ordering. However, we wanted to use as little human heuristics as possible and let the guidance figure out how to use the rewrite steps.

translated into the standard leanCoP proof format. In case of rewriting steps, plCoP references the relevant input clause (with an equational literal), its substitution, the goal before and after rewriting, the equation used and the side literals of the instantiated input clause. Using this information, `leancheck` replaces the rewriting step with finitely many instances of equational axioms (reflexivity, symmetry, transitivity, and congruence) and proceeds as if there were no rewriting steps.

The converted output from plCoP includes the problem's input clauses and a list of clauses that contributed to the proof. The proof clauses are either input clauses, their instances, or extension step clauses. Proof clauses may have remaining uninstantiated (existential) variables. However, for proof checking, we can consider these to be new constants, and so we consider each proof clause to be ground. To confirm we have a proof, it suffices to verify two assertions:

1. The proof clause alleged to be an input clause, or its instance is subsumed by the corresponding input clause (as identified in the proof by a label).
2. The set of such proof clauses forms a propositionally unsatisfiable set.

Each proof clause alleged to be an instance of an input clause is reported as a clause $B$, a substitution $\theta$ and a reference to an input clause $C$. Optimally, the checker should verify that each literal in $\theta(C)$ is in $B$, so that $\theta(C)$ propositionally subsumes $B$. In many cases this is what the checker verifies. However, Prolog may rename variables so that the domain of $\theta$ no longer corresponds to the free variables of $C$. In this case, the checker computes a renaming $\rho$ such that $\theta(\rho(C))$ propositionally subsumes $B$.[6] We could alternatively use first-order matching to check if $C$ subsumes $B$. This would guarantee a correct proof exists, although it would accept proofs for which the reported $\theta$ gives incorrect information about the intended instantiation. For the second property, we verify the propositional unsatisfiability of this ground clause set using PicoSat [7]. While plCoP is proving a theorem given in disjunctive normal form PicoSat is refuting a set of clauses. Hence we swap polarities of literals when translating clauses to PicoSat.[7] An example is given in Appendix A.

## 3   Evaluation

We use two datasets for evaluation. The first is the *M2k* benchmark that was introduced in [19]. The M2K dataset is a selection of 2003 problems [15] from the larger Mizar40 dataset [16], which consists of 32524 problems from the Mizar Mathematical Library that have been proven by several state-of-the-art ATPs used with many strategies and high time limits in the Mizar40 experiments [18]. Based on the proofs, the axioms were ATP-minimized, i.e., only those axioms were kept that were needed in any of the ATP proofs found. This dataset is by

---

[6] Note that this $\rho$ may not be unique. Consider $C = \{p(X), p(Y)\}$, $B = \{p(c), p(c)\}$ and $\theta = W \mapsto c, Z \mapsto c$.

[7] Technically swapping the polarities is not necessary since unsatisfiability is invariant under such a swap. However, there is no extra cost since the choice of polarity is made when translating from the plCoP proof to an input for PicoSat.

construction biased towards saturation-style ATP systems. To have an unbiased comparison with state-of-the-art saturation-style ATP systems such as E, we also evaluate the systems on the bushy (small) problems from the MPTP2078 benchmark [1], which contains just an article-based selection of Mizar problems, regardless of their solvability by a particular ATP system.

We report the number of proofs found using a 200000 step inference limit. Hyperparameters (described in Appendix E) were selected to be consistent with those of rICoP. A lot of effort has already been invested in tuning rICoP, furthermore, we wanted to make sure that the effects of our most important additions are not obfuscated by hyperparameter changes. As Tables 1 and 2 show, our baseline is weaker, due to several subtleties of rICoP that are not reproduced. Nevertheless, since our main focus is to make learning more efficient, improvement with respect to the baseline can be used to evaluate the new features.

**M2k Experiments:** We first evaluate the features introduced in Sect. 2 on the M2k dataset. Table 1 shows that both limited policy training and training from all proofsteps yield significant performance increase: together, they improve upon the baseline with 31% and upon rICoP with 3%. However, guided reduction does not help. We found that the proofs in this dataset tend to only use reduction on ground goals, i.e., that does not involve unification, which indeed can be applied eagerly. The rewrite step yields a 9% increase. Improved training and rewriting together improve upon the baseline by 42% and upon rICoP by 12%. Overall, thanks to the changes in training data collection, pICoP shows greater improvement during training and finds more proofs than rICoP, even without rewriting. Note that rICoP has been developed and tuned on M2k. Adding ten more iterations to the best performing (**combined**) version of pICoP results in 1450 problems solved, which is 17.4% better than rICoP in 20 iterations (1235).

**Table 1.** Performance on the M2k dataset: original **rICoP**, pICoP (**baseline**), pICoP with **guided reduction**, pICoP with **limited policy** training, pICoP trained using **all proofsteps**, pICoP using the previous two **improved training** pICoP with **rewriting** and pICoP with rewriting and improved training **combined**. **incr** shows the performance increase in percentages from iteration 0 (unguided) to the best result.

| Iteration | 0 | 1 | 2 | 3 | 4 | 5 | 6 | 7 | 8 | 9 | 10 | Incr |
|---|---|---|---|---|---|---|---|---|---|---|---|---|
| rICoP | 770 | 1037 | 1110 | 1166 | 1179 | 1182 | 1198 | 1196 | 1193 | **1212** | 1210 | 57% |
| Baseline | 632 | 852 | 860 | 915 | 918 | 944 | 949 | **959** | 955 | 943 | 954 | 52% |
| Guided reduction | 616 | 840 | 884 | 905 | 915 | 900 | 914 | 924 | **942** | 915 | 912 | 53% |
| Limited policy | 632 | 988 | 1037 | 1071 | 1080 | 1094 | 1092 | 1101 | 1103 | **1118** | 1111 | 77% |
| All proofsteps | 632 | 848 | 930 | 988 | 986 | 1018 | 1033 | 1039 | **1053** | 1043 | 1050 | 67% |
| Improved training | 632 | 975 | 1100 | 1154 | 1180 | 1189 | 1209 | 1231 | 1238 | 1243 | **1254** | **98%** |
| Rewriting | 695 | 913 | 989 | 995 | 1003 | 1019 | 1030 | 1030 | 1033 | 1038 | **1045** | 50% |
| Combined | 695 | 1070 | 1209 | 1253 | 1295 | 1309 | 1322 | 1335 | 1339 | 1346 | **1359** | 96% |

**MPTP2078:** Using a limit of 200000 inferences, unmodified leanCoP solves 612 of the MPTP2078 bushy problems, while its OCaml version (mlcop), used as a basis for rlCoP solves 502. E solves 998, 505, 326, 319 in *auto, noauto, restrict, noorder* modes[8] plCoP and rlCoP results are summarized in Table 2. Improved training and rewriting together yield 63% improvement upon the baseline and 7% improvement upon rlCoP. Here, it is plCoP that starts better, while rlCoP shows stronger learning. Eventually, plCoP outperforms rlCoP, even without rewriting. Additional ten iterations of the **combined** version increase the performance to 854 problems. This is 12% more than rlCoP in 20 iterations (763) but still weaker than the strongest E configuration (998). However it performs better than E with the more limited heuristics, as well as leanCoP with its heuristics.

**Table 2.** Performance on the MPTP2078 bushy dataset: original **rlCoP**, **baseline** plCoP, plCoP using **improved training** and **combined** plCoP. **incr** shows the performance increase in percentages from iteration 0 (unguided) to the best result.

| Iteration | 0 | 1 | 2 | 3 | 4 | 5 | 6 | 7 | 8 | 9 | 10 | Incr |
|---|---|---|---|---|---|---|---|---|---|---|---|---|
| rlCoP | 198 | 300 | 489 | 605 | 668 | 701 | 720 | 737 | 736 | 732 | **733** | **270%** |
| Baseline | 287 | 363 | 413 | 420 | 429 | 441 | 454 | 464 | 465 | **479** | 469 | 67% |
| Improved training | 287 | 449 | 544 | 611 | 640 | 674 | 692 | 704 | 720 | 731 | **744** | 140% |
| Combined | 326 | 460 | 563 | 642 | 671 | 694 | 721 | 740 | 761 | 775 | **782** | 140% |

# 4    Conclusion and Future Work

We have developed a reinforcement learning toolkit for experiments with guiding automated theorem proving in the connection calculus. Its core is the Prolog-based plCoP obtained by extending leanCoP with a number of features motivated by rlCoP. New features on top of rlCoP include guidance of reduction steps, the addition of the rewrite inference rule and its guidance, external proof checker, and improvements to the training data selection. Altogether, plCoP improves upon rlCoP on the M2K and the MPTP2078 datasets by 12% and 7% in ten iterations, and by 17.4% and 12% in twenty iterations. The system is publicly available to the ML and AI/TP communities for experiments and extensions.

One lesson learned is that due to the sparse rewards in theorem proving, care is needed when extracting training data from the prover's search traces. Another lesson is that new sound inference rules can be safely added to the underlying calculus. Thanks to the guidance, the system still learns to focus on promising actions, while the new actions may yield shorter search and proofs for some problems. An important part of such an extendability scheme is the independent proof checker provided.

---

[8] For a description of these E configurations, see Table 9 of [19].

Future work includes, e.g., the addition of neural learners, such as tree and graph neural networks [10, 26]. An important motivation for choosing Prolog is our plan to employ Inductive Logic Programming to learn new prover actions (as Prolog programs) from the plCoP proof traces. As the manual addition of the rewrite step already shows, such new actions can be inserted into the proof search, and guidance can again be trained to use them efficiently. Future work, therefore, involves AI/TP research in combining statistical and symbolic learning in this framework, with the goal of automatically learning more and more complex actions similar to tactics in interactive theorem provers. We believe this may become a very interesting AI/TP research topic facilitated by the toolkit.

# A    Proof Checking Example

As a simple example suppose we are proving a proposition $q(a)$ under two assumptions $\forall x.p(x)$ and $\forall x.p(x) \Rightarrow q(a)$. plCoP will start with three input "clauses" $p(X)^-$, $p(Y) \vee q(a)^-$ and $q(a)$. The connection proof proceeds in the obvious way and yields three proof clauses that are either input clauses or instances of input clauses: $p(X)^-$, $p(X) \vee q(a)^-$ and $q(a)$. Unification during the search makes the two variables $X$ and $Y$ the same variable $X$. For proof checking, we now consider $X$ to be constant (in the same sense as $a$). Switching from the point of view of proving a disjunction of conjuncts to the point of view of refuting a conjunction of disjuncts, we see these as three propositional clauses: $P$, $P^- \vee Q$ and $Q^-$ where $P$ stands for the atom $p(X)$ (now viewed as ground) and $Q$ stands for $q(a)$ (also ground). The set $\{P, P^- \vee Q, Q^-\}$ is clearly unsatisfiable. In terms of the connection method, the unsatisfiability of this set guarantees every path [2, 6] has a pair of complementary literals.

# B    Monte Carlo Tree Search in Prolog

We show the most important predicates that perform the MCTS. The code has been simplified for readability.

We repeatedly perform playouts, which consist of three steps: 1) find the next tree node to expand, 2) add a new child to this node and 3) update ancestor values and visit counts:

```
mc_playout(ChildHash,ParentHash,NodeHash,FHash):-

    % get the current bigstep node (root of exploration)
    rootnode(StartId), !,

    % find node to expand
    mc_find_unexpanded(StartId,ChildHash,NodeHash,
                    ExpandId,UnexpandedActionIds),
    nb_hashtbl_get(NodeHash,ExpandId,[State,_,_,_,ChProbs]),
```

```
    State=state(_,_,_,_,_,_,Result),
 (  Result ==  1 -> Reward = 1 % we found a proof
 ;  Result == -1 -> Reward = 0 % proof failed
 ;  get_largest_index(UnexpandedActionIds, ChProbs,
                              ActionIndex),
    flag(inference_cnt, X, X+1), % increase inference count

    % we expand the child with the largest prior probability
    mc_expand_node(ExpandId,ChildHash,ParentHash,NodeHash,
                   FHash,ActionIndex,Reward)
 ),

    % update ancestor visit counts and values
    mc_backpropagate(ExpandId,Reward,ParentHash,NodeHash).
```

We search for the node to expand based on the standard UCT formula:

```
% +Id: current node id
% -Id2: next node id to expand
mc_find_unexpanded(Id,ChildHash,NodeHash,
                   Id2,UnexpandedActionIds):-
    mc_child_list(Id,NodeHash,ChildHash,ChildPairs),
    nb_hashtbl_get(NodeHash,Id,[State,_,VisitCount,_,_]),
    action_count(State,ActionCount),
    length(ChildPairs,L),
 (  ActionCount == 0 -> % no valid moves
    Id2=Id, UnexpandedActionIds=[]
 ;  mc_ucb_select_child(VisitCount,ChildPairs,NodeHash,
                        SelectedId,UCBScore),
    ( L < ActionCount,
      mc_ucb_score_unexplored(VisitCount,ActionCount,
                              UCBUnexploredScore),
      UCBUnexploredScore > UCBScore -> %select current node
      Id2=Id,
      missing_actions(ActionCount,ChildPairs,
                      UnexpandedActionIds)
    ;
      % we move towards the child with the highest UCB score
      mc_find_unexpanded(SelectedId,ChildHash,NodeHash,
                         Id2,UnexpandedActionIds)
    ), !
 ;  % The current node is a leaf, so we select it
    Id2=Id,
    missing_actions(ActionCount,ChildPairs,
                    UnexpandedActionIds)
 ).
```

Once the node to expand has been selected, we pick the unexplored child that has the highest UCB score:

```
mc_expand_node(ParentId,ChildHash,ParentHash,NodeHash,FHash,
               ActionIndex,ChildValue):-
    nb_hashtbl_get(NodeHash,ParentId,[ParentState,_,_,_,ChProbs]),

    % we perform the inference step corresponding to the new child
    copy_term(ParentState,ParentState2),
    logic_step(ParentState2,ActionIndex,ChildState),

    % value estimate from the external (xgboost) model
    guidance_get_value(ChildState, FHash, ChildValue),

    % probability estimates for the children of the new node
    % from the external (xgboost) model
    guidance_action_probs(ChildState,FHash,ChChProbs),

    % store the new node in the hash tables
    nth0(ActionIndex, ChProbs, ChProb),
    flag(nodecount, ChildId, ChildId+1),
    nb_hashtbl_set(ChildHash,ParentId-ActionIndex,ChildId),
    nb_hashtbl_set(ParentHash,ChildId,ParentId),
    nb_hashtbl_set(NodeHash,ChildId,
                   [ChildState,ChProb,1,ChildValue,ChChProbs]).
```

Finally, we update ancestor nodes after the insertion of the new leaf:

```
mc_backpropagate(Id,Reward,ParentHash,NodeHash):-
    nb_hashtbl_get(NodeHash,Id,[State,Prob,VCnt,Value,ChProbs]),
    VCnt1 is VCnt + 1,
    Value1 is Value + Reward,
    nb_hashtbl_set(NodeHash,Id,[State,Prob,VCnt1,Value1,ChProbs]),
    ( nb_hashtbl_get(ParentHash, Id, ParentId) ->
      mc_backpropagate(ParentId,Reward,ParentHash,NodeHash)
    ; true
    ).
```

## C  leancop_step.pl Module

Below we provide the code for the most important predicates that handle leanCoP inference steps in such a way that the entire prover state is explicitly maintained. For better readability, we omit some details, mostly related to problem loading, logging, and proof reconstruction. The nondet_step predicate takes a prover state along with the index of an extension or reduction step and returns the subsequent state. Before returning, it repeatedly calls the det_steps predicate, which performs optimization steps that do not involve choice (loop elimination, reduction without unification, lemma, single action).

```
% init_pure(+File,+Settings,-NewState)
init_pure(File,Settings,NewState):-
    NewState = state(Goal,Path,Lem,Actions,Todos,Proof,Result),

    % store options
    retractall(option(_)),
    findall(_, ( member(S,Settings), assert(option(S)) ), _ ),

    % load tptp file and store contrapositives
    {...}

    % perform any potential optimizations
    det_steps([-(#)],[],[],[],[init((-#)-(-#))],
              Goal,Path,Lem,Todos,Proof,Result),

    % collect valid moves from this state
    % valid_actions(Goal, Path, Actions).


:- dynamic(alt/6).
% step_pure(+ActionIndex,+State,-NewState,-SelectedAction))
step_pure(ActionIndex,State,NewState,Action0):-
    State = state(Goal0,Path0,Lem0,Actions0,Todos0,Proof0,_),
    NewState = state(Goal,Path,Lem,Actions,Todos,Proof,Result),

    nth0(ActionIndex,Actions0,Action0),

    % if there were other alternative actions, store them
    (option(backtrack), Actions0=[_,_|_] ->
        select_nounif(Action0, Actions0, RemActions0), !,
        asserta(alt(Goal0,Path0,Lem0,RemActions0,Todos0,Proof0))
     ; true
    ),

    % perform any potential optimizations
    nondet_step(Action0,Goal0,Path0,Lem0,Todos0,Proof0,
                Goal1,Path1,Lem1,Todos1,Proof1,Result1),

    % if proof search fails, pop an alternative
    ( Result1 == -1, option(backtrack),
      pop_alternative(Goal,Path,Lem,Actions,Todos,Proof) ->
          Result=0,
     ; [Goal,Path,Lem] = [Goal1,Path1,Lem1],
```

```
        [Todos,Proof,Result] = [Todos1,Proof1,Result1],
        valid_actions(Goal,Path,Actions)
    ).


%%% make a single proof step from a choice point
% nondet_step(Action,Goal,Path,Lem,Todos,Proof,
%             NewGoal,NewPath,NewLem,NewTodos,NewProof,Result)
% reduction step
nondet_step(red(NegL),[Lit|Cla],Path,Lem,Todos,Proof,
            NewGoal,NewPath,NewLem,NewTodos,NewProof,Result):-
    neg_lit(Lit,NegL),
    Proof2 = {...}
    det_steps(Cla,Path,Lem,Todos,Proof2,
              NewGoal,NewPath,NewLem,NewTodos,NewProof,Result).
% extension step
nondet_step(ext(NegLit,Cla1,_),[Lit|Cla],Path,Lem,Todos,Proof,
            NewGoal,NewPath,NewLem,NewTodos,NewProof,Result):-
    neg_lit(Lit, NegLit),
    ( Cla=[_|_] ->
      Todos2 = [[Cla,Path,[Lit|Lem]]|Todos]
    ; Todos2 = Todos
    ),
    Proof2= {...}
    det_steps(Cla1,[Lit|Path],Lem,Todos2,Proof2,
              NewGoal,NewPath,NewLem,NewTodos,NewProof,Result).


% perform steps until the next choice point (or end of proof)
det_steps([],_Path,_Lem,Todos,Proof,
          NewGoal,NewPath,NewLem,NewTodos,NewProof,Result):-
    !,
    ( Todos = [] -> % nothing to prove, nothing todo on the stack
           [NewGoal,NewPath,NewLem,NewTodos,NewProof,Result] =
           [[success],[],[],[],Proof,1]
    ; Todos = [[Goal2,Path2,Lem2]|Todos2] ->
           % nothing to prove, something on the stack
           det_steps(Goal2,Path2,Lem2,Todos2,Proof,NewGoal,
                     NewPath,NewLem,NewTodos,NewProof,Result)
    ).
det_steps([Lit|_Cla],Path,_Lem,_Todos,Proof,
          NewGoal,NewPath,NewLem,NewTodos,NewProof,Result):-
    member(P,Path), Lit == P, !, % loop elimination
    [NewGoal,NewPath,NewLem,NewTodos,NewProof,Result] =
    [[failure],[],[],[],Proof,-1].
det_steps([Lit|Cla],Path,Lem,Todos,Proof,
```

```
                NewGoal,NewPath,NewLem,NewTodos,NewProof,Result):-
    member(LitL,Lem), Lit==LitL, !, % perform lemma step
    Proof2 = [lem(Lit)|Proof],
    det_steps(Cla,Path,Lem,Todos,Proof2,
                NewGoal,NewPath,NewLem,NewTodos,NewProof,Result).
det_steps([Lit|Cla],Path,Lem,Todos,Proof,
            NewGoal,NewPath,NewLem,NewTodos,NewProof,Result):-
    neg_lit(Lit,NegLit),
    ( option(eager_reduction(1)) ->
      member(NegL,Path),
      unify_with_occurs_check(NegL, NegLit), ! % eager reduction
    ; member(NegL,Path),
      NegL == NegLit, ! % reduction without unification is safe
    ),
    Ext = [NegL, NegL],
    Proof2 = [red(Ext-Ext)|Proof],
    det_steps(Cla,Path,Lem,Todos,Proof2,
                NewGoal,NewPath,NewLem,NewTodos,NewProof,Result).
det_steps(Goal,Path,Lem,Todos,Proof,
            NewGoal,NewPath,NewLem,NewTodos,NewProof,Result):-
    valid_actions(Goal,Path,Actions),
    ( option(single_action_optim), Actions==[A] ->
        % only a single action is available, so perform it
        nondet_step(A,Goal,Path,Lem,Todos,Proof,NewGoal,
                    NewPath,NewLem,NewTodos,NewProof,Result)
    ;Actions==[] ->              % proof failed
        [NewGoal,NewPath,NewLem,NewTodos,NewProof,Result] =
        [[failure],[],[],[],Proof,-1]
    ; option(comp(PathLim)), \+ ground(Goal), length(Path,PLen),
      PLen > PathLim -> % reached path limit
      [NewGoal,NewPath,NewLem,NewTodos,NewProof,Result] =
      [[failure],[],[],[],Proof,-1]
    ;[NewGoal,NewPath,NewLem,NewTodos,NewProof,Result] =
      [Goal,Path,Lem,Todos,Proof,0]
    ).
```

## D  Policy and Value Functions

Here we describe 1) the default value and policy functions used in the first iteration, 2) the training data extraction and 3) how the predicted model values are used in MCTS. All these formulae are taken directly from rlCoP [19] and have been highly hand-engineered. We currently use these solutions in plCoP without altering them; however, we believe some of these decisions are worth reconsidering.

## D.1  Value Function

In the first iteration, the default value $V_d$ is based on the total term size of all open goals. Given $s$ with total term size of open goals $t$, its value is

$$V(s) = \frac{1}{1 + e^{-3.7 * e^{-0.05t} + 2.5}} \tag{2}$$

After the MCTS phase, training data is extracted from the states in the bigstep nodes. If a state $s$ is $k$ steps away from a success node, its target value is $0.99^k$. If none of its descendants are success nodes, then its target value is 0. We can then build a model using logistic regression. However, the authors of rlCoP find that the xgboost model works better if standard regression is used, so the target value is first mapped into the range $[-3, 3]$. The value $V_t$ used for model training is

$$V_t(s, k) = min(3, max(-3, log(0.99^k / (1 - 0.99^k))))$$

In subsequent iterations, the prediction $V_p$ of the model is mapped back to the $[0, 1]$ range $(V_p')$:

$$V_p'(V_p) = \frac{1}{1 + e^{-V_p}}$$

This value is further adjusted to give an extra incentive towards states with few open goals. If the state has $g$ open goals, then the final value $(V_f)$ used in MCTS is

$$V_f(V_p', g) = (\sqrt{V_p'})^g$$

## D.2  Policy Function

The default policy $P_d$ is simply the uniform distribution, i.e., if a state $s$ has $n$ valid inferences, then each action $a$ has a prior probability of

$$P_d(n) = \frac{1}{n}$$

After the MCTS phase, training data is extracted from the (state, action) pairs in the bigstep nodes. Target probabilities are based on relative visit frequencies of child nodes. These frequencies are again mapped to a range where we can do standard regression. Given state $s$ with $n$ valid inferences, such that $s$ was expanded $N$ times and its $j$th child was visited $N_j$ times, then the policy $P_t$ used for model training is

$$P_t(s, n, N, N_j) = max(-6, log(\frac{N_j}{N} n))$$

The prediction $P_p$ is mapped back to the $[0, 1]$ range and normalized across all actions using the softmax function $softmax(x)_i = \frac{e^{\frac{x_i}{T}}}{\sum_j e^{\frac{x_i}{T}}}$, where $T$ is the temperature parameter that was set to 2. The final prior probabilities used in MCTS are

$$P_f(P_t) = \text{softmax}(P_t)$$

# E   Experiment Hyperparameters

plCoP is parameterized with configuration files (see examples in the `ini` directory of the distributed code), so the key parameters can be easily modified. Here we list the most important hyperparameters used in our experiments.

*Feature Extraction.* Our main features are term walks of length up to 3. We also add several more specific features: number of open goals, number of symbols in them, their maximum size and depth, length of the current path, and two most frequent symbols in open goals. The resulting feature vectors are sparse and long, so they are first compressed to a fixed size $d$: vector $f$ is compressed to $f'$, such that $f'_i = \sum_{\{j|j \bmod d=i\}} f_j$.

One difference from rlCoP is that they hash features to a fixed 262139 dimensional vector while plCoP uses a 10000 dimensional feature vector for faster computation. Over 5 iterations of our baseline on the M2k dataset, this even yields a small improvement (928 vs. 940), likely due to less overfitting.

*MCTS.* MCTS has an inference limit of 200000 steps and an additional time limit of 200 s. Bigsteps are made after 2000 steps. The exploration constant (cp) is 3 in the first iterations and 2 in later iterations.

leanCoP *Parameters.* leanCoP usually employs an iteratively increasing path limit to ensure completeness. We set path limit to 1000, i.e., we practically remove it, in order to allow exploration at greater depth.

*XGBoost Parameters.* To train XGBoost models, we use a learning rate of 0.3, maximum tree depth of 9, a weight decay of 1.5, a limit of 400 training rounds with early stopping if no improvement takes place over 50 iterations. We use the built-in "scale_pos_weight" XGBoost training argument to ensure that our training data is sign-balanced.

Furthermore, there is an option to keep or filter duplicate inputs with different target values. Our experiments did not show the importance of this feature, and all results presented in this paper apply duplicate filtering.

# References

1. Alama, J., Heskes, T., Kühlwein, D., Tsivtsivadze, E., Urban, J.: Premise selection for mathematics by corpus analysis and kernel methods. J. Autom. Reason. **52**(2), 191–213 (2013). https://doi.org/10.1007/s10817-013-9286-5
2. Andrews, P.B.: On connections and higher-order logic. J. Autom. Reason. **5**(3), 257–291 (1989)

3. Anthony, T., Tian, Z., Barber, D.: Thinking fast and slow with deep learning and tree search. arXiv preprint arXiv:1705.08439 (2017)
4. Bansal, K., Loos, S.M., Rabe, M.N., Szegedy, C., Wilcox, S.: HOList: an environment for machine learning of higher-order theorem proving (extended version). arXiv preprint arXiv:1904.03241 (2019)
5. Beckert, B., Posegga, J.: Leantap: lean tableau-based deduction. J. Autom. Reason. **15**, 339–358 (1995)
6. Bibel, W.: Automated Theorem Proving. Artificial Intelligence, 2nd edn. Vieweg, Braunschweig (1987)
7. Biere, A.: Picosat essentials. J. Satisf. Boolean Model. Comput. (JSAT) **4**, 75–97 (2008)
8. Browne, C., et al.: A survey of Monte Carlo tree search methods. IEEE Trans. Comput. Intell. AI Games **4**, 1–43 (2012)
9. Chen, T., Guestrin, C.: XGBoost: a scalable tree boosting system. In: Proceedings of the 22nd ACM SIGKDD International Conference on Knowledge Discovery and Data Mining, KDD 2016, pp. 785–794 (2016). https://doi.org/10.1145/2939672.2939785
10. Chvalovský, K., Jakubův, J., Suda, M., Urban, J.: ENIGMA-NG: efficient neural and gradient-boosted inference guidance for E. In: Fontaine, P. (ed.) CADE 2019. LNCS (LNAI), vol. 11716, pp. 197–215. Springer, Cham (2019). https://doi.org/10.1007/978-3-030-29436-6_12
11. Gauthier, T., Kaliszyk, C., Urban, J., Kumar, R., Norrish, M.: Learning to prove with tactics. arXiv preprint arXiv:1804.00596 (2018)
12. Goertzel, Z., Jakubův, J., Urban, J.: ENIGMAWatch: proofwatch meets ENIGMA. In: Cerrito, S., Popescu, A. (eds.) TABLEAUX 2019. LNCS (LNAI), vol. 11714, pp. 374–388. Springer, Cham (2019). https://doi.org/10.1007/978-3-030-29026-9_21
13. Jakubův, J., Urban, J.: ENIGMA: efficient learning-based inference guiding machine. In: Geuvers, H., England, M., Hasan, O., Rabe, F., Teschke, O. (eds.) CICM 2017. LNCS (LNAI), vol. 10383, pp. 292–302. Springer, Cham (2017). https://doi.org/10.1007/978-3-319-62075-6_20
14. Jakubuv, J., Urban, J.: Hammering Mizar by learning clause guidance. In: Harrison, J., O'Leary, J., Tolmach, A. (eds.) 10th International Conference on Interactive Theorem Proving, ITP 2019, Portland, OR, USA, 9–12 September 2019. LIPIcs, vol. 141, pp. 34:1–34:8. Schloss Dagstuhl - Leibniz-Zentrum für Informatik (2019). https://doi.org/10.4230/LIPIcs.ITP.2019.34
15. Kaliszyk, C., Urban, J.: M2K dataset. https://github.com/JUrban/deepmath/blob/master/M2k_list
16. Kaliszyk, C., Urban, J.: Mizar40 dataset. https://github.com/JUrban/deepmath
17. Kaliszyk, C., Urban, J.: FEMaLeCoP: fairly efficient machine learning connection prover. In: Davis, M., Fehnker, A., McIver, A., Voronkov, A. (eds.) LPAR 2015. LNCS, vol. 9450, pp. 88–96. Springer, Heidelberg (2015). https://doi.org/10.1007/978-3-662-48899-7_7
18. Kaliszyk, C., Urban, J.: MizAR 40 for Mizar 40. J. Autom. Reason. **55**(3), 245–256 (2015). https://doi.org/10.1007/s10817-015-9330-8
19. Kaliszyk, C., Urban, J., Michalewski, H., Olšák, M.: Reinforcement learning of theorem proving. In: NeurIPS 2018, pp. 8836–8847 (2018)
20. Kocsis, L., Szepesvári, C.: Bandit based Monte-Carlo planning. In: Fürnkranz, J., Scheffer, T., Spiliopoulou, M. (eds.) ECML 2006. LNCS (LNAI), vol. 4212, pp. 282–293. Springer, Heidelberg (2006). https://doi.org/10.1007/11871842_29

21. Kovács, L., Voronkov, A.: First-order theorem proving and vampire. In: Sharygina, N., Veith, H. (eds.) CAV 2013. LNCS, vol. 8044, pp. 1–35. Springer, Heidelberg (2013). https://doi.org/10.1007/978-3-642-39799-8_1

22. Letz, R., Stenz, G.: Model elimination and connection tableau procedures. In: Robinson, J.A., Voronkov, A. (eds.) Handbook of Automated Reasoning, vol. 2, pp. 2015–2114. Elsevier and MIT Press (2001)

23. Loos, S.M., Irving, G., Szegedy, C., Kaliszyk, C.: Deep network guided proof search. In: Eiter, T., Sands, D. (eds.) 21st International Conference on Logic for Programming, Artificial Intelligence, and Reasoning (LPAR), vol. 46, pp. 85–105 (2017)

24. Lukácsy, G., Szeredi, P.: Efficient description logic reasoning in prolog: the DLog system. TPLP 9(3), 343–414 (2009). https://doi.org/10.1017/S1471068409003792

25. Muggleton, S., Raedt, L.D.: Inductive logic programming: theory and methods. J. Log. Program. 19/20, 629–679 (1994). https://doi.org/10.1016/0743-1066(94)90035-3

26. Olsák, M., Kaliszyk, C., Urban, J.: Property invariant embedding for automated reasoning. arXiv preprint arXiv:1911.12073 (2019)

27. Otten, J.: leanCoP 2.0 and ileanCoP 1.2: high performance lean theorem proving in classical and intuitionistic logic (system descriptions). In: Armando, A., Baumgartner, P., Dowek, G. (eds.) IJCAR 2008. LNCS (LNAI), vol. 5195, pp. 283–291. Springer, Heidelberg (2008). https://doi.org/10.1007/978-3-540-71070-7_23

28. Otten, J.: MleanCoP: a connection prover for first-order modal logic. In: Demri, S., Kapur, D., Weidenbach, C. (eds.) IJCAR 2014. LNCS (LNAI), vol. 8562, pp. 269–276. Springer, Cham (2014). https://doi.org/10.1007/978-3-319-08587-6_20

29. Otten, J., Bibel, W.: leanCoP: lean connection-based theorem proving. J. Symb. Comput. 36, 139–161 (2003)

30. Ross, S., Gordon, G., Bagnell, D.: A reduction of imitation learning and structured prediction to no-regret online learning. In: Gordon, G., Dunson, D., Dudík, M. (eds.) Proceedings of the Fourteenth International Conference on Artificial Intelligence and Statistics. Proceedings of Machine Learning Research, 11–13 Apr 2011, vol. 15, pp. 627–635. PMLR, Fort Lauderdale (2011). http://proceedings.mlr.press/v15/ross11a.html

31. Schulz, S.: E - a brainiac theorem prover. AI Commun. 15(2–3), 111–126 (2002)

32. Silver, D., et al.: Mastering the game of go with deep neural networks and tree search. Nature 529(7587), 484–489 (2016). https://doi.org/10.1038/nature16961

33. Silver, D., et al.: Mastering chess and shogi by self-play with a general reinforcement learning algorithm. arXiv preprint arXiv:1712.01815 (2017)

34. Silver, D., et al.: Mastering the game of go without human knowledge. Nature 550(7676), 354 (2017)

35. Stickel, M.E.: A prolog technology theorem prover: implementation by an extended prolog computer. J. Autom. Reason. 4(4), 353–380 (1988). https://doi.org/10.1007/BF00297245

36. Sutton, R.S., Barto, A.G.: Reinforcement Learning: An Introduction, vol. 1. Cambridge University Press, Massachusetts (1998)

37. Urban, J.: MPTP 0.2: design, implementation, and initial experiments. J. Autom. Reason. 37(1–2), 21–43 (2006)

38. Urban, J., Sutcliffe, G., Pudlák, P., Vyskočil, J.: MaLARea SG1 - machine learner for automated reasoning with semantic guidance. In: Armando, A., Baumgartner, P., Dowek, G. (eds.) IJCAR 2008. LNCS (LNAI), vol. 5195, pp. 441–456. Springer, Heidelberg (2008). https://doi.org/10.1007/978-3-540-71070-7_37

39. Urban, J., Vyskočil, J., Štěpánek, P.: MaLeCoP machine learning connection prover. In: Brünnler, K., Metcalfe, G. (eds.) TABLEAUX 2011. LNCS (LNAI), vol. 6793, pp. 263–277. Springer, Heidelberg (2011). https://doi.org/10.1007/978-3-642-22119-4_21

40. Wielemaker, J., Schrijvers, T., Triska, M., Lager, T.: SWI-Prolog. Theory Pract. Log. Program. **12**(1–2), 67–96 (2012)

41. Zombori, Z., Urban, J.: Learning complex actions from proofs in theorem proving. Accepted to AITP 2020. http://aitp-conference.org/2020/abstract/paper_11.pdf

# Author Index

Printed in the United States
By Bookmasters